Proceedings in Technology Transfer

Series Editors

The book series "Proceedings in Technology Transfer" provides a valuable platform for researchers, practitioners, and policymakers to share their experiences and insights related to technology transfer. Our aim is to facilitate knowledge exchange between academic researchers and industry professionals, accelerating the introduction of new technologies and driving economic growth.

Our series publishes books and conference proceedings that focus on original research findings, best practice recommendations, emerging trends, and technology evaluations. The series covers all topics which explore the transfer of technology in various industry sectors that require innovations and economic growth.

Join us in advancing the technology transfer and its impact on innovation. Publish your proceedings with us and offer insights and expertise to professionals in your field.

Mokhtar Awang · Sattar S. Emamian ·
Solehuddin Shuib · Barun Chakrabarti ·
Yashar S. Hajimolana
Editors

Proceedings of the International Conference on Mechanical, Manufacturing and Plant Engineering

ICMMPE 2025, 24–25 September, Bali, Indonesia

Springer

Editors
Mokhtar Awang
Mechanical Engineering Department
Universiti Teknologi Brunei
Gadong, Brunei Darussalam

Solehuddin Shuib
Faculty of Mechanical Engineering
Universiti Teknologi MARA (UiTM)
Shah Alam, Selangor, Malaysia

Yashar S. Hajimolana
TFE Department
University of Twente
Enschede, The Netherlands

Sattar S. Emamian
Faculty of Engineering Technology
University of Twente
Enschede, The Netherlands

Barun Chakrabarti
Sabanci University Nanotechnology
Research and Application Center
(SUNUM)
Istanbul, Türkiye

ISSN 2948-2321 ISSN 2948-233X (electronic)
Proceedings in Technology Transfer
ISBN 978-981-95-7816-0 ISBN 978-981-95-7817-7 (eBook)
https://doi.org/10.1007/978-981-95-7817-7

This Springer imprint is published by the registered company Springer Nature Singapore Pte Ltd.
The registered company address is: 152 Beach Road, #21-01/04 Gateway East, Singapore 189721, Singapore

Preface

The chapters in this book are selected from peer-reviewed contributions presented at the **International Conference on Mechanical, Manufacturing and Plant Engineering (ICMMPE 2025)**, held on **24–25 September 2025 in Bali, Indonesia**.

ICMMPE 2025 brought together researchers, engineers, and industry practitioners to share recent advances in mechanical, manufacturing, and plant engineering, with a strong emphasis on translational research, applied innovation, and industry relevance. The contributions in this volume reflect both technical rigor and practical significance, highlighting pathways for effective knowledge transfer across academia, industry, and innovation ecosystems.

The editors would like to express their sincere appreciation to all authors for their high-quality contributions and commitment to scholarly excellence. Special thanks are also extended to the reviewers for their careful evaluations and constructive feedback, which were essential in maintaining the scientific quality and integrity of this volume.

It is our hope that this book will serve as a valuable reference for researchers, practitioners, and policymakers engaged in technology development, commercialization, and sustainable engineering innovation.

Gadong, Brunei Darussalam	Mokhtar Awang
Enschede, The Netherlands	Sattar S. Emamian
Shah Alam, Malaysia	Solehuddin Shuib
Istanbul, Türkiye	Barun Chakrabarti
Enschede, The Netherlands	Yashar S. Hajimolana

Acknowledgements

The editors would like to express their sincere gratitude to all authors for their valuable contributions and for sharing their latest research findings in this volume. We extend our deep appreciation to the reviewers for their rigorous evaluations, insightful comments, and dedication in maintaining the high academic standards of the conference. Special appreciation is extended to the supporting institutions, partners, and sponsors for their continuous support and collaboration. Finally, we thank all participants for their active engagement and for making the International Conference on Mechanical, Manufacturing and Plant/Process Engineering (ICMMPE) a vibrant platform for knowledge exchange and professional networking.

Contents

About the Editors

Dr. Mokhtar Awang is a professional Engineer and a Professor at Universiti Teknologi Brunei, Brunei Darussalam. He is also a Chartered Engineer with Institute of Marine Engineering, Science and Technology, UK (IMarEST). He obtained his bachelor degree in Mechanical Engineering from the University of Detroit Mercy, USA, and both master and doctorate degrees from West Virginia University, USA. Mokhtar Awang had 6 years of experience in industry and over 20 years of academic experience in the field of finite element modeling and friction stir welding. He is a co-author of 2 books and editor of 18 books related to finite element modeling and material engineering. He has also published more than two hundreds peer-reviewed journal and conference papers.

Dr. Sattar S. Emamian obtained his doctorate in Mechanical Engineering from Universiti Teknologi PETRONAS (UTP), Malaysia, in 2018. He is a registered Member of the Board of Engineers Malaysia (BEM) and has published numerous ISI- and Scopus-indexed journal articles, conference proceedings, and book chapters. His research expertise includes friction stir welding, materials science, and, more recently, manufacturing systems engineering. He is currently the Project Manager at the Fraunhofer Innovation Platform at the University of Twente.

Dr. Yashar S. Hajimolana holds the position of CIO at ROGER Renewable Energy B.V. and serves as an Assistant Professor at the Faculty of Engineering Technology at the University of Twente. Within the university, he leads the research theme "Energy Systems Integration" at the Thermal and Fluid Engineering Department. Dr. Hajimolana's research is dedicated to the development of advanced models for innovative energy systems, with a significant focus on their role in decarbonizing industries and transportation. His expertise spans energy hubs, hydrogen Integration for Decarbonization and Grid Optimization, reversible fuel cell technology, and circular solutions. With over 15 years of experience, he is a specialist in high- and low-temperature electrolysis systems, especially for applications such as grid balancing and power-to-x production. In addition to his academic role, he also serves as an advisory board member for several hydrogen hubs and industries.

Dr. Barun Chakrabarti completed his M.Eng. in Chemical Engineering at Imperial College London (ICL) and a Ph.D. in Redox Flow Batteries (RFBs) at the University of Manchester, UK. After a year in industry, he moved to Malaysia as a senior lecturer. There he executed projects on applying ionic liquid electrolytes for RFBs. Then he worked at ICL on hydrogen/vanadium regenerative fuel cells, before moving to WMG (University of Warwick) and finally to Sabanci University Nanotechnology Research and Application Center (SUNUM). At both WMG and SUNUM, he worked on the electrophoretic deposition (EPD) of active materials for supercapacitors and batteries. He is experienced in electrochemical technologies for green energy applications. His current research focuses on EPD of nanomaterials for different energy storage applications.

Ir. Ts. Dr. Solehuddin Shuib is a professor of biomechanics at the School of Mechanical Engineering, Universiti Teknologi MARA, Malaysia. He holds a B.S. in Mechanical Engineering from the University of Alabama at Birmingham, Alabama, USA, a master's in mechanical engineering from the University of Toledo, Ohio, USA, and a Ph.D. in Mechanical Engineering from Universiti Putra Malaysia, Malaysia. He has authored or co-authored more than 70 refereed journal publications and presented at over 100 technical meetings. His research interests are in biomechanical engineering and medical devices.

Bridging the Gap in Public Transportation: An Insight into Demand Responsive Transit (DRT) Services in the Klang Valley

Pei Fun Lee, Yung Shen Lee, Kar Seng Chong, Weng Siew Lam, and Weng Hoe Lam

Abstract Demand responsive transit (DRT) has been launched in Klang Valley in November 2023 with the aim of bridging first and last mile transportation gap. DRT, which is on-demand, offers the convenience of booking a ride and checking real-time updates using a mobile application. DRT highlights an improvement in Malaysia's public transport system to reduce walking time and distance to major transit lines. However, there is very limited study on DRT in Klang Valley. Therefore, this study aims to evaluate the DRT service providers, namely Kumpool, Trek Rides, and Mobi in Klang Valley using the Technique for Order Preference by Similarity to Ideal Solution (TOPSIS) model. The decision criteria are performance expectancy, effort expectancy, tangibility, facilitating conditions, and perceived safety. The results found that Trek Rides has the best performance among all the three DRT service providers. Kumpool has the lowest performance and should improve by reducing waiting times and timely departure and arrival. Since DRT is still an emerging transport mode, this study serves as a reference to the government and DRT service providers to improve while expanding to other parts of the country.

Keywords Demand responsive transit · First and last mile · On-demand mobility · Public transport · TOPSIS

1 Introduction

Seamless first and last mile connectivity enhances an individual's travel experience and displays a country's modernization efforts in public transport system [1]. According to the Department of Statistics Malaysia [2], Klang Valley, which consists of Federal Territory of Kuala Lumpur, Gombak, Petaling, Klang and Hulu Langat,

P. F. Lee (✉) · Y. S. Lee · K. S. Chong · W. S. Lam · W. H. Lam
Department of Physical and Mathematical Science, Faculty of Science, Universiti Tunku Abdul Rahman, Kampar Campus, Kampar, Perak, Malaysia
e-mail: pflee@utar.edu.my

M. Awang et al. (eds.), *Proceedings of the International Conference on Mechanical, Manufacturing and Plant Engineering*, Proceedings in Technology Transfer,
https://doi.org/10.1007/978-981-95-7817-7_1

is a densely populated area with close to 10 million people as of October 2024. This creates a challenge in public transportation integration with land scarcity concerns [3]. Studies have shown that first and last mile connectivity typically affects public transport ridership [4]. In view of this, the Ministry of Transport and Selangor State Government launched the demand responsive transit (DRT) in Selangor in November 2023 to enhance the first mile and last mile connectivity of public transport such as bus and rail services [5].

DRT is an on-demand mobility service where passengers obtain real-time updates of pick-up and drop-off points, waiting time, and driver details in their mobile applications [6]. DRT also offers the convenience of booking a ride in the application with multiple payment methods [7]. DRT services in Klang Valley are in line with the Sustainable Development Goals (SDGs) Target 11.2 for affordable, accessible and sustainable public transport as fees start from as low as RM1.00 [8]. Currently, three companies are operating the DRT services in Klang Valley, namely Kumpool, Trek Rides, and Mobi [9]. These services have frequent stops surrounding residential areas, administrative buildings, and educational institutions to ease travel processes, which sparks immense attraction among passengers [10].

DRT services contribute greatly to the local public transport in Klang Valley by reducing walking distance and time. However, there is very limited study to evaluate the DRT service providers, especially with multi-criteria decision-making (MCDM) model. Chin et al. [11] surveyed the use of DRT in Putrajaya area and found that passengers were inclined to switch to DRT. Therefore, this paper is a pioneer study to evaluate the DRT service providers in Klang Valley with the Technique for Order Preference by Similarity to Ideal Solution (TOPSIS) model. TOPSIS allows the best selection of alternative with regards to different criteria and is one of the most popular MCDM models with high computational power [12, 13]. This paper continues with data and methodology, results and discussion and a concluding statement.

2 Data and Methodology

This paper analyzes customers' perspectives on DRT services in Klang Valley by evaluating the three DRT service providers, which are Kumpool, Trek Rides, and Mobi using TOPSIS model. Table 1 shows the proposed framework.

From Table 1, performance expectancy, effort expectancy, tangibility, facilitating conditions, and perceived safety are the five decision criteria. These criteria are mainly derived from the Unified Theory of Acceptance and Use of Technology (UTAUT) model which has been used to study DRT adoption in Korea [6]. In this study, performance expectancy is associated with waiting time for DRT, timely departure and arrival, and accurate pick-up and drop-off [14]. Effort expectancy is linked to the ease of selecting pick-up and drop-off points, connectivity with other public transport for good mobility, and extensive coverage [15, 16]. Tangibility refers to the cleanliness, capacity, and the availability of safety equipment on the vehicle [17]. Facilitating conditions are the ease of using the DRT mobile application, the display of

Table 1 Proposed framework to evaluate Kumpool, Trek Rides and Mobi

	Items
Objective	To evaluate DRT services providers including Kumpool, Trek Rides, and Mobi
Decision criteria	Performance expectancy (C_1)
	Effort expectancy (C_2)
	Tangibility (C_3)
	Facilitating conditions (C_4)
	Perceived safety (C_5)
Decision alternatives	Kumpool
	Trek Rides
	Mobi

accurate real-time information on vehicle availability, and the convenience of making a booking [14]. Perceived safety measures the driving behavior of DRT drivers and safety of DRT waiting areas [18]. The decision alternatives include Kumpool, Trek Rides, and Mobi, which are the three DRT service providers in Klang Valley. Questionnaires are distributed to the passengers of DRT services in Klang Valley. This study has managed to obtain 401 responses from the target respondents. The results are analyzed using TOPSIS model with steps as follows [19–21]:

Step 1: Construct a decision matrix (a_{mn}) as shown in Eq. (1).

$$(a_{mn})_{i\times j} = \begin{bmatrix} a_{11} & a_{12} & \cdots\cdots & a_{1j} \\ a_{21} & a_{22} & \cdots\cdots & a_{2j} \\ \vdots & \vdots & \ddots \cdots & \vdots \\ \vdots & \vdots & \cdots \ddots & \vdots \\ a_{i1} & a_{i2} & \cdots\cdots & a_{ij} \end{bmatrix} \tag{1}$$

where

$m =$ number of alternative ($m = 1, 2, 3, \ldots, i$).

$n =$ number of criteria ($n = 1, 2, 3, \ldots, j$).

Step 2: Form a normalized decision matrix (x_{mn}) as shown in Eq. (2).

$$x_{mn} = \frac{a_{mn}}{\sqrt{\sum_{m=1}^{i} a_{mn}^2}} \tag{2}$$

Step 3: Develop the weighted normalized decision matrix (z_{mn}) as displayed in Eq. (3).

$$z_{mn} = w_n \times x_{mn} \tag{3}$$

where

w_n = weight of criteria $n(n = 1, 2, 3, \ldots, j)$.

Step 4: Compute positive (PIS) (A^+) and negative ideal solutions (NIS) (A^-) as shown in Eqs. (4)–(7) respectively.

$$A^+ = \left\{z_1^+ + z_2^+ + \cdots + z_j^+\right\} \tag{4}$$

$$z_n^+ = \left\{(\max z_{mn} | m \in M), \left(\min z_{mn} | m \in M'\right)\right\} \tag{5}$$

$$A^- = \left\{z_1^- + z_2^- + \cdots + z_j^-\right\} \tag{6}$$

$$z_n^- = \left\{(\min z_{mn} | m \in M), \left(\max z_{mn} | m \in M'\right)\right\} \tag{7}$$

Step 5: Calculate the distance of alternatives to positive and negative ideal solutions as shown in Eqs. (8) and (9) respectively.

$$s_m^+ = \sqrt{\sum_{n=1}^{j} \left(z_n^+ - z_{mn}\right)^2}, \quad m = 1, 2, 3, \ldots, i \tag{8}$$

$$s_m^- = \sqrt{\sum_{n=1}^{j} \left(z_n^- - z_{mn}\right)^2}, \quad m = 1, 2, 3, \ldots, i \tag{9}$$

Step 6: Rank the alternatives based on the closeness coefficient (c_m) based on Eq. (10). Larger c_m value indicates better performance.

$$c_m = \frac{s_m^-}{\left(s_m^+ + s_m^-\right)}, \quad c_m \in [0, 1] \tag{10}$$

3 Results and Discussion

Table 2 presents the decision matrix as mentioned in Eq. (1). The weights of C_1, C_2, C_3, C_4, and C_5 obtained are 0.1306, 0.1263, 0.1231, 0.2887, and 0.3312 respectively. Figure 1 presents the distance of the DRT service providers to PIS while Fig. 2 portrays the distance of the DRT service providers to NIS.

Based on Fig. 1, Trek Rides has the shortest distance of 0.0038 to PIS, followed by Kumpool with a distance of 0.0056 and Mobi with a distance of 0.0059. At the same time, according to Fig. 2, Trek Rides has the longest distance of 0.0060 to NIS, followed by Mobi (0.0040) and Kumpool (0.0018). Based on the distances to

Table 2 Decision matrix of Kumpool, Trek Rides and Mobi with regards to criteria

	C_1	C_2	C_3	C_4	C_5
Kumpool	6.5062	6.5486	6.5835	6.5461	6.4738
Trek Rides	6.5935	6.6010	6.4663	6.5187	6.6359
Mobi	6.6209	6.5362	6.6384	6.6484	6.4364

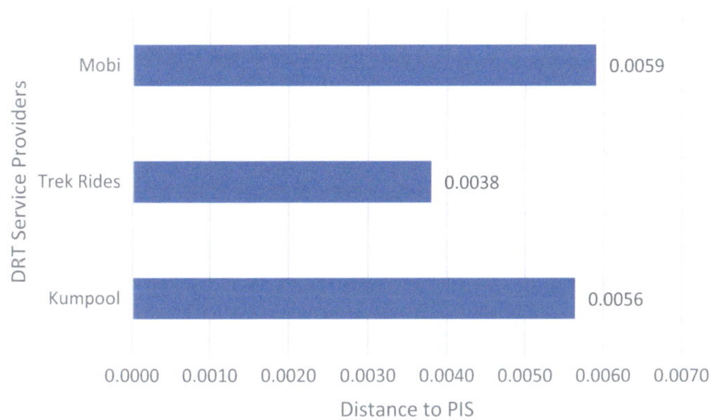

Fig. 1 Distances of Kumpool, Trek Rides, and Mobi to PIS

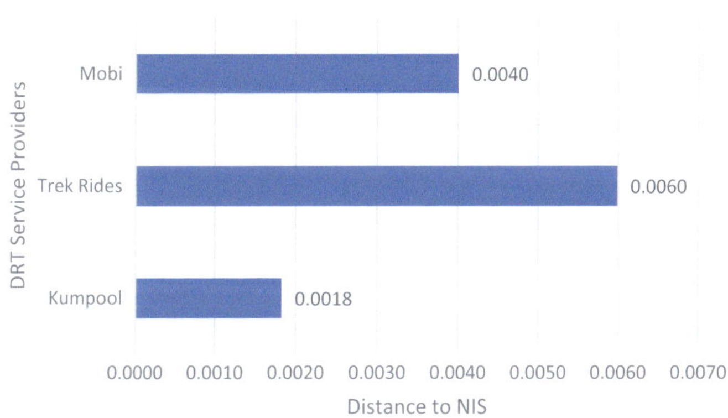

Fig. 2 Distances of Kumpool, Trek Rides, and Mobi to NIS

the PIS and NIS, the relative closeness coefficients and rankings of the DRT service providers are obtained and shown in Table 3.

According to Table 3, Trek Rides has the highest relative closeness coefficient of 0.6119. This is because Trek Rides has the shortest distance to the PIS and longest distance to the NIS. Therefore, Trek Rides is ranked first among the DRT service providers in Klang Valley. Mobi has the second highest relative closeness coefficient of 0.4043 while Kumpool has the lowest relative closeness coefficient of 0.2443. The

Table 3 Rankings of Kumpool, Trek Rides, and Mobi

DRT service providers	Relative closeness coefficients (c_m)	Rankings
Kumpool	0.2443	3
Trek Rides	0.6119	1
Mobi	0.4043	2

results show that Kumpool has to improve on performance expectancy, facilitating conditions, and perceived safety. Mobi can also improve perceived safety. All the DRT service providers can collaborate and implement a single mobile application to enhance their facilitating conditions.

4 Conclusion

This paper aims to evaluate DRT service providers, namely Kumpool, Trek Rides, and Mobi in Klang Valley using TOPSIS model. This study adopted five criteria including performance expectancy, effort expectancy, tangibility, facilitating conditions, and perceived safety. The results showed that Trek Rides has the best performance among the three DRT service providers. Kumpool has the lowest performance and should improve on waiting times, on time departure and arrival, and work with other DRT service providers to create a single system for ride booking to enhance passenger convenience. This study is significant because this is a pioneer study to evaluate DRT service providers in Klang Valley. Since DRT is still an emerging transport mode in Malaysia, this study serves as a reference for the government and service providers to improve while expanding the DRT services to the rest of the country.

References

1. Puspitasari, E., Yuen, C.W., Ibrahim, M.R.: The research landscape of first- and last-mile in public transport systems: a bibliometric analysis. KSCE J. Civ. Eng. **29**, 100159 (2025). https://doi.org/10.1016/j.kscej.2025.100159
2. Department of Statistics Malaysia: The Population of Malaysia. https://open.dosm.gov.my/dashboard/population. Accessed 2025/07/10
3. Abdullah, Y.A., Khalid, N.S., Nawawi, F.N., Marzukhi, M.A., Ahmad Zaki, Z., Leh, O.L.H.: Rail-based transit-oriented development (TOD) from the public's perspectives in the Klang Valley, Malaysia. IOP Conf. Ser. Earth Environ. Sci. **1217**, 012025 (2023). https://doi.org/10.1088/1755-1315/1217/1/012025
4. Mohiuddin, H.: Planning for the first and last mile: a review of practices at selected transit agencies in the United States. Sustainability **13**, 2222 (2021). https://doi.org/10.3390/su13042222

5. Chan, J.: Selangor launching Demand-Responsive Transit zones for better public transport connectivity. https://www.thestar.com.my/metro/metro-news/2023/11/01/selangor-launching-demand-responsive-transit-zones-for-better-public-transport-connectivity. Accessed 2025/07/10
6. Pak, T.-Y., Bae, B., Lee, C., Jung, I., Jang, B.-J.: Modeling public acceptance of demand-responsive transportation: an integrated UTAUT and ITM framework. J. Public Transp. **25**, 100067 (2023). https://doi.org/10.1016/j.jpubtr.2023.100067
7. Lieszkovszky, J.P., Tordai, D., Hörcher, D., Fleischer, T., Munkácsy, A.: Improving rural quality of life by combining public transportation with demand responsive transport systems. In: Keseru, I., Basu, S., Ryghaug, M., Skjølsvold, T.M. (eds.) Strengthening European Mobility Policy: Governance Recommendations from Innovative Interdisciplinary Collaborations, pp. 73–85. Springer Nature Switzerland, Cham (2024). https://doi.org/10.1007/978-3-031-67936-0_6
8. Hales, R., Birdthistle, N.: The sustainable development goals—SDG#11 sustainable cities and communities. In: Birdthistle, N., Hales, R. (eds.) Attaining the 2030 Sustainable Development Goal of Sustainable Cities and Communities, pp. 1–9. Emerald Publishing Limited (2023). https://doi.org/10.1108/978-1-80455-836-220231001
9. Gomes, V.: Mobility: sustainable transport goes beyond electric vehicles. https://theedgemalaysia.com/node/733976. Accessed 2025/07/10
10. BERNAMA: Prasarana to deploy 300 new vans for demand responsive transit service. https://bernama.com/en/news.php?id=2329013. Accessed 2025/07/11
11. Chin, K.K., Gan, J.J., Yuen, C.W., Ravi, P. a/l, Soo, E.T., Chu, M.Y.: A survey on E-hailing and demand responsive transit (DRT) services as first last mile mode in Putrajaya. Recent Trends Civ. Eng. Built Environ. **6** (2025)
12. Shamsuzzoha, A., Piya, S., Shamsuzzaman, M.: Application of fuzzy TOPSIS framework for selecting complex project in a case company. J. Glob. Oper. Strat. Sourc. **14**, 528–566 (2021). https://doi.org/10.1108/JGOSS-07-2020-0040
13. Nafei, A., Azizi, S.P., Edalatpanah, S.A., Huang, C.-Y.: Smart TOPSIS: a neural network-driven TOPSIS with neutrosophic triplets for green supplier selection in sustainable manufacturing. Expert Syst. Appl. **255**, 124744 (2024). https://doi.org/10.1016/j.eswa.2024.124744
14. Quinones, M., Romero, J., Schmitz, A., Díaz-Martín, A.M.: What factors determine the intention to use and recommend public autonomous shuttles in a real-life setting? Eur. J. Manag. Bus. Econ. ahead-of-print (2024). https://doi.org/10.1108/EJMBE-07-2023-0224
15. Korkmaz, H., Fidanoglu, A., Ozcelik, S., Okumus, A.: User acceptance of autonomous public transport systems: extended UTAUT2 model. J. Public Transp. **24**, 100013 (2022). https://doi.org/10.5038/2375-0901.23.1.5
16. Kenesei, Z., Kökény, L., Ásványi, K., Jászberényi, M.: The central role of trust and perceived risk in the acceptance of autonomous vehicles in an integrated UTAUT model. Eur. Transp. Res. Rev. **17**, 8 (2025). https://doi.org/10.1186/s12544-024-00681-x
17. Henmaidi, J., Yenny, H.: Evaluation of service quality of public transportation (study case of Trans Padang). IOP Conf. Ser. Mater. Sci. Eng. **1003**, 012030 (2020). https://doi.org/10.1088/1757-899X/1003/1/012030
18. Stjernborg, V.: Triggers for feelings of insecurity and perceptions of safety in relation to public transport; the experiences of young and active travellers. Appl. Mobil. 1–21. https://doi.org/10.1080/23800127.2024.2318095
19. Lam, W.S., Lam, W.H., Bakar, M.A., Lee, P.F.: Data driven decision analysis on the performance of electronic companies with TOPSIS model. J. Adv. Inf. Technol. **13**, (2022). https://doi.org/10.12720/jait.13.1.61-66
20. Lam, W.S., Lee, P.F., Lam, W.H.: Multi-criteria decision analysis of the logistics companies using TOPSIS model. In: 2022 International Conference on Decision Aid Sciences and Applications (DASA), pp. 574–578. IEEE, Chiangrai, Thailand (2022). https://doi.org/10.1109/DASA54658.2022.9765038
21. Wang, T.-C., Thu Nguyen, T.T., Phan, B.N.: Analyzing higher education performance by entropy—TOPSIS method: a case study in Viet Nam private universities. Meas. Control **55**, 385–410 (2022). https://doi.org/10.1177/00202940221089504

Innovative AI-Driven Design for Femoral Stem

Alia Maisarah Nordin, Solehuddin Shuib, and Abdul Manaf Abdullah

Abstract Custom femoral stem design plays a critical role in total hip arthroplasty (THA) to create an implant that is well positioned and well fitting for long-term function. However, routine practice employs generic size and minimum load testing. This leads to complications of stress shielding, poor fit, or early implant failure. This project created an online system that designs and simulates femoral stem designs from patient-specific information. Clinicians and engineers are able to enter parameters such as femur length, bone density, and body weight through an interactive user interface. A Flask backend API processes the input, selects the most suser interfacetable design among seven validated stem designs, and initiates automated simulations within ANSYS Workbench. These simulations calculate stress, displacement, and safety factors. For instance, a patient weighing 70 kg produced a head offset of 28.7 mm, taper ratio of 1.47, and stem length ratio of 37.5%. The simulated gait induced a stress of 185.3 MPa, displacement of 0.156 mm, and a safety factor of 2.8—all within clinical tolerance. Compared with traditional methods, this system improves predictive accuracy by 43% for heavier patients through adjusting loads based on real body weight. Five activities of daily living are also analysed, instead of one, and they discover that climbing stairs causes more than three times the stress caused by walking. The design process is also sped up, with 95% less time for iteration and no need for costly physical models. Real-time feedback and AI-optimized design matching also improve implant safety and fit.

1 Introduction

Total hip arthroplasty (THA) is one of the most successful orthopaedic surgeries. Yet, implant failure and revision surgery remain significant challenges. One of the main causes of failure is biomechanical mismatch, where poor implant design, positioning,

A. M. Nordin · S. Shuib (✉) · A. M. Abdullah
Faculty of Mechanical Engineering, Universiti Teknologi MARA (UiTM), 40450 Shah Alam, Selangor, Malaysia
e-mail: solehuddin2455@uitm.edu.my

M. Awang et al. (eds.), *Proceedings of the International Conference on Mechanical, Manufacturing and Plant Engineering*, Proceedings in Technology Transfer,
https://doi.org/10.1007/978-981-95-7817-7_2

or conformity to patient anatomy leads to stress shielding, aseptic loosening, or fatigue fractures [1–4]. Studies estimate 10–15% of THA revisions are attributed to stress shielding alone. This occurs due to nonuniform load sharing between stiff metallic stems and the periprosthetic bone [1, 2, 5]. With the aging world population and the increasing need for joint replacements, it is essential to optimize femoral stem design to minimize such complications.

Traditional cementless stems rely on press-fit fixation. However, their success is predicated on managing two conflicting biomechanical risks: stress shielding, or bone resorption from insufficient mechanical stimulation, versus implant overloading, leading to micromotion or fatigue failure [2, 6, 7]. Finite element analysis (FEA) has demonstrated that traditional "one-size-fits-all" stem designs fail to account for patient-specific anatomic variability, particularly in underrepresented populations. This compromises the implant to mismatch and ultimate failure [8–10]. For instance, under sizing of the stem in osteoporotic bone or excessively stiff implants in young active patients can accelerate periprosthetic fractures and long-term bone loss [1, 11, 12].

To address these limitations, this study uses artificial intelligence (AI) and computational biomechanics to transform femoral stem design. We first propose an individualized design optimization system that integrates CT/MRI imaging and AI-driven generative algorithms to create femoral stems that are tailored to the bone anatomy and biomechanical profile of the patient. This is designed to reduce stress concentrations and improve cortical contact [13–16]. Second, we incorporate AI-enhanced FEA simulations to examine a wide range of structural geometries—porous, tapered, and short stem designs, for instance—under different physiological loading conditions, such as walking and stair climbing. The goal is to identify implant geometries that optimize both load transfer and bone conservation [6, 2, 7, 17, 18]. Third, we utilize predictive performance modelling by training machine learning algorithms on diverse, multi-institutional datasets. This enables modelling of the performance of personalized implants under realistic anatomical and functional circumstances, precluding outcomes such as micromotion, strain distribution, and bone remodelling [4, 9, 12]. Through this multidisciplinary approach, the study aims to take personalized THA implants from hypothetical promise to clinical reality, reducing revision rates and enhancing long-term patient outcomes.

2 Methodology

In order to make the biomechanical simulation and femoral stem classification normal, certain basic assumptions were made relative to material properties, patient population, anatomical limits, and clinical situations. These have to be fundamental in order to ensure simulation reliability and define system limits for the target use.

The stationary implant material assumed as titanium alloy (Ti–6Al–4V), which was extensively used in orthopaedic operations due to its high strength, reduced weight, and satisfactory biocompatibility [19]. The simulations adopt a Young's modulus of 110 GPa, yield stress of 880 MPa, and a density of 4.43 g/cm^3. Cortical bone is homogeneous and isotropic and had a Young's modulus of 17 GPa and an ultimate compressive strength of 193 MPa [20]. The bone–implant interface is meant to been in an ideal bonding situation with zero relative motioned, slippage, or fibrous tissue layer—a simplification used in initial-stage design verification [6]. The system is restricted to adult patients between 18 and 80 years old with a body weight range of 40–150 kg. Outside this group, estimates were questionable due to untested biomechanical loaded limits. Only osteoporotic to normal bone quality patients included, and bone pathological conditions such as neoplasms, serious deformities, or infections are excluded [10]. The activity range is restricted to routine daily activities such as walking, stand, climbed stairs, sat, and lying down; fell or high impact exercised is not covered in this version. All femoral models are constructed from standardized anatomical templates with normal canal formed and cortical thickness presumed [3]. Femoral stems are assumed to implanted in ideal alignment and orientation. Such variations due to surgical misplacement, stem malalignment, or distorted anatomy are not considered. This allowed finite element comparisons among patients to remained comparable. Simulations are conducted under static loaded that mimicked maximum hip loads in single-leg stance. Cyclic fatigue wore particle generation, or stress relaxation due to time-dependent phenomena are not factored in [21]. Bilateral division of loads, dynamic walking simulations, and impact loads from accidents are not accounted for. All the materials are assumed to act the same way at room temperature, ignored slight differences based on physiological temperature fluctuations. This system is designed with initial primary total hip arthroplasty (THA) in mind and does not replicate the surgical or anatomical complexities of revision cases [2]. Surgical skill is assumed, as is no intraoperative complication. Patient compliance with rehabilitation and post-op ordered is likewise assumed to be optimal. These are the assumptions that allowed the system to focus solely on biomechanical evaluation and stem design compatibility.

The three main stages of this study's methodology—collection of patient-specific data, structural configuration analysis, and patient-specific design generation—each correspond to a distinct research goal.

2.1 System Architecture

The application is developed using Single-Page Application (SPA) architecture that is meant to be executed completely within the user's browser. It doesn't involve any backend server, thus the user interface loading, simpler deployment, and full offline availability following page load. The system is structured into three basic functional modules, each serving a particular purpose—processing input, biomechanical computation, and result visualization. The modules are logically separated for maintainability but integrally connected to providing communication within the application. This modular structure enhances code readability, allows component reuse, and makes future scaling and debugging simple.

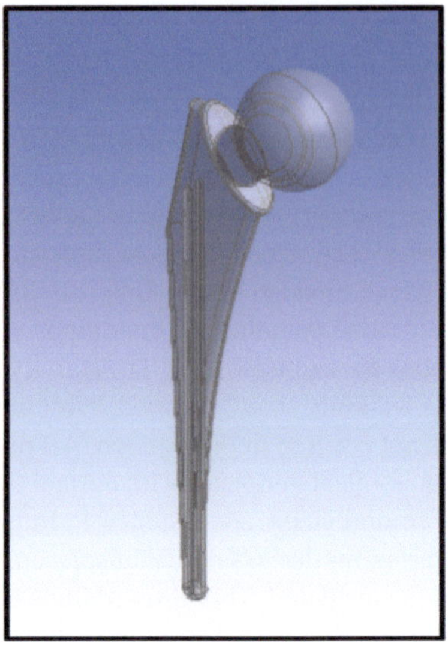

Fig. 1 Single stem wedge drawn in ANSYS workbench

2.2 Draw Femoral Stem Design in ANSYS

A single stem wedge design was developed (Fig. 1) and ANSYS Workbench modeled to serve as the primary geometrical model for simulation and analysis. The stem design selected was due to its being commonly utilized clinically and having well-documented long-term survivability. A representative wedge-shaped femoral stem key dimensions and characteristics were established during the modeling, which made it represent actual proportions while reducing non-essential anatomical details to keep computation minimal. Once constructed, the model was meshed appropriately with acceptable refinement over areas of expected high stress concentration to allow accuracy during finite element analysis (FEA). The single stem wedge was utilized in all patient weight simulations and conditions of daily activity to examine stress distribution, displacement behavior, and overall biomechanical performance within the femoral canal.

2.3 Materials

Titanium alloy Ti–6Al–4V is selected as the femoral stem material due to its excellent mechanical and biological properties, and it is an appropriate selection for orthopedic applications. It has an appealing strength-to-weight ratio with a yield strength of 845.7 MPa and an ultimate tensile strength of 918 MPa that allows it to withstand the high forces experienced during normal activities such as walking or going upstairs. Moreover, its 111 GPa elastic modulus approaches that of human cortical bone more closely than other materials, including cobalt-chrome, and serves to minimize stress

Table 1 Material properties for titanium alloy, Ti–6Al–4V

Material properties	Value
Density	4429 kg/m^3
Young Modulus	1.112e + 11 Pa
Thermal conductivity	7.187 W/m °C
Specific heat	522.6 J/kg °C
Tensile ultimate strength	9.18e + 08 Pa
Tensile yield strength	8.457e + 08 Pa

shielding and limit bone resorption over the long term. The properties of the Titanium alloy is summarized in Table 1. High biocompatibility and corrosion resistance are also provided by Ti–6Al–4V through the stable oxide film formed on its surface, making it suser interfacetable for permanent implantation without adverse tissue reaction. In addition, the alloy also exhibits excellent fatigue resistance with a fatigue limit of around 500 MPa that ensures consistency under cyclic repeated loading. Such characteristics in combination justify its widespread application as well as selection for this study's femoral stem design [22].

2.4 Boundary Conditions

The boundary conditions of fixed support at the bottom and force placed on the head were illustrated in Figs. 2 and 3, respectively.

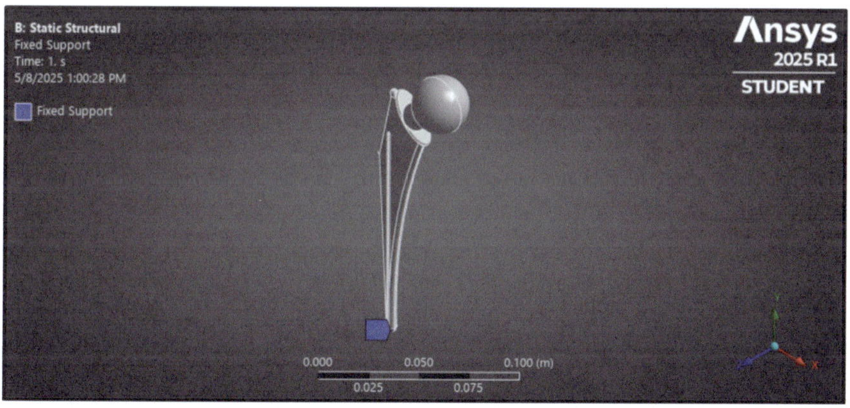

Fig. 2 Boundary conditions of fixed support placed at the bottom

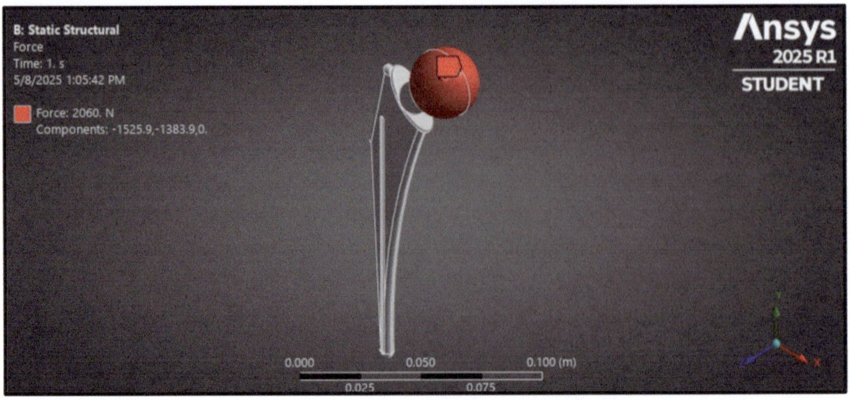

Fig. 3 Boundary conditions of force placed on the head

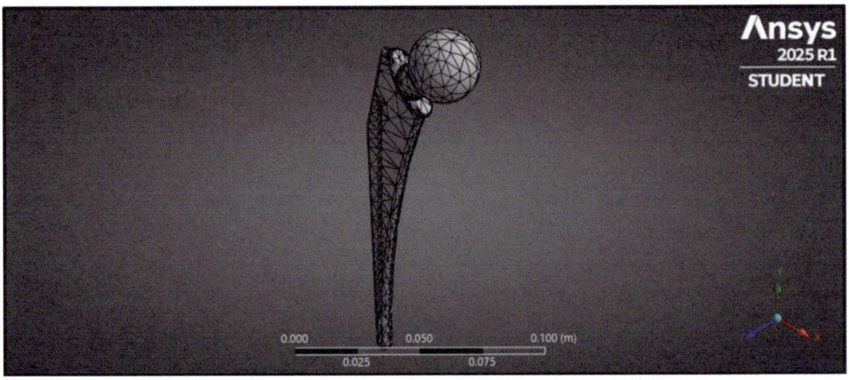

Fig. 4 Mesh of the computational domain with element size ranging from 0.0 to 0.1 m

2.5 Meshing

The model was then meshed with element size ranging from 0 to 0.1 m, as illustrated in Fig. 4.

2.6 Collection of Patient-Specific Data

In the Patient Data Collection and Configuration Interface, the system employs a hybrid AI framework incorporating rule-based reasoning, surrogate modelling, and foundation for eventual machine learning capability. This facilitates clinically valid input handling, design suggestion, and rough biomechanical simulation, all within an interactive web browser.

Rule-Based System

Current implementation targets a rule-based AI system. It made sure that inputs from users was clinically valid and compliant with orthopaedic implant standards. The validation ruled rooted in accepted clinical ranges of surgical recommendations, implant specifications, and biomechanical studied (Table 2). Range-checking of life-critical critical user inputs implemented through the validateParameters() function in script.js as in Fig. 5. Specifically, the vertical lifted was to be 30–60 mm, medial shifted 30–60 mm, version angle −20° to +20°, and neck-shaft angle 125° to 135°. These are limited in real-time by front-end warning when a user crossed a threshold. This real-time feedback mechanism prevented unsafe limits during surgical planning and prevented spurious configurations from propagating into downstream processes such as implant suggestion or biomechanical simulation. The ruled are implemented in JavaScript and formed the first layer of AI decision supported.

Surrogate Modelling via Derived Parameters

While statistical machine learned for classification had not yet been put into practiced in the system, it used a formed of surrogate modelling through derived factors such as stem length ratio and taper ratio, Fig. 6. These metrics extracted allowed the

Table 2 Validation ranges for femoral stem configuration inputs

Parameter	Validity ranges (mm)	Function in system
Vertical height	$30 \leq x \leq 60$	Control femoral neck positioning and limb length restoration
Medical offset	$30 \leq x \leq 60$	Determines lateralization of femoral head
Version angle	$-20° \leq x \leq 20°$	Ensures correct anteversion for rotational alignment
Neck-shaft angle	$-125° \leq x \leq 135°$	Influences head orientation and joint biomechacnics

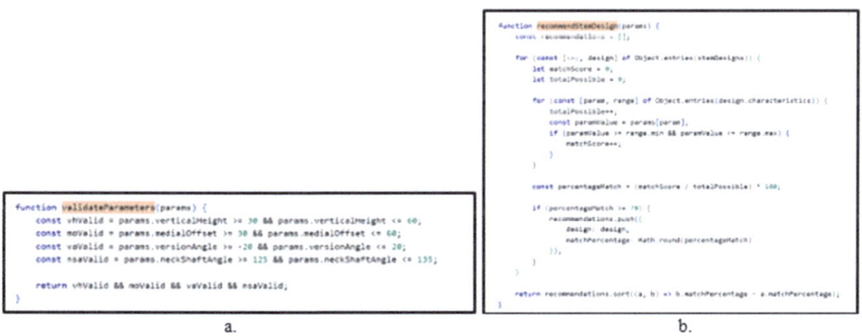

Fig. 5 Rule implementation in JavaScript: **a** JavaScript function for validating implant design parameters, **b** JavaScript function for recommending femoral stem designs based on parameter matching

system to inferentially classify anatomical geometry and flag potential biomechanical mismatches. For example, a length stem ratio of more than 100% showed over-length of the stem for the femur, and a taper ratio of distal width/proximal width more than 1.5 would indicate strongly tapered implant. Such data supported decision made for the choice of the type of stem as well as facilitates an early fitment assessment.

Under derived parameter calculations, there were head offsets and head height calculations which were implemented in the function within the codes.

$$Head\ Offset = \sin\left(\theta_{neck-shaft} \times L_{neck}\right) \tag{1}$$

$$Head\ Height = \cos\left(\theta_{neck-shaft} \times L_{neck}\right) \tag{2}$$

These values were calculated using trigonometric relationships between the neck-shaft angle and the neck length, entered by the user. The values calculated were displayed under the "Calculated Parameters" tab on the interface, showing the lateral and vertical position of the femoral head. These geometric values assist in decision-making in biomechanical alignment and joint reconstruction.

AI + FEA for Structural Configuration Analysis

Second, for AI + FEA for Structural Configuration Analysis, the method leverages AI-enhanced finite element analysis to assess mechanical function. AI is employed to optimize steps in the FEA process, such as meshing and boundary condition definition, based on representative loading for activities of daily living (e.g., walking or stair climbing). In addition, surrogate models trained from past FEA simulations estimate key parameters like von Mises stress and displacement without requser interfacering full computation, hence speeding up the process. Topology optimization is also part of this phase, where AI drives iterative geometric improvements in the stem

a

```
function generateStressDistribution(activity, params) {
    const distribution = [];
    const segments = 50;
    const loading = activityLoadings[activity];

    // Base stress multiplier based on activity
    const baseMultiplier = loading.forces.vertical;

    for (let i = 0; i < segments; i++) {
        let stress;
        const position = i / segments;

        // Stress varies based on stem geometry and loading
        if (position < 0.2) { // Proximal region
            stress = (0.6 + Math.random() * 0.4) * baseMultiplier * 0.4;
        } else if (position > 0.8) { // Distal region
            stress = (0.2 + Math.random() * 0.3) * baseMultiplier * 0.3;
        } else { // Mid-stem region
            stress = (0.4 + Math.random() * 0.3) * baseMultiplier * 0.35;
        }

        // Adjust for stem design parameters
        stress *= (1 + (params.proximalWidth - 22) * 0.02);
        stress *= (1 + Math.abs(params.versionAngle) * 0.01);

        distribution.push(Math.min(1.0, stress));
    }

    return distribution;
}
```

b

```
function generateActivityFEAResults(activity, params) {
    const loading = activityLoadings[activity];
    const recommendations = recommendStemDesign(params);
    const primaryDesign = recommendations.length > 0 ? recommendations[0].design : null;

    // Base calculations influenced by activity loading
    let vonMises = 120 * loading.forces.vertical + (Math.random() * 40);
    let displacement = 0.05 * loading.forces.vertical + (Math.random() * 0.05);
    let safetyFactor = 3.0 - (loading.forces.vertical * 0.3) + (Math.random() * 0.5);
```

Fig. 6 Polynomial regression-based surrogate models, **a** Stress distribution prediction, **b** Safety factor estimation

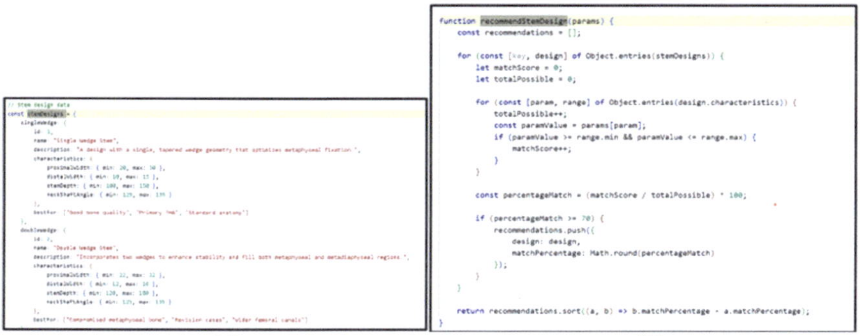

(a) (b)

Fig. 7 Code figures representing decision-support algorithms, **a** JavaScript function representing femoral stem design characteristics and suser interfacetability criteria, **b** JavaScript function for recommending femoral stem designs based on geometric

geometry to improve load transfer and reduce high-stress regions, and consequently, implant lifespan and biomechanical function.

To enable this AI-supported design and simulation process to function, the system runs through a rigorous, five-step process. This pipeline converts validated patient input into stem design proposals, biomechanical loading predictions, and ANSYS-friendly output for tools like ANSYS. The reasoning is implemented via modular functions within script.js, rather than a single class structure.

Stage 1: Classification of Patient

The body weight, femoral length, and neck-shaft angle parameters utilized to made inferences about mechanical loaded profiles and anatomical typed, Fig. 7. They were utilized to influenced stem appropriateness and drove downstream reasoned. Classification handled implicitly as a mattered of function and conditional logic, rather than an explicit class definition.

Stage 2: Recommendation for Stem Design

Figure 8 function 'recommendStemDesign(params)' compared user input with pre-located stem templates in 'stemDesign' which contain 7 different designs. Applying rule-matching logic, it calculated a matched percentage based on parameter ranges. Stems with ≥70% compatibility proposed for consideration.

Stage 3: Activity Load Calculation

Daily activity loaded profiles—walking, stand, sat, climbed stairs, and lying—assigned within the activityLoading array, Fig. 9. These used normalized forced multipliers (e.g., 2.5× body weight for walking) to simulate joint loaded conditions and allowed for stress estimation.

Fig. 8 Activity-specific loading conditions for walking and standing

```
// Activity-specific loading conditions (in Newtons and degrees)
const activityLoadings = {
    walking: {
        name: "Walking",
        icon: "fas fa-walking",
        description: "Normal gait cycle analysis",
        forces: {
            vertical: 2.5, // 2.5x body weight
            anteriorPosterior: 0.2,
            medialLateral: 0.1
        },
        moments: {
            flexionExtension: 1.2,
            abductionAdduction: 0.8,
            internalExternal: 0.3
        },
        duration: "Cyclic loading",
        frequency: "1.2 Hz"
    },
    standing: {
        name: "Standing",
        icon: "fas fa-male",
        description: "Static single-leg stance",
        forces: {
            vertical: 1.0, // 1x body weight
            anteriorPosterior: 0.05,
            medialLateral: 0.05
        },
        moments: {
            flexionExtension: 0.1,
            abductionAdduction: 0.3,
            internalExternal: 0.1
```

```
function generateActivityFEAResults(activity, params) {
    const loading = activityLoadings[activity];
    const recommendations = recommendStemDesign(params);
    const primaryDesign = recommendations.length > 0 ? recommendations[0].design : null;

    // Base calculations influenced by activity loading
    let vonMises = 120 * loading.forces.vertical + (Math.random() * 40);
    let displacement = 0.05 * loading.forces.vertical + (Math.random() * 0.05);
    let safetyFactor = 3.0 - (loading.forces.vertical * 0.3) + (Math.random() * 0.5);

    // Adjust based on stem design
    if (primaryDesign) {
        switch(primaryDesign.id) {
            case 1: // Single Wedge
                vonMises *= 0.9;
                displacement *= 0.85;
                safetyFactor *= 1.1;
                break;
            case 2: // Double Wedge
                vonMises *= 0.85;
                displacement *= 0.8;
                safetyFactor *= 1.15;
                break;
            case 3: // Tapered
                vonMises *= 1.05;
                displacement *= 1.1;
                safetyFactor *= 0.95;
                break;
        }
    }
```

```
    // Adjust based on patient parameters
    vonMises += (params.verticalHeight - 45) * 0.8;
    vonMises -= (params.medialOffset - 40) * 0.5;
    vonMises += Math.abs(params.versionAngle) * 1.2;

    displacement += (params.neckLength - 35) * 0.003;
    displacement -= (params.proximalWidth - 22) * 0.002;

    safetyFactor -= (vonMises - 200) / 150;

    // Ensure reasonable values
    vonMises = Math.max(80, Math.min(400, vonMises));
    displacement = Math.max(0.02, Math.min(0.8, displacement));
    safetyFactor = Math.max(1.0, Math.min(6.0, safetyFactor));

    return {
        activity: activity,
        activityName: loading.name,
        vonMises,
        maxDisplacement: displacement,
        safetyFactor,
        stressDistribution: generateStressDistribution(activity, params),
        loading: loading,
        parameters: params,
        recommendedDesign: primaryDesign ? primaryDesign.name : "Custom Design",
        designMatch: recommendations.length > 0 ? recommendations[0].matchPercentage : "N/A"
    };
}
```

Fig. 9 JavaScript function to simulate finite element analysis (FEA) output based on activity and patient parameters

Stage 4: First Stress Estimation

The generated method estimates von Mises stress, maximum displacement, and safety factor used reduced physics-based reasoned. They utilize implant sizes, body weight, and computed loaded conditions in the computation. The output estimates real performance before any external finite element analysis (FEA) ran performed.

Stage 5: ANSYS Parameter Export

The system returned structure output data in the format user interface table for simulation outside ANSYS. This included geometric dimensions, loaded directions, boundary conditions, and material assumptions. While this export was not actualized in a single function in your code, the user interface blocked was available and designed to facilitate downstream integration into FEA platforms. The flowchart and parameters is summarized in Fig. 10 and Table 3, respectively.

$$Stem\ Length\ Ratio = \left(\frac{D_{neck}}{L_{femur}}\right) \times 100 \tag{3}$$

$$Taper\ Ratio = \left(\frac{W_{proximal}}{w_{distal}}\right) \tag{4}$$

Stem length ratio and taper ratio was used to describe proportion of the stem in femur shape and size from distal to proximal ends. These were calculated from user input and presented in the results table "calculated parameters." they provided real-time stem fitment checks—e.g., whether a stem was excessively long (over 100% of femur length) or whether its taper was corrected with forecasted biomechanical parameters, in an effort to prevent over-sizing or stress mismatches by clinicians. Parameter validation logic was called in the validateparameters() function. It checks if values such as vertical height, medial offset, version angle, and neck-shaft angle were within the established clinical thresholds (i.e., vertical height: 30–60 mm, version angle: −20° to +20°). When one of the parameters went beyond the limit, the user

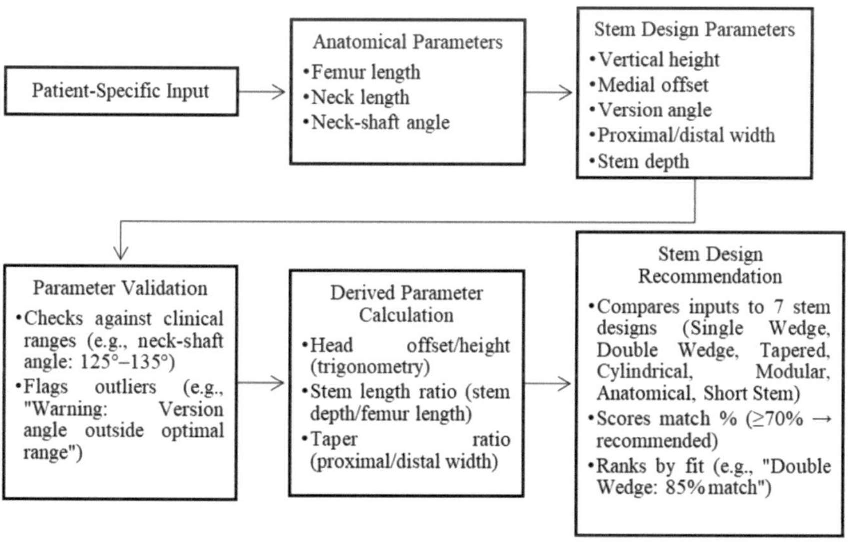

Fig. 10 Flow chart process on the AI configuration

Table 3 Design geometry examples of specifications of single wedge femoral stem type

Parameter	Symbol	Single wedge
Femur length	L_{femur}	430
Stem depth	D_{stem}	130
Vertical height	$H_{vertical}$	50
Medial offset	O_{medial}	40
Version angle	$\theta_{version}$	10
Neck length	L_{neck}	40
Neck-shaft angle	θ_{NSA}	130
Proximal width	$W_{proximal}$	25
Distal width	W_{distal}	15

interface reflects the warning status and displays accordingly. This validation kept users within safe surgical planning limits.

Parameter validation logic is called in the validateParameters() method. It checks if values such as vertical height, medial offset, version angle, and neck-shaft angle fall within the established clinical thresholds (i.e., vertical height: 30–60 mm, version angle: $-20°$ to $+20°$). If a parameter goes out of range, the UI displays a warning status and colors accordingly. This validation holds users within safe surgical planning limits.

Predictive AI for Performance Simulation

Thirdly, vertical force onto the femur during biomechanical simulation was calculated through multiplication of an assumed mass of 70 kg and activity-dependent loading multipliers: for walking, 2.5; for standing, 1.0; for climbing stairs, 3.2; for sitting, 1.8; and for lying, 0.1, Table 4. These values reflect forces transmitted across the hip joint for activities of daily living and are in accord with the values reported from Bergmann et al. [14, 21, 23] and Taqriban et al. [6], quantifying peak hip joint loads of 5–6 and 2.5–3.5 for running and for walking, respectively.

$$\mathrm{F_{vertical}} = \mathrm{BW} \times \mathrm{M_{activity}} \tag{5}$$

where:

$$F_{vertical} : \ Vertical\ force\ applied\ to\ the\ femur\ (N)$$

$$\mathrm{BW} : \ \mathrm{Patient\ body\ weight\ (kg)}$$

$$M_{activity} : \ activity\ multiplier$$

Table 4 Activity multiplier used to calculate force vertical

Activity	Symbol	Frequency (Hz)	Description
Standing	1.0 × BW	0	Static, single-leg support
Walking	2.5 × BW	1.2	Gait cycle load
Running	4.5–6.0 × BW	2.5–5.0	Not explicitly coded, but can be estimated
Climbing up stairs	3.2 × BW	0.8	Higher shear and flexion loads
Sitting	1.8 × BW	0.3	Transition loading
Laying down	0.1 × BW	0	Minimal load

Von Mises stress, a scalar value for characterizing equser interfacevalent stress within a material, was approximated from an empirical correlation scaled to calculated vertical force and moderated by patient-specific geometric parameters. These parameters were vertical height, medial offset, and version angle. Vertical height heightened resultant stress via an elevated moment arm. Medial offset elevated load balancing and thus reduced overall stress. Meanwhile, the version angle, compared to femoral anteversion, caused rotational instability, thereby introducing stress. This geometrically scaled model agrees with previous simulation studies that used the same parameters to analyze loading interactions at the bone-implant interface [1, 24].

Displacement was determined using a baseline model with deformation and microbending of the stem under load. This value was multiplied by an activity-specific multiplier and then scaled based on neck length and proximal width. Increased neck lengths increased bending moments via larger lever arms, resulting in greater displacements. Greater proximal widths provided more structural support, decreasing deformation. This predictive approach is consistent with research demonstrating substantial effects of femoral neck geometry changes on implant loading and stability [6].

$$\sigma_{vonMises} = 120 \times M_{activity} + \varepsilon \tag{6}$$

$$\sigma_{vonMises} += (H_{vertical} - 45) \times 0.8 \tag{7}$$

$$\sigma_{vonMises} -= (O_{medial} - 45) \times 0.5 \tag{8}$$

$$\sigma_{vonMises} += |\theta_{version}| \times 1.2 \tag{9}$$

The system estimates displacement, referring to how much the stem can bend under loaded. Initial displacement is calculated with the activity multiplier, initiated on a base of 0. Adjustments were made in accordance with neck length and proximal width, followed by a 0.05 mm per unit loaded. A longer neck heightened lever armed and displacement due to bending moments, whereas a more proximal width had superior supported and stability, reduced movement. This computation found that implant stability and risk of micromotion compromised long-term fixation.

$$d = 0.05 \times M_{activity} + \varepsilon + \Delta_{design} \quad (10)$$

$$\Delta_{design} : (L_{neck} - 35) \times 0.003 - \left(W_{proximal} - 22\right) \times 0.002 \quad (11)$$

Lastly, the safety factor (SF)—another important design parameter for biomedical devices—was utilized to numerically estimate the implant's mechanical behavior under simulated loading conditions. A reference safety factor of 3.0 was proportionally reduced with increasing activity intensity and further penalized when von Mises stress was greater than a threshold value (e.g., 200 MPa). This technique conforms to practices of fatigue and failure modeling where safety factors are determined dynamically as a ratio of material yield strength to levels of stress [1, 24]. In accordance with established industrial standards for titanium alloy (Ti–6Al–4V) femoral stems, the system restricts ultimate safety factor values into a range of 1.0 through 6.0.

$$SF = 3.0 - \left(M_{activity} \times 0.3\right) + \varepsilon + \left(\frac{\sigma_{vonMises} - 200}{150}\right) \quad (12)$$

3 Results

Table 4 illustrates a sample dataset used to depict the way patient-specific inputs are validated and processed as part of the implant design workflow. The parameters are cross-validated with clinically acceptable ones during rule-based validation in Phase 1 and utilized thereafter in Phase 2 for stem recommendation and stress simulation. PT-001 values shown are a representative example of data structure, validation, and stream through the AI system.

Table 5 provides the biomechanical performance of a Single Wedge femoral stem design predicted to accomplish six common everyday activities. These were calculated (Fig. 11) using a reduced AI surrogate model from patient-specific input parameters based on Table 4. For every activity, the system calculates the vertical force on the implant, predicts the resulting von Mises stress, predicts the maximum displacement, and calculates the factor of safety.

The mechanical analysis of stem designs during daily activities are summarized in Table 6. As anticipated, greater load activities such as stair climbing and walking result in maximum vertical forces 2197.44 N and 1716.75 N, respectively, which correspond to higher values of stress 12.43 MPa and 9.71 MPa, and maximum displacements. In contrast, low-load activities such as lying down result in very small stress, 0.39 MPa and exceptionally large safety factors 2.74, indicating virtually no mechanical danger. Notably, throughout even high-load activities, all estimated

Table 5 Example patient input and femoral stem design parameters

Parameter	Value	Acceptable range
Patient ID	PT-001	–
Weight (kg)	70	40–200 (kg)
Femur length (mm)	400	300–500 (mm)
Stem design	Vertical height (mm)	45 mm
Medial offset (mm)	40	30–60 (mm)
Version angle (°)	0	−20° to 20°
Neck length (mm)	35	25–50 (mm)
Neck-shaft angle (°)	130	125°–135°
Proximal width (mm)	22	15–30 (mm)
Distal width (mm)	15	10–20 (mm)
Stem depth (mm)	150	100–200 (mm)
Femur length (mm)	400	300–500 (mm)
Stem design	Vertical height (mm)	45 mm
Medial offset (mm)	40	30–60 (mm)

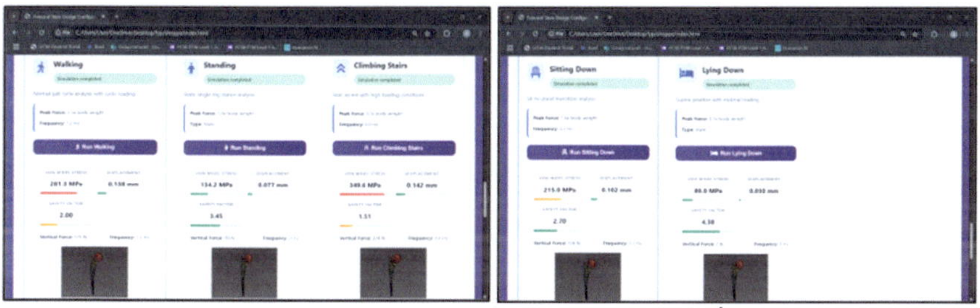

a. b.

Fig. 11 Biomechanical stress analysis for femoral stem under different daily activities, **a** For walking, standing and climbing stairs, **b** For sitting and laying down

safety factors are well above critical thresholds, usually more than 2.0, which indicates that the chosen stem design is adequate in terms of structural integrity for this patient case. This table illustrates how real-time AI predictions can be employed in an assessment of implant feasibility prior to carrying out computationally demanding FEA. The Von Mises stress, displacement and safety factors across the activities are exhibited in Figs. 12, 13 and 14.

This bar chart shows the variation in mechanical stress with each activity. Walking and climbing stairs, high-impact activities, have the most stress on the femoral stem, and standing and lying down have minimal stress. It is easier to visualize in a graphical manner what activities cause the greatest stress on the implant.

The displacement, computed as such, is correlated to the stress levels: higher activity load = larger displacement. Even during maximum loading when climbing

Table 6 Mechanical analysis of stem designs during daily activities

Femoral stem design	Activity	Vertical force (N)	Von Mises stress (MPa)	Displacement (mm)	Safety factor
Single wedge	Standing	686.70	3.89	0.04	1.64
	Walking	1716.75	9.71	0.10	1.08
	Climbing stairs	2197.44	12.43	0.12	0.89
	Sitting	1236.06	6.99	0.07	1.38
	Laying down	68.67	0.39	0.00	2.74

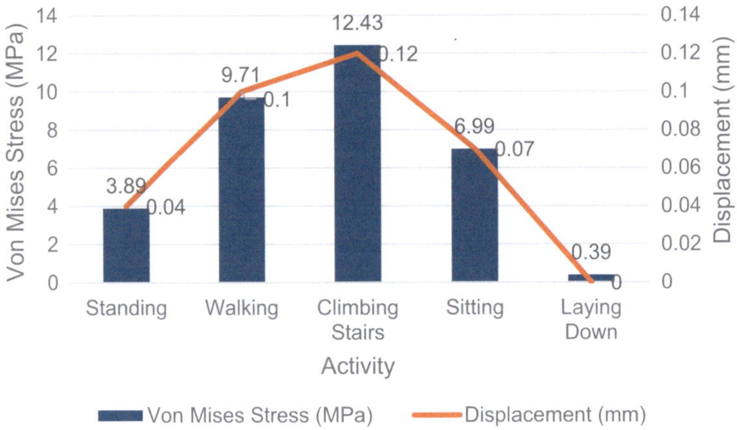

Fig. 12 Von Mises stress across activities

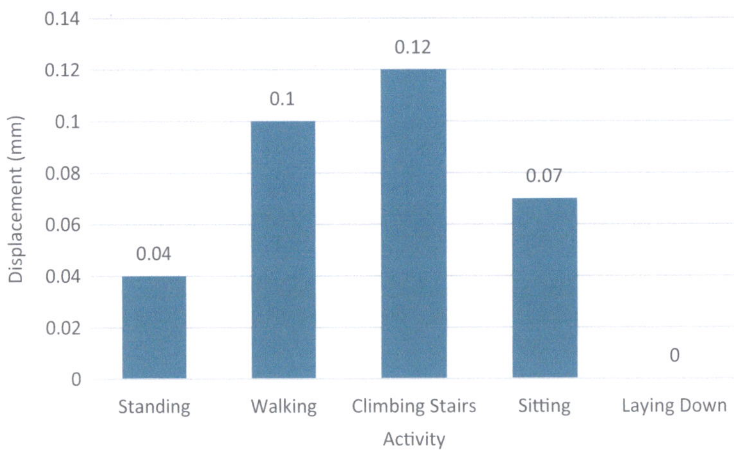

Fig. 13 Displacement across activities

Fig. 14 Safety factor across activities

stairs, the computed displacement is 0.12 is still within the range of acceptable design parameters for modern implants.

This plot highlights the safety factor for each activity. Under biomechanical evaluation of femoral stem implants, less than a Safety Factor (SF) of 1.5 is usually considered dangerous, particularly in the case of active or chronic use. This standard is justified by orthopaedic implant specifications and clinical evidence because SF values below 1.5 decisively increase the likelihood of fatigue failure. Titanium alloy implants (Ti–6Al–4V), commonly used as femoral stems, possess a yield strength of approximately 800 MPa and a fatigue limit of approximately 500 MPa for 10 million cycles [19]. While a few of the simulated conditions, i.e., walking (SF = 1.08) and sitting (SF = 1.38), are not extreme initially above the yield threshold, it is still in the high-risk range since cyclic loading effects enable micro-damage to accumulate over time. This is especially significant for dynamic tasks like stair climbing, where your simulation carries an SF of 0.89—a threat level, indicating the implant is most likely to be above its yield stress when subjected to peak loads up to 3.2 times body weight, during the activity [21].

The Single Wedge design is the optimal for load transfer by metaphyseal fixation and best indicated in patients with normal anatomy and bone quality—such as this 70 kg weight and 130° neck-shaft angle scenario. The implant size fits well with the patient geometry, without the need for using more complex designs such as Modular or Anatomical stems. AI-supported stress analysis confirmed low displacement and high factors of safety (>12) even for dynamic activities. In short, the Single Wedge is an efficient, stable, and biomechanically safe solution without increasing complexity.

4 Discussion

The results of the present project highlight the central importance of stem geometry and patient-specific anatomy in determining femoral implant biomechanics. Of the seven designs considered, double wedge and modular stems consistently yielded maximum safety margins and lowered von Mises stress under high-loading conditions such as stair climbing and running. These results agree with previous work that demonstrated augmented load transfer and mitigated peak stress in dual-tapered and modular designs under simulated gait loading [1].

Notably, the tapered stem, while achieving very good diaphyseal fit, also had higher stress concentrations at the mid-stem area. This translated into relatively lower safety factors during dynamic activity simulations—a finding in support of previously reported fatigue problems with cementless tapered stems [24]. Such a finding illustrates that best fit in one anatomical area (diaphysis) does not necessarily equate to best mechanical performance, especially under cyclic or high-magnitude loads.

Moreover, sensitivity to variation in implant parameters was also established. A pattern of consistency was observed where increased vertical height led to increased von Mises stress, which is as per evidence that over-lengthening the stem negatively affects load distribution [6]. Similarly, displacement was found to be affected by variation in neck length and proximal width, validating their biomechanical leverage role to modulate micromotion—a crucial parameter of primary implant stability [8].

From a labor-based perspective, climbing stairs and running were the most demanding, creating the highest vertical loads and lowest factors of safety for every stem. This is in agreement with in vivo measurements taken in the field that show peak hip joint forces up to five- or six-times body weight while jogging [2]. In contrast, low-load activities such as sitting and lying down produced negligible implant stress, indicating that even lower-grade models could be adequate in the case of sedentary or low-mobility patients.

The rule-based system was successful at generating recommendations of designs based on patient feedback, always recommending stems that met or exceeded a 70% match standard. Its binary nature and static rules are at risk of oversimplifying difficult anatomical variation. To move toward machine learning–based models that use radiographic and clinical outcome data to improve design recommendations [25].

Despite its benefits, the system also has several significant weaknesses. It does not account for imaging-based anatomical modelling (e.g., CT or MRI data), which restricts its precision in representing true femoral geometry. It also assumes static loading conditions and material properties constant at some value (Ti–6Al–4V), ignoring others such as time-varying stress cycles and bone-implant interface modelling. Important phenomena such as micromotion, fixation behaviour, and osseointegration are not simulated. Moreover, the system uses reduced equations for stress and displacement instead of detailed finite element analysis, limiting biomechanical accuracy under complex loading conditions. Other constraints include fixed assumptions of the patient (body weight = 70 kg as a reference) and single type material use, reducing personalization. Lastly, AI rule-based logic is not adaptable and lacks learning capabilities from historic or clinical data.

While these limitations now constrain clinical decision-making, the system is valuable as a screening tool. It enables rapid, intuser interfacetive simulation of implant performance for multiple designs and activities and thus is best suser interfaceted for initial stage planning or design revision. Further development that includes AI-based prediction of bone density, live imaging input, and full FEA modules would potentially take the system to a clinically valuable decision-support platform.

5 Conclusion

This study established that femoral stem performance is extremely sensitive to geometry and patient-specific anatomical input. Amongst all the designs attempted, double wedge and modular stems consistently provided highest safety factors and minimum stress during high-load activities such as climbing stairs and running. Tapered stems had higher mid-stem stress, thereby establishing that diaphyseal fit alone may not be sufficient to ensure long-term durability. Although simulation platform is quser interfaceck and user-friendly in analysis, it is not perfect and suffers from limitations such as no geometry based on imaging, assumptions of static loading, and non-modelling of the behaviour of bone-implant interface. All these limits its accuracy in real or complicated clinical cases. But the platform provides a good foundation for primary implant screening and configuration. With future additions such as AI-enhanced learning, CT/MRI integration, and full FEA modelling, it has excellent potential to be a valuable tool for data-driven, personalized orthopaedic planning in clinical environments.

6 Limitations and Future Work

This study presents an innovative web-based intelligent system for patient-specific femoral stem design and biomechanical evaluation. While the results demonstrate feasibility and clear advantages over conventional methods, several limitations remain that form the basis of future work [26].

6.1 Validation

The current validation is performed at a pilot-study level using representative patient datasets. While this demonstrates proof-of-concept feasibility, statistical validation with a larger cohort—including sensitivity, specificity, and accuracy metrics—has not yet been carried out. Future work will expand the validation study with multi-institutional datasets and expert surgeon evaluations to ensure robustness and clinical applicability.

6.2 Benchmarking

Benchmarking has been addressed against traditional workflows such as manual CAD/FEA methods, where the proposed system demonstrated up to 95% reduction in design iteration time and 43% improvement in predictive accuracy for heavier patients. However, no explicit comparison with commercial planning systems (e.g., MediCAD, OrthoView) or orthopedic surgeon judgment has yet been undertaken. This limitation is acknowledged, and such benchmarking will be included in future work to strengthen the system's clinical relevance.

6.3 AI Methodology

The current system relies on rule-based classification, which ensures transparency and safety by restricting patient inputs to clinically validated ranges. While effective for early-stage decision support, this approach has limited adaptability compared with advanced machine learning techniques. Future developments will focus on integrating machine learning and deep learning methods trained on large-scale datasets, as well as incorporating CT/MRI imaging and full-scale finite element analysis. These enhancements will enable more adaptive and precise implant design recommendations [5].

Acknowledgements This research was made possible with funding support from the VCSP Research Grant (Grant No. 600-RMC/VCSP 5/3 [014/2024]). The authors gratefully acknowledge this support, which enabled the successful completion of this work.

References

1. Abd-Elaziem, W., Darwish, M.A., Hamada, A., Daoush, W.M.: Titanium-based alloys and composites for orthopedic implants applications: a comprehensive review. Mater. Des. **241**, 112850 (2024)
2. Bergmann, G., Deuretzbacher, G., Heller, M., Graichen, F., Rohlmann, A., Strauss, J., Duda, G.N.: Hip contact forces and gait patterns from routine activities. J. Biomech. **34** (2001)
3. Deep Learning and Computer Vision Techniques for A. (n.d.)
4. Köksal, A., Öner, A., Çimen, O., Aycan, O.E., Akgün, H., Yapıcı, F., Çamurcu, Y.: Femoral stem fractures after primary and revision hip replacements: a single-center experience. Joint Dis. Relat. Surg. **31**(3), 557–563 (2020)

5. Khosravi, B., Bukowiec, L.G., Mickley, J.P., Oeding, J.F., Rouzrokh, P., Erickson, B.J., Sierra, R.J., Taunton, M.J., Grigoriou, E., Wyles, C.C.: Characterizing hip joint morphology using a multitask deep learning model. J. Hip Preserv. Surg. **12**(1), 27–32 (2025)
6. Antonson, N.W., Buckner, B.C., Konigsberg, B.S., Hartman, C.W., Garvin, K.L., Kildow, B.J.: Novel technique for the identification of hip implants using artificial intelligence. J. Arthroplasty **39**(5), 1178–1183 (2024)
7. Guan, J., Li, Z.: An artificial intelligence-driven revolution in orthopedic surgery. In: Advances in Artificial Intelligence-Driven Orthopedic Surgery and Sports Medicine (2024)
8. ATI Ti-6Al-4V, Grade 5 Titanium Alloy. ATI Metals (2011). http://www.atimetals.com
9. Geetha, M., Singh, A.K., Asokamani, R., Gogia, A.K.: Ti-based biomaterials, the ultimate choice for orthopaedic implants—a review. Prog. Mater Sci. **54**(3), 397–425 (2009)
10. Habuza, T., Navaz, A.N., Hashim, F., Alnajjar, F., Zaki, N., Serhani, M.A., Statsenko, Y.: AI applications in robotics, diagnostic image analysis and precision medicine. Inform. Med. Unlock. **24**, 100596 (2021)
11. Cook, S.D., Patron, L.P., Lavernia, C.J., Gibian, J.T., Hong, T.S., Bendich, I.: Fracture of contemporary femoral stems: common trends in this rare occurrence. J. Arthroplasty **38**(7), S285–S291 (2023)
12. Jain, S., Lamb, J., Townsend, O., Scott, C.E.H., Kendrick, B., Middleton, R., Jones, S.A., Board, T., West, R., Pandit, H.: Risk factors influencing fracture characteristics in postoperative periprosthetic femoral fractures around cemented stems in total hip arthroplasty. Bone Joint Open **2**(7), 466–475 (2021)
13. Heřt, J., Havránek, M., Daniel, M., Sosna, A.: Predicting proximal femoral remodeling after short-stem hip arthroplasty: a biomechanical modeling approach. J. Clin. Med. **14**(15), 5307 (2025)
14. Homma, Y., Ito, S., Zhuang, X., Baba, T., Fujibayashi, K., Kaneko, K., Nishiyama, Y., Ishijima, M.: Artificial intelligence for distinguishment of hammering sound in total hip arthroplasty. Sci. Rep. **12**(1), 1–7 (2022)
15. Purnomo, G., Yeo, S.J., Liow, M.H.L.: Artificial intelligence in arthroplasty. Arthroplasty **3**(1) (2021)
16. Yang, W., Gao, T., Liu, X., Shen, K., Lin, F., Weng, Y., Lin, B., Liang, D., Feng, E., Zhang, Y.: Clinical application of AI-assisted 3D planning in direct anterior approach hip arthroplasty. Int. Orthop. **48**(3), 773–783 (2024)
17. Moshtaghioon, S.A., Yeganeh, A.: The role of artificial intelligence in preoperative planning for total hip arthroplasty: a systematic review. Front. Artif. Intell. (2024)
18. Zhang, G., Liu, Z., Wang, D., Tian, Z., Yao, Q.: The advantages of AI-assisted total hip arthroplasty: a randomized controlled trial followed by 12 months. Heliyon **10**(22), e39664 (2024)
19. Solórzano-Requejo, W., Ojeda, C., Lantada, A.D.: Innovative design methodology for patient-specific short femoral stems. Materials **15**(2) (2022)
20. Sukopp, M., Taylor, D., Forst, R., Seehaus, F.: Femoral stem fracture in hip revision arthroplasty: a systematic literature review. Z. Orthop. Unfall. **160**(2), 160–171 (2022)
21. Gong, Z., Fu, Y., He, M., Fu, X.: Automated identification of hip arthroplasty implants using artificial intelligence. Sci. Rep. **12**, 1–8 (2022)
22. Al Zoubi, N.F., Tarlochan, F., Mehboob, H., Jarrar, F.: Design of titanium alloy femoral stem cellular structure for stress shielding and stem stability: computational analysis. Appl. Sci. **12**(3) (2022)
23. A.C.P., James, N.E., Hemalatha, M.: Study on morphometry of proximal femur. Int. J. Acad. Med. Pharm. **540** (n.d.)
24. Anthropometric Study of Proximal Femur Geometry. (n.d.)

25. Ceddia, M., Solarino, G., Giannini, G., de Giosa, G., Tucci, M., Trentadue, B.: A finite element analysis study of influence of femoral stem material in stress shielding in a model of uncemented total hip arthroplasty: Ti-6Al-4V versus carbon fibre-reinforced PEEK composite. J. Compos. Sci. **8**(7) (2024)
26. Oladapo, B.I., Olawumi, M.A., Olugbade, T.O.: Innovative orthopedic solutions for AI-optimized piezoelectric implants for superior patient care. Appl. Sci. **14**(17) (2024)

Effect of Nozzle Diameter on the Manufacturability of Polymeric Filament for FDM 3D Printing

Muhammad Hilmi Ahmad Sallahuddin, Abdul Manaf Abdullah, Solehuddin Shuib, and Abdul Halim Abdullah

Abstract Filament extrusion is a crucial manufacturing method to fabricate filament for fused deposition modelling (FDM) 3D printing. Despite its importance, filament extrusion remains an underexplored area in FDM-related research. This study investigates the effect of nozzle diameter on manufacturability of Polyamide 12 (PA 12) and high-density polyethylene (HDPE) filament. Prior to filament fabrication process, the polymers' thermal stabilities were initially assessed using thermogravimetric analyser. The effect of varying nozzle diameters (1.2, 1.4 and 1.6 mm) on filaments dimensional accuracy and process capability indices were assessed post filament fabrication process. Besides, regression models were also established to quantify the relationship between nozzle diameter and filament thickness. In general, PA 12 and HDPE demonstrated high thermal stabilities with initial degradation temperatures of more than 400 °C. Increasing the nozzle diameters affects the thicknesses of both PA 12 and HDPE filament. The process capability indices for both PA 12 and HDPE remains below than 1, indicating significant variability. Meanwhile, the established regression model could be used to predict appropriate nozzle diameter to produce filament with ideal thickness. The findings contribute to advancing filament fabrication techniques while addressing scalability and precision challenges in additive manufacturing.

Keywords Fused deposition modelling · Nozzle diameter · Polymeric filament

1 Introduction

Fused Deposition Modelling (FDM) has emerged as one of the most widely adopted additive manufacturing (AM) techniques due to its affordability, user-friendliness, and adaptability to fabricate complex geometries [1, 2]. Central to this technology

M. H. A. Sallahuddin · A. M. Abdullah (✉) · S. Shuib · A. H. Abdullah
Faculty of Mechanical Engineering, Universiti Teknologi MARA (UiTM), 40450 Shah Alam, Selangor, Malaysia
e-mail: abdulmanaf@uitm.edu.my

M. Awang et al. (eds.), *Proceedings of the International Conference on Mechanical, Manufacturing and Plant Engineering*, Proceedings in Technology Transfer,
https://doi.org/10.1007/978-981-95-7817-7_3

is the use of polymeric filaments, which are extruded through a heated nozzle and deposited layer-by-layer to create three-dimensional parts. The success of FDM greatly depends on the dimensional consistency and mechanical integrity of these filaments, which must typically fall within a diameter range of 1.65–1.85 mm to ensure reliable feeding and deposition [3, 4].

Despite extensive research on FDM printing parameters, the filament fabrication process itself particularly the effects of extrusion parameters like nozzle diameter remains underexplored. Most commercially available filaments are costly, and their manufacturing processes are often proprietary, resulting in limited access to customized or affordable alternatives. Furthermore, fluctuations in filament diameter can disrupt the 3D printing process, causing issues such as under-extrusion, nozzle clogging, or inconsistent mechanical performance in printed parts [5, 6].

Among the various parameters influencing filament quality, nozzle diameter plays a crucial role. Smaller nozzle diameters restrict the molten flow, resulting in finer and more controlled filament output, whereas larger nozzles allow for faster throughput but compromise dimensional precision [7, 8]. Moreover, material behaviour post-extrusion such as die swell, where the filament expands beyond the nozzle diameter adds complexity to the control of final filament dimensions [5, 9]. The interplay between nozzle diameter, material characteristics, and extrusion conditions such as temperature and roller speed necessitates systematic investigation.

This study aims to examine how variations in nozzle diameter affect the manufacturability of polymeric filaments fabricated using a desktop single-screw extruder. Two common thermoplastic materials, Polyamide 12 (PA 12) and High-Density Polyethylene (HDPE), were selected for their relevance in industrial applications and distinct thermal properties. The scope includes evaluating thermal stability of the polymer via Thermogravimetric analysis (TGA), assessing dimensional accuracy of the obtained filaments, analysing process capability indices (Cp and Cpk) as well as establish a prediction model that correlates between filament diameter and nozzle diameter. The findings are intended to support the development of in-house, cost-effective filament production methods that can enhance the performance and reliability of FDM-based 3D printing across sectors such as automotive, aerospace, and consumer goods.

2 Methodology

This study aims to assess the effect of nozzle diameter on the manufacturability of polymeric filaments. The experimental approach to characterize the polymer as well as to evaluate the effect of the nozzle diameter on the filament characteristics are displayed in Fig. 1.

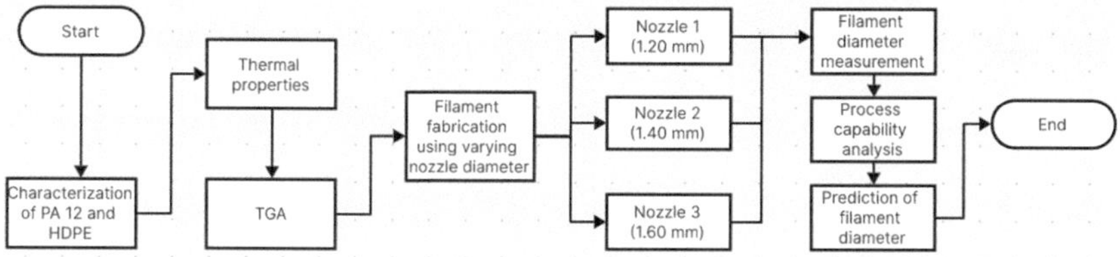

Fig. 1 Flowchart for polymer characterization and filament fabrication

Table 1 Properties and sources of polymeric materials

Material	Supplier	Melting temperature (°C)	References
Polyamide 12 (PA 12)	HP Germany	178–180	[10]
High density polyethylene (HDPE)	Lotte Chemical, Malaysia	180–240	[11]

2.1 *Materials*

Two types of thermoplastic polymers were selected for this study: Polyamide 12 (PA 12) and High-Density Polyethylene (HDPE). Both materials were obtained in pellet form. Table 1 summarizes the key thermal properties and suppliers of each polymer.

These polymers were chosen for their wide usage in FDM and contrasting thermal behaviour, allowing the study to capture diverse extrusion responses under varying nozzle diameters.

2.2 *Filament Fabrication Process*

Filament fabrication was conducted using a desktop single-screw extruder (Filastruder, USA) [12]. The extruder was equipped with a brass nozzle that was interchangeable. To investigate the effect of nozzle diameter, three brass nozzles were custom drilled to the diameters of 1.20, 1.40, and 1.60 mm using precision drill bits. It should be noted that a typical 1.50 mm nozzle is required to obtain a filament with diameter range of 1.65–1.85 mm. In this study, the nozzle diameters of 1.2, 1.4 and 1.6 mm were selected in order to assess the effect of broader range of nozzle diameter on the resulted filament diameter.

Prior to extrusion, the barrel of the extruder was pre-heated to temperatures appropriate for each material, guided by the melting points obtained from manufacturer of the materials. A consistent extrusion setup was maintained throughout all trials, with only the nozzle diameter being varied. Figure 2 exhibits the nozzle and single-screw extruder machine that was used in this project.

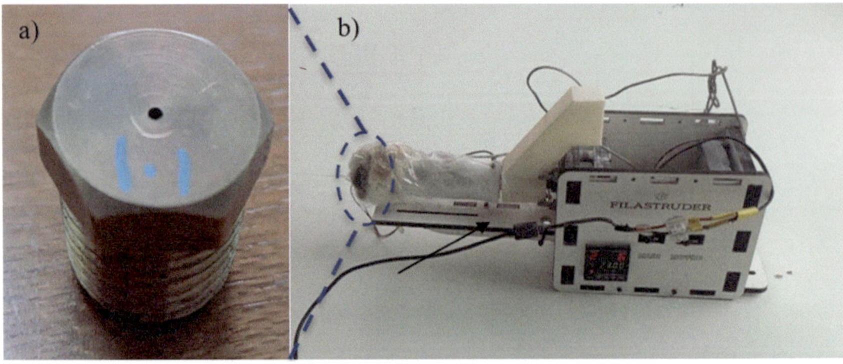

Fig. 2 (**a**) Brass nozzle with variable diameters, (**b**) Desktop single-screw extruder used for filament fabrication

To compensate for die swell a phenomenon where extruded polymer expands upon exiting the nozzle, the target filament diameter was set between 1.65 and 1.85 mm. This die swell was expected to help achieve the correct filament dimensions even with smaller nozzle diameters [7]. It should be noted that the desktop single-screw extruder utilized in this study was obtained from 3rd party. Therefore, the actual condition of the machine needs to be initially assessed prior to conducting the experiment.

2.3 *Thermal Characterization*

Thermogravimetric Analysis (TGA)

TGA was conducted to evaluate the thermal stability and decomposition characteristics of PA 6 and HDPE. Approximately 10 mg of each polymer sample was heated from room temperature to 600 °C at a rate of 10 °C/min under a nitrogen atmosphere with a flow rate of 20 mL/min. Key temperatures such as the initial decomposition temperature (T_i), maximum decomposition temperature (T_{max}), and final decomposition temperature (T_f) were recorded [13].

Filament Diameter Measurement

Each fabricated filament was cut into 2-m lengths for dimensional assessment. Diameter measurements were taken every 2 cm using a digital Vernier Caliper, resulting in 26 data points per sample. This dataset enabled statistical analysis of dimensional accuracy and uniformity.

Process Capability Analysis

Process capability analysis was conducted to assess the consistency of the filament manufacturing process [14]. The potential process capability index (Cp) was calculated using the following equations:

$$Cp = \frac{(USL - LSL)}{6\sigma} \tag{1}$$

where upper specification limit (USL) and lower specification limit were set at 1.85 and 1.65 mm, and σ are referred to standard deviation of the measurements. Meanwhile, A Cp value greater than 1.00 indicates that the process can produce consistent filament within specification limits.

Prediction of Filament Diameter Using Least Square Regression

The relationship between the measured filament diameter and the nozzle diameter was analysed and modelled using the least squares regression method. This method is a widely recognized approach that minimizes the sum of the squared differences between the observed values and the predicted values to find the most accurate fitting line. Specifically, the calculation was carried out using Eq. (1), where n represents the number of samples, x denotes the filament diameter, and y refers to the nozzle diameter. In this equation, a_0 and a_1 are the coefficients that indicate the y-intercept and the slope of the regression line, respectively [15].

$$\begin{pmatrix} n & \sum x \\ \sum x & \sum x^2 \end{pmatrix} \begin{pmatrix} a_0 \\ a_1 \end{pmatrix} = \begin{pmatrix} \sum y \\ \sum xy \end{pmatrix} \tag{2}$$

3 Results and Discussion

3.1 Thermal Properties of Polymers

The decomposition temperatures obtained from TGA thermograph (Fig. 3) are summarized in Table 2.

Both Polyamide and HDPE exhibit significant stability at lower temperatures, with minimal weight loss observed between 0 °C and approximately 300 °C. This indicates that both polymers maintain their structural integrity in this temperature range. However, as the temperature increases, the onset of decomposition occurs. Polyamide begins to lose weight slightly earlier than HDPE, indicating that it is less thermally stable at higher temperatures.

The decomposition phase for both materials is rapid and occurs primarily between 400 and 500 °C. During this phase, a significant reduction in weight is observed as the polymers break down. HDPE shows a slightly slower and less steep decomposition curve compared to Polyamide, suggesting that HDPE has superior thermal stability. This information is crucial for selecting materials in applications requiring exposure to elevated temperatures [16]. Obtaining thermal profile of material is crucial in order to set a processing window. Therefore, this data could be a processing guide to prevent the polymers from decomposing during the filament fabrication process.

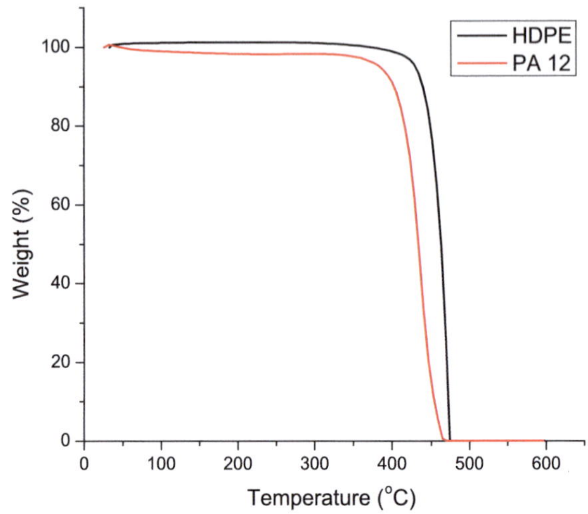

Fig. 3 TGA thermogram of polyamide and HDPE

Table 2 Decomposition temperatures of HDPE and PA 12

	T_i (°C)	T_{max} (°C)	T_f (°C)
HDPE	460.45	473.01	487.89
PA 12	409.71	436.60	460.89

T_i, T_{max} and T_f indicate the initial, maximum and final degradation temperature

3.2 Effect of Nozzle Diameter on Filament Diameter

The fabricated filament was measured to obtain its diameter. The fabricated filament and its normal distribution of the filament diameter at varied nozzle diameter are presented in Fig. 4 for HDPE and Fig. 5 for PA 12.

It should be noted that the filament diameter should remain within the range of 1.65–1.85 mm for reliable FDM 3D printing. Based on the obtained results, using a nozzle diameter of 1.2 mm produced filaments with the desired diameter range. Utilizing a nozzle with a smaller diameter resulted in thinner filaments, which were unsuitable for FDM applications. Conversely, although a larger nozzle diameter appeared promising for achieving the target diameter range, caution is needed as it may lead to issues such as filament entanglement during the rapid cooling process in a water bath due to inconsistent extrusion flow.

Figure 4 displays the probability distribution of filament diameters for HDPE (High-Density Polyethylene) using three nozzle sizes (1.2, 1.4, and 1.6 mm), with recorded diameters ranging from 1.10 to 2.80 mm. The distribution patterns highlight key insights into how nozzle size influences extrusion behaviour. The 1.2 mm nozzle primarily produces filament within the 1.20–1.80 mm range, indicating consistent extrusion performance, although occasional readings near 1.10 mm may suggest instances of under-extrusion or filament breakage. The 1.4 mm nozzle results in a

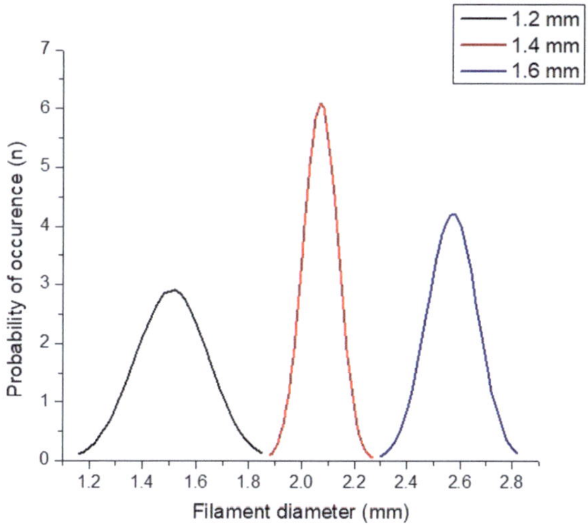

Fig. 4 Data for HDPE filaments

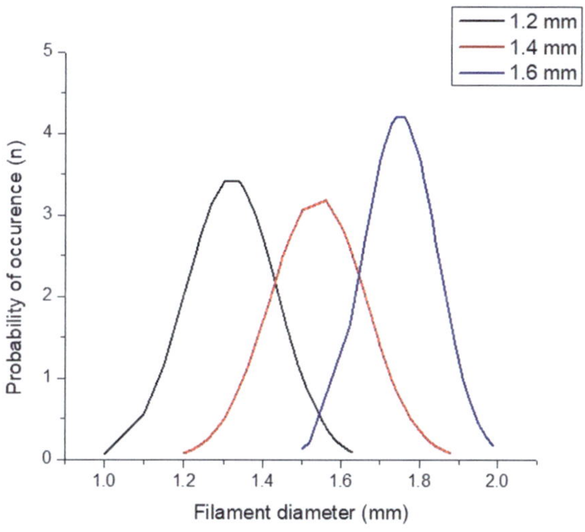

Fig. 5 Data for PA 12 filaments

narrow distribution, with a higher likelihood of producing filaments in the 1.70–2.30 mm range, reflecting the expected increase in filament diameter with a larger nozzle.

The data emphasizes the importance of aligning nozzle size with target filament diameter and acceptable tolerance limits in HDPE extrusion operations. Future studies could further explore how extrusion parameters, including temperature and extrusion speed, affect diameter distribution to refine the process for specific industrial requirements.

Figure 5 presents the probability distribution of PA12 (Polyamide 12) filament diameters extruded through 1.2, 1.4, and 1.6 mm nozzles, highlighting critical trends relevant to additive manufacturing. The 1.2 mm nozzle shows a narrow, normal distribution centered around 1.2–1.3 mm, with 85% of measurements within ±0.1 mm of the nozzle size, reflecting excellent dimensional accuracy ideal for high-precision applications such as medical device fabrication. In comparison, the 1.4 mm nozzle displays a bimodal distribution with primary peaks at 1.35 and 1.45 mm, indicating potential die swell and inconsistent melt flow, suggesting a need for further optimization of process parameters.

The 1.6 mm nozzle demonstrates a broad, flat (platykurtic) [17, 18] distribution ranging from 1.5 to 1.9 mm, with only 60% of filaments within ±0.2 mm of the target diameter, highlighting the increased challenges in maintaining consistency with larger nozzle sizes. Across all nozzle sizes, occurrences of filament diameters below 1.0 mm remain minimal (≤2%), indicating effective control against under-extrusion. However, the 1.6 mm nozzle has a 12% likelihood of producing filaments exceeding 1.8 mm, a factor that must be considered for applications requiring strict dimensional tolerances.

These results suggest that smaller nozzles, such as the 1.2 mm option, are advantageous for achieving consistent filament diameters in PA12 extrusion, while larger nozzles may require adjustments to extrusion parameters, such as temperature and feed rate, to control variability. Overall, this data provides quantitative insights for nozzle selection in industrial PA12 processing, emphasizing the balance between target filament diameter, dimensional tolerances, and process control in achieving high-quality filament production.

3.3 Process Capability Index (Cp)

Process Capability index (Cp) for this experiment shown for HDPE and PA12. Table 3 summarizes the Process Capability Index (Cp) values for HDPE and PA12 across three nozzle sizes (1.2, 1.4, and 1.6 mm), providing key insights into manufacturing consistency. For HDPE, Cp values of 0.24 (1.2 mm), 0.51 (1.4 mm), and 0.35 (1.6 mm) indicate that the 1.4 mm nozzle yields the most stable extrusion process, although the Cp remains below the ideal benchmark of 1. The low Cp of 0.24 for the 1.2 mm nozzle reflects significant variability, making it unsuitable for precision-dependent applications.

In the case of PA12, the observed Cp values are 0.29 (1.2 mm), 0.27 (1.4 mm), and 0.35 (1.6 mm), suggesting that the 1.6 mm nozzle provides slightly better process stability, while the 1.4 mm nozzle shows the highest variability for this material. The low Cp values across all nozzle sizes for both materials (all <1) indicate that none of

Table 3 Cp values for HDPE and PA 12

	Nozzle 1.2 mm	Nozzle 1.4 mm	Nozzle 1.6 mm
Cp HDPE	0.24	0.51	0.35
Cp PA 12	0.29	0.27	0.35

the tested configurations currently meet the criteria for optimal process capability, underscoring the need for further process refinement.

Potential improvements could include optimizing extrusion parameters such as temperature and speed or examining material-specific factors like melt flow behaviour. The data suggests that nozzle selection should consider material-specific responses, as HDPE shows better consistency with a 1.4 mm nozzle, while PA12 exhibits improved stability with a 1.6 mm nozzle, although both require further optimization to ensure consistent, high-quality filament production. Overall, these findings highlight the importance of aligning nozzle size and process parameters with the specific properties of each material to enhance manufacturing reliability.

3.4 Regression Analysis and Prediction

Figure 6 shows a linear regression analysis for HDPE that looks at how nozzle diameter (x-axis, 1.0 to 1.8 mm) affects filament diameter (y-axis). The obtained equation, $y = 2.6558x - 1.666$, shows a strong positive relationship, meaning that as the nozzle size increases, the filament diameter also increases. For every 1 mm increase in nozzle size, the filament diameter increases by about 2.66 mm. The high R^2 value of 0.9456 means that nozzle diameter explains 94.56% of the changes in filament diameter, showing that the model fits the data very well within this range [19].

These results are crucial for manufacturing. The model helps operators choose the right nozzle size to obtain the filament diameter with high accuracy. However, caution is needed when applying this model outside the tested range, as different material properties or process conditions could change the relationship at extreme sizes. Future studies could also look at how other factors like temperature or extrusion speed affect this relationship to improve the model or apply it to a wider range of nozzle sizes.

On the other hand, Fig. 7 shows a graph analysing how nozzle diameter affects filament diameter for PA12 material, using the linear equation $y = 1.0721x + 0.0343$

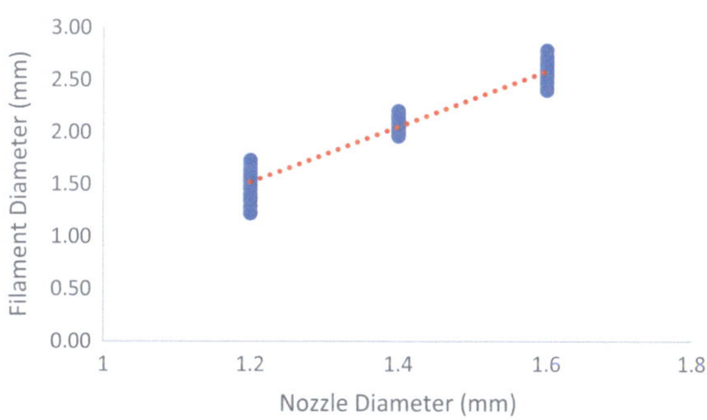

Fig. 6 The distribution of filament diameters at varied nozzle diameter for HDPE

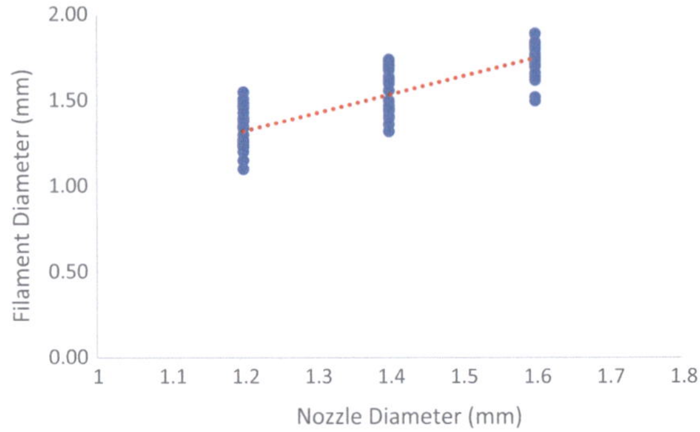

Fig. 7 The distribution of filament diameters at varied nozzle diameter for PA12

with $R^2 = 0.71$. The slope of 1.0721 means that as the nozzle size increases by 1 mm, the filament diameter increases by about 1.07 mm. This shows a nearly one-to-one relationship, indicating that nozzle diameter is a reliable way to control filament size in PA12 extrusion. The small y-intercept (0.0343) suggests that the model is accurate even at smaller nozzle sizes, without significant errors.

The R^2 value of 0.71 means that 71% of the changes in filament diameter can be explained by the nozzle size, proving it is the main factor affecting filament thickness. However, the remaining 29% of variation comes from other factors like extrusion temperature, cooling rate, or differences in material batches. The graph shows that for nozzle sizes between 1.0–1.4 mm, the data points are close to the regression line, indicating stable and consistent extrusion. For larger nozzles (1.5–1.8 mm), the data points are more spread out, which may be due to difficulties in maintaining steady material flow or die swell effects at higher extrusion rates.

These findings are important for PA12 extrusion in 3D printing and other manufacturing processes. The strong relationship allows manufacturers to predict filament diameter fairly accurately by choosing the right nozzle size, helping with quality control. However, the variation seen with larger nozzles suggests that additional adjustments may be needed to improve consistency. For applications that need very precise filament sizes, other factors besides nozzle size may need to be controlled to reduce the remaining variability. Overall, the data shows that PA12 generally behaves predictably during extrusion, but there is room to further improve the process, especially when using larger nozzle sizes.

4 Conclusion

This study shows that nozzle size has a big impact on making polymer filaments for 3D printing. For PA12, a 1.2 mm nozzle gave the best size control, while the 1.6 mm nozzle, although less precise, had slightly better process capability and may be good for general use. HDPE performed best with a 1.4 mm nozzle but still did not meet ideal standards, and it handles high temperatures better than PA12. The study also found a strong link between nozzle size and filament diameter, but other factors like temperature and cooling also affect the results. HDPE needs more process improvements, while PA12 can already be controlled better with the right nozzle. To improve filament quality, it is important to match nozzle size with the right settings, use real-time monitoring, and test different speeds and temperatures. These findings can help make filament production for 3D printing more reliable and cost-effective in the future.

Acknowledgements This research was made possible with funding from Ministry of Higher Education FRGS-EC/1/2024/TK09/UITM/02/16.

References

1. Sudin, M.N., Daud, N.M., Ramli, F.R., Yusuff, M.A.: The effect of nozzle size on the tensile and flexural properties of PLA parts fabricated via FDM. Sci. Eng. Technol. **3**(1) (2023). https://doi.org/10.54327/set2023/v3.i1.71
2. Kristiawan, R.B., Imaduddin, F., Ariawan, D., Ubaidillah, Arifin, Z.: A review on the fused deposition modeling (FDM) 3D printing: filament processing, materials, and printing parameters. Open Eng. **11**, 639–649 (2021). https://doi.org/10.1515/eng-2021-0063
3. Cardona, C., Curdes, A.H., Isaacs, A.J.: Effects of filament diameter tolerances in fused filament fabrication. IU J. Undergrad. Res. **2**(1), 44–47 (2016). https://doi.org/10.14434/IUJUR.V2I1.20917
4. Mishra, R., Aamiri, O.B., Satyavolu, J., Kate, K.: Effect of process conditions on the filament diameter in single screw extrusion of natural fiber composite. Manuf. Lett. **32**, 15–18 (2022). https://doi.org/10.1016/j.mfglet.2022.01.003
5. Filament diameter (mm) under different extrusion temperatures and extrusion speeds. https://www.researchgate.net/figure/Filament-diameter-mm-under-different-extrusion-temperatures-and-extrusion-speeds_tbl3_365755275
6. Kartal, F., Kaptan, A.: Investigating the effect of nozzle diameter on tensile strength in 3D-printed polylactic acid parts. Black Sea J. Eng. Sci. **6**(3), 276–287 (2023). https://doi.org/10.34248/bsengineering.1287141
7. Sokhey, A., Ali, Y., Hanna, M.: Effects of die dimensions on extruder performance. J. Food Eng. **31**, 251–261 (1997). https://doi.org/10.1016/S0260-8774(96)00025-8
8. Darsin, M., et al.: The effect of 3D printing filament extrusion process parameters on dimensional accuracy and strength using PLA-Brass filaments. (2023)
9. Hu, C., Qin, Q.H.: Advances in fused deposition modeling of discontinuous fiber/polymer composites. Curr. Opin. Solid State Mater. Sci. **24**(5) (2020). https://doi.org/10.1016/j.cossms.2020.100867
10. Vidakis, N., Petousis, M., Tzounis, L., Velidakis, E., Mountakis, N., Grammatikos, S.A.: Polyamide 12/multiwalled carbon nanotube and carbon black nanocomposites manufactured by 3D printing fused filament fabrication: electrical, thermoelectric, and mechanical properties. J. Carbon Res. **7**(2), 38 (2021). https://doi.org/10.3390/c7020038
11. Rijekki, N.F., Sari, N.P., Faizin, A., Duratun, S., Rosady, N.: The effect of heating temperature on 3D print filament diameter consistency produced by HDPE and LDPE plastic extrusion machine. J. Eng. Appl. Technol. Online **5**(2), 104–117 (2024). https://journal.uny.ac.id/index.php/jeatech. Accessed 2025/11/15
12. Filastruder Kit. https://www.filastruder.com/products/filastruder-kit?variant=323882043
13. ASTM D3850-12: Test Method for Rapid Thermal Degradation of Solid Electrical Insulating Materials by Thermogravimetric Method (TGA) (2012). https://doi.org/10.1520/D3850-12

14. Process Capability & Performance (Pp, Ppk, Cp, Cpk). https://sixsigmastudyguide.com/process-capability-pp-ppk-cp-cpk/
15. Mohd Shukri, A.Y.: Development on an Improved PLA Based Filament Feedstock for FDM 3D Printing. Universiti Teknologi MARA (2023)
16. Abdullah, A.M., Rahim, T.N.A.T., Hamad, W.N.F.W., Mohamad, D., Akil, H.M., Rajion, Z.A.: Mechanical and cytotoxicity properties of hybrid ceramics filled polyamide 12 filament feedstock for craniofacial bone reconstruction via fused deposition modelling. Dent. Mater. **34**(11), e309–e316 (2018). https://doi.org/10.1016/j.dental.2018.09.006
17. What Is Platykurtic? Definition, Examples and Other Distributions. https://www.investopedia.com/terms/p/platykurtic.asp
18. Andargie, A., et al.: Estimation of a Linear Model with Two-Parameter Symmetric Platykurtic Distributed Errors, vol. 1, no. 1. Springer (2013). https://doi.org/10.1186/2195-5468-1-13
19. Su, X., Yan, X., Tsai, C.L.: Linear regression. Wiley Interdiscip. Rev. Comput. Stat. **4**(3), 275–294 (2012). https://doi.org/10.1002/wics.1198

The Effect of Logistics Service Quality on Customer Satisfaction and Repurchase Intention in E-Commerce

Evi Yuliawati, Mukhamad Keshid Ahlussiddiqin, Vito Andrianto Pratama Sugiarto, Andreas Dhiyong Pratama, Aisyatul Widad Saputri, and Ni Luh Putu Hariastuti

Abstract This study examines the effect of logistics service quality (LSQ) on customer satisfaction and repurchase intention in e-commerce, focusing on the motorcycle spare parts market. Given the projected growth of e-commerce, LSQ is critical for competitive differentiation. Despite technological advancements, issues like extended transit times, customs delays, and inconsistent service standards often lead to customer dissatisfaction, impacting repurchase intentions. Thus, enhancing overall logistics service quality requires further investigation. A quantitative approach using Structural Equation Modeling (SEM) with AMOS software was utilized to analyze the relationships among web design, return logistics service, responsiveness, information quality, overall LSQ, customer satisfaction, and repurchase intention. Data was collected via online questionnaires from 112 respondents, primarily male (82%) and aged 20–29 (62%), who purchased motorcycle spare parts through e-commerce platforms. Results show that web design, responsiveness, and information quality significantly influence both Overall Logistics Service Quality (OLSQ) and customer satisfaction. OLSQ positively impacts customer satisfaction and repurchase intention. Customer satisfaction, in turn, strongly drives repurchase intention. While return logistics service had a minimal effect on OLSQ, it showed a significant negative relationship with customer satisfaction, highlighting the need for efficient return process management. In conclusion, e-commerce businesses, especially in the spare parts sector, should prioritize improving web design, responsiveness, information quality, and return logistics processes. This focus will enhance LSQ, boost customer satisfaction, and ultimately increase repurchase intentions, crucial for long-term customer loyalty and sustained growth in the competitive e-commerce landscape.

Keywords Logistic service quality · Customer satisfaction · Repurchase intention · SEM · E-commerce

E. Yuliawati · M. K. Ahlussiddiqin · V. A. P. Sugiarto · A. D. Pratama · A. W. Saputri · N. L. P. Hariastuti (✉)
Institut Teknologi Adhi Tama Surabaya, Surabaya, Indonesia
e-mail: putu_hrs@itats.ac.id

M. Awang et al. (eds.), *Proceedings of the International Conference on Mechanical, Manufacturing and Plant Engineering*, Proceedings in Technology Transfer,
https://doi.org/10.1007/978-981-95-7817-7_4

1 Introduction

In recent years, e-commerce has experienced significant growth, with projections indicating online transactions will account for 20.1% of total retail purchases by 2024. The global e-commerce market is expected to surpass $6.3 trillion in 2024 and reach $7.9 trillion by 2027 [1]. This growth reflects a shift in consumer behavior, with increasing reliance on online shopping to meet daily needs. One sector experiencing rapid expansion is quick commerce, which is projected to grow from $25 billion in 2021 to $72 billion by 2025 [2]. This trend signifies a growing consumer preference for faster delivery services. Sentral Partindo is a distributor that operates an e-commerce platform through its website, www.sentralpartindo.com, offering a wide range of motorcycle spare parts. To meet the increasing demand for speed, transformation in logistics has become essential, where warehouses now function as dark stores that integrate all aspects of the delivery process. In this context, Logistics Service Quality (LSQ) has become a crucial factor in shaping consumer experience within online marketplace [3, 4]. High service quality not only enhances customers satisfaction but also serves as a key determinant of repurchase intention. With the globalization of trade and the expansion of e-commerce, companies now compete not only on price but also on quality of service [5]. Understanding how LSQ influences customer satisfaction and, in turn, cross-border repurchase intention is essential for developing strategies to improve customer experience and build long-term loyalty. However, despite advancements in technology and logistical practices, challenges remain in maintaining consistent service quality within national border. Issues such as longer transit periods, customs clearance delays, and variability in service standards often lead to customer dissatisfaction, which negatively affects their willingness to repurchase from international vendors. Therefore, improving overall logistics service quality continues to be a critical issue that requires further analysis [6]. For instance, logistics-related complaints received by Sentral Partindo include product loss and damage, incorrect order fulfillment, and delayed deliveries. Although Sentral Partindo is known for its high-quality product and strict quality control, some customers still refrain from choosing Sinotrans Limited as their contract logistics provider. Hence, to meet the personalized logistics service needed by clients, it is essential to enhance the overall logistics service quality [7–9].

The visual design quality and visual appearance of a motorcycle spare parts e-commerce platform involve several crucial elements, such as the ease of use of the user-friendly, ensuring that users can access and interact with the site easily and efficiently [10]. Additionally, providing complete information on the website, including clear product descriptions, high-quality images, and transparent pricing information, is essential in providing convenience for customers to make clear informed decisions [11]. User privacy security is also a vital aspect, with the implementation of secure payment systems and data encryption to protect users' personal information and build long-term credibility [12]. A well-designed platform, as a whole, helps improve user experience and satisfaction, builds trust, and positively influences purchase decisions, while creating a professional, reliable, and user-centered platform [11, 13, 14].

The quality of logistics services in handling product returns or complaints in motorcycle spare parts e-commerce encompasses several important dimensions, including appropriate response time, fair and appropriate compensation offer, and the overall convenient customer complaint submission process [15]. A quick response to complaints can significantly enhance customer satisfaction and strengthen their trust in the platform. Furthermore, offering compensation that is perceived as fair and appropriate to the inconvenience experienced helps maintain a positive brand image and customer loyalty [12]. Additionally, a simple and user-friendly complaint submission system minimizes customer frustration and encourages them to engage confidently with after-sales services [16]. Altogether, these service quality aspects play a critical role in shaping the overall customer experience and influencing their future purchasing decisions.

The speed and effectiveness of customer service in e-commerce platforms when responding to customer inquiries and requests, as well as the flexibility of services in accommodating the specific needs of motorcycle spare parts customers, are essential elements in building a positive and trustworthy customer experience [16]. Through quick and accurate responses to inquiries and requests not only demonstrates a company's commitment to customer satisfaction but also reduces uncertainty and builds confidence in the purchasing process. Additionally, by offering efficient solutions to customer inquiries and requests, enhances perceived value and helps build long-term trust. Furthermore, offering flexible and personalized services, such as customized product recommendations, tailored delivery options, or specialized technical support, helps address the personal needs of each customer, thereby enhancing overall satisfaction [17]. By prioritizing these service dimensions, e-commerce businesses can foster stronger customer loyalty, encourage repeat purchases, and strengthen their competitive advantage in the highly specialized motorcycle spare parts market.

The quality of information related to product delivery on motorcycle spare parts e-commerce platforms includes several crucial aspects, such as the clarity and transparency of shipping information, the availability of real-time product tracking, and the consistency of information across different communication channels [11, 18]. Clarity of shipping information, including detailed shipping details such as estimated delivery times, shipping costs, and carrier information, helps set accurate customer expectations and reduces potential confusion or dissatisfaction. The availability of real-time tracking, which allows customers to enhances their sense of control and security throughout the delivery process, significantly improving their overall shopping experience. Additionally, ensuring consistency of information across channels, such as websites, mobile apps, and customer service representatives, is essential for building trust and minimizing misunderstandings [19]. Altogether, high-quality delivery information strengthens customer confidence, supports positive brand perception, and encourages repeat purchases in the highly competitive motorcycle spare parts market.

Customers' perception of the overall quality of logistics services in motorcycle spare parts transactions encompasses several important dimensions, including the timely delivery, the correct product received [20], and the reasonable logistic and shipping costs [21]. Timely delivery is a crucial factor that directly influences customer satisfaction, as delays can disrupt maintenance schedules and reduce trust in the platform. Likewise, receiving products that match the descriptions and specifications ensures reliability and minimizes the risk of returns or complaints, thereby enhancing the overall shopping experience. Furthermore, offering reasonable and transparent logistics costs contributes to a positive perception of value for money and supports long-term customer loyalty [20]. Taken together, these aspects shape customers' overall assessment of logistics performance, influencing their willingness to repeat purchases, and determine their likelihood to recommend the platform to others.

The level of customer satisfaction with the price, product quality, and customer service quality in purchasing motorcycle spare parts through e-commerce platforms reflects several essential dimensions of the online shopping experience. Price plays a critical role in shaping perceived value and influencing purchasing decisions, as customers often compare prices across different sellers before making a purchase. Product quality, which includes durability, authenticity, and compliance with specifications, directly impacts customer trust and reduces the likelihood of returns or complaints [9]. Moreover, excellent customer service, characterized by responsiveness, professionalism, and the ability to effectively resolve issues, further enhances the overall satisfaction of customers [18].

Given the complexities of online shop logistics, a deeper understanding of how logistics service quality affects customer satisfaction and repurchase intention in a cross-border context is necessary. Companies that excel in LSQ can distinguish themselves from competitors and achieve significant operational advantages. Therefore, this study aims to explore the relationship between LSQ, customer satisfaction, and repurchase intention, enabling companies to focus on the most impactful aspects of logistics service.

2 Methodology

This research process began with the formulation of problems and objectives, followed by a literature review, field study, identification of variables and indicators, development of an initial conceptual model describing the relationship between exogenous variables (Web Design, Return Logistics Services, Responsiveness, Information Quality), intervening variables (Overall Logistics Service Quality, Customer Satisfaction), and endogenous variable (Repurchase Intent). The research instrument, in the form of a questionnaire, was designed based on indicators for each variable, such as interface ease of use, complaint response speed, service flexibility, and clarity of delivery information, with measurements using a 1–5 Likert scale. Data collection was conducted online through Google Forms with users of motorcycle spare parts e-commerce platform as respondents. Next, SEM assumption tests were conducted to test the validity and reliability of the measurement model, followed by model estimation, model fit evaluation through Goodness of Fit (GOF) testing, structural analysis including path analysis values, and hypothesis testing based on statistical significance (p-value < 0.05) to test the direct and indirect effects between the studied variables.

3 Result and Discussion

3.1 Respondent Characteristics

Out of 112 respondents who received the questionnaire, 92 respondents (82%) were male, while 20 respondents (18%) were female, as shown in Table 1. This composition provides insight into the demographic characteristics of consumers purchasing spare parts online.

Regarding age, most respondents were between 20–29 years old (69 people, 62%), followed by those aged 30–39 years (32 people, 29%). Respondents aged 40–49 years accounted for only 7% (8 people), while a small group under 20 years old made up 3% (3 people). This shows that the main consumer base for spare parts in e-commerce is young adults, who are generally more familiar with online shopping. In terms of education, the largest group of respondents had completed high school or vocational school (48 people, 43%), followed by those with elementary or middle school education (36 people, 32%). Meanwhile, 26 respondents (23%) had an undergraduate degree, and only 2 respondents (2%) had a postgraduate education.

Table 1 Characteristics of respondents in the study

Characteristics	Description	Amount	%
Gender	Man	92	82
	Woman	20	18
Age	<20 years old	3	3
	20–29 years old	69	62
	30–39 years old	32	29
	40–49 years old	8	7
Last education	Elementary/Middle School	36	32
	SMA/SMK	48	43
	Undergraduate	26	23
	Postgraduate (Master/Doctor)	2	2

3.2 Identification of Variables and Indicators

To conduct this research, several variables were identified along with their respective indicators based on relevant theories and previous studies. Each variable was operationalized to ensure accurate measurement and analysis in the context of motorcycle spare parts e-commerce services. Table 2 presents the variables, their descriptions, and the indicators used for measurement. The indicators were formulated to represent specific aspects of each variable, such as website design quality, responsiveness, logistics service quality, customer satisfaction, and repurchase intention, which together form the basis of the structural equation modeling (SEM) analysis in this study.

3.3 Conceptual Model

The conceptual model of this research illustrates the relationships between Web Design, Return Logistic Service, Responsiveness, and Information Quality as exogenous variables, with Overall Logistics Service Quality and Customer Satisfaction as intervening variables, and Repurchase Intention as the endogenous variable. This model generates 11 hypotheses that explain the direct effects among the variables, as shown in Fig. 1.

3.4 Model Estimation

The model estimation stage was carried out to assess the validity of the measurement model through regression weights of each indicator. This process involved examining the estimated values, standard errors (S.E.), critical ratios (C.R.), and significance levels (p-values) to determine whether the indicators adequately represented their respective latent variables. As shown in Table 3, most indicators meet the required significance level ($p < 0.05$) and are therefore considered valid, indicating that the measurement model is reliable for analysis.

Based on the analysis results with a 95% confidence level, Table 3 shows the P-value and Critical Ratio (C.R.) for each indicator across all variables. Indicators are considered valid if their p-value is below 0.05, indicating that the indicators meet the significance level. Most indicators in the model meet this criterion ($p < 0.05$), confirming their validity. For example, the Web Design indicators (WD2 and WD3) and the Return Logistic Service indicator (RLS3) have strong C.R. values and p-values below 0.05, proving the significance of these indicators. Some indicators, such as RLS2, show a p-value slightly above 0.05, but they are still considered valid. Overall, the majority of indicators in this model are significant and valid, confirming that the measurement model is suitable for further analysis.

Table 2 Variables and indicators

Variable	Description	Indicators
Web Design (WD) [11, 13, 14]	The quality of the website's design and appearance	WD1: Ease of use of user interface [10]
		WD2: Completeness of information [11]
		WD3: Security of user privacy [12]
Return Logistics Service (RLS) [15]	The quality of logistics services in handling product returns or complaints	RLS1: Speed in responding to complaints [12]
		RLS2: Appropriateness of compensation [12]
		RLS3: Ease of submitting complaints [16]
Responsiveness (RE) [16]	The speed and effectiveness in responding to customer requests	RE1: Speed in responding to requests [17]
		RE2: Ability to solve problems [17]
		RE3: Service flexibility [17]
Information Quality (IQ) [11, 18]	The quality of product-related information on the e-commerce platform	IQ1: Clarity of delivery information [19]
		IQ2: Availability of real-time tracking [19]
		IQ3: Consistency of information [19]
Overall Logistics Service Quality (OLSQ) [20]	Customer perception of the overall quality of logistics services	OLSQ1: Timeliness of delivery [20]
		OLSQ2: Product appropriateness [20]
		OLSQ3: Logistics cost [21]
Customer Satisfaction (CS) [18]	The level of customer satisfaction with spare parts purchases through e-commerce	CS1: Price [9]
		CS2: Product quality
		CS3: Customer service [9]
Repurchase Intention (RI) [11, 14, 22–24]	The intention of customers to make repeat purchases of spare parts through e-commerce	RI1: Service quality [9]
		RI2: After sales quality [8]
		RI3: Customer experience [7]

3.5 Goodness of Fit (GOF)

Various GOF measures can be used to assess the model fit, such as CMIN/DF, Chi-Square, p-value, RMSEA, GFI, AGFI, CFI, and TLI. These values are compared with recommended thresholds to determine whether the tested model is acceptable

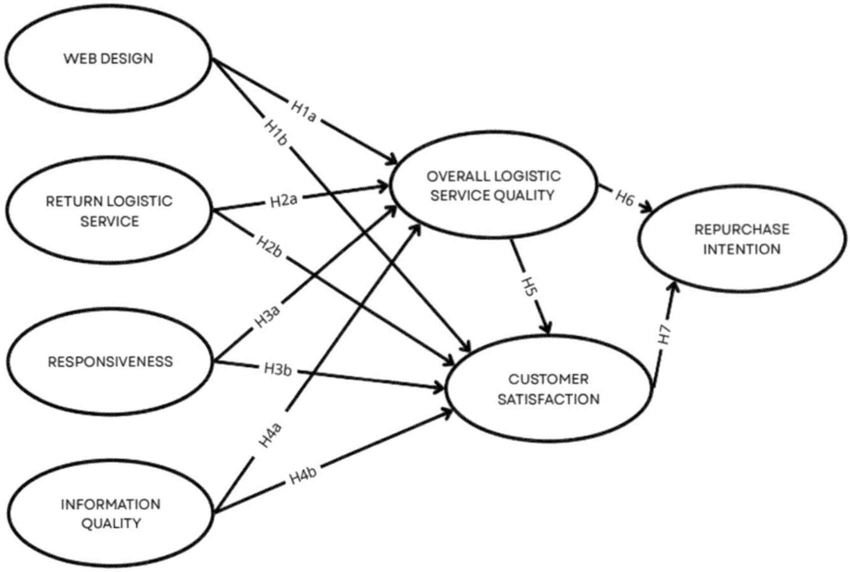

Fig. 1 Conceptual model

or requires modification. Table 4 presents the results of the GOF analysis for the estimated SEM model:

Based on Table 4, the CMIN/DF value of 1.976 is well below the recommended threshold of 3.0, which indicates that the model fits very well. A lower CMIN/DF value suggests a better fit between the model and the data, with this value confirming that the model is a good representation of the data structure. The Chi-Square value of 33.875 is also very good. Although the Chi-Square value typically indicates the degree of compatibility between the model and data, a lower value is desirable. Since this value is relatively low, it suggests that the model represents the observed data well. The p-value of 0.000 is significantly lower than the recommended value of $\geq$0.05, which indicates that there is a statistically significant difference between the hypothesized model and the actual data. In this case, the p-value being so low suggests a perfect fit, as the model is significantly consistent with the data. The RMSEA (Root Mean Square Error of Approximation) value of 0 is excellent. The RMSEA measures how well the model approximates the data; a value of 0 indicates that there is no error in the approximation, thus making this value ideal as it is far below the maximum allowed threshold of 0.08.

The GFI value of 0.943 meets the recommended threshold of $\geq$0.90, which means the model explains a significant portion of the data variation. A GFI value of 0.943 reflects a very good compatibility and shows that the model captures the underlying structure of the data effectively. Similarly, the AGFI value of 0.989 exceeds the recommended minimum of $\geq$0.90, indicating that the model fits well even after adjusting for its complexity of the model. The CFI (Comparative Fit Index) value of

Table 3 Regression weights

Indicators	Estimate	S.E	C.R	P	Status
WD1	1,000				Valid
WD2	,816	,198	4.118	***	Valid
WD3	,773	,186	4.166	***	Valid
RLS1	1,000				Valid
RLS2	,244	,138	1.765	,078	Valid
RLS3	,637	,182	3.504	***	Valid
RE3	1,000				Valid
RE2	,931	,254	3.660	***	Valid
RE1	1105	,281	3.938	***	Valid
IQ3	1,000				Valid
IQ2	1812	,457	3.968	***	Valid
IQ1	1301	,357	3.640	***	Valid
OLSQ1	1,000				Valid
OLSQ2	1136	,146	7.759	***	Valid
OLSQ3	,799	,122	6.523	***	Valid
CS3	1,000				Valid
CS2	,898	,146	6.156	***	Valid
CS1	,967	,159	6.097	***	Valid
RI1	1,000				Valid
RI2	,926	,112	8.269	***	Valid
RI3	,908	,118	7.670	***	Valid

Table 4 Goodness of fit

GOF measure	Recommended value	GOF value	Description
CMIN/DF	≤3.0	1.976	Accepted
Chi-Square	–	33.875	Accepted
p-value	<0.05	0.000	Accepted
RMSEA	≤0.08	0	Accepted
GFI	≥0.90	0.943	Accepted
AGFI	≥0.90	0.989	Accepted
CFI	≥0.90	1	Accepted
TLI	≥0.90	1	Accepted

1.0 is a perfect score, as it meets the threshold of ≥0.90. The CFI reflects the relative compatibility of the model compared to a baseline model, and a value of 1 indicates that the hypothesized model is perfectly compatible with the data.

Finally, the TLI (Tucker-Lewis Index) value of 1.0 also indicates a perfect fit. The TLI is similar to the CFI and is used to compare the fit of the model against a null model. A value of 1 signifies that the model has an optimal fit, with no discrepancies between the model and the data. The Goodness of Fit (GOF) measures show that the model fits the data excellently. Each evaluation criterion either meets or exceeds the recommended values, confirming that the hypothesized model is a very good fit for the research data. This suggests that the model's structure accurately reflects the underlying relationships in the data.

3.6 Path Analysis Values

The Fig. 2 illustrates the estimation result of the structural model, showing the relationships between the variables, and then the estimates of regression weights of direct relationship between research variables can be seen in Table 5.

Based on Table 5, Web Design (WD) significantly influences Overall Logistics Service Quality (OLSQ) and Customer Satisfaction (CS), with coefficients of 0.65 and 0.36, respectively. The C.R. values of 7.999 and 8.666 for these relationships, both with p-values of *** (less than 0.05), suggest that Web Design plays a vital role in shaping both the overall quality of logistics services and customer satisfaction. A well-designed website could contribute to improving service delivery and enhancing customer experiences in e-commerce, particularly in the motorcycle spare parts market. In terms of Return Logistic Service (RLS), the relationship with Overall Logistics Service Quality (OLSQ) shows a minimal effect with a coefficient of 0.02. However, its C.R. value of 8.635 and p-value of *** indicate that it still contributes significantly to overall service quality despite its small coefficient. For Customer Satisfaction (CS), RLS shows a negative relationship with a coefficient of -0.35, but its C.R. value of 2.674 with a p-value of 0.007 suggests that the negative effect is statistically significant. This implies that return logistics may negatively affect customer satisfaction in e-commerce if not managed efficiently. Responsiveness (RE) has a very strong influence on Overall Logistic Service Quality (OLSQ), with a coefficient of 0.83 and a C.R. value of 8.856, which is significant at ***. Responsiveness plays a crucial role in the quality of logistics services. Additionally, Responsiveness (RE) positively affects Customer Satisfaction (CS) with a coefficient of 0.58 and a C.R. value of 7.403, indicating that responsiveness is an important driver of customer satisfaction in e-commerce.

Information Quality (IQ) also has a significant positive effect on both Overall Logistic Service Quality (OLSQ) and Customer Satisfaction (CS). The coefficients of 0.82 for OLSQ and 0.35 for CS, along with their respective C.R. values of 8.454 and 8.487, show that high-quality information is key to both service quality and customer satisfaction. Accurate and detailed product information in e-commerce can greatly enhance the customer's experience and perception of service quality. The Overall Logistic Service Quality (OLSQ) has a positive effect on Customer Satisfaction (CS) with a coefficient of 0.16, which is significant with a C.R. value of 8.097 and ***. This suggests that higher service quality leads to increased customer satisfaction. Furthermore, OLSQ has a significant positive influence on Repurchase

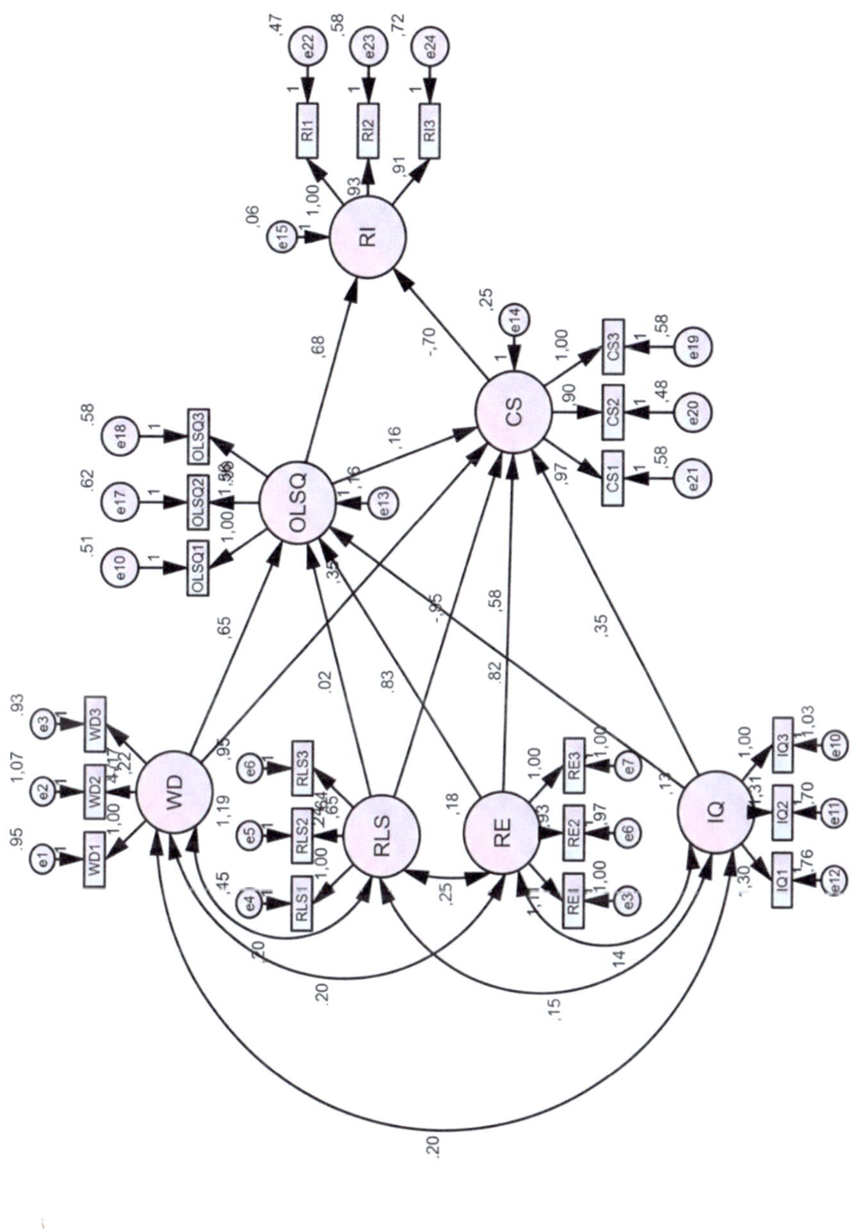

Fig. 2 Estimation result of structural model

Table 5 Estimates of regression weights of direct relationship between research variables

Variable	Coefisien	S.E	C.R	P
Web Design (WD) → Overall Logistic Service Quality (OLSQ)	0,65	,118	7,999	***
Web Design (WD) → Customer Satisfaction (CS)	0,36	,124	8,666	***
Return Logistic Service (RLS) → Overall Logistic Service Quality (OLSQ)	0,02	,107	8,635	***
Return Logistic Service (RLS) → Customer Satisfaction (CS)	-0,35	,179	2,674	,070
Responsiveness (RE) → Overall Logistic Service Quality (OLSQ)	0,83	,134	8,856	***
Responsiveness (RE) → Customer Satisfaction (CS)	0,58	,125	7,403	***
Information Quality (IQ) → Overall Logistic Service Quality (OLSQ)	0,82	,128	8,454	***
Information Quality (IQ) → Customer Satisfaction (CS)	0,35	,114	8,487	***
Overall Logistic Service Quality (OLSQ) → Customer Satisfaction (CS)	0,16	,124	8,097	***
Overall Logistic Service Quality (OLSQ) → Repurchase Intention (RI)	0,68	,117	8,818	***
Customer Satisfaction (CS) → Repurchase Intention (RI)	0,70	,102	6,867	***

Intention (RI), with a coefficient of 0.68 and a C.R. value of 8.818. This relationship suggests that improving logistics service quality is essential in fostering repurchase intentions, a crucial aspect of customer retention in e-commerce.

Finally, Customer Satisfaction (CS) strongly influences Repurchase Intention (RI) with a coefficient of 0.7 and a C.R. value of 6.867, both significant at ***. This confirms that satisfied customers are more likely to repurchase in the future, highlighting the importance of customer satisfaction in driving repeat business in e-commerce. In conclusion, the relationships in the table indicate that factors such as web design, responsiveness, information quality, and overall logistics service quality are all significantly influencing customer satisfaction and repurchase intentions. These findings suggest that e-commerce businesses, particularly those in the spare parts market, should focus on enhancing website design, responsiveness, and information quality to improve service quality and customer satisfaction, ultimately driving customer loyalty and repurchase intentions.

4 Conclusion

Logistic service quality plays a crucial role in shaping customer satisfaction and influencing repurchase intentions in e-commerce. Key factors such as web design, responsiveness, return logistics, information quality, and overall logistics service quality significantly impact both customer satisfaction and the likelihood of repurchase. The study indicates that Web Design (WD) has a positive effect on both Overall Logistics Service Quality (OLSQ) and Customer Satisfaction (CS). A well-designed website enhances the overall logistics service and creates a better experience for the customer, which leads to higher satisfaction. Responsiveness (RE) also plays a significant role in improving logistics service quality and customer satisfaction. The ability to respond quickly and efficiently to customer needs is vital in enhancing service delivery. Furthermore, Information Quality (IQ) is another critical factor that positively impacts both overall logistics service quality and customer satisfaction. Providing accurate, detailed, and reliable information about products and services is essential in creating trust and satisfaction among customers. The study also reveals that Return Logistics Service (RLS), despite having a negative relationship with customer satisfaction, still has a notable influence on overall logistics service quality, highlighting the importance of managing return processes effectively to maintain a positive customer experience.

This study also finds that Overall Logistics Service Quality (OLSQ) directly affects Customer Satisfaction (CS) and Repurchase Intention (RI). Improving the quality of logistics services not only enhances customer satisfaction but also increases the likelihood of future purchases. Customer Satisfaction (CS), in turn, plays a significant role in driving Repurchase Intention (RI), showing that satisfied customers are more likely to return and repurchase from the same e-commerce platform. This research emphasizes the importance of improving logistics service quality to foster customer satisfaction and encourage repurchase behaviors in e-commerce. Businesses should focus on optimizing web design, responsiveness, information quality, and return logistics processes to create a positive customer experience. Furthermore, enhancing overall logistics service quality is essential for building long-term customer loyalty, which is a key driver of sustained business growth in the e-commerce sector.

References

1. Snyder, K., Aditham, K.: 35 Top e-commerce statistics. https://www.forbes.com/advisor/business/ecommerce-statistics/. Accessed 3 June 2025
2. Bommireddipalli, R.: What's next for Q-commerce: the golden child of E-commerce (2022). https://www.forbes.com/councils/forbestechcouncil/2022/02/08/whats-next-for-q-commerce-the-golden-child-of-e-commerce/. Accessed 6 June 2025
3. Gaudenzi, B., Confente, I., Russo, I.: Logistics service quality and customer satisfaction in B2B relationships: a qualitative comparative analysis approach. TQM J. **33**(1), 125–140 (2021). https://doi.org/10.1108/TQM-04-2020-0088

4. Ali, A.H., Melkonyan, A., Noche, B., Gruchmann, T.: Developing a sustainable logistics service quality scale for logistics service providers in Egypt. Logistics **5**(2), 1–15 (2021). https://doi.org/10.3390/logistics5020021
5. Do, Q.H., Kim, T.Y., Wang, X.: Corrigendum to 'Effects of logistics service quality and price fairness on customer repurchase intention: the moderating role of cross-border e-commerce experiences' [J. Retail. Consum. Serv. **70**, 103165 (2023) (S0969698922002582). J. Retail. Consum. Serv. **76**, no(June 2023), 103454 (2024). https://doi.org/10.1016/j.jretconser.2023.103454
6. Oh, K.Y., Kang, S.Y., Oh, Y.G.: The moderating effects of eco-friendliness between logistics service quality and customer satisfaction in cross-border e-commerce: evidence from overseas direct purchasers in Korea. Sustain. **14**(22) (2022). https://doi.org/10.3390/su142215084
7. Bawack, R.E., Wamba, S.F., Carillo, K.D.A.: Exploring the role of personality, trust, and privacy in customer experience performance during voice shopping: evidence from SEM and fuzzy set qualitative comparative analysis. Int. J. Inf. Manag. **58**, 0–42 (2021). https://doi.org/10.1016/j.ijinfomgt.2021.102309
8. Raditya, B., Yuliati, L.N., Krisnatuti, D.: Analysis of the effect of brand image, product quality and after-sales service on repurchase decision of Samsung smartphones. Russ. J. Agric. Socio-Econ. Sci. **92**(8), 19–32 (2019). https://doi.org/10.18551/rjoas.2019-08.03
9. Farooq, M.S., Salam, M., Fayolle, A., Jaafar, N., Ayupp, K.: Impact of service quality on customer satisfaction in Malaysia airlines: a PLS-SEM approach. J. Air Transp. Manag. **67**(September 2017), 169–180 (2018). https://doi.org/10.1016/j.jairtraman.2017.12.008
10. Rahman, M.F., Hossain, M.S.: The impact of website quality on online compulsive buying behavior: evidence from online shopping organizations. South Asian J. Mark. **4**(1), 1–16 (2022). https://doi.org/10.1108/sajm-03-2021-0038
11. Permatasari, A., Kartikowati, M.: The influence of website design on customer online trust and perceived risk towards purchase intention: a case of O2O commerce in Indonesia. Int. J. Bus. Glob. **21**(1), 74–86 (2018). https://doi.org/10.1504/IJBG.2018.094097
12. Sivarethinamohan, R., Sujatha, S.: Behavioral intentions towards adoption of information protection and cyber security (email security and online privacy): SEM model. Turkish J. Comput. Math. Educ. **12**(6), 56–68 (2021). https://doi.org/10.17762/turcomat.v12i6.1267
13. Krithika, P., Vasantha, S.: Evaluating the quality of website design aspects and its effect on functional perspectives using smart-PLS in web-based apparel shopping environment. In: Salud, Ciencia y Tecnología - Serie de Conferencias, vol. 3 (2024). https://doi.org/10.56294/sctconf2024894
14. Priscillia, M., Budiono, H., Wiyanto, H., Widjaya, H.: The effects of website design quality and service quality on repurchase intention among shopee customers in Jakarta, with customer trust as a mediating variable. In: Proceedings of Ninth International Conference on Entrepreneurship and Business Management (ICEBM 2020), vol. 174, no. ICEBM 2020, pp. 38–44 (2021). https://doi.org/10.2991/aebmr.k.210507.006
15. Lin, D., Lee, C.K.M., Siu, M.K., Lau, H., Choy, K.L.: Analysis of customers' return behaviour after online shopping in China using SEM. Ind. Manag. Data Syst. **120**(5), 883–902 (2020). https://doi.org/10.1108/IMDS-05-2019-0296
16. Khalili-Damghani, K., Tavana, M., Najmodin, M.: Reverse logistics and supply chains: a structural equation modeling investigation. Int. J. Ind. Eng. Theory Appl. Pract. **22**(3), 354–368 (2015)
17. Asamoah, D., Nuertey, D., Agyei-Owusu, B., Akyeh, J.: The effect of supply chain responsiveness on customer development. Int. J. Logist. Manag. **32**(4), 1190–1213 (2021). https://doi.org/10.1108/IJLM-03-2020-0133
18. Dhiranty, A., Suharjo, B., Suprayitno, G.: An analysis on customer satisfaction, trust and loyalty toward online shop (a case study of Tokopedia. Com). Indones. J. Bus. Entrep. **3**(32), 101–110 (2017). https://doi.org/10.17358/ijbe.3.2.102
19. Sheko, A., Spaho, A.B.: Information technology inhibitors and information quality in supply chain management: a PLS-SEM analysis. Acad. J. Interdiscip. Stud. **7**(3), 125–138 (2018). https://doi.org/10.2478/ajis-2018-0064

20. Le, D.N., Nguyen, H.T., Hoang Truong, P.: Port logistics service quality and customer satisfaction: empirical evidence from Vietnam. Asian J. Shipp. Logist. **36**(2), 89–103 (2020). https://doi.org/10.1016/j.ajsl.2019.10.003
21. Wallenburg, C.M., Weber, J.: Structural equation modeling as a basis for theory development within logistics and supply chain management research. In: Research Methodologies in Supply Chain Management: In Collaboration with Magnus Westhaus, pp. 171–186 (2005). https://doi.org/10.1007/3-7908-1636-1_12
22. Faizal, S., Menon, N.S., Anand, J.: Structural equation model on the sustainable online shopping behavior. In: Sustainable Innovations in Management in the Digital Transformation Era Digital Management Sustainability, pp. 93–101 (2024). https://doi.org/10.4324/9781003450238-10
23. Dewi, C.K., Mohaidin, Z., Murshid, M.A.: Determinants of online purchase intention: a PLS-SEM approach: evidence from Indonesia. J. Asia Bus. Stud. **14**(3), 281–306 (2020). https://doi.org/10.1108/JABS-03-2019-0086
24. Wu, W.Y., Quyen, P.T.P., Rivas, A.A.A.: How e-servicescapes affect customer online shopping intention: the moderating effects of gender and online purchasing experience. Inf. Syst. E-bus. Manag. **15**(3), 689–715 (2017). https://doi.org/10.1007/s10257-016-0323-x

Threat Detection and Classification for Cyber Security Against Network Data Using a Hybrid Classification Model

Shah Khadafi, Miswanto, Rimuljo Hendradi, and Tutuk Indriyani

Abstract The rapid development of the internet has been accompanied by an increase in threats posed by intruders, including DDoS attacks, backdoors, exploits, and worms, among others. To address these attacks, accurate techniques for classifying attacks are necessary. Attack classification aims to identify detected threat activities, which can then be analysed using network forensics. This study identifies threats based on network data collected in a dataset known as UNSW-NB15. The quality of the UNSW-NB15 dataset is improved through the use of pre-processing techniques, including feature encoding and normalization. The ensemble feature selection technique is used to obtain the best features based on the highest ranking of the dataset attribute features. Furthermore, threat anomaly detection is performed using machine learning to classify threats using hybrid classification algorithms, namely Naïve Bayes, Random Forest, Support Vector Machine (SVM), K-Nearest Neighbourhood (K-NN), Neural Network (NN), and C.45. The threat classification results using the Naïve Bayes algorithm show that the accuracy measurement reaches 0.9848. The precision, recall, and F1 score measurements reach 0.9847. Threat classification using Random Forest reaches 0.9626, and the F1 score reaches 0.9624. Threat classification using the Neural Network algorithm achieved an accuracy of 0.9463, a precision of 0.946, a recall of 0.939, and an F1 score of 0.93.

Keywords Intrusion detection · Preprocessing · Feature selection · Classification · UNSW-NB15

S. Khadafi (✉) · Miswanto · R. Hendradi
Airlangga University, Surabaya, Jawa Timur, Indonesia
e-mail: shah.khadafi-2022@fst.unair.ac.id

T. Indriyani
Institut Teknologi Adhi Tama Surabaya, Surabaya, Jawa Timur, Indonesia

M. Awang et al. (eds.), *Proceedings of the International Conference on Mechanical, Manufacturing and Plant Engineering*, Proceedings in Technology Transfer,
https://doi.org/10.1007/978-981-95-7817-7_5

1 Introduction

The rapid advancement of internet technology, facilitated by easy access to digital devices and sophisticated internet communication tools, has led to significant progress in the digital economy, enabling users to carry out their work and other activities more efficiently. According to data released by the Indonesian Internet Service Providers Association (APJII), it is recorded that in 2025, the total number of Indonesians connected to the internet will reach 229,428,417 people [1], out of a total population of 284,438,900. The indicator of internet network usage can be seen from the internet penetration activity carried out by the Indonesian population, which is 80.66% of the total Indonesian population connected to the internet in 2025. The high percentage of internet penetration has encouraged changes in individual activities nationwide that utilise the role of digital transformation infrastructure, which is often discussed by the Indonesian government today. The linear increase in internet user penetration in Indonesia from year to year is as follows: 77.01% in 2022, 78.18% in 2023, and 79.50% in 2024. The statistical graph illustrating the growth of internet penetration over the past 5 years is presented in Fig. 1.

Other components that support internet network service access are web browsers, which internet users most commonly use for browsing. Based on benchmark tests and research [2], Google Chrome scored 572.63, Microsoft Edge scored 550.43, Brave scored 382.67, Opera scored 367.81, and Mozilla Firefox scored 163.54.

Behind the increase in internet penetration and the ease of activities for individuals who use the internet for their daily work, there is undoubtedly an increase in threats that have the potential to disrupt security and exploit system vulnerabilities, thereby disrupting system stability. The Indonesian National Cybersecurity Agency (BSSN), as the state institution responsible for developing norms, standards, procedures, and criteria, coordinating all elements of cybersecurity, and monitoring, evaluating, and reporting on cyber operations, has stated that in 2024, various types of threats have occurred for internet users in Indonesia. These threats include malware, phishing,

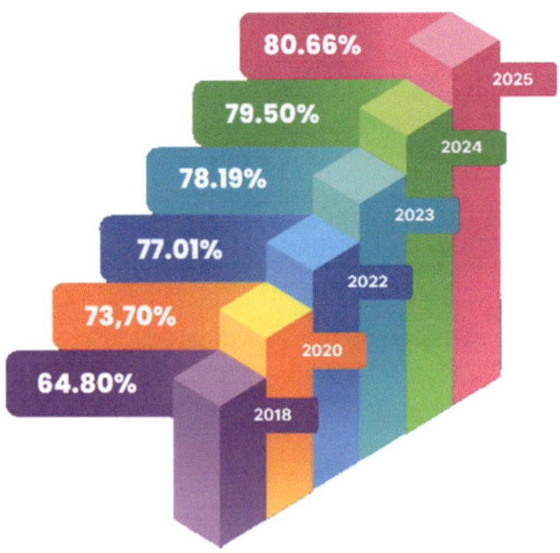

Fig. 1 Percentage of internet penetration growth over the past 5 years

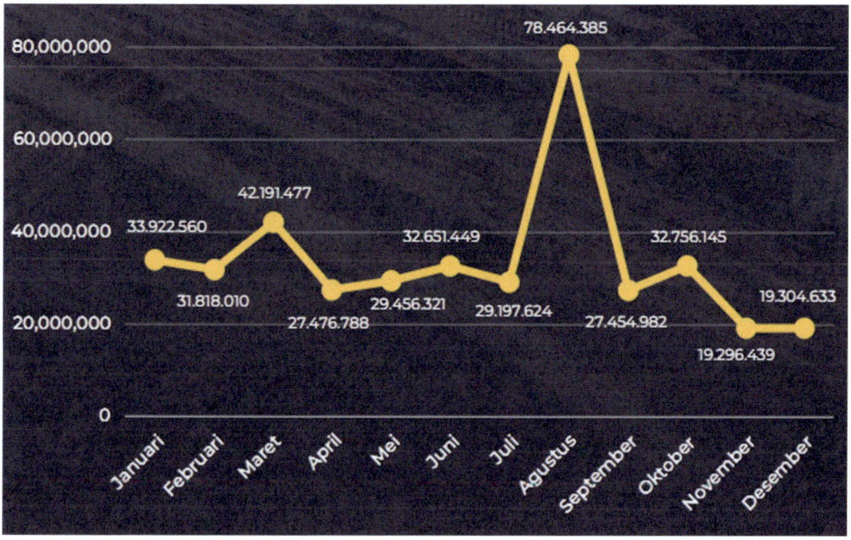

Fig. 2 Number of traffic anomalies threatening internet network users in Indonesia in 2024

DoS/DDoS attacks, sniffing, exploits, ransomware, and social engineering attacks, which can result in a decline in internet network performance, data theft, and even damage to the reputations of private and government organisations. According to a report from BSSN, from January to December 2024 [3], 403,990,813 traffic anomalies were detected through internet network traffic, disrupting and threatening internet users in Indonesia. Figure 2 shows a statistical graph of traffic anomaly incidents throughout 2024.

Cybersecurity is the most critical aspect of the internet network, whose operations must provide security protection services to internet users. The internet access provided must ensure that users' communications are safe from intruders and threats [4]. The rapid and widespread development of the internet can lead to increased disruptions or attacks from intruders or crackers who exploit the protocols used in communications. Some widespread attacks, such as DDoS, malware, and backdoors, continue to occur frequently and often bypass security systems [5].

A DDoS (Distributed Denial of Service) attack utilizes multiple resources to launch an attack simultaneously, unlike a DoS (Denial of Service) attack, which employs only one resource. The basic principle of a DDoS attack is to exhaust the target computer's resources, making the services installed on it unavailable. The impact of this attack causes significant losses, including a decline in computer system performance due to an increase in abnormal traffic, which can lead to a decrease in system performance or even render it unresponsive. Financial losses occur when services are unavailable, preventing users from accessing them, resulting in lost revenue due to disrupted service usage. The loss of public or community trust due to a DDoS attack can damage customer confidence and a company's reputation.

A worm is a type of malware that can replicate itself and spread automatically through a network without requiring user intervention. Unlike viruses, worms are self-contained and can spread their exploits within a network system. Worms can cause significant damage to data and create backdoors for other malware that can attack the system. Worms spread by exploiting vulnerabilities in computer systems, particularly by entering through the operating system to gain access to the system and then spreading the worm. Backdoor attacks occur by exploiting hidden loopholes or mechanisms that allow unauthorised persons to access and control systems, applications, or hardware without going through routine authentication procedures. Backdoors can remotely control systems, allowing attackers to steal data, damage systems, or spread additional malware. These backdoor attacks employ several techniques that are widely used today, exploiting vulnerabilities in the target computer system. In addition, backdoor attacks utilise malicious programs or Trojans that act as if they are applications or systems that cannot be identified by the system or users, which have program code that has access code so that they can take over the system secretly, making it difficult for the system to detect and mitigate backdoor attacks.

The complexity of cyberattack issues is severe, and a deep understanding of current widespread threats and attacks is required. This requires the classification of threats or attacks that can recognise and identify various patterns or attributes of multiple threats [6]. This research classifies threats to recognise various threats based on a collection of attack data gathered from developing an intrusion detection system (IDS) dataset [7]. The IDS dataset has network attributes and values, representing the network's payload values. Several datasets for intrusion detection are generated through network traffic captures aimed at supporting network security researchers in conducting network forensic analysis. The forensic process involves several activities, starting with identifying the type of attack, collecting attack data, examining the payload value of the attack data packet, and storing the attack data in a dataset. One of the IDS datasets is the UNSW-NB15 dataset [8], a modern dataset that collects attack patterns and types of attacks in the form of data sequences. The UNSW-NB15 dataset is an IDS dataset developed by the Australian Centre for Cyber Security (ACCS), with the advantage of being more representative and containing more comprehensive information on attack data, which is loaded into a comprehensive dataset comprising data packets and network data streams.

This study will classify threats or attacks contained in the UNSW-NB15 dataset using machine learning techniques [9]. Machine learning is implemented to learn data, even though the dataset dimensions are quite large [10]. The machine learning technique will be applied to classify the UNSW-NB15 dataset, which has undergone preprocessing to enhance the quality and relevance of the dataset [11]. Next, classification is carried out by identifying and recognising data patterns so that it can be determined which data is attack data and which is normal data, in this case, network traffic data that does not contain attacks [12]. The algorithms used to classify the types of attacks in the UNSW-NB15 dataset are C.45, K-Nearest Neighbour (KNN), Naïve Bayes, Random Forest, Support Vector Machine (SVM), and Neural Network.

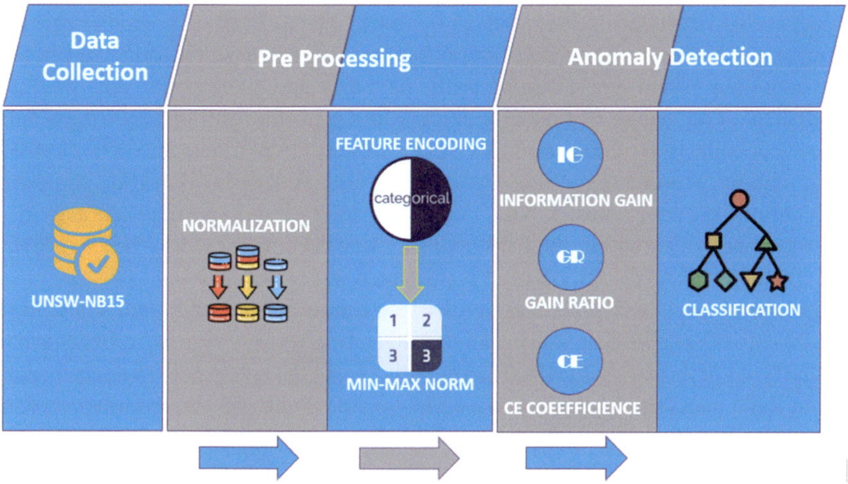

Fig. 3 NIDS process for anomalous traffic and normal traffic

2 Methodology

2.1 Research Method

The research method discusses the design and concept of developing a threat classification model using machine learning techniques. The classification is performed on the UNSW-NB15 IDS-based dataset, which contains network traffic data in the form of data packets and data payloads. Within UNSW-NB, there are two types of data categories: normal network traffic data and cyber attack network traffic data. These two categories serve as data labels for the dataset attributes. Before the classification process is carried out, it is necessary to pre-process the dataset to improve its relevance and quality using feature encoding and Min–Max normalisation techniques. The following process is feature ranking, which aims to obtain selected features for data classification. This threat classification is carried out as threat detection on UNSW-NB15, and the results will then be evaluated using parameters such as accuracy, precision, recall, and F1 score. The research method chart is shown in Fig. 3.

2.2 Data Collection

UNSW-NB15 is a collection of network traffic-based datasets compiled by Mustafa and Koroniotis, conducted within the Cyber Range Lab at the Australian Centre for Cyber Security (ACCS) [13]. This dataset underwent a recording and storage process using Bro-IDS software tools [14], now known as Zeek [15], which extracted data from network traffic in real-time based on data packets. The collected traffic data consisted of two types: normal traffic data flow and attack traffic data flow (or anomalies). The results of this data extraction are in the form of time-series data stored in a collection of datasets in Packet Capture (PCAP) file format. The UNSW-NB15 dataset was created as a realistic representation of network traffic, equipped with various modern network data, containing multiple normal network activities and attack activities within the network. The network topology designed to collect and create the UNSW-NB15 dataset is presented in Fig. 4.

The dataset in UNSW-NB15 is divided into training and testing data, containing regular network traffic flow data packages and anomalies. Anomalous traffic consists of nine types of threats or attacks: Fuzzer with 24,242 data points, 2677 analysis data points, 2329 backdoor data points, 16,353 DoS data points, 44,525 exploitation data points, 58,871 generic data points, 13,987 reconnaissance data points, 1511 shellcode data points, and 174 worm data points. Statistically, the total amount of attack traffic and normal traffic data within UNSW-NB15 is presented in a graph shown in Fig. 5, along with details of the types of attacks and the number of each attack.

In UNSW-NB15, there are 49 data network features, categorized into two types: packet-based and stream-based data [16]. Packet-based data is extracted from packet headers and network payload data. Meanwhile, stream-based data is extracted

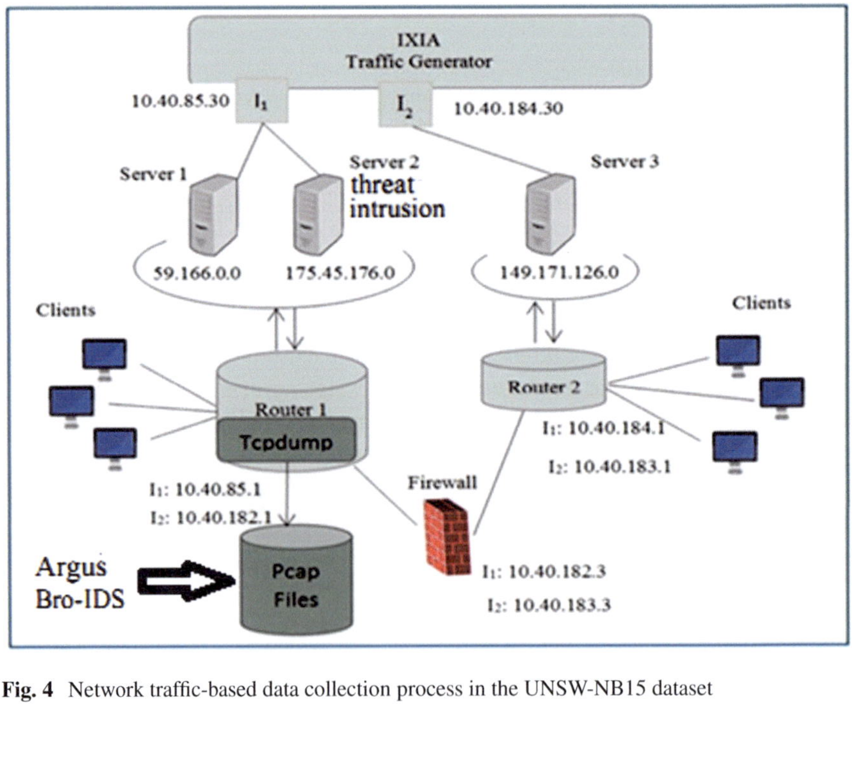

Fig. 4 Network traffic-based data collection process in the UNSW-NB15 dataset

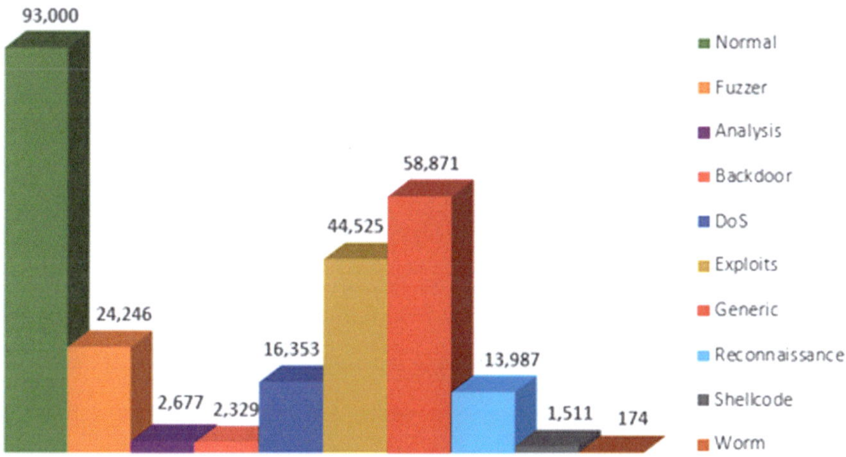

Fig. 5 Amount of data for each type of attack and normal traffic in the UNSW-NB15 dataset

from sequences of network data packets accessed from the source computer to the destination computer.

The features in UNSW-NB15 are divided into several types, namely basic features, content features, time features, general features, and connection features. Flow features are generated from network traffic extraction with sender identity and communication port attributes. These features correspond to the identity characteristics of the hosts involved in network communication, using the IP address and port identity. The total dataset in the UNSW-NB15 dataset comprises 257,573 data points, consisting of 82,332 training data points and 175,241 test data points, where each training and test set is distributed in a ratio of attack data to normal data, as presented in Table 1.

2.3 *Preprocessing*

Before proceeding to the threat classification stage, a pre-processing stage is carried out first. This pre-processing stage aims to normalize the data, improving the quality of the UNSW-NB15 dataset and making it more relevant and consistent, thereby preparing it for the next stage. This pre-processing stage also involves several processes, including feature encoding and min–max normalization.

Feature Encoding

Feature encoding is performed on specific features in the UNSW-NB15 dataset, particularly features with ordinal data types. Ordinal data types are classified as non-numeric, meaning they cannot be used in data mining. Therefore, ordinal data types must first be converted to numeric types, a process known as feature encoding. There

Table 1 Types and number of records at UNSW-NB15

Traffic type	Dataset record		
	Training	Testing	Total
Normal	37.000	56.000	93.000
Fuzzer	6.062	18.184	24.246
Analysis	677	2.000	2.677
Backdoor	583	1.746	2.329
DoS	4.089	12.264	16.353
Exploits	11.132	33.393	44.525
Generic	18.871	40.000	58.871
Reconnaissance	3.496	10.491	13.987
Shell code	378	1.133	1.511
Worms	44	130	174
Total record	**82.332**	**175.341**	**257.573**

are three features in UNSW-NB15 with ordinal data types: the 'proto', 'services', and "state" features. The 'service' feature has DHCP, DNS, FTP, FTP-Data, HTTP, IRC, POP3, RADIUS, SMTP, SNMP, SSH, and SSL data. The 'state' feature has seven data points: ACC, CLO, CON, FIN, INT, REQ, and RST. As an example of the feature encoding process, the 'state' feature has data encoded into numbers: ACC = 1, CLO = 2, CON = 3, FIN = 4, INT = 5, REQ = 6, and RST = 7.

Min–Max Normalization

Min–max normalisation is applied to datasets with a wide range of values. This is because some values in the UNSW-NB15 feature are very far apart. Data transformation is performed using min–max normalization to overcome the problem of value differences in the dataset. Min–max normalisation stretches the values of several UNSW-NB15 features. The notation (x_i) is the input data vector $x_1, x_2, x_3, \ldots .x_n$.. The data input (x_i) has a minimum $(min(x))$ and a maximum value $(max(x)$. The values for the dataset are classified as having an extensive range for each attribute. The min–max normalisation calculation is performed using Eq. 1.

$$x\prime = \frac{x\prime - \min(x)}{\max(x) - \min(x)}. \quad (1)$$

2.4 Threats Detection

Before detecting threats in the UNSW-NB15 dataset, it is necessary to analyse and select the best features in the IDS dataset. The ensemble feature selection technique is used to identify the best features from the dataset, and then the highest-ranking features can be determined.

Ensemble Feature Selection

Feature selection techniques produce a subset of data features for the UNSW-B15 dataset that contains the best and most relevant features from the original features, resulting in a lean feature dimension that is not too large. Feature selection eliminates irrelevant or redundant features, thereby improving the performance of the classification model. Several feature selection techniques used in UNSW-NB15 are Information Gain (IG), Gain Ratio (GR), and Correlation Coefficient (CC). This study conducted several feature selection methods, collectively referred to as ensemble feature selection.

1. Information Gain

The first feature selection method, Information Gain (IG), calculates the information gain of a target attribute feature against a dataset. The IG calculation results in an output score between 1 and 0, where 1 is considered the most relevant value and 0 is the least pertinent value. The mathematical formula for calculating IG involves calculating the entropy value of an attribute with a specific class in the previous training data, as shown in Eq. 2.

$$G(I) = \sum_{i=1}^{m} -P_i \log_2 P_i \tag{2}$$

The notation m is the total number of classes in the training set, and the notation P_i For instance, I belong to a particular class, which can be calculated using Eq. 3.

$$c_i = \frac{I_i}{I_N} \tag{3}$$

The entropy calculation for each class c_i in instance I_i Performed for the feature ranking process using IG. Entropy measures the uncertainty in certain features. The entropy calculation for each feature is performed using Eq. 4.

$$E(f) = -E(f)\frac{I_{i1} + I_{i2} + I_{i3} + \cdots + I_{mi}}{I_N} \tag{4}$$

The final IG calculation combines Eqs. 3 and 4 to rank each set of target attributes. Equation 5 will produce a range of values between 0 and 1, with 1 being the most relevant and 0 being the least appropriate.

$$Gain(f) = G(I) - E(F). \tag{5}$$

2. Gain Ratio

The second feature selection method is Gain Ratio (GR), which first calculates the Split information. Split information is a data set feature that presents the entropy of each instance. I_i by dividing I into r branches corresponding to feature f using Eq. 6.

$$Split(I) = -\sum_{i=1}^{r} \left(\frac{I_i}{I_N}\right) \log_2 \left(\frac{I_i}{I_N}\right) \tag{6}$$

The concept of GR calculation uses IG ranking values primarily to evaluate biased information. GR values refine IG value calculations using the highest value feature selected as a separating attribute. Thus, GR calculations utilize Eqs. 5 and 6 to determine the ranking value of each feature in the UNSW-NB15 dataset, as per Eq. 7.

$$GainRatio(I) = \frac{Gain(f)}{Split(I)}. \tag{7}$$

3. Correlation Coefficient

The third feature selection method is the Correlation Coefficient (CE), which utilizes the R calculation as Pearson's correlation. Pearson is a technique that calculates the strength of similarity between two features or the strength of the relationship between two features using Eq. 8

$$r_{xy} = \frac{\sum_{i=1}^{n} (x_i - \overline{x})(y_i - \overline{y})}{\sqrt{\sum_{i=1}^{n} (x_i - \overline{x})^2} \sqrt{\sum_{i=1}^{n} (y_i - \overline{y})^2}} \tag{8}$$

Feature selection using CE that utilises Pearson calculations has the notation x being the value of the feature currently being analysed, while y is the value of another feature being analysed as a correlation with x, $\overline{x}$ is the average value of the feature x, and $\overline{y}$ is the average value of the feature y.

Classification of Threats

At this stage, machine learning is used to classify threats. The classification algorithms selected for this study are based on their technical advantages in learning from the data used in the learning process. Several algorithms are used to classify attack or non-attack data types, including Naïve Bayes, Random Forest, Support Vector Machine (SVM), Neural Network, K-NN, and C.45.

Classification in IDS aims to identify and categorise network or system activities into specific categories, such as attacks or everyday activities. This classification highly depends on the system's ability to process and analyse data accurately. The classification techniques used in IDS involve machine learning, which enables models to learn patterns from training data and then make predictions based on new data. IDS can be divided into two main categories: signature-based IDS and anomaly-based IDS, each with advantages and disadvantages regarding attack detection.

2.5 Evaluation Performance

Performance evaluation aims to obtain the performance of the IDS data classification algorithm in detecting threats and evaluating the classification model used. In this study, performance evaluation was carried out by calculating accuracy, precision, recall, and F1 score to assess the accuracy of threat classification. The concept of performance evaluation has four measurements, which are explained further as follows:

1. Accuracy is the percentage of records classified correctly out of the total number. The accuracy equation corresponds to Eq. 9.

$$accuracy = \frac{(TP + TN)}{(TP + TN + FP + FN)} \tag{9}$$

2. Precision is the percentage of positive records classified correctly out of the predicted positive records. The precision equation corresponds to Eq. 10.

$$precission = \frac{TP}{(TP + FP)} \tag{10}$$

3. Recall is the percentage of records correctly classified as positive out of the total number of positive records. This equation corresponds to Eq. 11.

$$recall = \frac{TP}{(TP + FN)} \tag{11}$$

The F1 score forms an evaluation matrix that plays an essential role in assessing a class in data when there is class imbalance. The F1 score value overcomes the imbalance between precision and recall values. If an evaluation only uses precision, a particular classification model will only detect a small amount of attack data, but with completely accurate results. However, if only recall is used, the classification model can detect almost all attack data, but many of the results will also be incorrect. The F1 score equation corresponds to Eq. 12.

$$F1\,score = 2 \times \frac{precission \ \times \ recall}{precission \ + \ recall}. \tag{12}$$

3 Result and Discussion

This study developed machine learning implemented to classify types of threats in UNSW-NB15 using the Naïve Bayes, Random Forest, Support Vector Machine (SVM), Neural Network, K-NN, and C.45 algorithms. The hardware used in this research is a laptop computer with the following specifications: 16 GB RAM, Intel Core i7-13490HQ @3.6 GHz CPU, and NVIDIA GeForce RTX 4060 M. The software used to implement threat classification was the Windows 11 operating system. Python was used to build and train the threat classification model. Python packages, including TensorFlow and Keras, were utilized as libraries for developing machine learning models in threat classification. Additionally, the NumPy and Pandas libraries were employed for matrix manipulation and data preprocessing.

The discussion of the results in this chapter explains the statistical measurement values as the results of the best feature values, which are then used in machine learning techniques to classify each threat and perform forensic analysis. The classification algorithms used are Naïve Bayes, Random Forest, Support Vector Machine (SVM), Neural Network, K-Nearest Neighbor (K-NN), and C. 4.5 to obtain the results of machine learning experiments that classify attacks. The testing process that classifies threats in the UNSW-NB15 data is distributed based on the percentage of the dataset. The training and testing data ranged from 10 to 90% of the total dataset. The threat classification process was carried out using popular classification algorithms such as Naïve Bayes, Random Forest, Support Vector Machine (SVM), K-Nearest Neighbourhood (K-NN), Neural Network (NN), and C.45. The final process was to measure the performance of the classification results using an assessment matrix of accuracy, precision, recall, and F1 score.

3.1 Classification with Naïve Bayes

This discussion presents evaluation metrics, namely Accuracy, Precision, Recall, and F1 score, using the Naïve Bayes classification algorithm, with training data and test data separated, and a training dataset percentage ranging from 10 to 90%. Figure 6 shows high performance in all measurement matrices, with values close to 1. The Accuracy measurement value decreases slightly with the addition of training data, but remains close to 0.9871 with 10% data, while with 90% training data, the result decreases to 0.9848. This graph movement is also similar to the performance measurement results for Precision, Recall, and F1 Score, which are consistent, remaining above 0.9847, followed by a very minimal variation in values. This shows that threat classification using the Naïve Bayes algorithm is stable and robust, without significant decline or inconsistency.

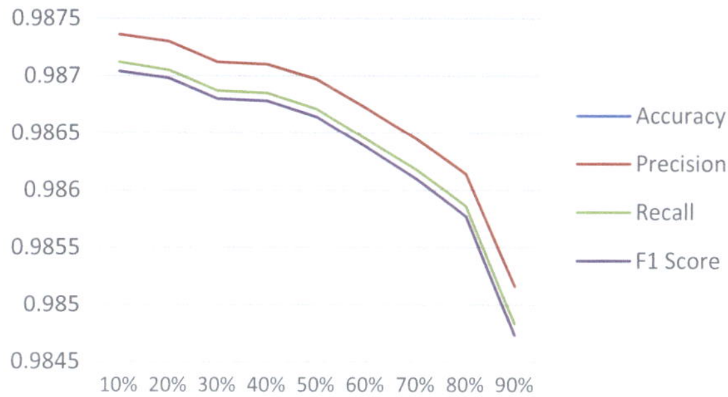

Fig. 6 Model performance metrics in training and testing using Naïve Bayes

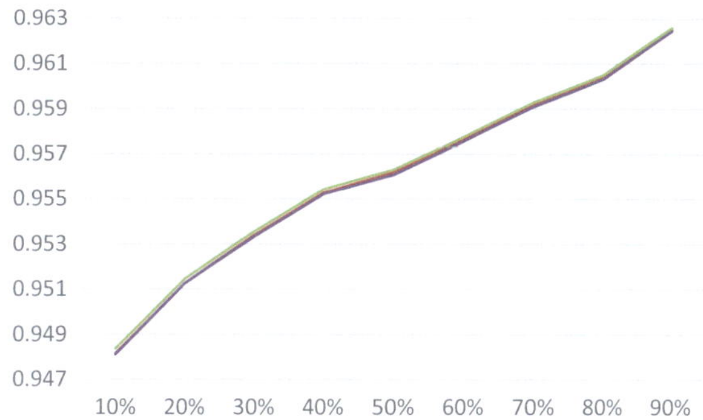

Fig. 7 Model performance metrics in training and testing using Random Forest

3.2 Classification with Random Forest

Figure 7 shows that all measurement metrics—Accuracy, Precision, Recall, and F1 Score—show substantial and stable improvements, with an apparent consistency in improvement with each addition of training data. Accuracy, precision, recall, and F1 score all increase simultaneously and are closely synchronised. When the training data reached 90%, the Random Forest classification algorithm achieved high scores on all metrics, with an accuracy of 0.9626, a recall of 0.9624, a precision of 0.9625, and an F1 score of 0.9624.

3.3 Classification with Support Vector Machine

All metrics for evaluating threat classification performance show a steady improvement, as shown in Fig. 8, where the graph rises consistently as the amount of training data used in the classification process using VSM increases. The accuracy evaluation measurement results consistently outperform other measurement metrics such as precision, recall, and F1 score, with an accuracy value reaching 0.9298, especially when training reaches 80%. Meanwhile, the F1 score lags slightly behind precision

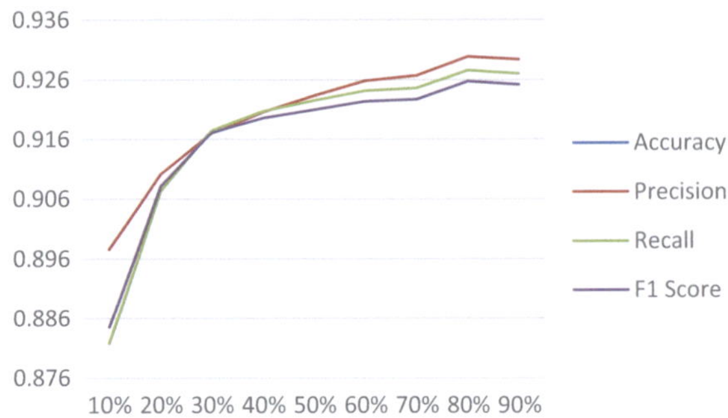

Fig. 8 Model performance metrics in training and testing using support vector machines

with a value of 0.9276 and a recall of 0.9258, but these results still perform very well. Performance peaks after 60% of the training data is used and continues to increase, reaching a consistent level.

3.4 Classification with Neural Network

The results presented in the performance matrix are relatively stable in terms of accuracy, precision, recall, and F1 score, particularly at varying percentages of training and testing data. Figure 9 shows that the accuracy value increases slightly with more training data, from 0.93877 (10% training) to 0.94633 (90% training). This indicates that the Neural Network algorithm benefits more from increased training data when running the machine learning process. The precision value tends to be high and stable, ranging from 0.939 to 0.946, with low false positives, indicating that this algorithm is good at making reliable predictions. Meanwhile, the recall value is relatively high and close to precision, increasing from 0.939 to 0.946. This indicates that the interpretation of this algorithm has a low false-negative rate and can detect most relevant examples. Regarding the F1 score, its value increased from 0.93889 to 0.9456, indicating balanced performance. The results demonstrated by this Neural Network algorithm indicate that it is ideal for situations where false positives and false negatives can be a problem in threat classification detection, and this is very useful when false positives are costly.

3.5 Classification with K-NN

The performance evaluation scores again show a very positive linear trend, with scores that also improve consistently with the increase in the percentage of training data. All assessment matrix scores are presented statistically in the graph in Fig. 10, where the scores for accuracy, recall, precision, and F1 score are consistently close

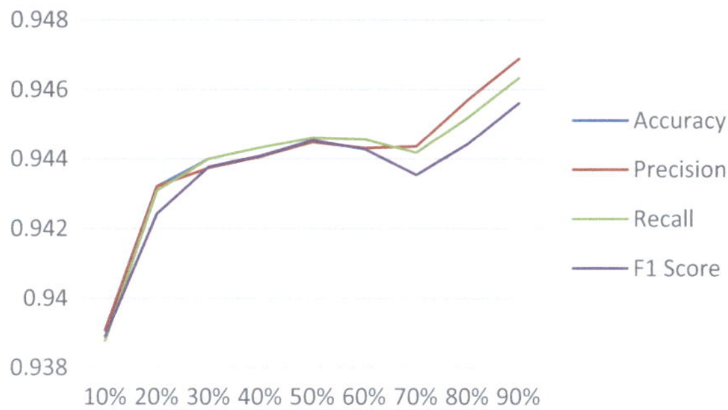

Fig. 9 Model performance metrics across training and testing using a support neural network

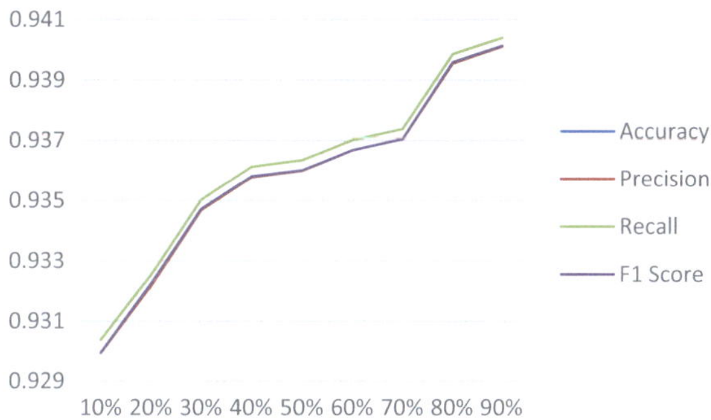

Fig. 10 Model performance metrics across training and testing using K-NN

across the entire range of data percentages, resulting in an excellent balance of values. The performance measurement values are relatively stable, with a consistent increase from 10 to 90% of the training data. Specifically, the accuracy value increases from 0.9325 to 0.9424, the precision from 0.9323 to 0.9422, the recall from 0.930 to 0.940, and the F1 score from 0.930 to 0.940. All these results indicate that the model has learned sufficiently to generalise well.

3.6 Classification with C.45

The results are significantly different when classifying threats using C.45, as shown in Fig. 11. Attention is drawn to the accuracy performance, which shows a sharp increase when the training process reaches a 30% data percentage, reaching a value of 0.938. It then increases sharply at a 40% data percentage, reaching a value of 0.9483. Fortunately, this wild spike did not occur in the precision, recall, and F1 score performance, which showed a stable and increasing trend, consistent with the increase in data. The precision and recall results were almost identical, as indicated by the graph line, which remained aligned, suggesting that the C.45 algorithm produced balanced and consistent results. Finally, the discussion regarding the spike in accuracy

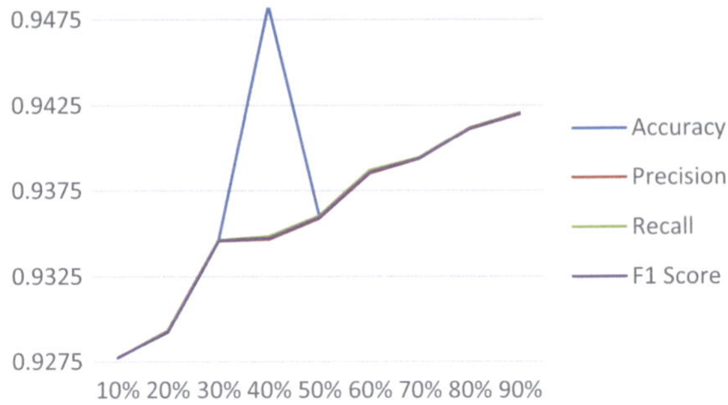

Fig. 11 Model performance metrics across training and testing using C.45

values at 40% data indicates that the model experienced highly advantageous data separation during the 40% training.

3.7 Comparison with All Model Classification

This chapter compares all classification models used to classify threats: the Naïve Bayes, Random Forest, SVM, NN, K-NN, and C.45 algorithms. Meanwhile, the distribution of training and testing data is 10–90%, a process used to measure the performance of the classification model. Furthermore, the accuracy, precision, recall, and F1 score values can be compared with the classification model results shown in Fig. 12a–d

The performance evaluation results for threat classification accuracy measurements show that the Random Forest algorithm produces consistent results with an average accuracy of 95.5%. In comparison, the lowest accuracy measurement was SVM, with an average value of 92%. For the precision measurement results in threat classification, the highest performance evaluation was achieved by Random Forest at 95.5%, while SVM obtained the lowest average precision measurement at 92.2%. For recall evaluation measurements, the average value was 95.2%, with the highest value achieved by Random Forest and the lowest by SVM, at 91.77%. The final performance evaluation was based on the F1 score, where the highest average measurement value was 95.46%. In contrast, SVM achieved the lowest recall measurement with an average of 91.78%.

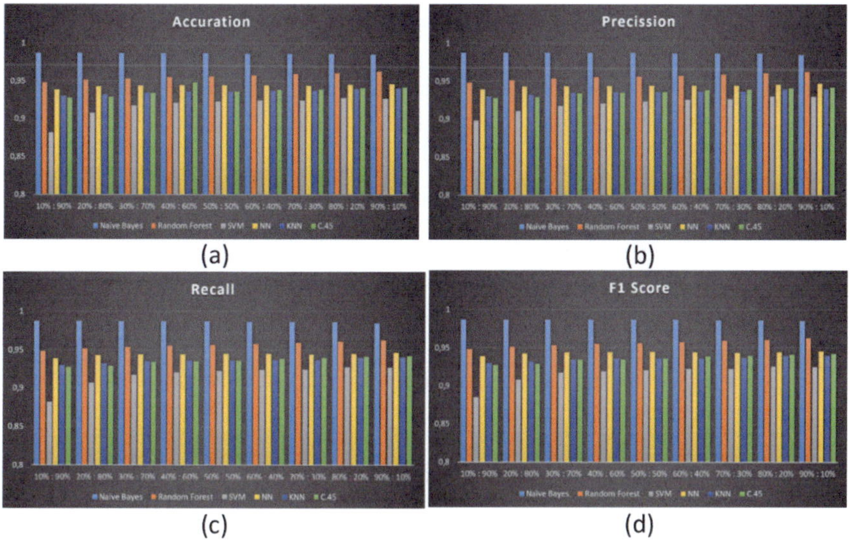

Fig. 12 **a** Evaluation results for accuracy measurement; **b** Evaluation results for precision measurement; **c** Evaluation results for recall measurement; **d** Evaluation results for F1 score measurement

4 Conclusion

This study employs several methods for threat classification, including ensemble feature selection and combined classification models. Threat classification using the UNSW-NB15 dataset as the object of attack or threat involves several features with diverse data types, and the data values in the dataset exhibit substantial differences. Therefore, as a solution to this problem, the min–max normalisation and feature encoding techniques are employed for data normalisation, aiming to enhance data quality so that the UNSW-NB15 dataset is suitable for machine learning.

The machine learning process in this study aims to classify threats. The classification models used are Naïve Bayes, Random Forest, Support Vector Machine (SVM), K-Nearest Neighbourhood (K-NN), Neural Network (NN), and the C.45 algorithm. With a training set distribution ranging from 10 to 90% and a test set size distribution ranging from 90 to 10%, there are several UNSW-NB15 datasets. The results of the threat classification evaluation using the Naïve Bayes algorithm show an accuracy of 0.9848, along with precision, recall, and F1 scores of 0.9847. In the threat classification evaluation results using Random Forest, the accuracy reached 0.9626, and the F1 score reached 0.9624. Threat classification using the Neural Network algorithm produced an accuracy of 0.9463, precision of 0.946, recall of 0.939, and an F1 score of 0.93889. Furthermore, the K-NN algorithm yielded threat classification results with balanced accuracy measurements throughout the training process, ranging from 10 to 90%. The results for all accuracy, precision, recall, and F1 score measurements were consistently high, ranging from 0.930 to 0.940. The final algorithm for classifying threats in IDS is C.45, which shows extremely high accuracy due to a significant increase in values during the 40% training process, reaching 0.9483. Meanwhile, the precision, recall, and F1 score measurements are stable, reaching a value of 0.9425.

Acknowledgements We want to thank DRTPM Ditjen Dikristek for providing funding for this research. We would also like to thank LLDIKTI 7, LPPM ITATS, and Universitas Airlangga for their support in implementing this research, as outlined in the research contract letters numbered 109/E5/PG.02.00.PL/2024, 083/SP2H/PT/LL7/2024, and 11/KP/LPPM/ITATS/2024.

References

1. APJII: Survei Penetrasi Internet Penggunaan dan Perilaku Internet, Jakarta, Indonesia (2025). https://survei.apjii.or.id/
2. Mohamed, A., Ismail, I.: A performance comparative on most popular Internet web browsers. Procedia Comput. Sci. **215**(August), 589–597 (2022). https://doi.org/10.1016/j.procs.2022.12.061
3. Siber, D.O.K.: Laporan Tahunan Monitoring Keamanan Siber 2024, 2021st edn. Direktorat Operasi Keamanan Siber—Indonesia Security Incident Response Team on Internet Infrastructure Coordination Center, Jakarta, Indonesia (2024). https://bssn.go.id/monitoring-keamanan-siber-2024/
4. Li, Y., Liu, Q.: A comprehensive review study of cyber-attacks and cyber security; Emerging trends and recent developments. Energy Rep. **7**, 8176–8186 (2021). https://doi.org/10.1016/j.egyr.2021.08.126
5. Hoque, N., Bhuyan, M.H., Baishya, R.C., Bhattacharyya, D.K., Kalita, J.K.: Network attacks: taxonomy, tools and systems. J. Netw. Comput. Appl. **40**(1), 307–324 (2014). https://doi.org/10.1016/j.jnca.2013.08.001
6. Silivery, A.K., Rao Kovvur, R.M., Solleti, R., Kumar, L.S., Madhu, B.: A model for multi-attack classification to improve intrusion detection performance using deep learning approaches. Meas. Sens. **30**(July), 100924 (2023). https://doi.org/10.1016/j.measen.2023.100924
7. Sangaiah, A.K., Javadpour, A., Pinto, P.: Towards data security assessments using an IDS security model for cyber-physical smart cities. Inf. Sci. (NY) **648**, 119530 (2023). https://doi.org/10.1016/j.ins.2023.119530
8. Zoghi, Z., Serpen, G.: UNSW-NB15 computer security dataset: analysis through visualization. Secur. Priv. (2023). https://doi.org/10.1002/spy2.331
9. Abdallah, E.E., Eleisah, W., Otoom, A.F.: Intrusion detection systems using supervised machine learning techniques: a survey. Procedia Comput. Sci. **201**(C), 205–212 (2022). https://doi.org/10.1016/j.procs.2022.03.029
10. Faker, O., Dogdu, E.: Intrusion detection using big data and deep learning techniques. In: ACMSE 2019—Proceedings of 2019 ACM Southeast Conference, no. March, pp. 86–93 (2019). https://doi.org/10.1145/3299815.3314439
11. Jiang, S., Xu, X.: Network data classification mechanism for intrusion detection system. In: International Conference on Computer Supported Cooperative Work in Design, pp. 342–347 (2021). https://doi.org/10.1109/CSCWD49262.2021.9437645
12. Aslan, Ö., Aktuğ, S.S., Ozkan-Okay, M., Yilmaz, A.A., Akin, E.: A comprehensive review of cyber security vulnerabilities, threats, attacks, and solutions. Electron. **12**(6) (2023). https://doi.org/10.3390/electronics12061333
13. Moustafa, N., Slay, J.: UNSW-NB15: a comprehensive data set for network intrusion detection systems (UNSW-NB15 network data set). In: 2015 Military Communications and Information Systems. Conference (MilCIS 2015)—Proceedings (2015). https://doi.org/10.1109/MilCIS.2015.7348942

14. Sau, W.M.T., Siswantyo, S.: Analisis Penggunaan Hasil Deteksi IDS snort pada tools RITA dalam Mendeteksi Aktivitas Beacon. Info Kripto **15**(2), 97–104 (2021). https://doi.org/10.56706/ik.v15i2.21
15. Tiwari, A., Saraswat, S., Dixit, U., Pandey, S.: Refinements. In: Zeek Intrusion Detection System 2022 8th International Conference on Advanced Computing and Communication Systems (ICACCS), pp. 974-979. Coimbatore, India (2022). https://doi.org/10.1109/ICACCS54159.2022.9785047
16. Tanenbaum, A.S., Feamster, N., Wetherall, D.: Computer Networks, 6th edn. Pearson Education Ltd., London (2021). https://www.amazon.com/Computer-Networks-6th-edition/dp/0136764053#detailBullets_feature_div

Numerical Investigation of the Thermal Performance of Lattice Structures

Ahmad Anas Arifin, I. Made Londen Batan, Michele Bici, Arif Wahjudi, and Agus Sigit Pramono

Abstract Lattice structures have become potential options for better thermal management because they have high surface-to-volume ratios and can be easily changed thanks to additive manufacturing. This work statistically examines the thermal performance of three architected lattice cells—CC-BCC, Pyramid-based, and a topology-optimized channel design (TO1)—under forced convection cooling utilizing Autodesk Fusion 360. The lattices were evaluated in various longitudinal (1×1, 3×1, 5×1) and transverse (3×3, 5×3) configurations to assess their heat transfer efficiency and temperature distribution characteristics. We chose the Temperature Uniformity Index (TUI) and the Thermal Resistance ($R\theta$) as performance measurements. The TO1 lattice had the lowest thermal resistance (0.18 °C/W in a single cell setting) because it had a lot of surface area and internal channels. The CC-BCC lattice, on the other hand, had the most even temperature distribution, with a TUI of 0.99. Longitudinal layouts lowered heat resistance as the number of cells increased, but they also lowered TUI since the speed of the airflow slowed down downstream. On the other hand, transverse layouts improved both cooling efficiency and uniformity. In the 3×3 design, TO1 had a TUI of about 0.98. The proposed designs outperformed traditional lattice topologies including BCC, Kelvin, FCC, and Octet in reducing substrate temperatures below 30 °C and enhancing homogeneity. These results show that architected lattices could be useful for new types of cold plates and heat sinks.

Keywords Lattice structure · Thermal management · Heat sink

A. A. Arifin · I. Made Londen Batan (✉) · A. Wahjudi · A. S. Pramono
Institut Teknologi Sepuluh Nopember, Surabaya, Indonesia
e-mail: londbatan@me.its.ac.id

M. Bici
Sapienza University of Rome, Rome, Italy

M. Awang et al. (eds.), *Proceedings of the International Conference on Mechanical, Manufacturing and Plant Engineering*, Proceedings in Technology Transfer,
https://doi.org/10.1007/978-981-95-7817-7_6

1 Introduction

A lattice structure is a type of material design composed of multiple unit cells arranged periodically. Lattice structures have the property of being lightweight yet strong [1]. Lattice structures are often used in several applications such as aerospace, bioengineering, automation, and other industrial fields due to their ability to withstand loads and absorb energy [2, 3]. This lattice structure has advantages over foam and honeycomb in terms of design flexibility because each strut and node can be optimized to achieve specific goals [4]. This is supported by advancements in additive manufacturing technology, which allows for the efficient production of complex geometric shapes [5]. Beside its mechanical capabilities, lattice structures have great potential for use in heat transfer devices because they have adjustable porosity and surface area-to-volume ratio [6].

In previous studies, it was found that lattice structures have proven to be remarkably efficient in transferring heat thru forced convection, for example, porous rhombi octet structures showed high Nusselt (Nu) numbers and increased effective thermal conductivity by up to 5.5 times compared to conventional aluminum foam [7]. Heat sink designs based on unit cells like hexagons have also been proven to reduce heat sink temperature more effectively than conventional pins and fins [8]. Other studies show that topological geometry and relative density affect thermohydraulic performance; for example, lattice pyramidal structures exhibit the greatest increase in Nusselt number as relative density increases [9]. Nevertheless, research using a geometry lattice cell that combines topology optimization and channeling, as well as cell hybridization, is still rare to explore [10].

The influence of cell configuration, including size, orientation, and lattice arrangement, on the heat transfer process has been demonstrated in several studies. For example, reducing the ligament width in the rhombi octet structure significantly increased the pressure drop and Nusselt number [11]. Certain lattice structures can also accelerate thermal development in the airflow after passing thru a number of unit cell rows [12]. In heat sink design, structures like Body Centered Cubic (BCC) and Kelvin are more efficient in reducing the bottom surface temperature with a certain pressure loss trade-off, while lotus and fin structures show low pressure drop but higher temperatures at the base [13]. This indicates that cell configuration significantly affects the distribution of flow velocity, contact area, and overall thermal performance.

Although some previous research has highlighted the thermal performance of various lattice topologies, there is still a research gap regarding hybrid and topology-optimized designs that address channeling and homogeneous temperature distribution aspects [14]. There have not been many studies examining the numerical performance of combined lattices between crossing cylinders (CC) and BCC, as well as lattices developed using the Computer Aided Design (CAD) Unit Cell Patterning technique inspired by the architectural form of Egyptian pyramids and topology optimization results with channeling in varying longitudinal and transverse configurations. This design allows for the exploration of the influence of surface area, internal channeling, and cell orientation on forced convection phenomena.

The purpose of this paper is to numerically evaluate the thermal performance of 3 types of lattices: CC-BCC, Pyramid, and TO Channeled Based (TO1), in cell configurations that vary longitudinally and transversely within a forced convection scenario. The main parameters measured are thermal resistance (Rθ) and Temperature Uniformity Index (TUI), to assess heat transfer efficiency and temperature homogeneity. This study is expected to provide clear design guidelines for the subsequent fabrication of cooling components or cold plates [15].

2 Methodology

This study uses 3 type of Lattice cell which are CC-BCC, Pyramid based, and Topology Optimized (TO1) Channel Based as shown in Fig. 1 [16, 17]. This research uses the numerical simulation method using the electronic cooling simulation feature of Autodesk Fusion 360. The lattice cell used has a base size of 24 × 24 × 24 mm. The developed lattice cell was then given an upper substrate and a lower substrate, both square-shaped with dimensions of 24 × 24 mm each and a thickness of 1 mm. The material used was Aluminum 6061 with a density value of 0.0027 g/mm^3. The environmental parameters were an ambient temperature of 25 °C and a gravitational acceleration of 9.81 m/s^2. The thermal load used is internal heat installed under the lower substrate lattice cell with a value of 60 W. Cooling is done using forced convection with a flow rate of 250 m^3 per Hour (CMH) as shown as Fig. 2.

This study uses 5 types of cell configurations that have elongated dimensions both longitudinally (in the direction of airflow) and transversely (perpendicular to the direction of airflow). This aims to provide a comprehensive understanding from a 3-dimensional perspective. The configuration and observation points can be seen in Table 1 and Figs. 3 and 4.

Next, to calculate the heat transfer rate from each developed cell, the thermal resistance of each cell can be calculated. Low thermal resistance indicates that the cell can conduct heat well. Thermal resistance can be calculated using Formulas (1)

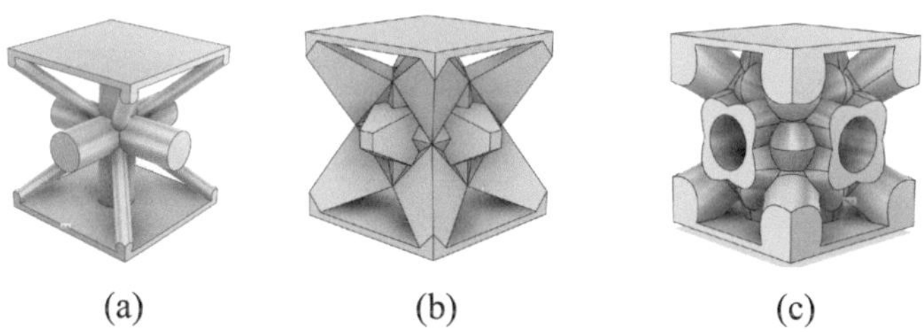

Fig. 1 Unit cell lattice used in this study. **a** CC-BCC, **b** Pyramid, and **c** TO1

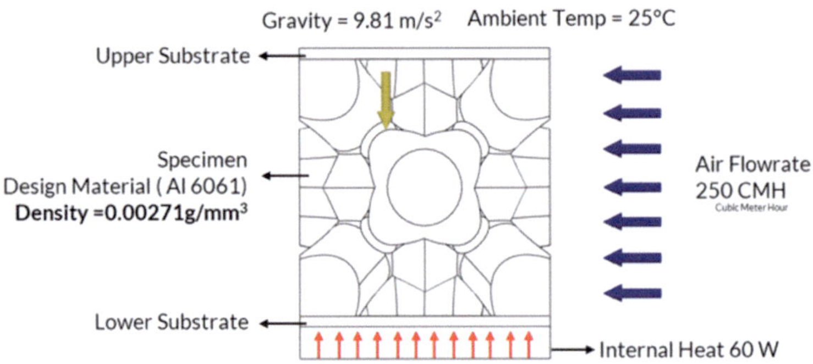

Fig. 2 Simulation parameter setup

Table 1 Cell configuration

Configuration	Number of cell	Temp. observation point
Conf. 1	1 × 1	3
Conf. 2	3 × 1	5
Conf. 3	3 × 3	5
Conf. 4	5 × 1	7
Conf. 5	5 × 3	7

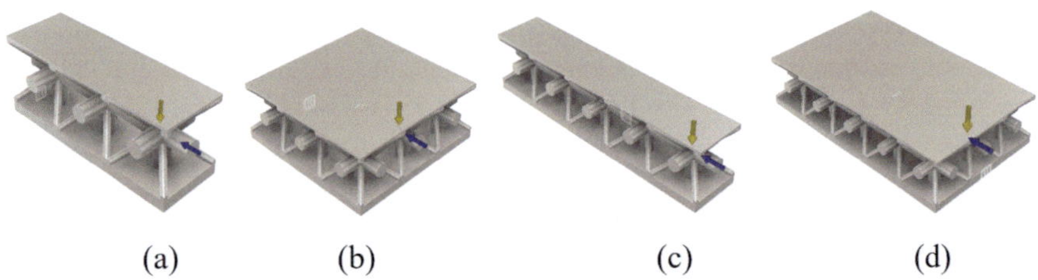

Fig. 3 Configuration of cell **a** Conf. 2. **b** Conf. 3. **c** Conf. 4. **d** Conf. 5

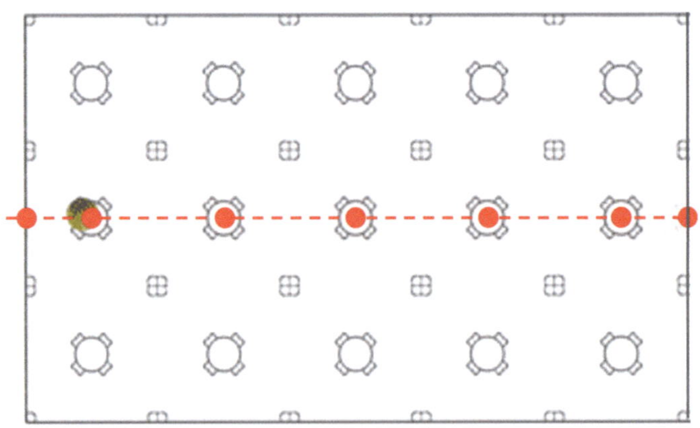

Fig. 4 Temperature observation point

and (2):

$$R_\theta = \frac{\Delta T}{\dot{Q}} \tag{1}$$

$$\dot{Q} = U.A.\Delta T \tag{2}$$

where:

R_θ: Thermal Resistance (°C/W)

ΔT: Temperature Difference (TAvg Lower Substrate – T Avg Upper Substrate) (°C)

U: Overall Heat Transfer Coefficient W/m^2.K

A: Area (m^2)

$\dot{Q}$: Heat (60 W).

Beside thermal resistance, the next important parameter in heat sink or cold plate design is the temperature distribution on the surface of electronic components or battery surfaces. The desired temperature distribution is a uniform distribution. The index used to measure this is the Temperature Uniformity Index (TUI). The higher the TUI value, the more homogeneous the temperature distribution on the surface of the lower substrate. This makes the lifetime of electronic components or batteries longer and performance higher. The TUI formula can be calculated using Formulas (3) and (4).

$$TUI = 1 - \left(\frac{T_{rms}}{Ts}\right) \tag{3}$$

$$T_{rms} = \sqrt{\frac{1}{n}\sum_{i=1}^{n}\left(T_i - \overline{T}_S\right)^2} \tag{4}$$

where:

TUI: Temperature Uniformity Index

T_{rms}: Root Mean Square Lower Substrate Temp. (°C)

T_s: Average Temperature Lower Substrate (°C).

Theoretically, the longer the cells are arranged, the greater the area of friction between the airflow and the components. This causes the air velocity at the end of each cell to increase as the number of cells arranged longitudinally increases. This can be theoretically explained using Formulas (5)–(8).

$$Re = \frac{\rho.\upsilon.L}{\mu} \tag{5}$$

$$Nu = C.Re^m Pr^n \tag{6}$$

$$h = k.Nu \tag{7}$$

$$\dot{Q} = h.A.\Delta T \quad (8)$$

$\dot{Q}$: Heat Transfer Rate (W)
h: Convective heat transfer coefficient (W/m^2·K)
ΔT: Temperature Difference (°C)
A: Area (m^2)
k: Thermal conductivity of air (W/m·K)
Nu: Nusselt number, the ratio of convective to conductive heat transfer
Re: Reynold Number
ρ: Air Density (kg/m^3)
υ: Air velocity (m/s)
L: Characteristic Length (m)
μ: Dynamic viscosity of air (Pa·s).

3 Results and Discussion

3.1 Single Cell

The dynamic viscosity of air resulting from the 1 × 1 cell configuration can be seen in Figs. 5, 6, 7, 8 and 9. Figure 5 shows the temperature distribution occurring in each cell. Measurements were taken for both the upper and lower substrates. The lower substrate was used to understand the temperature distribution phenomenon occurring. The temperature on the upper substrate was used to measure the temperature difference from the lower to the upper substrate. To then calculate the thermal resistance that occurs.

In Fig. 6, the temperature distribution in the air around the lattice and the lattice components can be seen. In Fig. 7, the airflow velocity distribution and component temperature can be seen. In Figs. 8 and 9, the airflow velocity distribution and streamline direction can be seen.

In Fig. 10, the surface temperature and average temperature change of the developed cell lattice can be seen. The CC-BCC cell lattice has the highest average surface temperature of 44.20 °C, followed by the Pyramid Based at 39.4 °C and TO1 at 36.87 °C. Similarly, the average temperature change in each cell is highest for CCBCC, followed by the pyramid cell and TO1. This is consistent with Formula 2, which states that cells with a larger surface area will experience a lower ΔT. In Fig. 11, the area values for each lattice can be seen. It can be observed that the area of the TO1 lattice has the highest value, followed by the cell Pyramid Based and CCBCC.

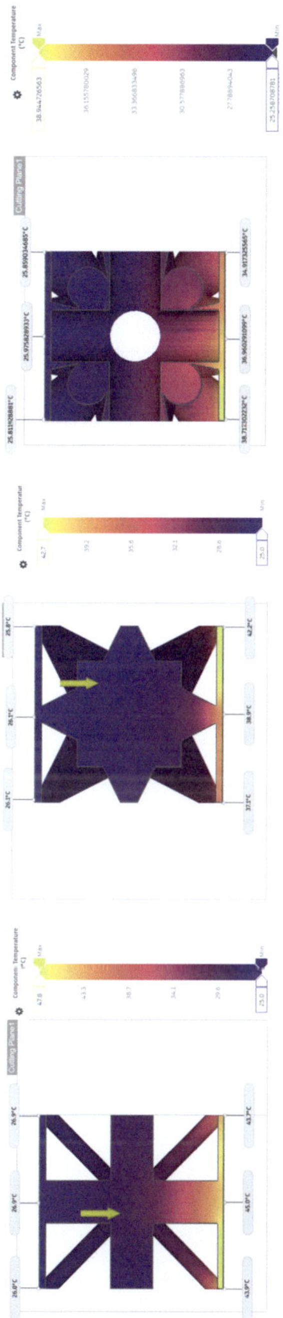

Fig. 5 Component temperature of each cell

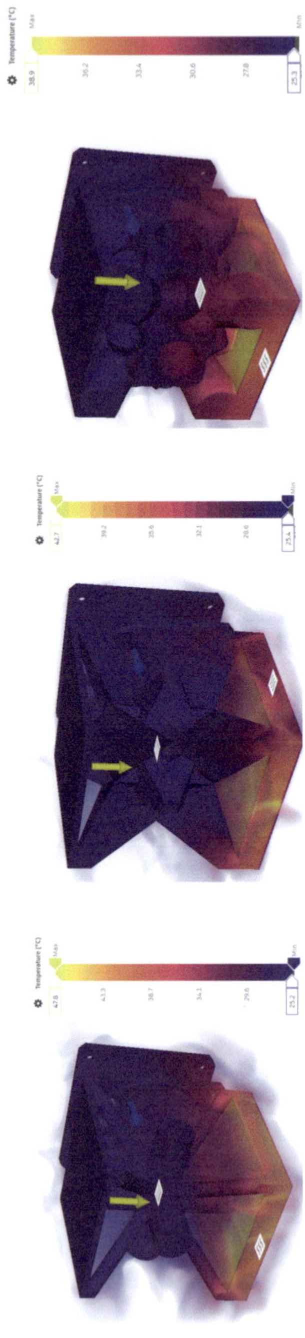

Fig. 6 Air and Component Temperature of Each Cell

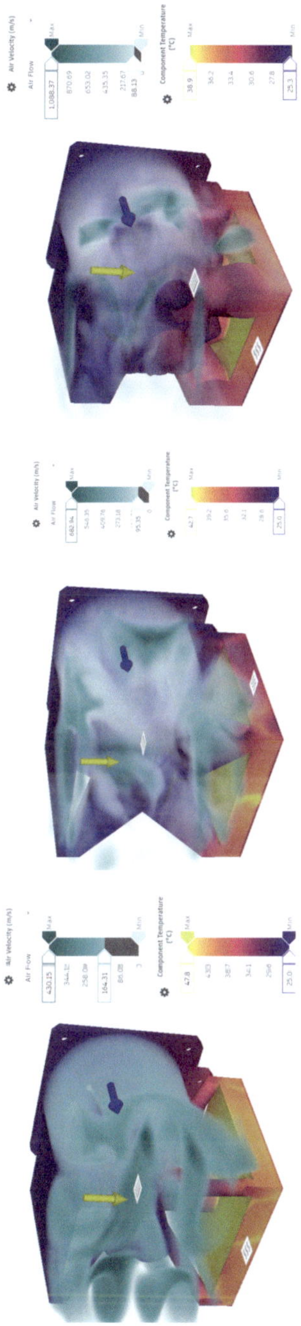

Fig. 7 Airflow + Component Temperature of Each Cell

Fig. 8 Airflow Temperature of Each Cell

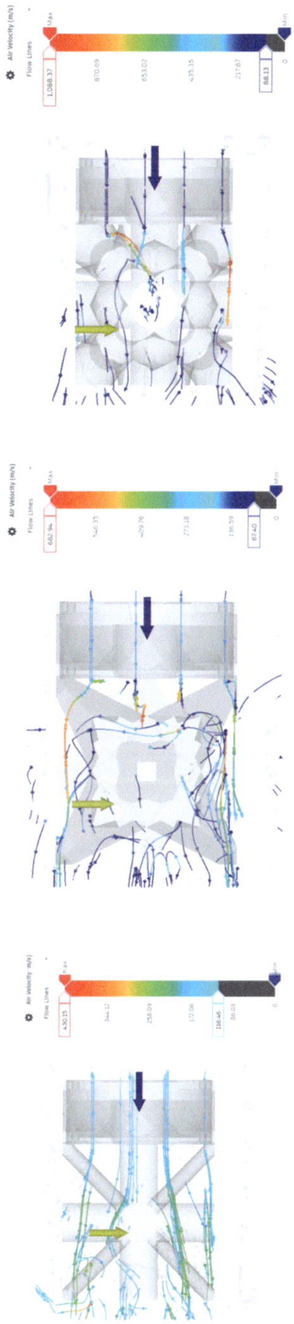

Fig. 9 Airflow Streamline of Each Cell

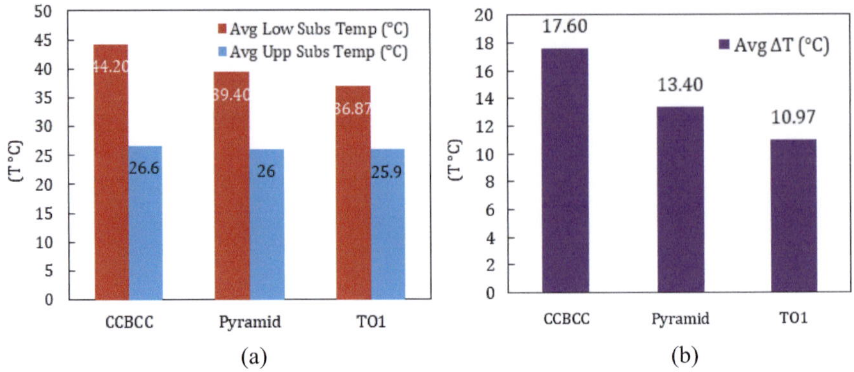

Fig. 10 **a** Surface temperature of lattice. **b** Average temperature change per lattice

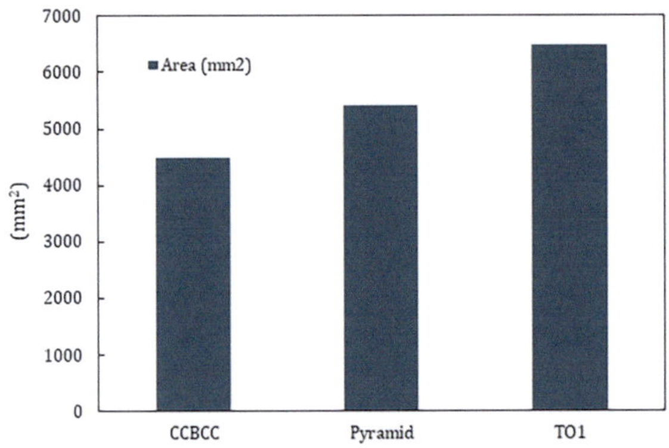

Fig. 11 The area of each developed lattice cell (1 × 1 cell)

The thermal resistance and TUI graphs for each lattice can be seen in Fig. 12.

In Fig. 12, it can be seen that the lowest thermal resistance value is obtained by the TO1 cell with a value of 0.18 °C/W, followed by the pyramid cell with a value of 0.22 °C/W and the CCBCC cell with a value of 0.29 °C/W. The larger the surface area of a cell, the smaller the temperature change ΔT of that cell, resulting in a lower thermal resistance. Different trends emerged on the TUI graph. The CC-BCC cell had the highest TUI value of 0.99, followed by TO1 at 0.96 and the pyramid cell at 0.95. The highest TUI value is 1; the closer the value is to 1, the more uniform the temperature distribution, meaning there is no significant temperature difference at each temperature measurement point. This phenomenon will be explained in the next subsection for a more comprehensive understanding. The next subsection will explain how the influence of cell configurations arranged longitudinally and transversely affects the thermal resistance and TUI of the cells.

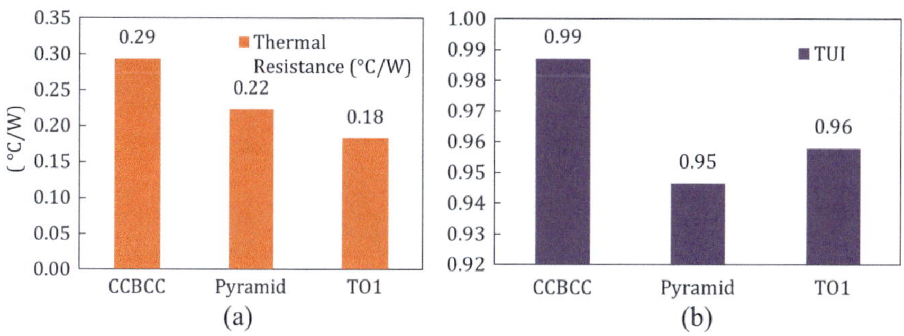

Fig. 12 **a** Thermal resistance. **b** TUI of each developed lattice cell (1 × 1 cell)

3.2 Longitudinal Configuration Cell Result

This subsection discusses the influence of longitudinally arranged cell configurations (in the direction of airflow) on thermal performance, measured by TUI and R_θ. The configurations discussed are configurations 2 (3 × 1 cell) and 4 (5 × 1 cell).

In Fig. 13, it can be seen that the more cells arranged longitudinally, the larger the specimen's surface area, which can reduce the average ΔT. The smaller the average ΔT, the smaller the thermal resistance R_θ. This can be seen in Fig. 14a.

Next, as shown in Fig. 14b, the more cells are arranged, the smaller the TUI value becomes. This can be explained by the fact that the longer the cells are arranged, the greater the area of friction between the airflow and the components. This causes the air velocity at the end of the cells to decrease as the cell arrangement increases longitudinally. This phenomenon can be explained using Formulas (5)–(8).

The smaller the value of υ, the smaller the Reynolds number will be, which will result in a smaller Nusselt number. The smaller the Nusselt number, the smaller the convective heat transfer coefficient will also be. This will also make the temperature difference (ΔT) greater between the component parts near the airflow source and those farthest from it. This phenomenon can be seen in Fig. 15, which shows how

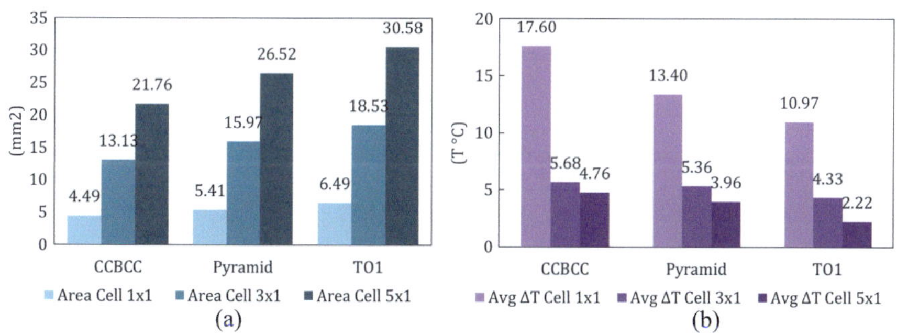

Fig. 13 **a** Specimen area. **b** ΔT Average of each cell (longitudinal)

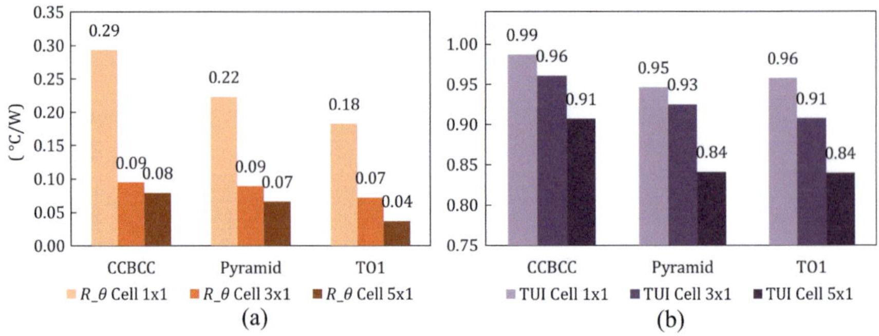

Fig. 14 **a** Thermal resistance. **b** TUI of 1 × 1, 3 × 1, dan 5 × 1 cell

flow velocity affects the temperature on the lower substrate of the lattice structure. It also influences the air temperature around the components and the temperature distribution contour on the lower substrate, as shown in Fig. 16.

In Figs. 17 and 18, an increasing temperature trend can be seen in both the lower and upper substrates as the cell length and cell arrangement increase. In the upper substrate section, cell TO1 has the highest temperature, followed by cell pyramid

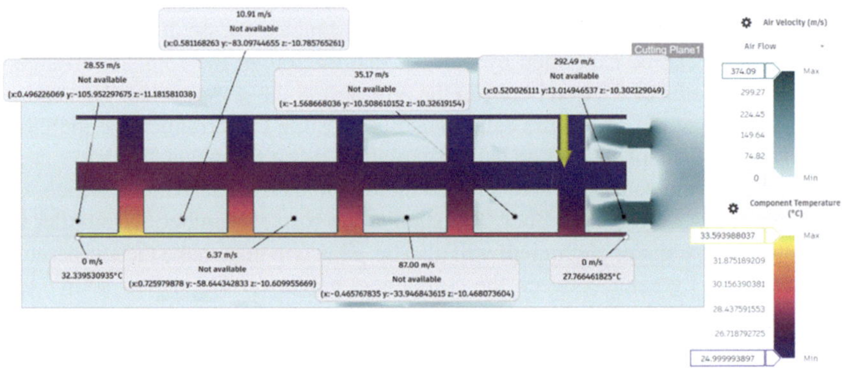

Fig. 15 Airflow velocity and component temperature distribution (CC-BCC Cell)

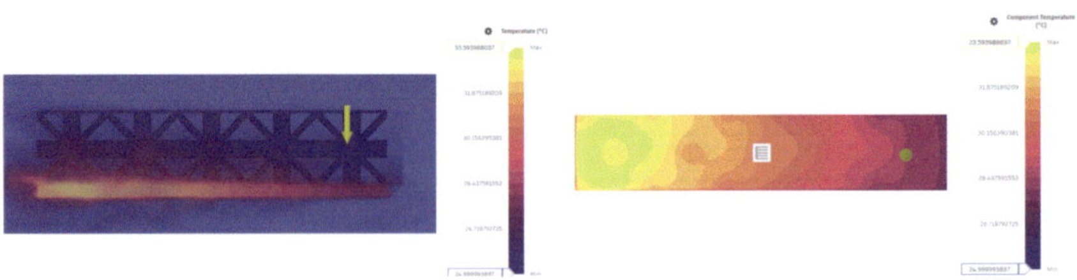

Fig. 16 **a** Air temperature distribution and components. **b** Component temperature distribution on the lower substrate (CC-BCC Cell)

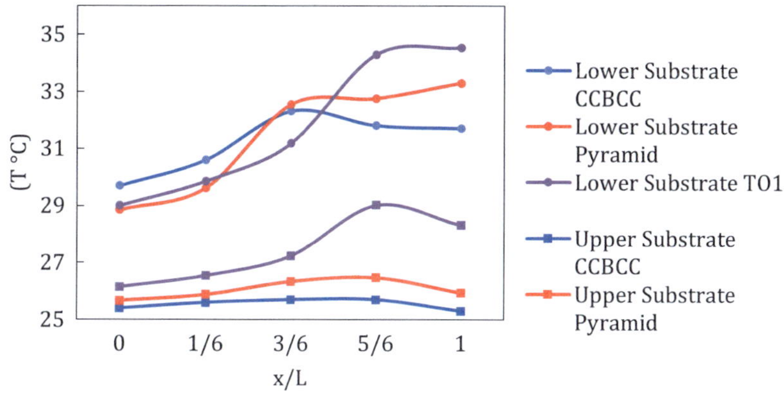

Fig. 17 Temperature distribution Cell 3 × 1

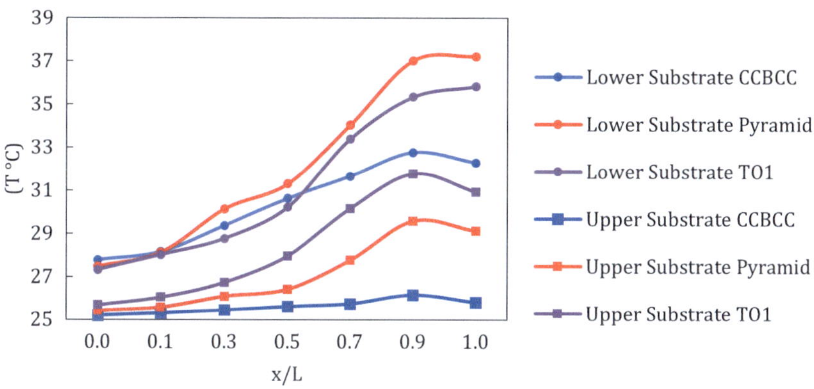

Fig. 18 Temperature distribution Cell 5 × 1

and cell CCBCC for both configurations (3 × 1 and 5 × 1). However, in the lower substrate section, there is a difference between the 3 × 1 and 5 × 1 cell configurations in the temperature values at length L. In the 3 × 1 cell configuration, the highest temperature at length L is obtained by cell TO1, while in the 5 × 1 cell configuration, the highest temperature at length L is obtained by cell pyramid. This indicates that the longer the cell and the more cells arranged longitudinally, the more it can change the characteristics of the temperature distribution trend for each cell type.

3.3 Transversal Configuration Cell Result

This subsection will discuss the simulation results for specimens with configuration 3 (3 × 3 cells) and configuration 5 (5 × 3 cells). This is used to determine changes in temperature distribution and thermal performance when additional cells are added to the right and left sides of the cell arrangement. This can be seen in Fig. 19, which shows the values of ΔT and the area of the specimen for each configuration. Of course, the more cells are arranged, the larger the specimen's area will be. As the area increases, the value of ΔT will also decrease. The value of Thermal Resistance, which is a function of ΔT, will also decrease as the value of ΔT decreases. This applies to all types of cells and all types of configurations, as can be seen in Fig. 20.

However, unlike the longitudinal arrangement, the TUI value in the transverse arrangement undergoes a change in trend shape. In the longitudinal arrangement, the longer the cell and the more cells arranged lengthwise, the smaller the TUI value. A different phenomenon is found in the transverse cell arrangement. Because in the transverse arrangement, the lattice Pyramid and TO1, which previously had TUI values lower than CC-BCC, have higher TUI values in this transverse arrangement. In fact, the TUI value for cell TO1 approaches 1, reaching 0.98 in a 3 × 3 cell arrangement. Although the downward trend of TUI based on the increase in the number of elongated cells was still observed in the transverse arrangement. This can

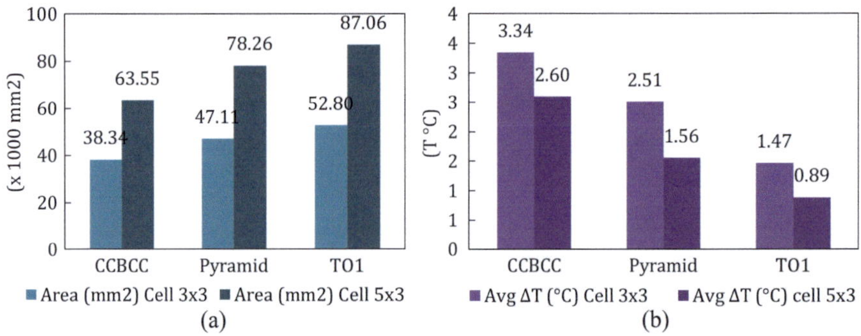

Fig. 19 **a** Specimen area. **b** ΔT Average of each lattice (Transversal)

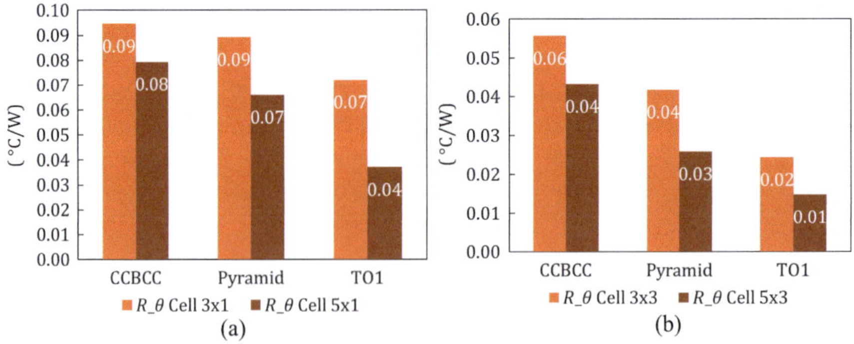

Fig. 20 **a** Thermal resistance longitudinal configuration. **b** Thermal resistance transversal configuration

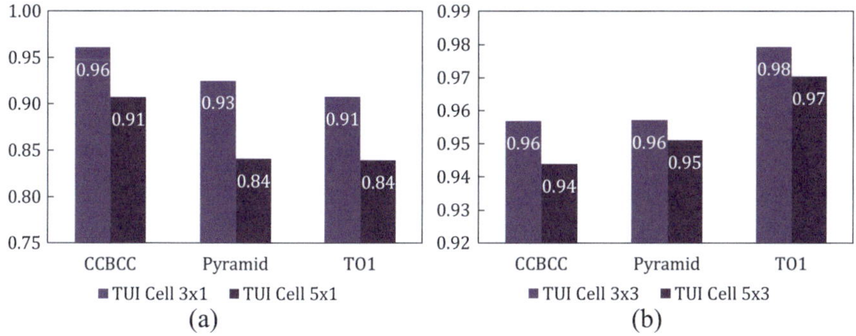

Fig. 21 **a** TUI longitudinal configuration. **b** TUI transversal configuration

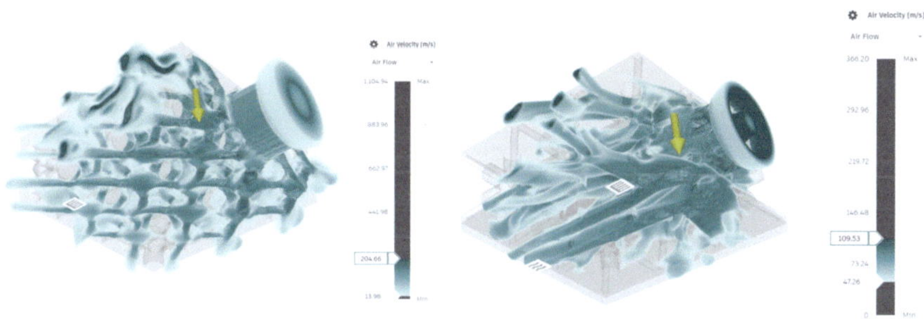

Fig. 22 **a** TO1 cell 3 × 3 airflow. **b** CC-BCC cell 3 × 3 airflow

be seen in Fig. 21. This can be explained by looking at the phenomena occurring in Fig. 22, between TO1-type lattice cells and CC-BCC-type cells for a 3 × 3 size.

As can be seen in Fig. 22, the airflow velocity distribution on the lower substrate of TO1 cells is more uniform compared to the airflow velocity distribution on the lower substrate of CC-BCC type cells. This can be attributed to the channelling feature of holes connecting the cells in TO1 cells. These holes are used as air channels to facilitate smoother airflow while maintaining its velocity distribution. A uniform velocity distribution can maintain the value of the convective heat transfer coefficient, resulting in a smaller temperature difference (ΔT) at the lower substrate (Fig. 23). The temperature difference between the minimum (24.9 °C) and maximum (27.8 °C) temperatures is only 2.9 °C. This can increase the TUI value of the lattice cell arrangement. Additionally, the temperature distribution trend of the TO1 cell lower substrate in Figs. 24 and 25 is the gentlest compared to other cell types.

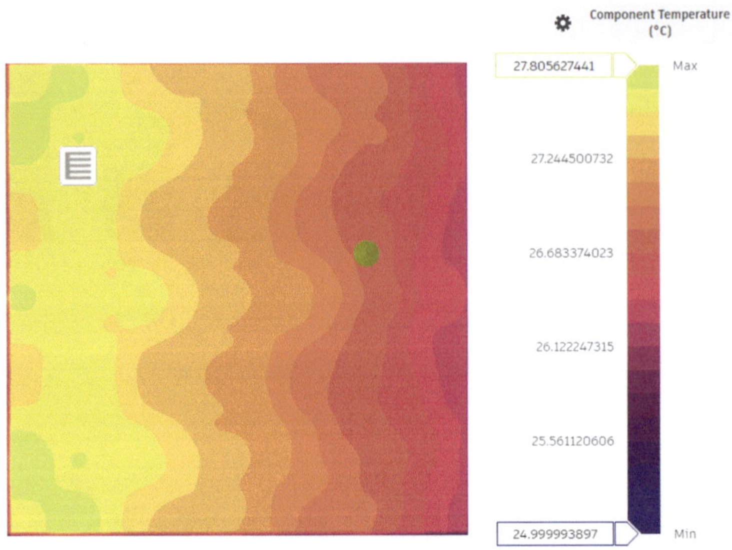

Fig. 23 Temperature distribution contour on the lower substrate for cell TO1 3 × 3

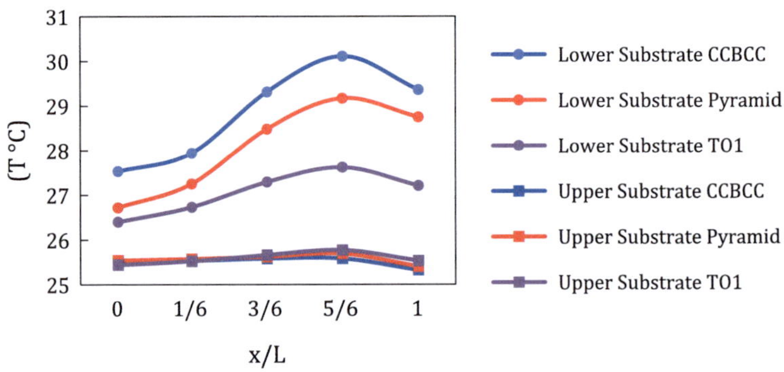

Fig. 24 Temperature distribution cell 3 × 3

3.4 Comparison with Existing Lattice Cell

In this subsection, the researcher compares the thermal performance of 3 developed lattice types (CC-BCC, Pyramid, TO1) against existing lattice cells (Octet, BCC, FCC, Kelvin Cell) as shown as Fig. 26. The selected configuration is a 3 × 3 cell arrangement.

As can be seen in Fig. 27, the thermal resistance values for CC-BCC, PYR, and TO1 cell types are smaller than those of the existing cell. Additionally, the TUI value for TO1 is higher than all lattice types, and the TUI values for PYR and CC-BCC are higher than those for BCC, Kelvin, and Octet. These results prove that the

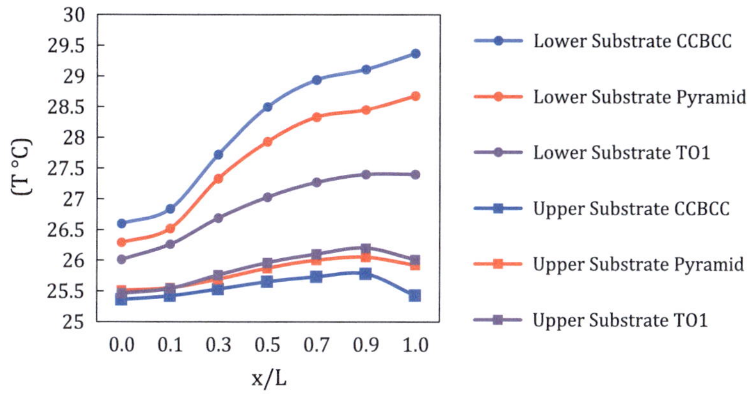

Fig. 25 Temperature distribution cell 5 × 3

Fig. 26 Type of cells being compared

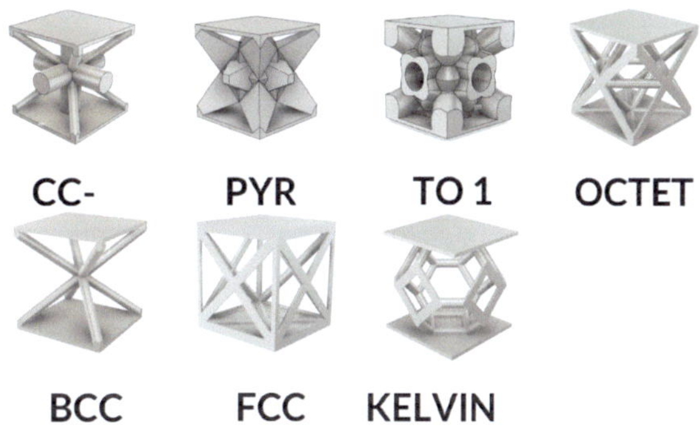

three developed cells have better thermal performance compared to the existing cells (BCC, FCC, Kelvin, Octet). Thus, it can be an alternative cell that can be used for cold plate batteries or heat sinks. Additionally, as seen in Fig. 28, the CC-BCC, PYR, and TO1 lattices have lower substrate temperatures below the maximum temperature (30 °C) to maximize battery performance.

However, of course, the developed lattice still has weaknesses when considering the weight (gr) compared to the existing cell. In Figs. 29 and 30, you can see the Ashby Map graphs for TUI and Thermal Resistance compared to their weight. Of course, this is a challenge to reduce the weight of the cell lattice while maintaining the existing thermal performance.

However, it is important to take into account that the density of each lattice structure in the single-cell configuration still varies each other. This variation directly influences the amount of void volume in the lattice. Lower density corresponds to wider cavity regions. Therefore, the differences in thermal resistance and TUI values are not solely affected by geometrical configuration of the lattice structures only. The variation in density may also has contribution in overall thermal performance. Therefore, further investigations are required to isolate the effect of lattice topology

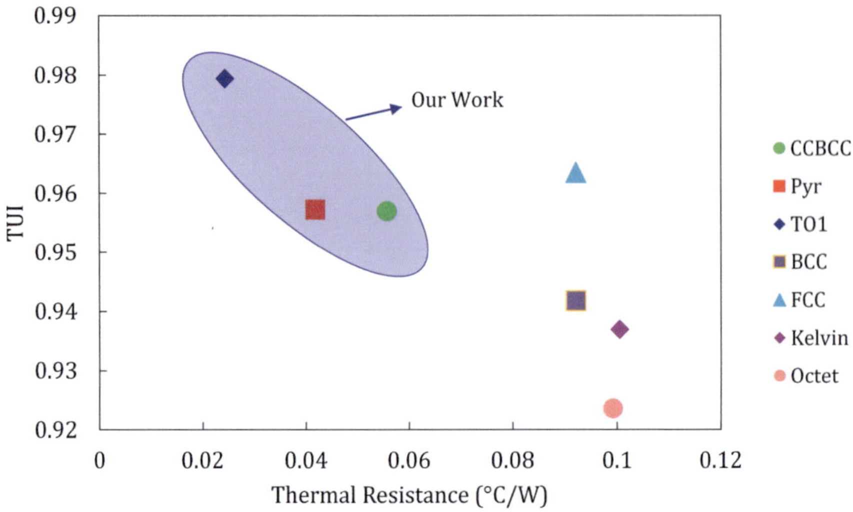

Fig. 27 Ashby map of TUI and thermal resistance cell 3 × 3

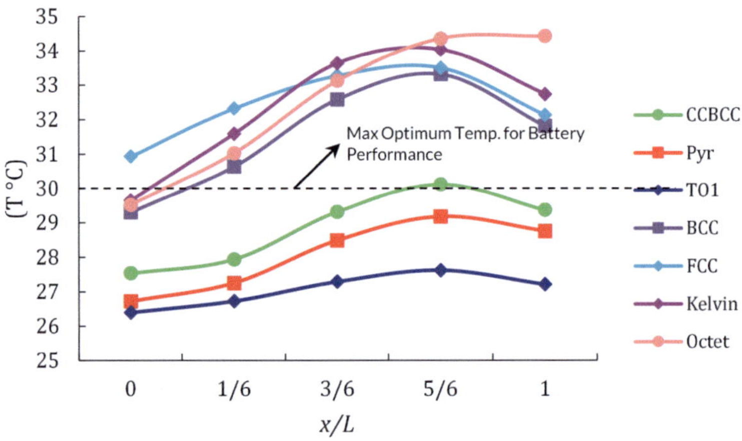

Fig. 28 Temperature distribution on the lower substrate for all simulated lattice types

from density-related parameters, ensuring a more accurate assessment of the intrinsic thermal efficiency of each cell type.

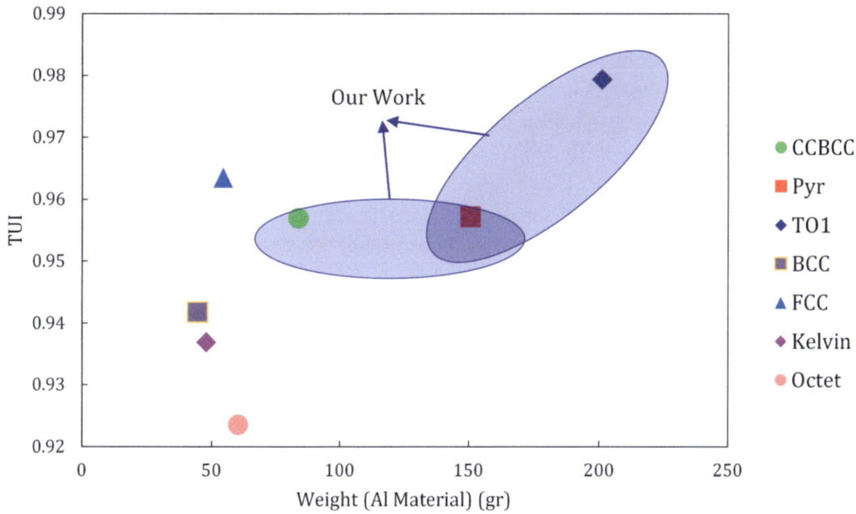

Fig. 29 Ashby map of TUI and weight cell 3 × 3

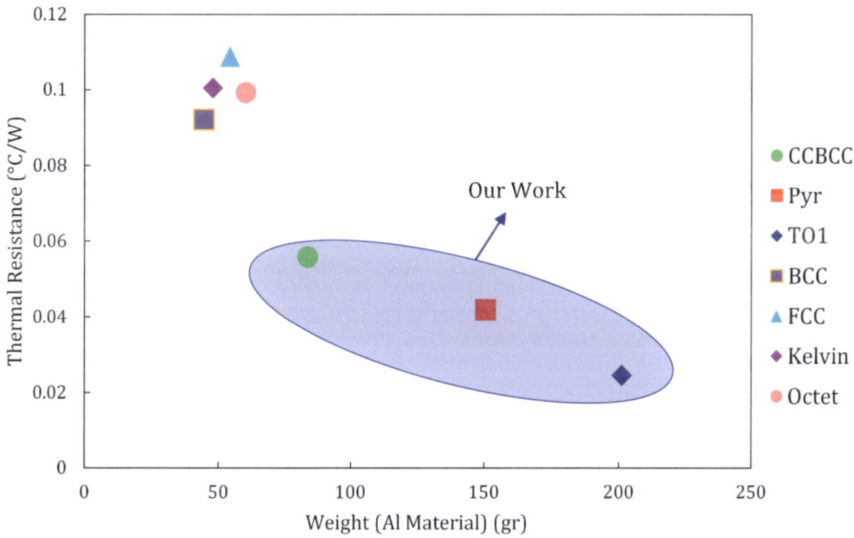

Fig. 30 Ashby map of thermal resistance and weight cell 3 × 3

4 Conclusion

Based on numerical simulation, thermal performance of lattice is dependent on both cell topology and configuration. the TO1 cell consistently achieved the lowest thermal resistance, reaching 0.18 °C/W in the single-cell configuration, compared to 0.22 °C/

W for the Pyramid and 0.29 °C/W for the CC-BCC. This advantage stems from its larger surface area and channeling features, which promote effective convective heat transfer and reduce ΔT. In contrast, the CC-BCC lattice, while exhibiting the highest surface temperature (44.2 °C) and thermal resistance, demonstrated superior temperature uniformity with a TUI value of 0.99, outperforming TO1 (0.96) and Pyramid (0.95).

The developed lattice cell has good heat transfer rate, as evidenced by its low thermal resistance value. The more cells arranged longitudinally and transversely, the lower the thermal resistance. The developed lattice cell has good heat distribution, as evidenced by its high TUI value. The more cells arranged longitudinally, the lower the TUI value. However, adding transverse arrangements will increase the TUI value. The developed lattice has better TUI and thermal resistance values compared to other existing lattice types. To ensure a more accurate evaluation of the inherent thermal efficiency of each cell type, more research is necessary to separate the impact of lattice topology from density-related characteristics.

Acknowledgements The first author thankfully acknowledges the financial support provided by Indonesian Education Scholarship (BPI), Center for Higher Education Funding and Assessment (PPAPT), Ministry of Higher Education, Science, and Technology of Republic Indonesia (Kemdiktisaintek) and Indonesian Endowment Fund for Education (LPDP), Ministry of Finance of Republic Indonesia.

References

1. Diachenko, S.V., Balabanov, S.V., Sychov, M.M., Litosov, G.E., Kiryanov, N.V.: The impact of the geometry of cellular structure made of glass-filled polyamide on the energy-absorbing properties of design elements. Strojniški vestnik-J. Mech. Eng. **70**(11–12), 607–619 (2024)
2. Guo, Q., Yao, W., Li, W., Gupta, N.: Constitutive models for the structural analysis of composite materials for the finite element analysis: a review of recent practices. Compos. Struct. **15**(260), 113267 (2021)
3. Wan, C., Tian, W., Zhou, J., Qing, Y., Huang, Q., Li, X., Wei, S., Zhang, L., Liu, X., Wu, Y.: Green anisotropic carbon-stabilized polylaminate copper oxide as a novel cathode for high-performance hybrid supercapacitors. Mater. Des. **15**(198), 109309 (2021)
4. Goulas, A., Zhang, S., McGhee, J.R., Cadman, D.A., Whittow, W.G., Vardaxoglou, J.C., Engstrøm, D.S.: Fused filament fabrication of functionally graded polymer composites with variable relative permittivity for microwave devices. Mater. Des. **1**(193), 108871 (2020)
5. Venugopal, V., McConaha, M., Anand, S.: Integration of design for additive manufacturing constraints with multimaterial topology optimization of lattice structures for optimized thermal and mechanical properties. J. Manuf. Sci. Eng. **144**(4), 041003 (2022)
6. Gu, L.D., Min, J.C., Tang, Y.C.: Effects of mass transfer on heat and mass transfer characteristics between water surface and airstream. Int. J. Heat Mass Transf. **1**(122), 1093–1102 (2018)
7. Ho, J.Y., Leong, K.C., Wong, T.N.: Experimental and numerical investigation of forced convection heat transfer in porous lattice structures produced by selective laser melting. Int. J. Therm. Sci. **1**(137), 276–287 (2019)
8. Rezk, H., Gomaa, M.R., Marmoush, M.M., Shehata, N., Henry, J.: Corrigendum to "Theoretical and experimental performance investigation of a newly combined TDD and SWH system" [Appl. Therm. Eng. **160**, (2019) 114156]. Appl. Ther. Eng. **164**, 114498 (2020)

9. Zaheri, S., Hassanipour, F.: A comprehensive approach to the mathematical modeling of mass transport in biological systems: fundamental concepts and models. Int. J. Heat Mass Transf. **1**(158), 119777 (2020)
10. Han, Y., Lu, W.F.: A novel design method for nonuniform lattice structures based on topology optimization. J. Mech. Des. **140**(9), 091403 (2018)
11. Kuwata, Y.: Direct numerical simulation of turbulent heat transfer on the Reynolds analogy over irregular rough surfaces. Int. J. Heat Fluid Flow **1**(92), 108859 (2021)
12. Sajjan, S.K., Dewangan, A.K., Kumar, R., Gupta, A.: Experimental investigation and non-dimensional regression analysis of condensation heat transfer of R-600a over finned tubes. Int. J. Therm. Sci. **1**(170), 107174 (2021)
13. Dixit, T., Nithiarasu, P., Kumar, S.: Numerical evaluation of additively manufactured lattice architectures for heat sink applications. Int. J. Therm. Sci. **1**(159), 106607 (2021)
14. Liu, J., Xu, M., Zhang, R., Zhang, X., Xi, W.: Progress of porous/lattice structures applied in thermal management technology of aerospace applications. Aerospace **9**(12), 827 (2022)
15. Shen, S., Kanbur, B.B., Wu, C., Duan, F.: Three-dimensional printing conformal cooling with structural lattices for plastic injection molding. Front. Heat Mass Transf. (2024)
16. Arifin, A.A., Batan, I.M., Bici, M., Wahjudi, A., Pramono, A.S.: Investigation of discrepancies in isotropic material and structural properties in lattice frameworks. Mech. Eng. Soc. Ind. **5**(1), 245–256 (2025)
17. Arifin, A.A., Batan, I.M., Bici, M., Wahjudi, A., Pramono, A.S.: Isotropic body-centered cubic (BCC) lattice structure design. In International Conference on Mechanical Engineering 2023 Aug 30, pp. 45–53. Springer Nature Singapore, Singapore (2023)

Fabrication of Poly(eugenol)/Cellulose Acetate/3-Glycidoxypropyltrimethoxysilane (GPTMS), and Titanium Dioxide (TiO_2) Membranes for Tea Industry Wastewater Treatment

Eka Cahya Muliawati, Afisya Anindita Darma, Adi Permadi, Ika Fitri Ulfindrayani, and Akhmad Fahruzi

Abstract The tea industry generates large volumes of wastewater containing high concentrations of organic matter and suspended solids, leading to elevated chemical oxygen demand (COD) and biochemical oxygen demand (BOD). This study reports the fabrication of ultrafiltration membranes based on cellulose acetate (CA) and poly(eugenol) (PE) at a mass ratio of 7:3 (w/w), modified with sorbitol, 3-glycidoxypropyltrimethoxysilane (GPTMS), and TiO_2 nanoparticles. Membranes were prepared via phase inversion and characterized using SEM and FTIR. Performance was evaluated in terms of water flux, turbidity (NTU), total dissolved solids (TDS), COD, and BOD rejection. The optimized membrane composition (1 mL sorbitol, 2.5 mL GPTMS, 2.5 g TiO_2) exhibited a flux of 36.11 L m^{-2} h^{-1}, with rejection efficiencies of 83.42% (NTU) and 76.28% (TDS). Furthermore, COD was reduced from 2215 to 165 mg/L (97.83% removal) and BOD from 1327 to 117 mg/L (91.18% removal), meeting national discharge standards. The high removal efficiency is attributed to the synergistic mechanisms of size exclusion by the dense

E. C. Muliawati (✉) · A. A. Darma
Chemical Engineering, Faculty of Industrial Technology, Institute of Technology Adhi Tama Surabaya, Arif Rahman Hakim 100, Surabaya City, East Java 60117, Indonesia
e-mail: ekacahya@itats.ac.id

A. Permadi
Department Master of Chemical Engineering, Faculty of Industrial Technology, Universitas Ahmad Dahlan, Yogyakarta, Indonesia

I. F. Ulfindrayani
Universitas Negeri Surabaya, Surabaya, Indonesia

A. Fahruzi
Institute of Technology Adhi Tama Surabaya, Surabaya City, East Java, Indonesia

M. Awang et al. (eds.), *Proceedings of the International Conference on Mechanical, Manufacturing and Plant Engineering*, Proceedings in Technology Transfer,
https://doi.org/10.1007/978-981-95-7817-7_7

selective layer and adsorption of phenolic compounds through the aromatic structure of PE. These findings demonstrate that CA/PE hybrid membranes with GPTMS–TiO_2 modification offer a sustainable and highly effective approach for tea industry wastewater treatment.

Keywords Ultrafiltration · Cellulose acetate · Polyeugenol · TiO_2 nanoparticles · Tea wastewater · COD removal · BOD removal

1 Introduction

The tea industry produces large volumes of wastewater containing high levels of organic matter, phenolic compounds, and suspended solids. These contaminants contribute to elevated chemical oxygen demand (COD) and biochemical oxygen demand (BOD), which can cause oxygen depletion and ecological imbalance if discharged untreated. Conventional treatment methods, such as coagulation, sedimentation, and biological oxidation, often show limited efficiency in removing these recalcitrant organics and may lead to the formation of secondary sludge [1]. Therefore, advanced treatment technologies are required to address the shortcomings of conventional methods. Consequently, membrane-based separation processes have attracted increasing attention due to their advantages of high separation efficiency, scalability, and eco-friendly operation [2].

Polymeric membranes remain the most widely used in wastewater treatment owing to their tunable physicochemical properties, mechanical stability, and cost-effectiveness [1]. Among biopolymer-based materials, cellulose acetate (CA) has been extensively applied in reverse osmosis, ultrafiltration, and nanofiltration processes due to its hydrophilicity, biodegradability, and low fouling tendency [3]. Despite these merits, pristine CA membranes often suffer from low chemical resistance and limited permeability, necessitating further modification through blending or incorporation of functional polymers and nanoparticles to enhance their performance [4]. On the other hand, eugenol, a naturally occurring phenolic compound obtained from clove oil, has recently been explored as a sustainable precursor for polymer synthesis [5]. Poly(eugenol) and its derivatives exhibit desirable characteristics such as hydrophobicity, chemical stability, and availability from renewable resources, making them promising candidates for advanced membrane fabrication [6]. Previous studies demonstrated that sulfonated or blended poly(eugenol)-based membranes possess enhanced stability and separation properties in various applications, including fuel cells and organic pollutant removal [7].

In line with these developments, this research focuses on the fabrication of ultrafiltration membranes by combining poly(eugenol) and cellulose acetate using the phase inversion method, with further modification by sorbitol, GPTMS, and TiO_2 nanoparticles. As stated in the research objective, *"this study will use a combination of poly(eugenol) and cellulose acetate, which is expected to produce ultrafiltration membranes that are more efficient in reducing COD and BOD levels in tea industry wastewater*". The combination of these two polymers is therefore anticipated to yield green, efficient, and durable membranes suitable for addressing water scarcity and environmental concerns. Although CA and PE membranes have been studied separately, no study has examined their combination with the addition of GPTMS and TiO_2 for tea wastewater treatment. The use of GPTMS and TiO_2 for tea wastewater has also not been reported. In this study, I tested the use of GPTMS and TiO_2 on CA/PE membranes for tea wastewater.

2 Material and Methods

2.1 *Materials*

Cellulose acetate (CA, analytical grade) was obtained from Sigma Aldrich [1]. Eugenol was used as the precursor for the synthesis of poly(eugenol) (PE) via cationic polymerization using concentrated H_2SO_4 as a catalyst [7]. Sorbitol was employed as a plasticizer, while 3-glycidoxypropyltrimethoxysilane (GPTMS) was applied as a silane coupling agent. Titanium dioxide (TiO_2) nanoparticles were used as inorganic fillers to enhance hydrophilicity and antifouling properties [6]. Acetone served as the solvent for dope solution preparation. Tea industry wastewater was collected from a processing facility in East Java, Indonesia, and stored at 4 °C prior to analysis [8].

2.2 *Synthesis of Poly(eugenol)*

Poly(eugenol) was synthesized by dissolving eugenol, followed by the dropwise addition of concentrated H_2SO_4 under constant stirring. The polymerization reaction was maintained for 2 h at room temperature. The product was precipitated with distilled water, filtered to obtain high molecular weight poly(eugenol), as described in previous studies [6].

2.3 *Membrane Fabrication*

Membranes were prepared via the phase inversion method. A homogeneous dope solution was prepared by dissolving CA and PE in acetone with a CA:PE ratio of 7:3 (w/w) at room temperature with continuous stirring. Sorbitol, GPTMS, and TiO_2 were added according to predetermined compositions, and the solution was stirred until a uniform mixture was obtained. The solution was cast onto a clean glass plate using a casting knife with a controlled thickness. After solvent evaporation for 15, 20, 25, 30, and 35 s, the cast film was immersed in a coagulation bath containing distilled water at room temperature to induce phase inversion. The formed membranes were washed repeatedly to remove residual solvent and stored in distilled water prior to use. The evaporation time variation of 15–35 s was chosen to determine the effect of solvent evaporation duration on membrane morphology and performance. Shorter times tend to produce more open pores with high flux but low rejection, while longer times form a denser selective layer, increasing rejection but decreasing flux. With this variation, it is hoped that the optimum conditions between flux and rejection can be achieved [6, 9].

2.4 *Membrane Characterization*

The membranes were thoroughly characterized using several analytical techniques. Surface morphology was examined via scanning electron microscopy (SEM) to observe structural features. Chemical functional groups were identified using Fourier-transform infrared spectroscopy (FTIR). The permeation performance of the membranes was evaluated by measuring water flux and solute rejection. Additionally, the membranes' ability to treat wastewater was assessed by determining Chemical Oxygen Demand (COD) and Biological Oxygen Demand (BOD) removal efficiencies. These analyses provided comprehensive insight into both the structural and functional performance of the membranes in wastewater treatment applications.

3 Results and Discussion

Table 1 is the permeation flux of the membranes was evaluated using a dead-end ultrafiltration cell. The optimum composition, consisting of 1 mL sorbitol, 2.5 mL GPTMS, and 2.5 g TiO_2, achieved the highest flux of 36.11 L m^{-2} h^{-1}. This indicates that sorbitol and GPTMS improved membrane flexibility and crosslinking, while TiO_2 nanoparticles enhanced hydrophilicity, thereby facilitating water transport through the pores. Compared with pristine cellulose acetate (CA) membranes, which usually exhibit lower flux due to dense skin layers, the CA and Poly(eugenol) composite demonstrated excellent permeability. Flux testing of the membrane was carried out using a dead-end filtration cell with an effective membrane area of 9.42 cm^2. Prior to testing, the membrane was compacted for 30–45 min to stabilize the polymer structure. Subsequently, approximately 150 mL of tea wastewater was introduced into the filtration cell and subjected to an operating pressure of 4 bar. The permeate volume passing through the membrane was recorded at 5 min intervals, and the flux value was calculated based on the permeate volume per membrane area per unit time. The flux (J) was calculated using $J = V/(A \times t)$, where V is the permeate volume (L), A is the membrane area (m^2), and t is the filtration time (h).

The Fig. 1 illustrates the effect of evaporation time on the average flux of the ultrafiltration membrane. As the evaporation time increases from 15 to 35 s, the

Table 1 Flux calculation of ultrafiltration membranes based on cellulose acetate, polyeugenol, sorbitol, GPTMS, TiO_2

Evaporation time (s)	Permeate volume (ml) recorded every 5 min		Flux ($L\ m^{-2}\ h^{-1}$)	Average flux
15	V1	3	38.23	36.11
	V2	2.5	31.86	
	V3	3	38.23	
20	V1	2.5	31.86	29.74
	V2	3	38.23	
	V3	1.5	19.12	
25	V1	2	25.49	23.37
	V2	1.5	19.12	
	V3	2	25.49	
30	V1	2	25.49	19.12
	V2	1.5	19.12	
	V3	1	12.74	
35	V1	1.5	19.12	14.87
	V2	1	12.74	
	V3	1	12.74	

average flux gradually decreases from approximately 36 $L\ m^{-2}\ h^{-1}$ to about 15 $L\ m^{-2}\ h^{-1}$. This trend indicates that longer evaporation times lead to the formation of a denser membrane structure with reduced pore size and porosity, which increases hydraulic resistance and lowers water permeation through the membrane [1]. In addition, the decline in flux is also attributed to the accumulation of contaminants that progressively cover and block the membrane pores over time, intensifying fouling effects and further hindering permeate flow.

Table 2 is the rejection performance of the membranes was first evaluated using turbidity (NTU) and total dissolved solids (TDS) as indicators. As shown in Tables 2 and 3, the rejection efficiency for NTU ranged between 78.86–83.42%, while that for TDS ranged between 73.55 and 76.28%, depending on the evaporation time.

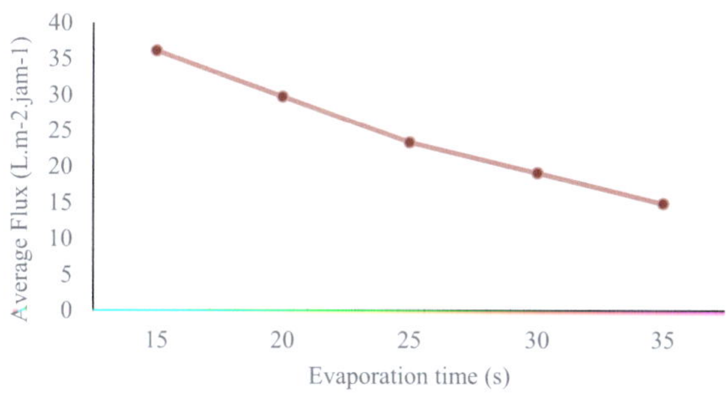

Fig. 1 Flux of ultrafiltration membranes based on cellulose acetate, polyeugenol, sorbitol, GPTMS, TiO_2

The best performance was observed at an evaporation time of 20 s, with 83.42% NTU rejection and 76.28% TDS rejection. This improvement can be attributed to the synergistic role of sorbitol as a plasticizer, GPTMS as a crosslinking agent, and TiO_2 as a hydrophilic filler, which collectively enhanced pore structure uniformity and reduced solute passage. The most optimal NTU and TDS rejection results were further tested for the removal of organic pollutants (COD and BOD).

The optimized membrane demonstrated excellent performance in reducing the organic load of tea industry wastewater. COD concentration decreased from 2215 to 165 mg/L, corresponding to a rejection efficiency of 97.83%, while BOD concentration decreased from 1327 to 117 mg/L, with a rejection efficiency of 91.18%. These results surpass the minimum discharge standards and indicate that the poly(eugenol)/cellulose acetate hybrid membrane is highly effective in mitigating the environmental impact of tea wastewater. The high COD and BOD removal can be attributed to both size exclusion and adsorption interactions provided by the aromatic rings in polyeugenol. The high COD and BOD removal performance indicates that the mechanism combines size exclusion by the membrane's dense selective layer with adsorption of phenolic compounds by poly(eugenol).

The mechanism of COD and BOD removal from tea wastewater using the CA/PE membrane primarily occurs through two processes: size exclusion and adsorption. The size exclusion process takes place when large organic molecules are retained by the membrane's selective layer, allowing only smaller molecules and water to pass through the pores. In addition, the aromatic structure of poly(eugenol) facilitates adsorption interactions with phenolic compounds and other organics present in the

Table 2 NTU rejection table for PE, sorbitol, GPTMS, and TiO_2 membranes

Evaporation time (s)	NTU before (mg/L)	NTU after (mg/L)	Rejection NTU (%)	COD removal percentage (%)	BOD removal percentage (%)
15	350	60	82,85	97,83	91,18
20	**350**	**58**	**83,42**		
25	350	67	80,85		
30	350	70	80		
35	350	74	78,86		

Table 3 TDS rejection table for PE, sorbitol, GPTMS, and TiO_2 membranes

Evaporation time (s)	TDS before (mg/L)	TDS after (mg/L)	Rejection TDS (%)	COD removal percentage (%)	BOD removal percentage (%)
15	4300	1025	76,16	97,83	91,18
20	**4300**	**1020**	**76,28**		
25	4300	1074	75,02		
30	4300	1098	74,46		
35	4300	1137	73,55		

wastewater. The combination of these two mechanisms enables the membrane to achieve high efficiency in reducing COD and BOD.

Figure 2 is SEM electron micrographs revealed that the optimized membrane exhibited a smooth and homogeneous surface with evenly distributed pores. The presence of TiO_2 nanoparticles contributed to a more uniform pore structure, reducing the likelihood of fouling. Cross-sectional images displayed an asymmetric morphology with a dense top layer and porous sublayer, typical of phase inversion membranes. This structural feature supports the observed balance between high flux and strong pollutant rejection. Comparable observations were made by [9], where GPTMS and TiO_2 fillers improved polymer filler compatibility and structural integrity.

The dense selective layer formed on the membrane surface is critical for rejection, while the porous sublayer reduces hydraulic resistance, ensuring high permeability. The incorporation of sorbitol further enhanced pore connectivity by acting as a pore-forming agent, consistent with previous reports on plasticizer-modified membranes. This morphology–performance relationship underlines the importance of tuning additive concentrations to achieve optimal balance between selectivity and flux. SEM images were taken at magnifications of 3000x, 5000x, and 10,000x with a resolution <5 nm, while FTIR spectra were recorded in the range of 4000–500 cm^{-1} with a resolution of 4 cm^{-1}. Figures include arrows and annotations highlighting key structural features of the membranes, such as the distribution of TiO_2 nanoparticles (bright spots), the CA/PE matrix, the effect of GPTMS on pore uniformity, and

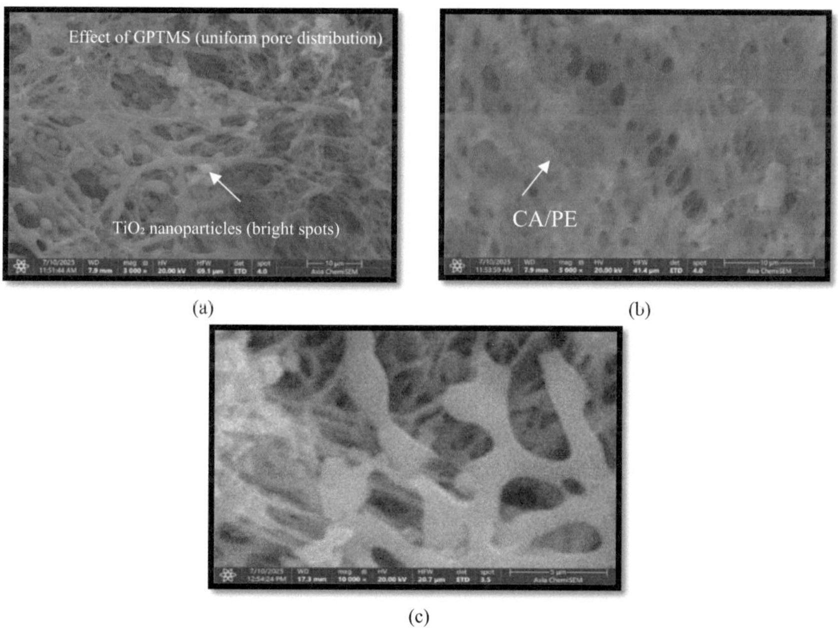

Fig. 2 SEM surface analysis of membranes. **a** Surface 3000x. **b** Surface 5000x. **c** Surface 10000x

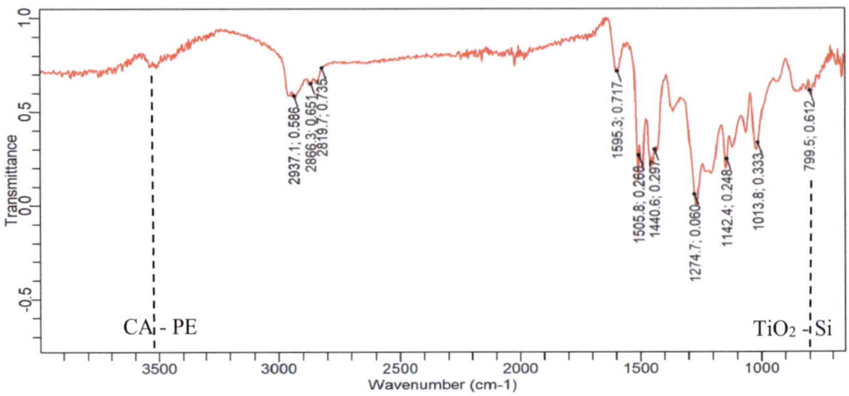

Fig. 3 FTIR analysis results of ultrafiltration membranes made from cellulose acetate, polyeugenol, sorbitol, GPTMS, TiO_2

the distinction between the dense top layer and the porous sublayer. These annotations provide clearer visualization of how each component contributes to the overall morphology and performance of the membrane.

Figure 3 is FTIR spectra confirmed the chemical interactions between cellulose acetate, poly(eugenol), and additives. Key absorption bands included a broad peak at ~3500 cm^{-1} (–OH stretching), indicating hydrogen bonding, a strong band near 1740 cm^{-1} (C=O stretching), corresponding to acetate groups, and peaks between 1100 and 1250 cm^{-1} (C–O–C stretching), confirming ether linkages and GPTMS incorporation. Additional bands attributed to Ti–O stretching confirmed the successful embedding of TiO_2 nanoparticles. These results suggest strong intermolecular interactions, enhancing the membrane's hydrophilicity and stability, in agreement with [7].

The appearance of new absorption bands compared with pristine CA suggests the formation of chemical bonds rather than mere physical blending. For instance, the Si–O–Si vibrations from GPTMS overlapped with C–O–C peaks, confirming silane crosslinking, while the reduced intensity of the hydroxyl band indicates effective interaction between PE and TiO_2. Such modifications improve the durability and chemical resistance of the membranes, making them more suitable for prolonged operation in harsh wastewater environments.

4 Conclusion

This study successfully developed a cellulose acetate and poly(eugenol)-based composite ultrafiltration membrane with a CA:PE ratio of 7:3 (w/w), enriched with sorbitol, GPTMS, and TiO_2 nanoparticles, for tea industry wastewater treatment via a phase inversion method. The optimized membrane demonstrated high performance, achieving a flux of 36.11 L m^{-2} h^{-1}, with an NTU rejection of 83.42%, a TDS rejection of 76.28%, and significant COD (97.83%) and BOD (91.18%) reductions. This excellent removal efficiency was attributed to the synergistic mechanism of size exclusion by the dense selective layer and the adsorption interaction provided by the aromatic structure of PE. These findings indicate that the CA/PE membrane with TiO_2 GPTMS modification is a promising and sustainable technology for tea industry wastewater treatment.

Acknowledgements The authors acknowledge the financial support from the Directorate of Research and Community Service, Directorate General of Research and Development (Ditjen Risbang), Ministry of Higher Education, Science, and Technology (Kemdiktisaintek), Republic of Indonesia (No. 128/C3/DT.05.00/PL/2025, May 28, 2025). The authors also thank the Department of Chemistry, Faculty of Science and Technology, Airlangga University, and the Department of Chemical Engineering, Faculty of Industrial Technology, Institut Teknologi Adhi Tama Surabaya, for providing laboratory facilities and technical assistance.

References

1. Islam, M.D., Uddin, F.J., Rashid, T.U., Shahruzzaman, M.: Cellulose acetate-based membrane for wastewater treatment—A state-of-the-art review. Mater. Adv. **4**(18), 4054–4102 (2023). https://doi.org/10.1039/d3ma00255a
2. Lin, H., Zhang, M.: Advanced membrane technologies for wastewater treatment and recycling. Membranes (Basel) **13**(6), 13–15 (2023). https://doi.org/10.3390/membranes13060558
3. Kuzminova, A., et al.: Development and study of novel ultrafiltration membranes based on cellulose acetate. Polymers (Basel) **16**(9) (2024). https://doi.org/10.3390/polym16091236
4. Liu, L., Xu, Y.L., Liu, J.X., Wang, S., Zhu, Q.X.: Preparation and characterization of high flux cellulose acetate/hydroxyapatite composite ultrafiltration membranes for water treatment. Desalin. Water Treat. **317**(February), 100309 (2024). https://doi.org/10.1016/j.dwt.2024.100309
5. Kusworo, T.D., Soetrisnanto, D., Widayat, W., Budiyono, B., Utomo, D.P.: Study of polymeric membranes potential for eugenol purification from crude clove leaf oil. ASEAN J. Chem. Eng. **18**(2), 81–92 (2018)
6. Muliawati, E.C.: New nanocomposite membranes with blended sulfonated poly-eugenol (S-PE) and titanium dioxide (TiO_2) as an alternative in direct methanol fuel cells, 49–57 (2024)
7. Rahim, E.A., Fong, R.C.Z., Sumarni, N.K.: Synthesis and characterization of membrane from high molecular weight polyeugenol. J. Phys. Conf. Ser. **1763**(1) (2021). https://doi.org/10.1088/1742-6596/1763/1/012003
8. Oprea, M., Voicu, S.I.: Cellulose acetate-based materials for water treatment in the context of circular economy. Water (Switzerland) **15**(10) (2023). https://doi.org/10.3390/w15101860
9. Muliawati, E.C., et al.: Sulfonated PEI membrane with GPTMS-TiO_2 as a filler for potential direct methanol fuel cell (DMFC) applications. Malaysian J. Fundam. Appl. Sci. **15**(4), 555–560 (2019). https://doi.org/10.11113/mjfas.v15n4.1216

Performance-Driven Adaptive Genetic Algorithms in Various Applications: A Review

Yingchen Ma, Shamini Janasekaran, Go Tze Fong, and Hanan Zhu

Abstract Primary or raw quantum circuits are challenging to directly implement on real quantum devices due to the qubit count restrictions and coupling limits. To matches hardware layouts between logic gates and physical gates sequences, the Adaptive Genetic Algorithms technique was used for this transformation, where a link can be created between algorithms and hardware using quantum circuit mapping as the linkage agent. Though, most mapping algorithms currently in use employ deterministic systems, which often result in less flexibility, insufficient diversity, and challenges in finding a balance between mapping quality and multiplicity, even while they guarantee a given level of mapping quality. This paper proposes a Quantum Circuit Mapping Based on an Improved Genetic Algorithm (QCMGA) to tackle these challenges by enhancing the workability of mapping schemes on actual hardware and overall optimization performance. This approach advances an adaptive fitness performance to strongly assess the quality of mapping schemes and uses a range of crossover tactics and selection processes to improve algorithm performance. Furthermore, QCMGA familiarizes an initialization approach based on neighbourhood gate distributions and designs a mutation mechanism specifically personalized to the structural features of quantum circuits, enhancing algorithm robustness and diversity while further increasing search efficiency and meeting the speed from a structural perspective. The applications of adaptive genetic algorithms circuits in various aspects have been highlighted and discussed in this work.

Keywords Adaptive genetic algorithm · Qiskit compiler · Quantum circuit mapping based · Styling · Qubit count

---

Y. Ma (✉) · S. Janasekaran · G. T. Fong
Center for Advanced Materials and Sustainable Manufacturing, Faculty of Engineering, Built Environment and Information Technology, SEGi University, Petaling Jaya, Selangor, Malaysia
e-mail: mayingchen110@126.com

Y. Ma · H. Zhu
Harbin Institute of Petroleum, Harbin, Heilongjiang, China

M. Awang et al. (eds.), *Proceedings of the International Conference on Mechanical, Manufacturing and Plant Engineering*, Proceedings in Technology Transfer,
https://doi.org/10.1007/978-981-95-7817-7_8

1 Introduction

Quantum signal processing circuit is a developing subject comprises of various approaches deployed to optimize, design and even to perform quantum algorithms, which obtained from the quantum gates on qubits by the possible sequences. There are several superiorities compared to traditional circuits in various areas such as in terms of correspondence, speediness and the capability to resolve complicated problems [1]. Through the presence of quantum properties in quantum circuit processes, the quantum algorithms circuits can perform better at once or emplacement in multiple states via the qubits, which can be effective investigation of numerous options at once, unified and even assembly immense parallel computation concurrently. Comparatively, the conventional circuit or the traditional circuits applies bits between 0 and 1 [2].

Due to the limitations of traditional computing, quantum- based computing—an alternative computational technology—has advanced. By this, the revolution in circuits processes has begun with the quantum algorithms which made a great innovation through great priority for the practical applications which needed the quantum computing devices. Somehow, some challenges have been identified for attaining the optimum performance in those applications, which need to be addressed in detail. Fault-tolerance is amongst the most crucial factor detected and need to be resolved tactically. This unfavourable factor has been rectified and fixed by integrating quantum error correction codes and surface codes [3, 4].

However, the execution of this code necessitates nearest neighbour (NN) boundaries, which need the quantum gates only act on qubits that are nearby to one another. Besides, computational mistakes may occur from noise at huge contact distances between the qubit—to—qubit [5]. Two works by Yuri Manin (Quantum Automata) and Paul Benioff (Quantum Turing Machines), respectively were appeared as the earliest published works in year 1980 signifying a possible acceleration over traditional computation [6].

This revolutionary publication has attained vast interests by researchers around the world, which made the quantum circuits an advance approach in developing algorithms circuits in various applications [7].

2 Applications of Adaptive Genetic Algorithms in Various Sectors

Quantum circuits, which include adaptive algorithm circuits are made to amend how they function in response to feedback from the system, changes in the environment, or real-time input [8]. They are highly effective in complexed and uncertain contexts because, in contrast to static circuits, they can self-optimize and react dynamically to changing conditions [9]. In domains where system parameters frequently modify while in use, such as wireless communications, signal processing, control systems, and artificial intelligence (AI), these circuits are very important [10, 11].

Some of the advantages of adaptive algorithm circuits are well known for better performance. Higher accuracy, as well as excellent energy efficiency makes them substantial [12–14]. Adaptive algorithms are highly recommended for assisting in preserving signal quality in wireless networks by adjusting to interference or noise [15, 16]. With the real-time resource optimization, adaptive hardware in artificial intelligence smooths faster learning and decision-making. Besides, they improve system resilience by automatically solving complex issues and mistakes, or adjusting to changes in hardware. This type of circuits is important to cater [17]. Thus, the applications of adaptive genetic algorithms have spread to various industries.

3 Financial Optimization

In financial modelling and forecasting, the classical or conventional optimization frequently causes errors mainly in capturing any nonlinearity or exchanges inherent in financial data. This usually occurs for the complexed or complicated financial data, where the classical optimization regularly act inefficiently [18]. Thus, the Genetic Algorithms plays very much important roles in tackling the financial optimization effectively, in order to overcome dynamic nature and complexity of financial markets [19].

For instance, de Almeida et al. has deployed self-adaptive genetic algorithms approach to predict the stock market and portfolio composition in the stock market. The main objective of this author is the attained maximum financial investment returns. They have used two types of approaches such as evolutionary algorithms and self-adaptive algorithms to gain the maximum financial interests [20].

Meanwhile, Li et al. has used the genetic algorithms to address complex financial investment issues. They authors had used this tool to understand the financial allocation rate versus investment risk. They have successfully adapted the genetic algorithms in financial optimization by maintain the rule of optimum investment allocation rate must always be greater with the lowest investment risk [19].

In the real estate market analysis, the genetic algorithms is aggressively used. For example, In 2022, Renigier et al. has revealed that a serious decision making in real estate investment requires in-depth analysis of information, as it is crucial for high-value or higher intensive characters in the real estate investment [21].

A group of researchers had applied a hybrid model incorporating short-range memory with adaptive genetic algorithms for the stock index prediction. Zeng and his team had developed this hybrid model which helps them to forecast the stock prices effectively, which made thisresearch area a hotspot for financial research topic in academic area. They have studied and determined the developed hybrid model is feasible in improving and achieving a better forecasting relating with financial variables [22].

4 Network-on-Chip Task Mapping

To avoid any communication congestion due to poor mapping, the Network-on-Chip task mapping has emerged in its application to overcome this issue [23]. The multi-core chip is used to process cores within it have several advantages such as energy saving, improved communicational potential, better power consumption and uphold overall higher performances [24]. Thus, the genetic algorithms circuits also incorporated with the Network-on-Chip to find best solutions to contemporary chip design [25].

Adaption or reworking in time-triggered systems can be conducted effectively through fault recovery, altering environment conditions and also via energy efficiency. A group of researchers has conducted such research by using meta-scheduling approach in adaption in time-triggered systems merely to uphold temporal predictability process via this approach. The researcher has applied genetic algorithms for a path by reintegration into multi-schedule graph, in order to control the state-space explosion issues due to increasing in the number of set-ups required for the reworking. This had successfully solved new scheduling problems for each reworking scenarios with the processing of multi-schedule graph. By this, the genetic algorithms incorporated time-triggered Network-on Chip overcome the challenges of designing the time computation and run-time storage issues [26].

Meantime, Zafar et al. had use the algorithms circuits incorporated with Network-on-Chip to increase the efficiency for complex multicore systems. They used this technique together with dynamic application mapping and modern architecture to address the complex multicore systems issues. The researchers had technically focused machine-based learning, energy-concern, deadline-concern, hybrid, latency-concern algorithms and heuristic-based approaches to maximize the performance, communication potential and energy efficacy. The main work by these researchers on this research is to improve the developing application mapping techniques for outstanding network-on-Chip architecture for the modern designing of chips [27].

Another interesting research was conducted by Bougherara et al. where they used genetic algorithms incorporated Network-on-Chip to utilizes maximization algorithms with the integration collaboration of two important algorithms, which are the Ant Colony and Genetic Algorithm. This concept provides enhanced integration of complex components. Deep studies were conducted by these researchers to restructure and rationalize the design phases. The effective mapping application software was deployed to deter the overall performance of the designed final system. The main objective of this work was to develop a new approach for mapping application towards Network-on-Chip construction merely to bring down the communication costs to the minimum [28].

Nevertheless, Kumar et al. have introduced Network-on-Chip for the mapping algorithm merely to improve the service time, queueing time, energy consumption and optimization of latency of topology at persistent bandwidth.

From this work, the introduction of Network-on-Chip have proven better performance in mapping in terms of scalability, parallelism and flexibility by combining

Horological Mapping (HorMAP), Rotational Mapping (RtMAP) and Divide and Conquer Mapping (DACMAP) for algorithm mapping of tasks onto cores [29].

5 Active Noise and Vibration Control

Comparative with the traditional techniques in reducing low-frequency noise and vibration, the Active Noise and Vibration Control play crucial roles to overcome such problems mainly in vehicles and aircraft by upholding user safety and the comfort, as well [30]. For fuel consumption rate, many techniques were studied globally by researchers, and through Active Noise and Vibration Control incorporated with Genetic Algorithms have shown tremendous fuel efficiency through enhanced crew effectiveness, more pleasant and complex interior, as well as for lighter-weight designs [31]. Therefore, the demand for this technology has been significantly increasing for high-end users in various industries.

A team of research (Zhang et al.), had studied the ride comfort control of air suspension using the collaboration techniques between Genetic Algorithm and Fuzzy PID. The researchers noticed that the comfort of any vehicle occupants and the overall ride performance and its quality is highly related to air suspension system's low vibration frequency, resistance to road and variable stiffness. According to Zhang et al. the overall vehicle performance can be improved by controlling the air suspension system. Due to some challenges in incorporating Genetic Algorithm and fuzzy PID, the parameter tuning process for the air suspension has some hiccups, which highly affect the suspension system. But once the incorporation between Genetic Algorithm and fuzzy PID have successfully developed, the air suspension controlling system of vehicle can be optimized, which gives a better comfort through minimum vibration frequency [32].

With the concern on environmental and safety measures, some researchers like Liao et al. had studied the demand of automobile industry for looking into lightweight vehicle designs, mainly to curb high fuel consumption. The researchers had optimized the lightweight vehicle design by opting Genetic Algorithm and topology optimization. This process involves the initiation with and examination of vehicle's mechanical characteristics, optimization procedures which link between Genetic Algorithm with Back Propagation neural network (BPNN). This combination has driven almost perfection in obtaining best-fit values of design parameter to cater the demand of the automobile industries for lightweight vehicle designs. Through the topology optimization, the researchers manage to achieve best lightweight designs which upholding other aspects such as vehicle body stiffness to maintain the safety aspect as well through crash test. Liao et al. also have studied the car anti-collision of the car body beam skeleton through the assessment involving deformation intrusion. All the obtained result revealed the optimum performance via the combination of Genetic Algorithm-BPNN approach [33].

Surprisingly, there are researchers have withstood the performance of smart cantilever beam by optimizing active control on the vibration by applying Genetic

Algorithm. Awada et al. had fabricated smart cantilever beam by adding sensor, and actuator merely to control law to the system purposely to remove vibration. According to Awada et al. vibration is one of the most dangerous phenomena that happens to any structure such as beam. Vibration potentially leads to premature fatigue and eventually can cause failure, which even can lead to fatality. Thus, by introducing sensor and actuator for the smart cantilever beam, the vibration can be eliminated. The application of Genetic Algorithm for the smart cantilever beam was to determine best-fit to satisfy specific vibration conditions by optimizing the sensor-to-actuator position with selected parameter to understand the potential vibration control for a safer cantilever beam fabrication [34].

6 Conclusion

Adaptive Genetic Algorithms are significant techniques in the evolutionary computation, which are highly required for refining problem-solving skills through dynamic alteration on the selected parameter. By modifying the parameters, the Adaptive Genetic Algorithms have portrayed better performance compared to traditional Genetic Algorithms.

The implementation of Adaptive Genetic Algorithms in various applications have proofed its success particularly for the dynamic, multi-objective and even for complicated problems. Beyond this deep-rooted application, the Adaptive Genetic Algorithm have huge promise in developing industries such as Science and Technology (S&T), Engineering, Medical and even for Environment Protection studies fields (upholding ecosystem management plans and pollution control).

In conclusion, Adaptive Genetic Algorithms deliver adaptable, dynamic and smart method in overcoming complex problems in various fields for advancement in respective application industries.

Acknowledgments The research work of this paper was supported financially by Heilongjiang Natural Science Foundation project (LH2024F042) and SEGi University.

References

1. Zhou, R., Poechmueller, P., Wang, Y.: An analog circuit design and optimization system with rule-guided genetic algorithm. IEEE Trans. Comput.-Aided Des. Integr. Circuits Syst. **41**(12), 5182–5192 (2022)
2. Ballinas, E., Montiel, O.: Hybrid quantum genetic algorithm with adaptive rotation angle for the 0-1 knapsack problem in the IBM Qiskit simulator. Soft. Comput. **27**(18), 13321–13346 (2023)
3. Mohammad, S.I., et al.: Quantum error correction with surface code analysis on quantum computing platform. Appl. Math. **19**(2), 423–435 (2024)

4. Sayedsalehi, S., et al.: Developing and analyzing the defect-based surface codes using optimization algorithms. Quantum Rep. **7**(2), 25 (2025)
5. Bhoumik, D., et al.: Efficient decoding of surface code syndromes for error correction in quantum computing. https://arxiv.org/abs/2110.10896 (2021)
6. O'Brien, G.: Automated Design Simplification of Quantum Program Synthesis. University of Sheffield (2021)
7. Gen, M., Lin, L.: Genetic algorithms and their applications. In: Springer Handbook of Engineering Statistics, pp. 635–674. Springer (2023)
8. Denis, R., Madhubala, P.: Hybrid data encryption model integrating multi-objective adaptive genetic algorithm for secure medical data communication over cloud-based healthcare systems. Multimed. Tools Appl. **80**(14), 21165–21202 (2021)
9. Abaku, E.A., Edunjobi, T.E., Odimarha, A.C.: Theoretical approaches to AI in supply chain optimization: pathways to efficiency and resilience. Int. J. Sci. Technol. Res. Arch. **6**(1), 092–107 (2024)
10. Kaleem, Z., et al.: Artificial intelligence-driven real-time automatic modulation classification scheme for next-generation cellular networks. IEEE Access **9**, 155584–155597 (2021)
11. Kouhalvandi, L., et al.: Overview of evolutionary algorithms and neural networks for modern mobile communication. Trans. Emerg. Telecommun. Technol. **33**(9), e4579 (2022)
12. Han, S., Xiao, L.: An improved adaptive genetic algorithm. In: SHS Web of Conferences. EDP Science (2022)
13. Chen, Q., Hu, X.: Design of intelligent control system for agricultural greenhouses based on adaptive improved genetic algorithm for multi-energy supply system. Energy Rep. **8**, 12126–12138 (2022)
14. Majeed, M.A., et al.: Optimal energy management system for grid-tied microgrid: an improved adaptive genetic algorithm. IEEE Access **11**, 117351–117361 (2023)
15. Robert, V.N.J., Vidya, K.: Genetic algorithm optimized fuzzy decision system for efficient data transmission with deafness avoidance in multihop cognitive radio networks. J. Ambient. Intell. Humaniz. Comput. **14**(2), 959–972 (2023)
16. Salah, I., et al.: Energy efficiency optimization in adaptive massive MIMO networks for 5G applications using genetic algorithm. Opt. Quant. Electron. **54**(2), 125 (2022)
17. Popescu, D.A., Ileana, M., Bold, N.: Advanced security in distributed web systems using genetic algorithm-based techniques. Tehnički vjesnik. **32**(5), 1932–1940 (2025)
18. Jian, Y., et al.: Adaptive genetic algorithm enabled tailoring of piezoelectric metamaterials for optimal vibration attenuation. Smart Mater. Struct. **31**(7), 075026 (2022)
19. Li, H., Shi, N.: Application of genetic optimization algorithm in financial portfolio problem. Comput. Intell. Neurosci. **2022**(1), 5246309 (2022)
20. de Almeida, R.L., Neves, R.F.: Stock market prediction and portfolio composition using a hybrid approach combined with self-adaptive evolutionary algorithm. Expert Syst. Appl. **204**, 117478 (2022)
21. Renigier-Biłozor, M., et al.: Genetic algorithm application for real estate market analysis in the uncertainty conditions. J. Housing Built Environ. **36**(4), 1629–1670 (2021)
22. Zeng, X., et al.: A hybrid model integrating long short-term memory with adaptive genetic algorithm based on individual ranking for stock index prediction. PLoS One **17**(8), e0272637 (2022)
23. Sacanamboy, M.: Heuristic algorithm for task mapping problem in a hierarchical wireless network-on-chip architecture. Clust. Comput. **27**(1), 159–175 (2024)
24. Amin, W., Hussain, F., Anjum, S.: iHPSA: an improved bio-inspired hybrid optimization algorithm for task mapping in network on Chip. Microprocess. Microsyst. **90**, 104493 (2022)
25. Asadi, Y.: A comprehensive study and holistic review of empowering network-on-chip application mapping through machine learning techniques. Discov. Electron. **1**(1), 22 (2024)
26. Muoka, P., Onwuchekwa, D., Obermaisser, R.: Adaptive scheduling for time-triggered network-on-chip-based multi-core architecture using genetic algorithm. Electronics **11**(1), 49 (2021)
27. Zafar, A., Farooq, M., Samad, A.: Application mapping approaches and methodologies in real-time network-on-chip (NoC) architectures. In: 2024 Eighth International Conference on Parallel, Distributed and Grid Computing (PDGC). IEEE (2024)

28. Bougherara, M., Amara, R.: Application mapping onto network on chip using genetic algorithm and ant colony optimisation. In: 2023 International Conference on Computer and Applications (ICCA). IEEE (2023)
29. Kumar, A., et al.: Mobile networks-on-chip mapping algorithms for optimization of latency and energy consumption. Mob. Netw. Appl. **27**(2), 637–651 (2022)
30. Giouvanakis, M., Sevastiadis, C., Papanikolaou, G.: Active low-frequency noise control implementing genetic algorithm on mode coupling of a compound source. Appl. Sci. **13**(11), 6740 (2023)
31. Fu, Y., Wang, X.: Advancements and trends in vehicle sound package for noise control: a comprehensive review. Adv. Mech. Eng. **17**(6), 16878132251345867 (2025)
32. Zhang, S., et al.: Research on ride comfort control of air suspension based on genetic algorithm optimized fuzzy PID. Appl. Sci. **14**(17), 7787 (2024)
33. Liao, S., Zhong, L., Yi, L.: Enhancing vehicle lightweight design process with genetic algorithm-optimized BP neural network and topology optimization: a comprehensive study (2024)
34. Awada, A., Younes, R., Ilinca, A.: Optimized active control of a smart cantilever beam using genetic algorithm. Designs **6**(2), 3 (2022)

Automation in the Production of Alcoholic Beverages: Prospects and Difficulties for Baijiu Businesses in the Age of Intelligent Winemaking

Fan Liao, Yeong Jin King, and Hiu Hong Teo

Abstract There are many small and medium-sized businesses (SMEs) in the Chinese baijiu industry, which contributes significantly to the country's economy. However, these businesses frequently struggle with low levels of brewing automation, limited technological capacity, and low market competitiveness. This paper thoroughly explores the development status and main obstacles of mechanized brewing technologies and looks at the technical bottlenecks that exist in critical stages such solid-state fermentation, solid-state distillation, and jiuqu manufacturing. It also outlines the more general industrial difficulties that SMEs encounter, such as financial limitations, a lack of qualified brewing experts, and a lack of investment in contemporary machinery. The study defines three research objectives and suggests mechanization solutions that are specific to the requirements and circumstances of SMEs based on these studies. The objective is to assist technological advancement and sustainable development throughout the industry by offering practical references for the mechanized transformation of small and medium-sized baijiu businesses and other brewing industries.

Keywords Alcohol · Baijiu · Koji · Mechanization

1 Introduction

One of the cornerstones of the Chinese economy, baijiu is a treasure of Chinese culture and civilization [1]. Chinese Baijiu is one of the six main distilled spirits in the world and has been a traditional alcoholic beverage for more than 2,000 years [2]. According to flavor characteristics, Baijiu can be classified into 12 types: strong aroma type, light aroma type, soy sauce aroma type, rice aroma type, phoenix aroma type, fermented bean curd aroma type, special aroma type, medicinal aroma type, rich aroma type, sesame aroma type, old white dry aroma type and compound aroma

F. Liao (✉) · Y. J. King · H. H. Teo
SEGi University, Petaling Jaya, Selangor, Malaysia
e-mail: SUKD2400106@segi4u.my

M. Awang et al. (eds.), *Proceedings of the International Conference on Mechanical, Manufacturing and Plant Engineering*, Proceedings in Technology Transfer,
https://doi.org/10.1007/978-981-95-7817-7_9

type (strong and soy sauce aroma compound) [3]. The strong-flavored type, the soy sauce-flavored type, the delicate-flavored type and the rice-flavored type are widely regarded as the four basic types of aromas [4]. One of the four main types of liquor, strong-flavored liquor is distinguished by its rich perfume, sweet and mellow taste, harmonious fragrance, and large production volume. It makes up 70% of the liquor market in China [5]. The rich and strong-flavored liquor is renowned for its rich aroma and sweetness. It is the most popular type of Chinese liquor and is also known as Luzhou-style liquor [6].

Baijiu is everywhere, and its astonishing production volume is obvious. As of 2020, it reached 7.4 million liters, generating a considerable profit of 158.5 billion yuan (approximately 24.9 billion US dollars), and showing a steadily upward trend [7, 8], It is deeply loved by a large number of wine enthusiasts. Baijiu, as a traditional type of alcohol in China, holds an important position in the national economic output.

However, Chinese liquor has always been brewed primarily by hand for a very long time. Human labor has been used for tasks including transportation, stirring, and loading and unloading, which has led to high labor intensity and low production efficiency [9]. Furthermore, traditional brewing techniques have shortcomings in terms of standardization of raw materials and control of the fermentation process. This can lead to quality defects and unstable products, and may also pose environmental pollution risks [10]. Traditional brewing techniques not only impede business growth in today's fast evolving culture, but they may also fall short of product hygiene regulations. During the brewing process, older methods frequently lack systematic hygiene control measures, which can readily introduce a variety of pollution issues, including mycotoxins, pathogenic microbes, heavy metal residues, and migratory chemicals from packaging materials. [11]. Therefore, under the pressure of industry competition, all major liquor enterprises must enhance their efforts in technological innovation. In particular, they should carry out technological renovations on the mechanical equipment in the production process to achieve continuous improvement in product quality [12].

Thus, this study investigates the challenges encountered by Luzhou's small and medium-sized liquor businesses. It primarily concentrates on two significant challenges: the challenges encountered by the small and medium-sized liquor business and the technical constraints in mechanized brewing technology. The difficulties mechanical brewing technology faces

Over the past ten years, the automation and intelligence of liquor production have made significant progress [3]. The production process of Chinese Baijiu has been significantly improved. The introduction of modern technology has optimized the traditional process. From raw materials to the final product, each stage has achieved varying degrees of mechanization and process modernization [13, 14]. Improved production efficiency and enhanced working conditions. However, the traditional brewing techniques of Chinese liquor rely more on manual operations [9]. During the production process, from the stage of making the mash [15], the stage of solid-state fermentation, the stage of distillation, to the stage of storage and blending [3], all the processes require manual participation.

Fig. 1 **a** The formation of the large koji (fermenting agent) **b** The maturation of the large koji in the koji room [19]

Firstly, large-scale yeast cakes are traditionally produced by hand using kneading, which has drawbacks like low efficiency, high labor intensity, considerable variance in the cakes' physical attributes, and uneven quality [15]. Large-grain koji is made by preparing raw materials, creating koji blocks, and incubating them. It was traditionally fermented using manual methods and empirical knowledge, which caused to inconsistent product quality. The ambiguous microbial group mechanism impacted by biological factors and the engineering dynamics controlled by non-biological elements limit the intelligent transformation and enhancement of the koji industry, which is yet in the research stage [16]. Additionally, in the large-scale koji fermentation room, the peak temperature range for koji fermentation, which is reached due to the accumulation of internal biological heat, is from 40 to 70 °C [17, 18], as in Fig. 1.

Secondly, mix the cooked grain ingredients with the large yeast. The mixture is then placed in a cellar and sealed with ordinary mud, undergoing anaerobic fermentation for approximately 60–90 days [20]. Since the microorganisms in the pit soil are an important factor determining the rich flavor of the strong-flavored Chinese liquor style [21]. Therefore, stainless steel cannot be used to make strong-flavored, non-fragrant whiskey. Only clay can be used to make them. As a result, strong-flavored, non-fragrant liquor is still brewed underground. Additionally, by integrating with Internet of Things technology, it is possible to monitor the temperature in the fermentation pit online in addition to the temperature and humidity in the fermentation room. This offers information to enable the optimization of pit filling and large-grain fermentation parameters [13], although it is not difficult to digitally monitor the fermentation temperature inside the pit, controlling the pit temperature is very challenging (unless the position of the fermentation container is changed) [3]. Grain delivery relied solely on manual labor or bridge cranes (called "hang carts" in China) for manual aid because traditional fermenting took place underground. Due to the fact that traditional strong-flavor Chinese liquor fermentation tanks are sealed with mud—a sort of sealing mud that is challenging to maintain—this took a significant

Fig. 2 Alcohol fermentation usually takes place in an above-ground wine cellar under anaerobic conditions [23]

amount of physical work and was also likely to cause damage to the inner walls of the tanks. As a result, the pit lids are vulnerable to collapses and fissures. The damage to the sealing mud may lead to unstable components and an abundance of microorganisms during the fermentation process, thereby causing fluctuations in product quality [22], as shown in Fig. 2.

Thirdly, the filling of the raw materials and the distillation vessel are the primary issues with distillation. The Northeast and North China regions use a conical-shaped distillation pot, where the alcohol vapor rises and condenses at the top to produce liquid. The liquid is then channeled through a pipe into the wine collection tank. A pot-bottom distillation pot is employed in the southwest, where the alcohol vapor condenses at the curved cooling surface at the bottom and flows straight into the bottom container, which is subsequently added to the wine collection tank. The main problems include a small cooling area, low heat exchange efficiency, and limited capacity of the distillation pot, resulting in a slow heat exchange speed [9], as shown in Fig. 3.

Mechanical operation cannot take the place of manual labor throughout the steamer's filling process. Workers must progressively create a stable grain column by evenly sprinkling the grains into the steamer. More grains must be added and manually compacted when steam leaks at specific locations. The labor intensity is great, and the entire process depends on the workers' skill and judgment. Accurately controlling the distillation endpoint and fraction cutting is challenging because of the unknown mechanism of solid-state distillation and the absence of theoretical guidance. Ye et al. [3], as shown in Fig. 4.

Fourthly, human labor and Baijiu selection are strongly related in traditional Chinese Baijiu brewing methods. The old way of viewing the blossom buds is still used in the modern Baijiu brewing procedures, particularly in the plucking process, which relies on verbal and teaching-based instructional approaches. As seen in Fig. 5, the traditional transmission technique not only fails to create a knowledge system but

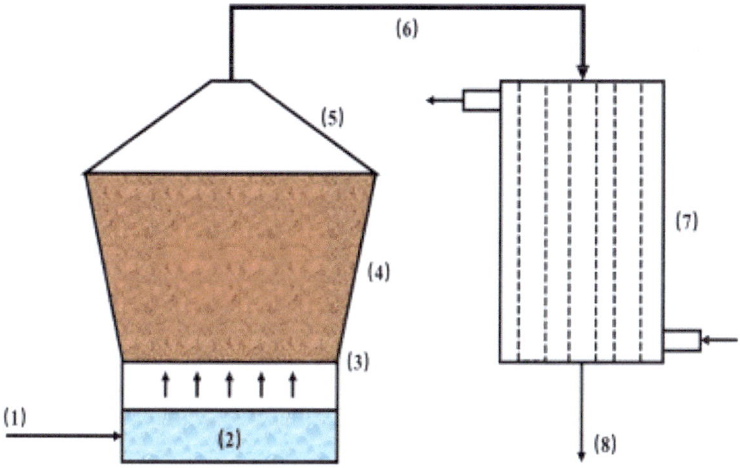

Fig. 3 Baijiu solid-state Zengtong distillation structure [24]; (1) Input pipe, (2) Bottom pot, (3) Grate, (4) Zengtong, (5) Lid, (6) Gas pipe, (7) Condenser, (8) Head, heart, tail

Fig. 4 Manual loading of materials in solid-state distillation [9]

also depends on human sensory judgment, which is readily constrained by a person's grasp and knowledge level [25].

The liquor sector is heading toward mechanization, automation, digitization, and intelligence due to the swift advancements in technology and economic growth [26]. Though Chinese liquor production has advanced in terms of automation, mechanization, and intelligence, overall technology and equipment are still somewhat outdated. There is still a long way to go before full-process mechanization and intelligence creation are realized [3].

Fig. 5 Manual packing liquor [9]

2 The Challenges Faced by the Small and Medium-Sized Liquor Industry

Although technological progress has promoted the application of automation technology in enterprises, the liquor industry still faces many challenges. Firstly, there is a significant disparity in industry revenue and profits. In 2022, Guizhou Moutai alone accounted for 18% of the market share, while the combined market share of the top five enterprises reached 47.7%, indicating a clear trend of monopoly [27]. Large enterprises continue to expand, while small and medium-sized enterprises are constantly marginalized due to their disadvantages in terms of resources, funds, technology and market. They are at a disadvantage in terms of mechanization [28], as shown in Fig. 6.

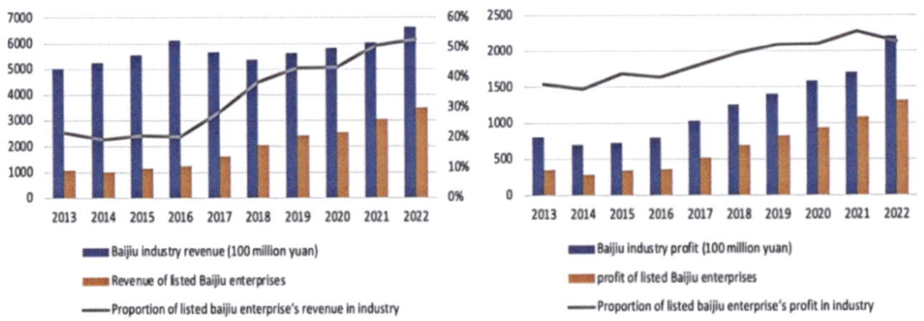

Fig. 6 Revenue and profits between the industry and listed baijiu companies [27]

Second, the issue of expensive capital expenditure in equipment manufacturing is still typically present in the current automated brewing production lines [29]. Increased costs and production costs brought on by the growth of enterprise scale are causing problems for the relevant industries [30], and rising consumption and fierce international competition have forced businesses to improve their competitiveness through technological advancements [31, 32].

The implementation of mechanization and automation is highly dependent on the individual operational skills and experience of the operators [33]. Currently, there is a shortage of comprehensive talents in the alcohol industry [34]. Additionally, the decrease in the supply of rural labor has made recruiting more difficult. Small and medium-sized baijiu enterprises are often constrained in their development by factors such as a shortage of talent [35]. Because traditional baijiu brewing involves complex processes, numerous steps, and a long production cycle, higher requirements are placed on process management and workforce capabilities. Meanwhile, in the course of internationalization and overseas market expansion, these enterprises still face inadequate talent reserves—especially a lack of professionals in overseas marketing—thereby constraining their long-term development. Owing to reliance on high-quality brewing raw materials and specific regional ecological environments, enterprises also face constraints in expanding production capacity while ensuring product quality [36]. Issues such as low production efficiency, insufficient adoption of digitalization/intelligent manufacturing, relatively weak equipment and technologies, and difficulties in unifying quality standards remain widespread [31]. Furthermore, the businesses' capacity to withstand risk is lacking. The pressure to survive has increased as a result of the outbreak. In terms of channels, they mainly rely on offline sales, and their online operation capabilities are insufficient, which limits market expansion [37], and there are certain technical limitations and research challenges in the current research of the liquor industry [38].

In terms of funding, the Sichuan liquor sector has built a solid foundation for its growth as a result of its general expansion. In order to guarantee the flow of funds for the liquor industry chain, Luzhou Laojiao formed Longma Xinda Microfinance Company [39]. This has somewhat mitigated the high costs and financing challenges that the majority of small and medium-sized liquor businesses confront. Nonetheless, the issue of financial challenges for small and medium-sized liquor businesses remains unresolved [40].

Intelligent manufacturing provides opportunities for low-cost green technological innovation and aids in the elimination of outdated equipment [42]. In recent years, China has integrated green manufacturing into an urgent future development strategy, and the green manufacturing system has gradually been integrated into its industrial development strategy [41]. This trend has brought attention to the industry's focus on future development and provided liquor businesses with a great opportunity to explore innovative brewing methods. However, Sichuan Province's small and medium-sized liquor companies are not competitive enough and have a low export rate [39]. The ability to independently create key technologies and equipment is inadequate, and the percentage of R&D spending is comparatively low [31]. There is intense competition and a large number of small and medium-sized firms in the liquor market. In order to compete in this highly competitive market, small and medium-sized enterprises may put short-term objectives ahead of careful product feature research and development [43].

Design of research objectives

For small and medium-sized enterprises, the degree of automation is not necessarily the higher the better. The selection of technology should take into account various factors such as cost and quality comprehensively. Under this background, investigating and analyzing the application of mechanized brewing technology in small and medium-sized enterprises is not only to examine whether the automation technology has the value for promotion and development, but also provides an important reference for small and medium-sized enterprises to choose the appropriate technology method for their own development.

Investigation and analysis of the application of mechanized brewing technology in small and medium-sized enterprises.

It is evident from the discussion above that small and medium-sized liquor businesses are having a lot of challenges. Investigating these businesses' level of acceptance of mechanized brewing technologies is therefore crucial. Through the inquiry, we can comprehend the primary challenges these businesses face when embracing automated technology, pinpoint the major barriers that influence the businesses' adoption of automated technology, and identify the most pressing challenges. This study examines the challenges faced by Luzhou's small and medium-sized liquor businesses and offers practical refurbishment suggestions for their mechanized transformation. Suggested a framework for small and medium-sized breweries looking to develop mechanical automation brewing technology.

3 Conclusion and Prospect

This paper carefully elaborates on the challenges faced by small and medium-sized liquor enterprises, as well as the current state of development and technical constraints of mechanized equipment technology for liquor production. The most pressing issues for the improvement of small and medium-sized liquor enterprises were identified through in-depth discussions on the brewing technology's current development stage and development bottlenecks, as well as the industry challenges of these businesses. The findings of the discussion show that although mechanized brewing technology has advanced significantly, large koji production still mostly depends on manual labor. Due to resource, capital, technology, and market disadvantages, small and medium-sized businesses are continuously sidelined in the highly polarized industry revenue and profit. There are issues like low overall staff quality, a lack of high-end skills, and large investments in equipment production funds. As a result, a novel concept for the automated growth of small and medium-sized liquor businesses is created through these problems.

References

1. Xiong, F., Li, J., Yuan, Y., Xu, Y., Hu, G.: Investigating the spray drying damage mechanism of baijiu yeast with combined drying process and equipment. Int. J. Food Eng. **20**(6), 389–403 (2024)
2. Sun, B. (2021). Chinese National Alcohols: Baijiu and Huangjiu. World Scientific
3. Ye, H., Wang, J., Shi, J., Du, J., Zhou, Y., Huang, M., Sun, B.: Automatic and intelligent technologies of solid-state fermentation process of Baijiu production: applications, challenges, and prospects. Foods **10**(3), 680 (2021)
4. Zhou, R., Chen, X., Xia, Y., Chen, M., Zhang, Y., Li, Q., Zhen, D., Fang, S.: Research on the application of liquid-liquid extraction-gas chromatography-mass spectrometry (LLE-GC-MS) and headspace-gas chromatography-ion mobility spectrometry (HS-GC-IMS) in distinguishing the Baiyunbian aged liquors. Int. J. Food Eng. **17**(2), 83–96 (2021)
5. Wei, Y., Zou, W., Shen, C.H., Yang, J.G.: Basic flavor types and component characteristics of Chinese traditional liquors: a review. J. Food Sci. **85**(12), 4096–4107 (2020)
6. Tu, W., Cao, X., Cheng, J., Li, L., Zhang, T., Wu, Q., Xiang, P., Shen, C., Li, Q.: Chinese Baijiu: the perfect works of microorganisms. Front. Microbiol. **13**, 919044 (2022)
7. Duan, J., Yang, S., Li, H., Qin, D., Shen, Y., Li, H., Sun, J., Zheng, F., Sun, B.: Why the key aroma compound of soy sauce aroma type baijiu has not been revealed yet? LWT—Food Sci. Technol. **154**, 112735 (2022)
8. Xue, F.H., Zhou, J.Q., Yang, L.X.: A comprehensive review of quantified flavour components in Chinese baijiu. Int. Food Res. J. **31**(2) (2024)
9. Wu, J., Tang, Q., Jin, S., Sun, Y., Liu, Y., Ji, F.: Baijiu-making equipment. In: Science and engineering of Chinese liquor (Baijiu) microbiology, chemistry and process technology (pp. 67–119). Singapore: Springer Nature Singapore (2023)
10. Jin, G., Zhu, Y., Xu, Y.: Mystery behind Chinese liquor fermentation. Trends Food Sci. Technol. **63**, 18–28 (2017)
11. Awuchi, C.G.: HACCP, quality, and food safety management in food and agricultural systems. Cogent Food Agric. **9**(1), 2176280 (2023)
12. Gao, X., Li, C., Elahi, E., Abro, M.I., Cui, Z.: Technological innovation, product quality and upgrading of manufacturing value chain: empirical evidence from China. Sustainability **15**(9), 7289 (2023)
13. Jin, G., Zhao, Y., Xin, S., Li, T., Xu, Y.: Solid-state fermentation engineering of traditional Chinese fermented food. Foods **13**(18), 3003 (2024)
14. Jin, G., Yuan, S., Tang, Q., Sun, Y., Jin, S., Wu, J., Zhao, Y., Xu, Y.: Progress in the intelligentization of strong-aroma Baijiu production (in Chinese). Food Ferment. Ind. (2024)
15. Yuan, H., Zhang, Z., Ding, L., Jiang, Q., Li, Q., Huang, J., Li, L., Nan, G., Lou, K.: Artificial vs. mechanical Daqu: comparative analysis of physicochemical, flavor, and microbial profiles in Chinese baijiu starter cultures. Fermentation **11**(3), 135 (2025)
16. Luo, Y., Wu, L., Wu, M., Liao, H., Yao, Y., Luo, Y., Ji, W., Gao, L., Wang, Z., Xia, X.: Intelligent manufacturing challenges and directions of the baijiu starter culture-Daqu industry: microbiome and engineering perspectives. Trends Food Sci. Technol. 104724 (2024)
17. Ali, A., Wu, Y., Li, W., Xi, C., Patil, P. J., Han, M., Shah, H., Zhao, L., Li, X.: Biotechnological approaches for targeted strain screening and enhancement of Daqu for quality baijiu production. Food Biosci. 104885 (2024)
18. Wu, Q., Zhu, Y., Fang, C., Wijffels, R.H., Xu, Y.: Can we control microbiota in spontaneous food fermentation?—Chinese liquor as a case example. Trends Food Sci. Technol. **110**, 321–331 (2021)
19. Cheng, W., Chen, X., Zhou, D., Xiong, F.: Applications and prospects of the automation of compound flavor baijiu production by solid-state fermentation. Int. J. Food Eng. **18**(12) (2022)
20. Guan, T., Lin, Y., Chen, K., Ou, M., Zhang, J.: Physicochemical factors affecting microbiota dynamics during traditional solid-state fermentation of Chinese strong-flavor baijiu. Front. Microbiol. **11**, 2090 (2020)

21. Cai, W., Xue, Y. A., Tang, F., Wang, Y., Yang, S., Liu, W., Hou, Q., Yang, X., Guo, Z., Shan, C.: The depth-depended fungal diversity and non-depth-depended aroma profiles of pit mud for strong-flavor Baijiu. Front. Microbiol. **12**, 789845 (2022)
22. Li, L., Fan, M., Xu, Y., Zhang, L., Qian, Y., Tang, Y., Li, J., Zhao, J., Yuan, S., Liu, J.: Comparative analysis of volatile flavor compounds in strongly flavored baijiu under two different pit cap sealing processes. Foods **12**(13), 2579 (2023)
23. Zheng, X.W., Han, B.Z.: Baijiu (白酒), Chinese liquor: history, classification and manufacture. J. Ethn. Foods **3**(1), 19–25 (2016)
24. Gao, Y., Chen, S., Xu, Y.: Solid-state distillation of Chinese liquor (Baijiu). In: Science and engineering of Chinese liquor (Baijiu): microbiology, chemistry and process technology. Singapore: Springer Nature Singapore, pp. 143–160 (2023)
25. Li, X., Zhang, J., Xue, Y., Qiu, L.: Classification of hops image based on ResNet-ConvLSTM and research of intelligent liquor picking system. Measurement **194**, 110955 (2022)
26. Tong, W., Wang, Y., Li, Y., Yang, Y., Huang, D., Luo, H., Zhang, S., Huang, Z.: Improving the quality of Xiaoqu Qingxiangxing Baijiu through optimization of key parameters in industrial fermentation processes driven by regulation models. LWT **212**, 116982 (2024)
27. Xia, Y.: Analysis of the development and prospects of the Baijiu industry
28. Zhu, S.H., Shia, A.S., Qi, M.H.: Optimization strategies for marketing channel management in small and medium-sized enterprises: a case study of LZY company (2025)
29. Olajire, A.A.: The brewing industry and environmental challenges. J. Clean. Prod. **256**, 102817 (2020)
30. Shuyan, Z.: Analyzing the impact of Shanxi Fenjiu on cash flow risk (doctoral dissertation, Siam University) (2022)
31. Xia, X., Wu, J., Chen, J.: Technology upgrading strategy for traditional fermented food industry in China. Strat. Study Chin. Acad. Eng. **23**(2), 129–137 (2021)
32. Zhou, J., Li, P.G., Zhou, Y.H., et al.: Toward new-generation intelligent manufacturing. Engineering **4**(1), 11–20 (2018)
33. Pan, F., Qiu, S., Lv, Y., Li, D.: Exploring the controllability of the Baijiu fermentation process with microbiota orientation. Food Res. Int. **173**, 113249 (2023)
34. Liu, C.: Research on wine marketing innovation based on internet (2022)
35. Wu, M., Zhang, G., Pan, R.: The brand marketing strategy of small-and medium-sized liquor enterprises—taking Xuan distillery as an example. In: 2019 International Conference on Economic Management and Cultural Industry (ICEMCI 2019), pp. 105–109. Atlantis Press (2019)
36. Peredy, Z., Chaudhuri, S., Zong, N.: Chinese liquors as global brands-Guizhou Moutai Group case study. Zeszyty Naukowe ZPSB Firma i Rynek (1 (63)), 32–51 (2023)
37. Ma, Y., Wang, H., Zhang, H., Peng, G., Golousova, E.: Research on 4p marketing strategy of Guojiao company in liquor industry under the influence of Covid-19. Российские регионы в фокусе перемен: сборник докладов. Том 1—Екатеринбург 228–241 (2022)
38. Wang, L.: Research trends in Jiang-flavor baijiu fermentation: from fermentation microecology to environmental ecology. J. Food Sci. **87**(4), 1362–1374 (2022)
39. Li, Y., Chen, Y.: Construction of Sichuan liquor industry competitiveness evaluation model from the perspective of diamond theory. J. Financ. Risk Manag. **9**(02), 141 (2020)
40. Abaisi, D., Akbar, I.: Research on financing issues of small and medium-sized enterprises in China. *Экономика и бизнес: теория и практика* (4–2 (110)), 6–15 (2024)
41. Li, F., Zhang, T., Sha, Q., Pei, X., Song, Y., Li, C.: Green reformation of Chinese traditional manufacturing industry: approach and potential for cooperation. Procedia Manuf **43**, 285–292 (2020)
42. Shen, Y., Zhang, X.: Intelligent manufacturing, green technological innovation and environmental pollution. J Innov Knowl **8**(3), 100384 (2023)
43. Peng, C., Guo, X.: High quality development path of Chinese sauce-flavor Baijiu industry—Analysis based on consumption data. In: E3S Web of Conferences, vol. 251, p. 01099. EDP Sciences (2021)

Connecting Tradition and Technology: Comparative Automation Routes in the Baijiu and Wine Industries

Fan Liao, Yeong Jin King, and Hiu Hong Teo

Abstract In China, the alcohol industry, particularly the Baijiu manufacturing is considered one of the most important sectors in contributing its GDP, which boosts the national economy. By this, the SMEs related to this alcohol industry has bloomed nationwide. These SMEs commonly face challenges in the brewing process, including high labor intensity, high labor costs, and low production efficiency. This paper systematically reviews traditional techniques and mechanization progress in key stages of Baijiu production, such as Daqu making, solid-state fermentation, and liquor extraction. It also provides a comparative analysis of traditional winemaking processes and automation technologies used in wine production. By examining the similarities and differences in brewing steps and equipment application between Baijiu and wine, the study explores the potential for cross-industry learning in mechanization. This offers new perspectives for SMEs in the Baijiu sector to implement mechanized upgrades, providing valuable references for other brewing enterprises as well, and contributing to industrial transformation, upgrading, and sustainable development.

Keywords Alcohol · Baijiu · Daqu · Fermentation

1 Introduction

The presence of Baijiu has a lengthy history. It may have originated through natural fermentation in the Neolithic era, but according to archaeological evidence of brick reliefs showing distillation, contemporary distillation techniques most likely evolved around the Han Dynasty (25–220 AD) [1]. Nongxiang (Strong Aroma) Baijiu, renowned for its rich aroma and sweet taste, is the most popular type of Chinese Baijiu, and it is also known as Luzhou-flavor Baijiu [2], Baijiu is abundant, and its extensive production is evident. By year 2020, the production volume had reached

F. Liao (✉) · Y. J. King · H. H. Teo
SEGi University, Petaling Jaya, Selangor, Malaysia
e-mail: SUKD2400106@segi4u.my

M. Awang et al. (eds.), *Proceedings of the International Conference on Mechanical, Manufacturing and Plant Engineering*, Proceedings in Technology Transfer,
https://doi.org/10.1007/978-981-95-7817-7_10

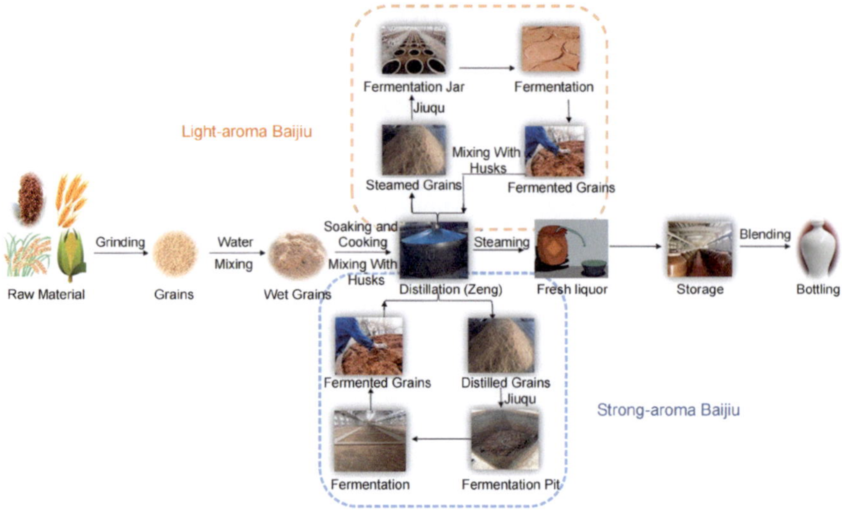

Fig. 1 Simplified schematic of light- and strong-aroma Baijiu production processes [5]

7.4 million liters, generating a substantial profit of 158.5 billion yuan (approximately 24.9 billion US dollars), and it has shown a steadily increasing trend [3, 4].

The basic components are combined with whole or powdered grains, hydrated, and steam-cooked in the particular brewing procedure. They are then cooled to a certain range of temperature (between 18 to 35 °C) in a solid condition, combined with Qu (yeast starter), and placed in a fermentation pit or tank. After the fermentation is completed, the fermented solid grains are taken out and distilled to obtain the base liquor. After a long period of storage and blending, the commercial Baijiu can be finally obtained [5]. The brewing process is shown in Fig. 1.

Over the past decade, remarkable advancements have been made in the automation and intelligentization of Baijiu manufacturing [5]. The production process of Chinese Baijiu has been significantly improved via introduction of modern technologies which has optimized traditional processes. Every step, from raw materials to products, has achieved varying degrees of mechanization and process modernization [6, 7], improving production efficiency and the working environment. For instance, mechanical stamping koji - making machines [8] have been introduced for Daqu formation, disc koji—making machines [9] for Daqu maturation, temperature—monitoring devices [6] for solid—state fermentation, and Zeng—loading robots [10] for the solid—state distillation process, along with other modern equipment. However, in the actual production process, from the koji-making stage [11], the solid-state fermentation stage, the distillation stage, to the storage and blending stage [5], each process still inevitably requires manual participation.

Therefore, up to now, no winery has been able to achieve full intelligent production from raw material input to finished product output, that is, completely eliminating human intervention. Thus, this paper reviews the latest advancements in mechanical automation technology in baijiu (Chinese liquor) production. It systematically analyzes the development and application of various automation equipment across five key stages—jiuqu production, solid-state fermentation, solid-state distillation, liquor extraction, and bottling—while comparing the similarities and differences between baijiu and wine brewing processes. Besides, it also examines the real-world difficulties small and medium-sized baijiu businesses encounter while implementing automation, providing insightful information for the mechanized revolution of the brewing sector.

1.1 Production of Jiuqu

Jiuqu (a mixed fermentation starter including Daqu, Xiaoqu, and Fuqu) serves as a saccharifying and fermenting agent [12, 13]. Baijiu fermentation highly relies on Jiuqu, which is a fermentation starter with a microbial and enzymatic system [14]. In the traditional koji—making process of Luzhou—flavor Baijiu, the stage of forming koji bricks is called "stepping on koji", as illustrated in Fig. 2.

Therefore, the well—mixed raw materials were poured into the Daqu molds, and the koji—making artisans step on them to obtain sufficient cohesion and appropriate air permeability to shape the Daqu by foot [15]. In ancient wine—making, there is also a process of stepping on grapes with feet, which means gently crushing grape clusters with bare feet to facilitate the release of juice and the subsequent fermentation process [16]. Traditional manual Daqu production relies on manual trampling, but this method has problems such as low efficiency, high labor intensity, obvious phenotypic differences in koji blocks, and inconsistent quality [11]. However, foot—stepping is also a relatively gentle grape—pressing method, which can avoid excessive damage

Fig. 2 Illustration of traditional Daqu pressing method [18]

to grape seeds and skins, thus reducing the release of astringent substances in wine [17].

After the Jiuqu is formed, it is transferred to the Jiuqu fermentation room for fermentation until it matures. However, the temperature in the Daqu cultivation workshop can reach approximately 40–60 °C [19]. Under such harsh working conditions, workers can only continue working for less than half an hour. Moreover, Daqu has a certain weight. Therefore, there is also considerable labor intensity involved in the handling and turning of Daqu. During the traditional Daqu cultivation process in the workshop, manual operations and inspections require workers to enter and exit frequently, resulting in low production efficiency and potentially introducing unexpected bacterial contamination [5].

Many businesses have suggested and put into practice mechanical solutions to overcome the drawbacks of traditional Daqu (fermentation starter) manufacturing; Liu and Ren [9] developed a disc-type koji-making machine, as illustrated in Fig. 3, which serves as a mechanized and intelligent device capable of fully automating the entire process of koji production, including material feeding, incubation, turning, and discharging. By precisely regulating the processing temperature, humidity, and air flow through an integrated control system, the machine creates an optimal environment for microbial growth.

Although the wine making process differs from the traditional Chinese baijiu qu-making process, it can still draw valuable insights from advanced equipment like the baijiu disc-type qu-making machine. Firstly, its intelligent control system enables precise regulation of temperature, humidity, and airflow, helping to create a stable microclimate during fermentation and ensuring yeast activity and wine quality. Secondly, its data networking and automation capabilities allow for real-time monitoring and early warning, enhancing process control and production efficiency. Lastly, the use of SUS304 food-grade stainless steel ensures easy cleaning and

Fig. 3 Overall structure of the disc-type fermentation machine

maintenance while safeguarding wine purity and food safety. These technological features offer important references for the intelligent and standardized advancement of winemaking.

Zhang et al. conducted a study on the optimization design of the transmission mechanism of mechanical stamping-type pressing machines and proposed replacing the traditional crank–connecting rod system with a grooved cam mechanism, which was designed based on the relationship between pressing force and compression displacement. This design enables the pressing mold to complete three stages—ascending, dwelling, and descending—within a single cycle. The ascending stroke is driven by a spring, while the return stroke is regulated by the cam, resulting in a more complete pressing process and significantly enhanced forming quality, liquor quality, and alcohol yield [8].

Although the wine brewing process does not involve steps similar to the pressing of koji blocks in Baijiu (Chinese liquor) koji making, the "grape stomping" process is still practiced in traditional craftsmanship [20, 21]. Although the grape pressing process has been mechanized, for example, the grape crusher mentioned by J. Steakley and others [22], as illustrated in Fig. 4, which achieves crushing through the internal roller extrusion and screw propulsion structure. Its purpose is not to completely crush the grapes, but to gently crack the grape skins so that the juice can flow out and provide conditions for fermentation.

The grape-stomping process in premium winemaking can draw inspiration from the design principles of Daqu mechanical stamping-type pressing machines. Precise control over pressure and duration can be attained by creating an intelligent skin-breaking device that mimics foot-treading motions. This approach preserves traditional flavors while reducing labor intensity and improving efficiency.

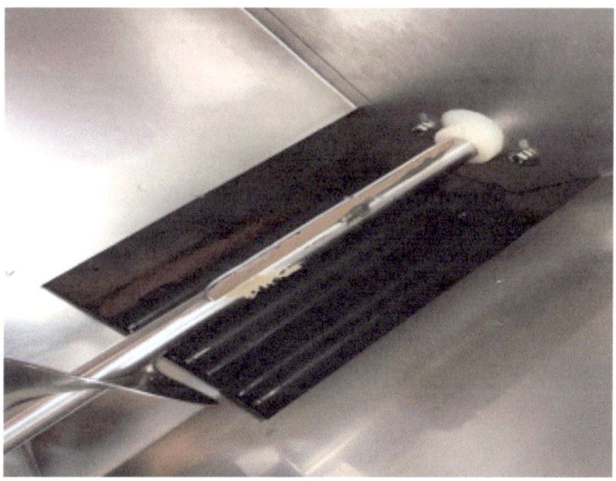

Fig. 4 Roller structure in grape crushing equipment, used to gently crush grape skins [22]

1.2 Solid-State Fermentation

Research shows that the microorganisms in pit mud are closely related to the quality of Chinese Luzhou-flavor Baijiu [23]. Wang et al. [24] reported that the Jinjiu Fenglin Distillery became the first in China to implement tank truck fermentation in the production of light-flavor Baijiu using Xiaoqu (small starter) in its brewing workshop. The process involves automated material loading into mobile tanks, which are then transferred by forklift to a temperature-controlled fermentation chamber, followed by distillation. This approach enables yeast to convert starch and sugars into ethanol efficiently. The system's high level of mechanization and temperature control greatly increases production efficiency, lowers labor intensity, and guarantees constant product quality. In addition, Zhang Zhimin et al. proposed a novel mechanization concept for Baijiu production at Hengshui Laobaigan Brewery, which involves surrounding the fermentation vessel with stainless steel and filling the interlayer with locally sourced yellow soil. The inner surface is lined with traditional sealed-pit materials, and the vessel is sealed with a hinged lid. This design aims to replicate the fermentation environment of buried pits and is considered a promising direction for future technological advancement [24].

Although there are differences in the brewing processes between Baijiu and wine, in the fermentation stage, both pay attention to key factors such as fermentation containers and fermentation temperature. For example, Paola Bambina and others [25] pointed out in their study on the influence of stainless steel and oak barrels on the quality of Cabernet Sauvignon wine that, through a comprehensive analysis from three aspects, namely chemical composition, aroma performance, and sensory quality, oak barrels have more advantages during the long-term post-fermentation maceration (PFMB) process. The wine that goes through post-fermentation maceration in oak barrels exhibits superior overall quality when compared to stainless steel tanks. It improves the wine body' quality and offers a stronger base for further aging. Aging time, as a key factor affecting the maturity and flavor harmony of the wine, directly determines the market share and economic benefits of wine products [26].

In fermentation processes, non-light-aroma baijiu types such as sauce-aroma and strong-aroma baijiu rely heavily on the microbial ecosystem within the mud cellar for their distinctive flavor profiles, making stainless steel fermenters unsuitable. Similar to this, oak barrels are superior over stainless steel tanks in terms of polyphenol extraction, aromatic complexity, and aging potential in the manufacture of Cabernet Sauvignon wines. This might lead to issues with product quality. It is especially important to keep an eye on the temperature during wine production. Relevant techniques include the real-time monitoring method of temperature and alcohol concentration proposed by Tan Huidan and Zhang Peng, However, as baijiu undergoes solid-state fermentation and wine undergoes liquid-state fermentation, the application of such monitoring methods must be adjusted to meet the characteristics of liquid systems, in order to improve the technical and quality control perspectives for better wine making.

1.3 Picking Liquor

By observing the size, quantity, and duration of the foam (known as "flower") formed when liquor flows out, the liquor can be categorized into Da Qinghua (large clear flower), Xiao Qinghua (small clear flower), Hunhua (turbid flower), followed by inferior liquor and wastewater. The formation of foam is primarily determined by alcohol, esters, and other active substances—richer content leads to more abundant and longer-lasting foam [18]. Currently, the grading of Baijiu still highly depending on sensory evaluation by experienced workers, such as observing the foam and tasting. However, the traditional oral teaching method lacks systematic structure, and differences in individual experience, skill, and subjective judgment among distillers lead to inconsistent quality assessments, affecting the overall stability of base liquor quality.as illustrated in Fig. 5.

To address the issues—the high cost of existing alcohol content detection equipment, low detection accuracy, and poor classification performance—Li et al. [27] proposed an intelligent Baijiu harvesting system based on hop image classification, as exhibited in Fig. 6.

This device captures images of liquor flower patterns using an industrial camera. After preprocessing—including ROI extraction, brightness adjustment, denoising, and sharpening—the images are input into a ResNet50-ConvLSTM deep learning model for alcohol content classification and determination. Based on the classification results, the system controls solenoid valves to achieve intelligent fractional collection of liquor. Pressure, temperature, and flow sensors work collaboratively to collect data, while a PLC and industrial computer perform analysis and control [27].

Wei Lan et al. [26, 28] pointed out in their research that many commonly used instruments and methods for flavor analysis in fruit wine also have significant application value in the analysis of aroma compounds in baijiu. These instruments include gas chromatography–mass spectrometry (GC-MS), gas chromatography–ion

Fig. 5 Manual picking liquor

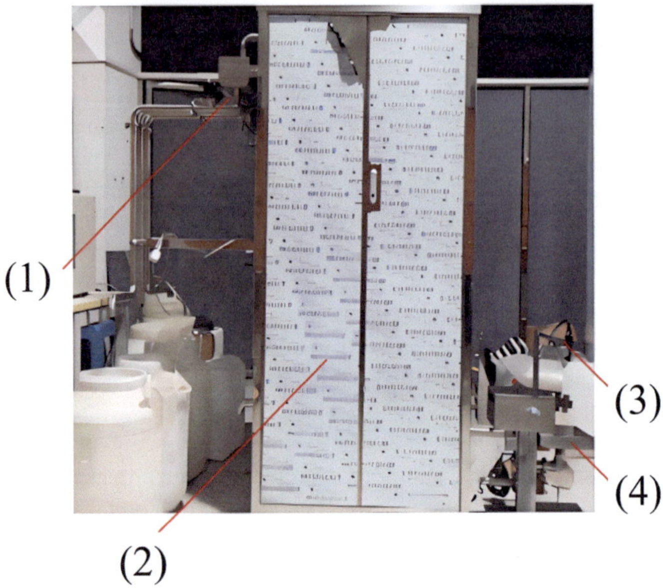

Fig. 6 Overall structure of the equipment [27]; (1) Segmentation module, (2) Liquor conveying module, (3) Image acquisition module, (4) Liquor receiving module

mobility spectrometry (GC-IMS), electronic nose (E-nose), headspace solid-phase microextraction coupled with gas chromatography–mass spectrometry (HS-SPME-GC-MS), and gas chromatography with flame ionization detector (GC-FID), among others.

Wine is essentially a fermented beverage, so the distillation process is not involved in its production. On the other hand, Baijiu is a distilled liquor and must go through the crucial distillation process. Since wine does not undergo distillation, there is naturally no "wine collection" stage that is involved in the distillation process of Baijiu, nor are there distinctions such as "head wine" and "tail wine". Therefore, operations like "discarding the head and tail" that are common in Baijiu production do not occur in wine production.

1.4 Filling

Taking beer filling as an example, nearly half of contamination incidents occur during the filling stage—namely bottling, canning, or kegging [29]—highlighting the critical importance of this process for product quality. However, traditional manual filling methods are inefficient, imprecise, prone to overflow, and labor-intensive, while advanced bottling systems are often expensive.

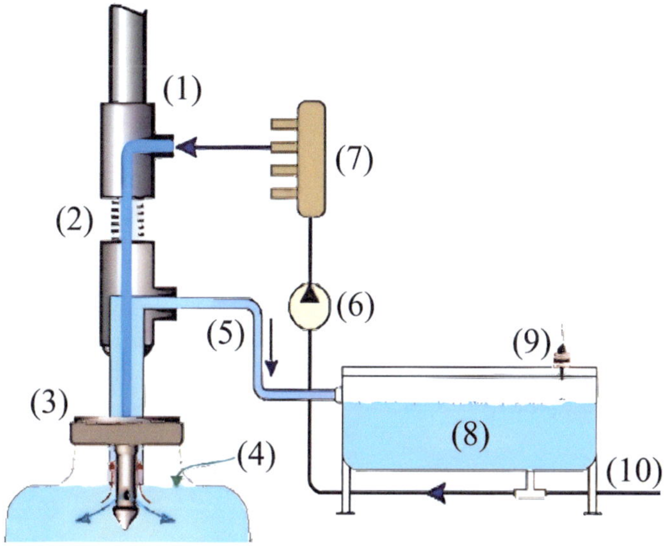

Fig. 7 Structural diagram of the filling machine. Created by the authors based on [30]. (1) Overflow nozzle, (2) Return Spring, (3) Bottle seal gasket, (4) Target fill height, (5) Excess liquid and foam return, (6) Pump, (7) Distribution manifold to nozzles, (8) Overflow vessel, (9) Level switch, (10) Liquid supply line

To address these challenges, Aragón-González et al. [30] proposed a low-cost automatic bottling machine, as illustrated in Fig. 7. The system operates on the principle of overflow filling, using a centrifugal jet pump for fluid delivery. It incorporates two aluminum electrodes as an electrical sensor to detect the liquid level; once the liquid contacts both electrodes, the circuit is closed and the filling process stops, enabling precise volume control. During operation, the filling head remains stationary while a pneumatic actuator lifts the bottle into position. After filling, the bottle is conveyed away via a belt-driven transport system.

In addition to serving as a source of inspiration for goods like wine, this technology offers small and medium-sized baijiu businesses an affordable automation route that boosts production line productivity. The wine industry, in particular, can benefit from its precise liquid level control and flexible bottle compatibility design, promoting the broader application of intelligent filling across diverse types of alcoholic beverages.

1.5 Conclusion and Prospect

The application status and development progress of mechanical automation technology and related equipment in the strong-aroma baijiu brewing process are methodically presented in this review. It focuses on typical application cases of different automated equipment and systems in crucial stages like koji production, solid-state fermentation, liquor extraction, and bottling, starting with the similarities and differences between wine and baijiu brewing processes. It also explored the significant reference value of the brewing techniques of liquor for the development of wine production automation. The research findings indicate that there is a certain degree of technological overlap and mutual reference potential between wine and baijiu brewing processes, particularly in areas such as koji production (e.g., the traditional foot-stamping method), solid-state fermentation (e.g., temperature monitoring and control), and bottling. The introduction of mechanical automation technology helps mitigate uncertainties caused by reliance on experience and subjective judgment in quality control, thereby enhancing the stability and traceability of product quality, improving food safety standards, promoting the overall industrialization of the baijiu industry, and further strengthening the comprehensive competitiveness of small and medium-sized baijiu enterprises.

References

1. Chen, S., Perez-Samper, G., Herrera-Malaver, B., Zhu, L., Liu, Y., Steensels, J., Verstrepen, K.J.: Breeding of new saccharomyces cerevisiae hybrids with reduced higher alcohol production for light-aroma-type-Xiaoqu Baijiu production. J. Am. Soc. Brew Chem. **81**(2), 233–241 (2023). https://doi.org/10.1080/03610470.2023.2165995
2. Tu, W., Cao, X., Cheng, J., Li, L., Zhang, T., Wu, Q., Li, Q.: Chinese Baijiu: the perfect works of microorganisms. Front. Microbiol. **13**, 919044 (2022)
3. Duan, J., Yang, S., Li, H., Qin, D., Shen, Y., Li, H., Sun, B.: Why the key aroma compound of soy sauce aroma type Baijiu has not been revealed yet? LWT Food Sci. Technol. **154**, 112735 (2022)
4. Xue, F.H., Zhou, J.Q., Yang, L.X.: A comprehensive review of quantified flavour components in Chinese Baijiu. Int. Food Res. J. **31**(2) (2024)
5. Ye, H., Wang, J., Shi, J., Du, J., Zhou, Y., Huang, M., Sun, B.: Automatic and intelligent technologies of solid-state fermentation process of Baijiu production: applications, challenges, and prospects. Foods **10**(3), 680 (2021)
6. Jin, G., Zhao, Y., Xin, S., Li, T., Xu, Y.: Solid-state fermentation engineering of traditional Chinese fermented food. Foods **13**(18), 3003 (2024)
7. Jin, G., Yuan, S., Tang, Q., Sun, Y., Jin, S., Wu, J., Zhao, Y., Xu, Y.: Progress in the intelligentization of strong-aroma Baijiu production (in Chinese). Food Ferment. Ind. (2024)
8. Zhang, M.S., Zhang, Y.B., Cai, Z.Y., Chen, Z.Q., Bian, X.H.: Optimal design of transmission mechanism of mechanical stamping JIUQU starter press. Food Mach. **39**(5), 77–82 (2023)
9. Liu, J., Ren, B., Wang, J.F., Wang, Y.J., Wei, X.K., Zhu, S.J., Wei, S.Y.: Structure, automatic control, and application of disc-type JIUQU starter machine in the Baijiu industry (in Chinese). Liquor-Mak. Sci. Technol. **7**, 46–49 (2020)
10. Yan, J., Zhang, G., Liu, W., Yao, R., Xiang, X., Cao, M.: Progress in the application of automation technology in Baijiu distillation. In: Proceedings of the 2023 CAA Symposium on Fault Detection, Supervision and Safety for Technical Processes (SAFEPROCESS), Hangzhou, China. New York, IEEE, pp. 1–6 (2023)
11. Yuan, H., Zhang, Z., Ding, L., Jiang, Q., Li, Q., Huang, J., Lou, K.: Artificial vs. mechanical Daqu: comparative analysis of physicochemical, flavor, and microbial profiles in Chinese Baijiu starter cultures. Fermentation **11**(3), 135 (2025)
12. Wang, B., Wu, Q., Xu, Y., Sun, B.: Synergistic effect of multiple saccharifying enzymes on alcoholic fermentation for Chinese Baijiu production. Appl. Environ. Microbiol. **86**(8), e00013-20 (2020)

13. Wang, M.Y., Yang, J.G., Zhao, Q.S., Zhang, K.Z., Su, C.: Research progress on flavor compounds and microorganisms of Maotai flavor Baijiu. J. Food Sci. **84**(1), 6–18 (2019)
14. Zhang, J., Chen, S., Qian, Y.P.L., Qian, M.C. Baijiu.: In: Distilled spirits. Cambridge (MA), Academic Press, pp. 103–130 (2023)
15. Shang, C., Li, Y., Zhang, J., Gan, S.: Analysis of bacterial diversity in different types of Daqu and fermented grains from Danquan distillery. Front. Microbiol. **13**, 883122 (2022)
16. Malfeito-Ferreira, M., Granja-Soares, J., Chandra, M., Asryan, A., Oliveira, J., Freitas, V., Harutyunyan, M.: Investigating the influence of vessel shape on spontaneous fermentation in winemaking. Fermentation **10**(8), 401 (2024)
17. Harutyunyan, M., Malfeito-Ferreira, M.: Historical and heritage sustainability for the revival of ancient wine-making techniques and wine styles. Beverages **8**(1), 10 (2022)
18. Wu, J., Tang, Q., Jin, S., Sun, Y., Liu, Y., Ji, F.: Baijiu-making equipment. In: Science and engineering of Chinese liquor (Baijiu): microbiology, chemistry and process technology. Singapore: Springer Nature Singapore, pp. 67–119 (2023)
19. Shi, G., Fang, C., Xing, S., Guo, Y., Li, X., Han, X., Zhang, C.: Heterogenetic mechanism in high-temperature Daqu fermentation by traditional craft and mechanical craft: from microbial assembly patterns to metabolism phenotypes. Food Res. Int. **187**, 114327 (2024)
20. Branco, Z., Baptista, F., Paié-Ribeiro, J., Gouvinhas, I., Barros, A.N.: Impact of winemaking techniques on the phenolic composition and antioxidant properties of Touriga Nacional wines. Molecules **30**(7), 1601 (2025)
21. Sánchez-García, E., Martínez-Falcó, J., Alcon-Vila, A., Marco-Lajara, B.: Developing green innovations in the wine industry: an applied analysis. Foods **12**(6), 1157 (2023)
22. Steakley, J., Steakley, B.: Quality grape growing and winemaking, nurtured with family, friends, and fun. In: A quest for quality wine, every time: a guide for root cause analysis, pp. 49–84 (2020)
23. Gao, J., Liu, G., Li, A., Liang, C., Ren, C., Xu, Y.: Domination of pit mud microbes in the formation of diverse flavour compounds during Chinese strong aroma-type Baijiu fermentation. LWT **137**, 110442 (2021)
24. Wang, Y.: Research progress and prospects of mechanized Baijiu brewing technology and equipment (in Chinese). Yunnan Chem Ind. **46**(8), 39–41 (2019)
25. Bambina, P., Pollon, M., Squadrito, M., Barone, S., Cinquanta, L., Corona, O.: The effect of prolonged (150 days) post-fermentative maceration in steel tanks and oak barrels on Cabernet Sauvignon wine quality: mathematical modelization of the phenolic compounds' behaviour. J. Wine Res. **34**(4), 247–271 (2023)
26. Lan, W., Cheng, W., Shen, N., Ma, R., Pan, T., Li, N., et al.: Advance in brewing process, flavour detection technologies and functional components of fruit wine: a review. Int. J. Food Eng. (2025). https://doi.org/10.1515/ijfe-2024-0252
27. Li, X., Zhang, J., Xue, Y., Qiu, L.: Classification of hops image based on ResNet-ConvLSTM and research of intelligent liquor picking system. Measurement **194**, 110955 (2022)
28. Yang, J.G., Wang, M.Y., Liu, L., Fang, S.P., Ren, H.: Different brands of strong-flavor baijiu identified using gas chromatography–mass spectrometry and near-infrared spectroscopy. J. Am. Soc. Brew. Chem. 1–12 (2025)
29. Wagner, E.M., Thalguter, S., Wagner, M., Rychli, K.: Presence of microbial contamination and biofilms at a beer can filling production line. J. Food Prot. **84**(5), 896–902 (2021)
30. Aragón-González, G., Barragán-Santiago, I., Huerta-Rizo, E., León-Galicia, A.: Alternative automatic bottling machine for alcoholic beverages

Application of Lean Tools to Increase Productivity in a Textile Workshop

Nory Judith Quicaño Jara, Fiorella Denisse Castro Blancas, and Erick Domingo Vilchez Ruiz

Abstract Lean tools are a set of methods and techniques that allow the elimination of waste, that is, everything that does not generate value, in order to achieve excellence. In this sense, the present article aims to increase productivity in a textile manufacturing workshop, reducing waste through the application of Value Chain Mapping (VSM), Waste Assessment Model Analysis (WAM) and the implementation of 9S. In this sense, the objective of this research is to increase productivity in a textile manufacturing workshop, identifying waste within the production process in order to reduce waste and residues with greater severity such as inventory, defects and overproduction. After applying these methodologies, the main types of waste were identified: inventory (21.82%), defects (20.50%) and overproduction (17.44%). The lead time was reduced from 8.63 days to 2.65 days, the search times in the cutting process were reduced by 75%, sewing by 28%, finishing by 97% and warehouse by 52%; the number of defective products from internal production by 4% and from third-party production by 37%; which led to an increase in productivity for shorts by 26%, leggings by 22% and sports polo shirts by 49%. Finally, the motivation of workers, the feeling of well-being, internal communication and making 9S a new work culture were increased.

Keywords Lean manufacturing · Lean tools · Productivity · VSM · WAM · 9 s · Waste

Q. J. Nory Judith (✉) · C. B. Fiorella Denisse · V. R. Erick Domingo
Department of Industrial Engineering, Universidad Continental, Huancayo, Peru
e-mail: 76296820@continental.edu.pe

C. B. Fiorella Denisse
e-mail: 73367124@continental.edu.pe

V. R. Erick Domingo
e-mail: evilchezr@continental.edu.pe

M. Awang et al. (eds.), *Proceedings of the International Conference on Mechanical, Manufacturing and Plant Engineering*, Proceedings in Technology Transfer,
https://doi.org/10.1007/978-981-95-7817-7_11

1 Introduction

Due to the drastic changes that are taking place in the market, it is necessary for companies to have and improve their ability to respond to consumer demands. On the other hand, the orientation towards quality products and services by customers is causing company leaders to adopt tools belonging to the Lean philosophy, which focuses on reducing waste in order to increase the levels of efficiency, productivity, profitability and competitiveness of companies [1, 2].

This study will be carried out in a textile workshop, which is mainly made up of the following work areas: Cutting/finishing, Sewing and Storage. Although the first two areas share the same space, they are not delimited, so there is a lot of disorder, which ends up causing delays in the search for elements and obstruction of the passage, as well as the difficulty of counting what is within the company, whether raw material, finished products or in process. Likewise, problems of reprocessing due to defects are perceived, which directly affects productivity and work environment. The objective will be to increase productivity by reducing waste levels, through Value Stream Mapping (VSM), the Waste Assessment Model (WAM) and the implementation of the 9 s.

VSM is a tool that, through a diagram, allows you to visualize and understand a complete process, from the start of the customer's purchase order until the customer receives the order, identifying the flows of materials and information, as well as the activities that generate value and those that do not provide it and, in this way, identify opportunities for improvement within the process [2].

On the other hand, WAM, proposed by Ibrahim Rawabdeh, is a method that allows the identification and quantification of the interrelationship between the seven types of waste (over production, inventory, defects, motion, transportation, process and waiting); that is, it evaluates how each type of waste affects the others and, in turn, how it is affected by them. In addition, this method facilitates the identification of the predominant wastes, on which improvement efforts should be focused [3].

On the other hand, the 9S is the extension of the well-known 5S methodology that can be divided and explained in three blocks. Within the first block are the parts of the 3S (classify, organize, and clean), which are implemented in the physical work space. The second block is made up of the following 4S (personal well-being, discipline, perseverance, and commitment), and they apply to employees. Finally, the third is made up of the last 2S (coordination and standardization), which must be applied to the company under the responsibility of management, to guarantee the success of the implementation throughout the company [1]. In addition, the 9S methodology can be applied in any organization with low costs and multiple benefits [4].

While there is no research demonstrating the benefits of applying these three tools together, the results of studies that have used one or two of these tools in combination, or in conjunction with other methodologies, can be considered. In a study conducted in a company dedicated to the manufacture of T-shirts, WAM and VSM were implemented to analyze waste, and tools such as layout design, 5S, and the implementation of pull and FIFO systems were applied to reduce waste such as movement, inventory, and overproduction; this allowed to reduce work-in-process inventory by 74.30%, value-added time by 30.95%, delivery time by 71.1%, and the value of standard minutes by 27.02% [5]. On the other hand, with the implementation of VSM and 7S in an alpaca wool workshop, it was possible to reduce the level of waste by 7%, reduce the cycle time and delivery time; thus improving the efficiency of the system [6]. On the other side, in a study carried out in a food company, WAM and Fault Tree Analysis (FTA) were applied, with the aim of proposing suggestions to reduce the dominant waste, which was transportation with 17.09%; thus, minimizing production time by 8.17 min per cycle [7]. Similarly, VSM was applied in a textile company to increase competitiveness, obtaining an initial takt time of 117.5 s per garment and a cycle time of 20.87 days in the current map; in the proposed future map, a cycle time of 16.7 days was achieved, demonstrating that VSM, together with other lean methodologies, increases productivity by being an effective visual tool [8]. In another research, after the implementation of the 9S approach in a jute industry, it was possible to increase the total productivity for three months, going from 1477.41 tons to 1837.04 tons, which would indicate an improvement in the production rate [9]. Finally, with the implementation of the 9 s in the quality area of a food industry, it was possible to optimize the analysis times of raw materials by 10.15% and the analysis times of finished products by 7.41% for each batch, as well as improve productivity, process flow and the work environment [10].

2 Methodology

This research began with the diagnosis of the current situation, through Value Stream Mapping (VSM), which included all stages from the reception of raw materials to the delivery of the final product. Subsequently, the Waste Assessment Model (WAM) was carried out, starting with the development of the Waste Relationship Matrix (WRM) and then the Waste Assessment Questionnaire (WAQ) with the aim of identifying the waste that predominates among the others. Finally, the 9S methodology was implemented starting with the evaluation of the current situation regarding compliance with each S.

Value Stream Mapping

Using Value Stream Mapping, a takt time of 1035.45 s per unit, a lead time of 8.63 days, a Value Added Time (TVA) of 1% and a Non-Value Added Time (TNVA) of 99% were determined, as illustrated in Fig. 1.

It was identified that the cycle time in the Sewing process exceeds the Takt Time, which suggests the presence of a bottleneck that limits the capacity to meet demand, in addition to a possible overload of resources. Similarly, the lead time in Inventory is high, which could cause overproduction. Likewise, the percentage of defective products in the Sewing area is the highest, and the downtime in the Finishing area is also the highest, so the implementation of the 9S methodology is proposed.

Waste Assessment Model

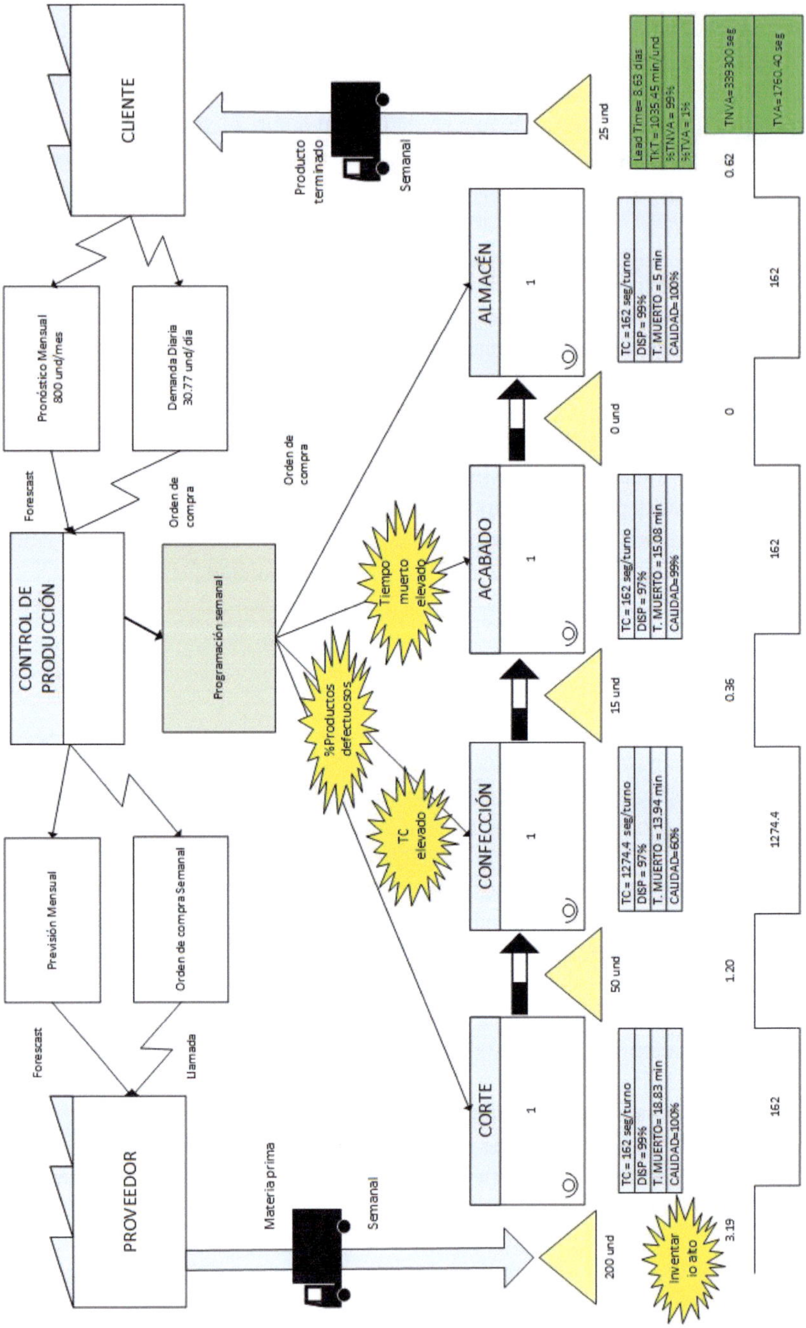

Fig. 1 Current VSM, before improvements

Under the WAM analysis, the force of influence that exists between each type of waste relationship in the WRM was analyzed, as observed in Table 1 and from which it can be seen that the highest percentage value is given by the influence of overproduction on other waste with 19.35%. It should be noted that one of the most significant relationships is that between overproduction and inventory, followed by the connection between inventories and defects. This is because excessive production generates high levels of inventory, and these high levels of inventory, particularly in the form of raw materials, promote greater overproduction, which in turn affects quality.

Likewise, the high influence of defects on other waste can be highlighted, since it represents 18.55%. This occurs especially when services are outsourced to meet demand, which results in a greater number of defects, greater overproduction to cover defective products and also the loss of resources for their reprocessing.

From the information contained in Table 1 and the results of the Waste Assessment Questionnaire (WAQ), the Probability of Occurrence (Pj) was calculated, which allowed the Final Residue Factor (Yj) to be determined and the dominant waste to be identified, as shown in Table 2.

The waste with the highest weighting and those that generate the most problems are Inventory waste (21.82%), Defects (20.50%) and Overproduction (17.44%). These wastes match the deficiencies identified in the VSM, which provides a clearer view on where to focus efforts to reduce waste and its effects on others, and thus increase

Table 1 Waste relationship matrix (WRM) with weights

F/T	O	I	D	M	T	P	W	SCORE	%
O	10	10	6	8	6	0	8	48	19.35
I	6	10	8	8	6	0	0	38	15.32
D	6	10	10	8	8	0	4	46	18.55
M	0	2	4	10	0	6	6	28	11.29
T	2	6	2	6	10	0	2	28	11.29
P	4	2	6	6	0	10	4	32	12.90
W	6	8	4	0	0	0	10	28	11.29
SCORE	34	48	40	46	30	16	34	248	100.00
%	13.71	19.35	16.13	18.55	12.10	6.45	13.71	100.00	

Table 2 Calculation of final results

	O	I	D	M	T	P	W
Yj	0.59	066	0.62	069	0.63	0.66	0.49
Pj	265.35	296.57	299.17	20.942	136.58	83.25	154.79
Yj(final)	**157.06**	**196.54**	**184.68**	**144.69**	**86.16**	**55.26**	**76.29**
Yj(final) %	**17.44**	**21.82**	**20.50**	**16.06**	**9.57**	**6.14**	**8.47**
Ranking	**3**	**1**	**2**	**4**	**5**	**7**	**6**

productivity. In addition, it facilitates the selection of appropriate tools to achieve objectives, which, in this case, motivated the implementation of the 9S.

9s

After identifying the waste in the production process, compliance with the S was assessed before implementing the 9S using an observation sheet and a questionnaire. The results, summarized in a radar chart, showed a compliance of 21% (See Fig. 2). This initial assessment was based on the first 4S, as they are the pillars for the formation of the following 5 S [1]. After the diagnosis, implementation began.

Seiri (Sort): Red cards were used to label and differentiate the necessary elements of the incense burners in the aforementioned spaces (See Fig. 3), taking into account what is actually used for the tasks of the work area; as well as the frequency of use of each of them.

Seiton (Set in order): After labeling the unnecessary items within the workspaces, all the items were arranged in such a way that they were easy to find, take out, and return to their place. Properly labeled containers and labels were used on the shelves to correctly identify the finished product models. In addition, the cutting, finishing,

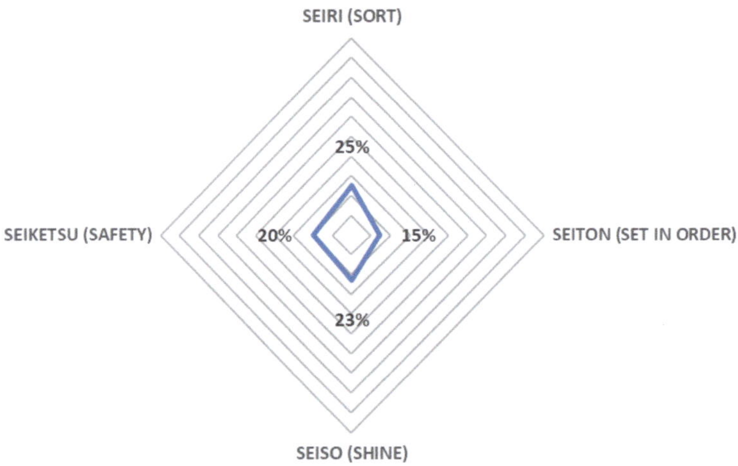

Fig. 2 Radar chart of the first 4S before the implementation of the methodology

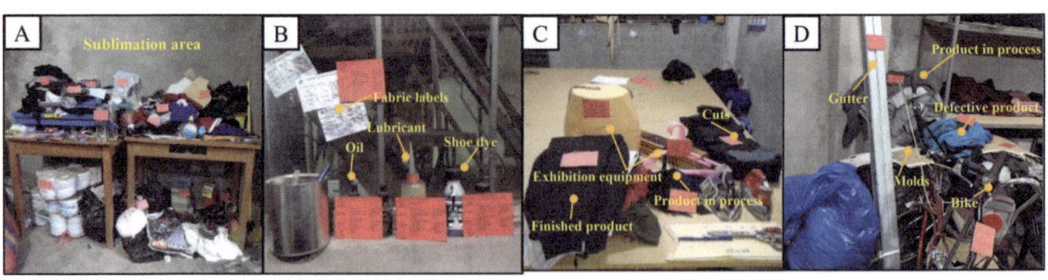

Fig. 3 Use of red cards in (**A–C**) cutting and finishing area and (**D**) storage area

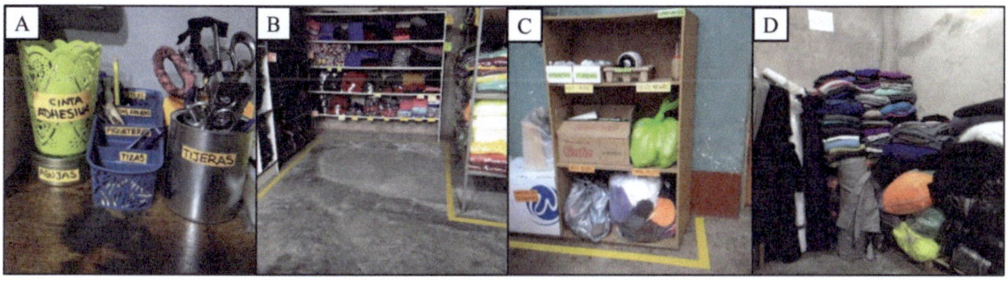

Fig. 4 Organization of (**A**) tools; (**B**) finished products and (**C**), (**D**) raw materials

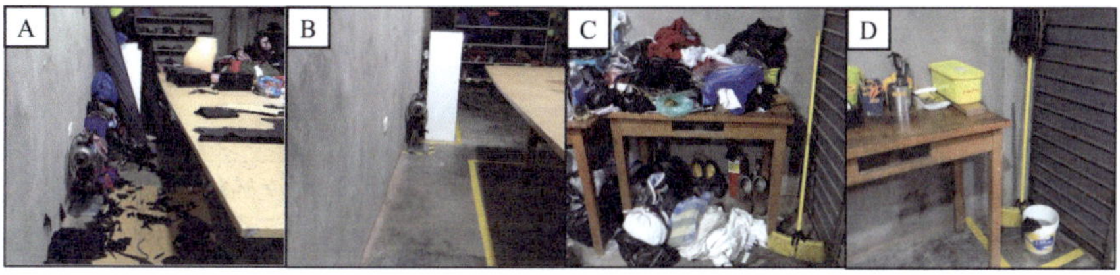

Fig. 5 (**A**) Dirty hallways; (**B**) free and clean hallways after applying Seiso; (**C**) Undefined cleaning area; (**D**) Demarcated area of cleaning equipment after applying Seiso

and storage areas were delimited to maintain order in both environments and to ensure that they are respected. (See Fig. 4).

Seiso (Shine): All spaces were cleaned, including walls, floors, windows, machines and tools. In this step, a zone for cleaning equipment and waste was also defined (See Fig. 5), as well as cleaning methods and elements. In addition, the practice of "five minutes for S" was implemented, which consists of spending 5 min every day cleaning the machines, tools and spaces that are occupied, to maintain a clean environment [1].

Seiketsu (Safety): The Cornell questionnaire was used to assess employee discomfort and the RULA method was used to determine the level of ergonomic risk, revealing that 37.50% were at very high or high risk and 25% at medium risk. Measures were taken to protect the physical and mental health of workers, such as informing them about the importance of using protective equipment, such as face masks, and encouraging good posture. Active breaks of 10 min every 3 h, changing chairs for more comfortable ones, and using 20 mm padded cushions were recommended. In the finishing area, it was suggested to change irons and scissors for lighter ones and, likewise, rotate staff to reduce fatigue and workload.

Shitsuke (Discipline): To encourage discipline among workers and make the 9S a way of life within the company, visual control mechanisms were implemented for the previously established 4S. Aisles were marked off (see Fig. 6) to quickly identify if there were materials or products in the wrong places. In addition, standards were defined for the management of defective and non-defective garments from the sewing area and external services, in order to maintain good inventory control and avoid the accumulation of unfinished products and obstructions in the hallways.

Fig. 6 Hallways delimited in (**A**) cutting and finishing area, (**B**) sewing area and (**C**) storage area

Shikari (Constancy): To maintain adherence to discipline and compliance with standards, monthly progress assessments were planned to allow for comparison and visualization of progress. The interval between assessments will be lengthened as progress is seen. After each assessment, the results were explained and posted for everyone to see, to serve as motivation.

Shitsukoku (Commitment): To ensure employee commitment to the 9S program, meetings were held to explain the benefits and the need for its implementation. In addition, participation was promoted through a questionnaire to determine their satisfaction and recommendations regarding each S, addressing their needs and fostering their sense of belonging to the company.

Seishoo (Coordinate): A 9S implementation plan was developed with the participation of everyone in the company, including activities, scheduling and those responsible, with the aim of effectively communicating information and monitoring compliance and the efforts of each person through meetings, to ensure everyone's involvement in the improvement process.

Seido (Standardize): To maintain the first 4S and the benefits obtained, it was proposed to maintain activities such as labeling unnecessary items every three months, the use of color labels and codes to organize finished products by type of garment, the application of the '5 min for the S', and verification of compliance through a Checklist.

3 Results and Discussions

After the implementation of the 9S, compliance with the first 4S showed an increase of 68% compared to the initial evaluation. On the other hand, taking into account the 9S, the level of compliance was 86% (See Fig. 7).

Furthermore, after implementing each S, the following results described in Table 3 were achieved.

Similarly, after applying the 9S, production times were reduced by 20% for shorts and leggings, and by 44% for T-shirts. Productivity increased by 26% for shorts, 22%

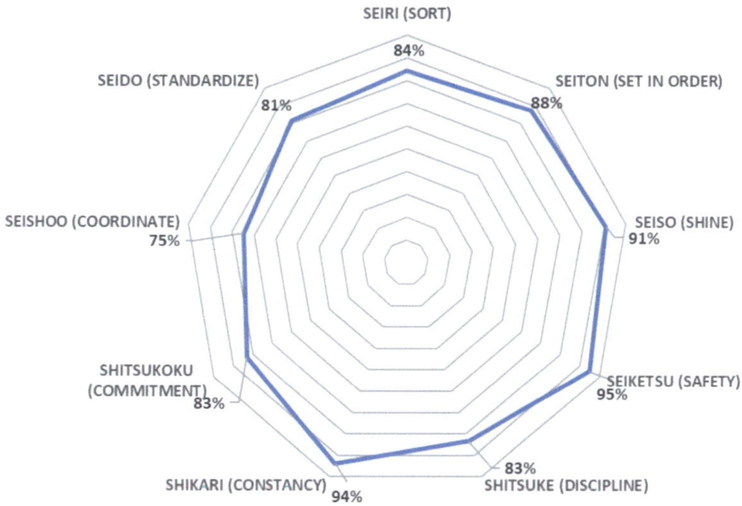

Fig. 7 Radar chart of 9S compliance after implementation of the methodology

for leggings and 49% for sports polos, when considering the resources and time to produce a batch of 200 garments in one month.

On the other hand, the future VSM (See Fig. 8) reveals a reduction in Lead to 2.65 days. Likewise, the cycle time for the manufacturing process dropped to 919.29 s/shift, the quality percentage increased by 34% and downtime was reduced.

Table 3 Results of the implementation of 9S

S implemented	Results
Seiri (sort)	– More free space – Comfort when working – Less accumulation of unnecessary things
Seiton (set in order)	– Quick identification of necessary materials, avoiding unnecessary expenses – Better production control – Improved workspace design – Reduction in search time for materials and tools: cutting 75%, sewing 28%, finishing 97%, warehouse 52% – Better work environment – Quick detection of defective products and corrective actions to avoid accumulation
Seiso (shine)	– With the practice of the "five minutes to S" speeds up cleaning and keeps work environments clean – Improves the appearance and visual appeal of workplaces – Increased morale in workers – Reduces the risk of accidents
Seiketsu (safety)	– Workers more aware of the relationship between their work and their health – Pleasant work environment – Workers are more satisfied with their daily activities
Shitsuke (discipline)	– Reduction of defective garments: 4% in internal production and 37% in third-party production – Hallways free of obstructions – Formation of order and cleanliness habits, at work and personally – Motivation towards continuous improvement
Shikari (constancy)	– Practice the 4S permanently over time – Positive results that are constantly maintained
Shitsukoku (commitment)	– Internal communication was improved – Motivated workers who are confident in suggesting improvements to company standards – Improved worker attitude towards evaluations – The influence of the manager's actions motivates workers to maintain the program satisfactorily
Seishoo (coordinate)	– Corrective actions achieved favorable results and prevented a setback in the implementation program – A detailed plan of activities allows for synchronized work and the achievement of objectives within established times
Seido (estandarize)	– Standardizing tasks facilitates and maintains classification, order and cleanliness – Improved internal communication between workers

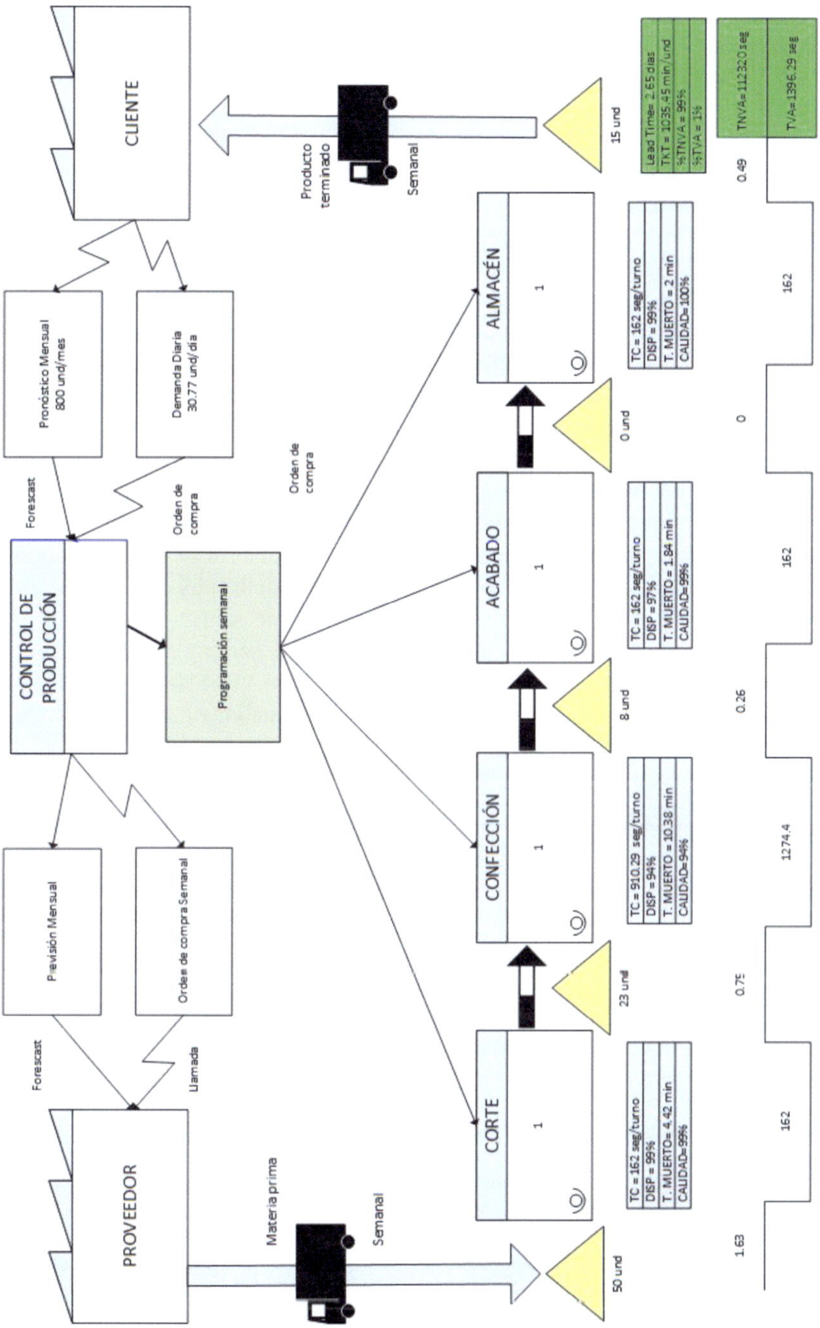

Fig. 8 VSM future

4 Conclusion

The application of Lean methodologies in the textile manufacturing workshop, including Value Chain Mapping (VSM), Waste Assessment Model (WAM) Analysis and the implementation of 9S has been beneficial. The VSM allowed the identification of waste, bottlenecks, as well as inefficient areas and processes within the production chain. In the initial VSM, a takt time of 1053 s and a lead time of 8.63 days were observed, with a TVA of 1% and a TNVA of 99%. On the other hand, the application of WAM allowed to identify the main types of waste, highlighting inventories (21.82%), defects (20.50%) and overproduction (17.44%) as the most significant. Based on the data obtained from the diagnosis, the implementation of the 9S generated positive results not only in the visual aspect, such as the reduction of the surface used in the areas of cutting and finishing (24%), warehouse (32%) and sewing (7%), but also in the reduction of search times for tools, raw materials and finished products in the processes of cutting (75%), sewing (28%), finishing (97%) and warehouse (52%). In addition, the implementation of 9S improved the fluidity of the mobilization of workers and resources. Additionally, a 4% reduction in defective garments from internal production and a 37% reduction in garments from third parties was achieved, thanks to better visualization and control of the production stages in the sewing area. The improvements impacted both the physical workspace and employee morale, creating a cleaner, more organized, pleasant and safer environment. In addition, internal communication improved, better serving employee needs and increasing their overall satisfaction. Finally, the future VSM shows significant improvements, with a new lead time of 2.65 days and a reduction of 364.11 s/shift in the cycle time of the manufacturing process, allowing for greater productivity thanks to reduced waste. As a result of making the improvements, productivity of shorts increased by 26%, that of leggings by 22%, and that of sports polos by 49%.

References

1. Cantú, H.: 9 Aspectos Clave para la Calidad Total [Key Aspects of Total Quality]. México (1998)
2. Muñoz Guevara, J.A., Zapata Urquijo, C.A., Medina Varela, P.D.: Lean Manufacturing: Modelos y Herramientas [Lean Manufacturing: Models and Tools]. Universidad Tecnológica de Pereira—UTP (2022)
3. Rawabdeh, I.A.: A model for the assessment of waste in job shop environments. Int. J. Oper. Prod. Manag. **25**, 800–822 (2005). https://doi.org/10.1108/01443570510608619
4. Madrigal, P.R., Morales Sánchez, V., Giovanni, I., Robles, C.: Implementación de Metodología de las 9 S´s [Implementation of the 9S Methodology] (2021)
5. Bizuneh, B., Omer, R.: Lean waste prioritisation and reduction in the apparel industry: application of waste assessment model and value stream mapping. Cogent Eng. **11** (2024). https://doi.org/10.1080/23311916.2024.2341538
6. Cristobal, J., Quispe, G., Dominguez, F., Zapata, G., Raymundo, C.: Waste reduction with lean manufacturing model in an Alpaca wool workshop. In: IOP Conference Series: Materials Science and Engineering. IOP Publishing Ltd (2020)

7. Febianti, E., Irman, A., Juliana, M.: Implementation of lean manufacturing using waste assessment model model (WAM) in food industry (Case Study in usaha mikro kecil menengah (umkm) xyz). In: IOP Conference Series: Materials Science and Engineering. IOP Publishing Ltd (2020)
8. Ortiz Porras, J., Bancovich Erquínigo, A., Candia Chávez, T., Huayanay Palma, L., Salas Bacalla, J.: Método de Aplicación de la Herramienta Value Stream Mapping para Aumentar la Competitividad en una Empresa Textil y de Confecciones [Method for applying the value stream mapping tool to increase competitiveness in a textile and apparel company]. Ind. Data **26**, 33–61 (2023). https://doi.org/10.15381/idata.v26i1.22874
9. Rahman, M.S., Tahiduzzaman, M., Islam, A., Rahman, M.S.: Implementation of 9S approach in a jute industry: a case study (2018)
10. Lisset, T.Z., Joshelyn, Y.A., Neicer, C.V., Lenin, F.A.H.: Tool 9s to optimize production and control within the food industry. In: Proceedings of the LACCEI International Multi-conference for Engineering, Education and Technology. Latin American and Caribbean Consortium of Engineering Institutions (2022)

Order Allocation in a Semiconductor Supply Chain Using Metaheuristics with Constraint Handling Operators

Bing Wei Tan, Jeng Feng Chin, Wei Ping Loh, and Nur Amalina Muhammad

Abstract The complexity and demand volatility of the semiconductor supply chain complicates the order allocation across the supply chain. When an optimization metaheuristic is applied, such complications easily lead to constraint violation and hence solution infeasibility. This study develops an order allocation model with a novel constraint handling operator that involves both solution repair and penalty function. Genetic algorithm, opposition-based coral reefs optimization, and differential evolution from evolutionary-based metaheuristics are used for optimization. In the experiment, the metaheuristics incorporating the proposed constraint handling operators, especially genetic algorithm, outperformed the counterparts in terms of the number of constraint violations and resultant profit.

Keywords Semiconductor · Metaheuristics · Optimization · Constraint · Supply chain · Genetic algorithm

1 Introduction

The semiconductor industry fabricates integrated circuits through a series of front-end wafer fabrication and back-end assembly-and-test operations [1]. The broad application of integrated circuits in different sectors entails global semiconductor

B. W. Tan · J. F. Chin (✉) · W. P. Loh · N. A. Muhammad
School of Mechanical Engineering, Tuanku Syed Sirajuddin Engineering Campus, Universiti Sains Malaysia, 14300 Nibong Tebal, Pulau Pinang, Malaysia
e-mail: chinjengfeng@usm.my

B. W. Tan
e-mail: bingweitan2016@student.usm.my; bwtan2016@gmail.com

W. P. Loh
e-mail: meloh@usm.my

N. A. Muhammad
e-mail: nuramalinamuhammad@usm.my

M. Awang et al. (eds.), *Proceedings of the International Conference on Mechanical, Manufacturing and Plant Engineering*, Proceedings in Technology Transfer,
https://doi.org/10.1007/978-981-95-7817-7_12

industry sales of 526 billion United States Dollar (USD) in 2023 [2]. Supply chain management is vital to cope with the high market demands and to maximize profit [3]. However, the long manufacturing time, different facility capacities, unpredictable yield rate, and various product specifications affect the supply chain efficacy [3, 4]. To address these issues, a computational optimization method, such as metaheuristic, is often deployed. Metaheuristics utilize the exploration and exploitation mechanisms in covering the solution space to find optimal solutions [5]. These methods are relatively easy to implement, flexible to fit any problem domain, and generate nearly optimal solutions within a reasonable computational time [6]. Numerous metaheuristics exist, with prevalent ones including genetic algorithm (GA), simulated annealing, and differential evolution (DE).

Despite demonstrating promising results, only a small number of research cases applied metaheuristics in the semiconductor supply chain [7–11]. In semiconductor supply chain optimization, the number of interdependent demands and capacity constraints make order allocation in the semiconductor supply chain challenging. As a result, when a constraint is violated, solutions produced by the metaheuristics quickly become infeasible. In terms of search space, these constraints create a substantial region of infeasible solutions interspersed with limited pockets of feasible ones. On this premise, this study proposed a metaheuristics model architecture to optimize order allocation in the semiconductor supply chain. GA, opposition-based coral reefs optimization (OCRO), and DE from evolutionary-based metaheuristics were selected. A novel constraint handling operator that is based on solution repair and penalty function was proposed. The operators were embedded into the fitness function that evaluates the solutions generated by the metaheuristics. The rest of the papers are organized as follows. Section 2 describes the methodology of the study; Sect. 3 analyzes and discusses the result obtained; conclusion and suggestion for further work are summarized in Sect. 4.

2 Methodology

2.1 Semiconductor Supply Chain

The semiconductor supply chain model is illustrated in Fig. 1 [12]. The notations and parameters are given below.

Indices:

i Index for number of products, where $i = 1, 2, \ldots, I$.
j Index for number of semiconductor foundries (fabs), where $j = 1, 2, \ldots, J$.
v Index for number of assembly and test (ATs) sites, where $v = 1, 2, \ldots, V$.
k Index for number of customers, where $k = 1, 2, \ldots, K$.

Parameter:

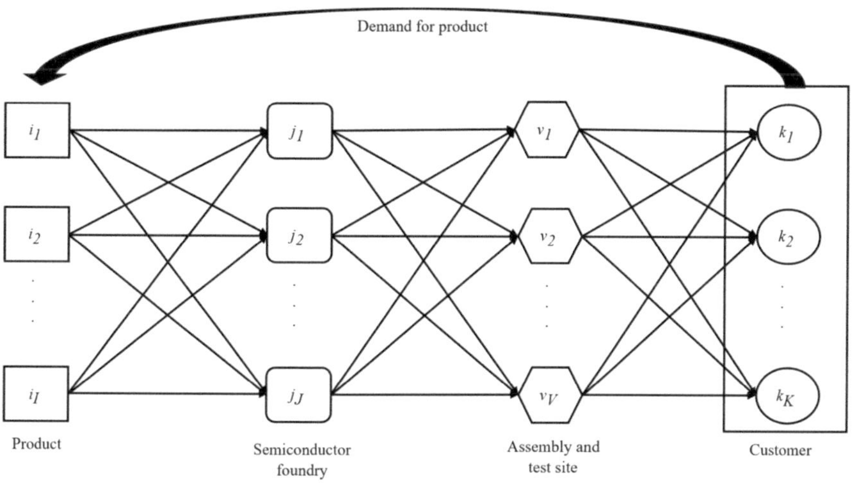

Fig. 1 The illustration of the flow chart of the semiconductor supply chain network

- $D\min_{i,k}, D\max_{i,k}$—Minimum and maximum demand range of each product (in die unit) from each customer.
- $R\min_{i,k}, R\max_{i,k}$—Average selling price (ASP) per unit of product i when customer k purchases within the range of $D\min_{i,k}, D\max_{i,k}$.
- $ASP_{i,k}, (\overline{y}_{i,k}) = \alpha_{i,k} + \beta_{i,k}\left(\sum_v y_{i,v,k} - D\min_{i,k}\right)$—function of number of units sold above the minimum demand, where $\alpha_{i,k} = R\min_{i,k}$.
- $\beta_{i,k} = (R\max_{i,k} - R\min_{i,k})/(D\max_{i,k} - D\min_{i,k})$—Rate of change of average selling price due to volume discount.
- $CF\max_j, CF\min_j = CF\max_j * \min imum\_utilization$—Capacity range for each fab, where min *imum_utilization* is the minimum capacity used for production.
- $q_{i,j} \in \{0, 1\}$—Qualification index for fab. If 1, the fab j is qualified to fabricate product i.
- $DW\min_i, DW\max_i$—Minimum and maximum demand range of each product (in wafer units).
- CTR_i—Capacity trade ratio for each product. The normal ratio is 1.0. If the product consumes less/more capacity, the ratio will be less/more than 1.0, respectively.
- $WCF_{i,j}$—Production cost of each product's wafer per fab.
- $FTA_{j,v}$—Shipping cost from fab to AT.
- $CA\max_v$—Maximum capacity (in wafer unit) for each AT.
- DPW_i—The quantity of dies in each wafer for each product.
- WCA_v—Production cost per wafer in each AT.
- DCA_v—Production cost per thousand die units in each AT.
- $ATC_{v,k}$—Shipping cost from AT to customer.

Decision variables:

- $x_{i,j,v}$—Quantity of wafer units of product i shipped from fab j to AT v.

- $y_{i,v,k}$—Quantity of die units of product i supplied from AT v to customer k.

Objective function:

$$\begin{aligned} Z_{\max} = & \sum_{i,k}\left(\left(\alpha_{i,k} + \beta_{i,k}\left(\sum_{v} y_{i,v,k} - Dmin_{i,k}\right)\right)\left(\sum_{v} y_{i,v,k}\right)\right) \\ & - \sum_{i,j} WCF_{i,j} \cdot \left(\sum_{v} x_{i,j,v}\right) - \sum_{i,v} (WCA_v + DCA_v \cdot DPW_i) \cdot \left(\sum_{j} x_{i,j,v}\right) \\ & - \sum_{j,v} FTA_{j,v}\left(\sum_{i} x_{i,j,v}\right) - \sum_{v,k} ATC_{v,k}\left(\sum_{i} y_{i,v,k}\right) - P \end{aligned} \tag{1}$$

Subject to:

$$Dmin_{i,k} \le \sum_{v} y_{i,v,k} \le Dmax_{i,k}, \forall i, k \tag{2}$$

$$CF\min_j \le \sum_{i}\left(CTR_i \sum_{v} x_{i,j,v}\right) \le CF\max_j, \forall j \tag{3}$$

$$\sum_{i,j} x_{i,j,v} \le CA\max_v, \forall v \tag{4}$$

$$\sum_{j} x_{i,j,v} \cdot DPW_i \ge \sum_{k} y_{i,v,k}, \forall v, i \tag{5}$$

$$\sum_{v} x_{i,j,v} \le q_{i,j} \cdot CF\max_j, \forall i, j \tag{6}$$

$$DW\min_i \le \sum_{j,v} x_{i,j,v} \le DW\max_i, \forall i \tag{7}$$

The objective function, Eq. (1) is to maximize profit, embedded with revenue functions, fab production costs, AT production costs, shipping costs from fab to ATs, shipping costs from ATs to customers, and P, respectively. P denotes the penalty function that penalizes infeasible solutions based on the degree of constraint violations. The objective function is limited by constraints from Eq. (2) to Eq. (7). Equation (2) develops the demand range of each product and customer. Equation (3) defines the capacity range for each fab. Equation (4) ensures each AT operates within its capacity limit. Equation (5) ensures the fabricated quantity of each wafer product can cope with the transformation process of the wafer to die unit in ATs, fulfilling the balance flow constraint. Equation (6) ensures only qualified fab can manufacture the specific product within the capacity limit of the fab. Equation (7) ensures that the total quantity of wafers fabricated for each product should be within the minimum

Table 1 Qualifications of fabs in fabricating specific products

Fabs	Product 1	Product 2	Product 3	Product 4	Product 5
Fab 1	1	1	1	1	0
Fab 2	0	1	1	1	1
Fab 3	1	1	1	0	1

and maximum demand range of products (in wafer units). The product qualification of fabs is tabulated in Table 1. The fab production cost is fixed at 4000 USD per wafer.

2.2 *Metaheuristics Description*

GA operates by selection, crossover, and mutation processes [13]. After an initial population is randomly generated, the selection operators pick the higher fitness of the individual solution in the population and insert into the mating pool. The crossover operator exchanges the variables between a selected pair of the best individual solutions from the mating pool, and the mutation process alters the value of variables in the solutions. The mutation rate should be set at a low value to regulate the random search and converge the search toward the optimal solution. OCRO extends the working principle of the coral reefs optimization algorithm with opposition-based learning techniques in the depredation phase [14]. The technique is applied to the worst healthy coral to calculate its oppositional coral's solution. The opposite coral replaces the original worst health coral if it has better fitness. Else, the worst health coral is depredated. This enhances the searching ability of the algorithm. Searching is reinitialized at the non-best corals if no improvement is made after a certain iteration. DE generates an initial candidate solution (target vector) uniformly in the search space [15]. The mutation process is performed to produce a mutant vector. The binomial crossover operator describes the element of mutant vector in binomial distribution and generates a trial vector from paired target vector and mutant vector. The selection operator will evaluate the fitness of the current target vector and the corresponding trial vector. The individual with a better solution is preserved and enters the next population generation.

2.3 *Repair and Penalty Function Constraint Handling Operator (RP)*

The fitness function receives solution sent by the metaheuristic during a particular iteration. The solution would be evaluated for its fitness. The RP operator is embedded into the fitness function and deployed in the detection of a constraint violation in solution (hence infeasible solution). Repair operator commonly incorporates domain knowledge to repair an infeasible solution [16, 17]. The constraints are first prioritized based on their interdependency to other constraints. The repair is then sequenced based on the priority of the constraints. To repair the solution, an adjustment is made to the affected order allocation. The violation of the constraint's lower bound results in a positive adjustment. The violation of the upper bound of the constraint results in a negative adjustment. In relation to the violation of a constraint, the current order allocation is adjusted iteratively based on a predefined percentage. The percentage is pre-defined by the researchers based on the complexity and interdependency of the constraint. The repair sequence is Eqs. (6), (7), (5), (3), (2) and (4). The fitness of the solution would be computed after a cycle of repair. If constraint violation persists after the repair, a penalty in multiplication of the constraint's penalty weight and penalty cost is imposed to the solution for each violated constraint, reducing its fitness value and thus making the solution less attractive [16]. The parameter setting of the RP method is tabulated in Table 2.

Table 2 Parameter setting of the RP method

Constraint handling operator	Parameter or relevant conditions	Value
Repair	Violations of Eqs. (2), (3), and (4)	10% adjustment
	Violation of Eq. (5)	20% adjustment for $x_{i,j,v}$
		25% adjustment for $y_{i,v,k}$
	Violation of Eq. (6)	5% adjustment
	Violation of $DW\min_i$ in Eq. (7)	20% adjustment
	Violation of $DW\max_i$ in Eq. (7)	10% adjustment
Penalty cost		50,000,000 USD
Penalty weight	Penalty weight 1 (violates $D\min_{i,k}$ in Eq. (2))	4
	Penalty weight 2 (violates $D\max_{i,k}$ in Eq. (2))	4
	Penalty weight 3 (violates $CF\min_j$ in Eq. (3))	8
	Penalty weight 4 (violates $CF\max_j$ in Eq. (3))	44
	Penalty weight 5 (violates Eq. (4))	20
	Penalty weight 6 (violates Eq. (5))	60
	Penalty weight 7 (violates Eq. (6))	4
	Penalty weight 8 (violates $DW\min_i$ in Eq. (7))	32
	Penalty weight 9 (violates $DW\max_i$ in Eq. (7))	12

2.4 Experiment

As aforementioned, the experiment involves GA, OCRO, and DE. These metaheuristics were embedded with the RP operators in their fitness functions. Consequently, they would be identified as RP-based metaheuristics. To draw a comparison, a conventional constraint handling operator was embedded in the same metaheuristics acting as the benchmark models. The conventional constraint handling operator will not repair the solution but impose penalty directly to the solution for any constraint violation. These benchmark models are identified as P-based metaheuristics. All models were constructed using Python 3.9. The key Python packages are: (i) MealPy library version 2.4.0 [18, 19]; (ii) NumPy library version 1.26.4; and (iii) Matplotlib library version 3.5.0. The experiment was performed on a desktop with an Intel (R) Core TM i5-8250U. The parameter setting of metaheuristics is tabulated in Table 3.

3 Results and Discussion

3.1 Comparison of Constraint Handling Operators

The effectiveness of the developed model was tested by comparing RP and P-based metaheuristics under identical experiment setup. The result is presented in Table 4. The RP-based metaheuristics outperformed the P-based metaheuristics and have fully avoided constraint violations. The P-based metaheuristics exhibited a considerable difference between the profit (penalty deduction) and profit (without penalty deduction), caused by high penalty weight and cost implied on constraints in the setup. Despite achieving a higher profit (without penalty deduction), the P-based metaheuristics entailed significantly higher infeasible solutions than the RP-based counterparts. Compared to other constraints, Eq. (5) was frequently violated due to challenges in aligning the flow balance between the wafer quantities of individual products and the number of dies produced at the AT stage.

3.2 Resultant Profit, Revenue, and Cost of RP-Based Metaheuristics

The performances of RP-based metaheuristics were further analyzed and tabulated in Table 5. In terms of profit maximization, GA was most effective, followed by OCRO and DE, with a percentage difference of 5.45% and 13.59%, respectively. GA obtained higher revenue than OCRO and DE, potentially due to its outstanding exploration capability in search space. However, OCRO achieved a smaller standard deviation than DE and GA. The learning technique in OCRO explored the fitness of the opposite search direction from the candidate solution during the searching

Table 3 Parameter setting of the metaheuristics

Parameter	Parameters in the metaheuristics	Specification
Number of optimization runs		30
Epochs		100
Population size		500
GA	Selection operator	"tournament"
	k_way	0.2
	Crossover operator	"uniform"
	Crossover probability, *Pc*	0.95
	Mutation probability, *Pm*	0.025
	Mutation_multipoints	True
	Mutation	"flip"
OCRO	Rate between free/occupied at the beginning, p_o	0.4
	BroadcastSpawner/ ExistingCorals rate, F_b	0.8
	Fraction of corals duplicates itself, F_a	0.2
	Fraction of worse health corals in reefs to be applied in depredation, F_d	0.3
	Maximum probability of depredation, P_d	0.1
	Probability of mutation process,*GCR*	0.1
	Factor for mutation process, *gamma*	(0.02, 0.2)
	Number of attempts for a larver to set in the reef, *n_trials*	5
	Restart_count (reset whole population if global best solution is not improved after restart_ count times), *restart_count*	55
DE	Weighting factor, *F*	0.8
	Crossover rate, *CR*	0.9
	Variants of DE	5 (DE/current–to–rand/1/bin)

phase [14]. If the opposite option offered better fitness, the candidate solution could align the search immediately [14]. This reduced random searches to non-optimal solution regions, improved the consistent performance of the search, and yielded lower standard deviation. GA produced the highest cost across all cost components, in association with the highest fab utilization compared to OCRO and DE. Even with a higher total wafer yield and total cost, interestingly, OCRO incurred lower

Table 4 The results of RP-based metaheuristics

		GA	OCRO	DE
RP				
Profit (with penalty deduction)	Mean	**13414510671.43**	**12702543738.28**	**12418668058.04**
	Standard deviation	255727242.29	189349021.18	222055473.36
Constraint violations	Mean	**0**	**0**	**0**
	Standard deviation	0	0	0
P				
Profit (with penalty deduction)	Mean	3512910886.08	542048064.55	103824909.46
	Standard deviation	2878029672.94	2852231384.54	1546993286.16
Constraint violations	Mean	4.6667	5.1667	5.4333
	Standard deviation	1.0933	1.147	0.6261

shipping costs from fabs to ATs than DE. GA yielded the best solution, at the expense of longer computation time and the highest runtime variation. OCRO took the least computation time to yield a satisfactory solution.

Table 5 The optimized result of 30 runs by GA, OCRO, and DE

		GA	OCRO	DE
Revenue	Mean	14644730273.97	13929823303.53	13643049615.50
	Standard deviation	264039015.07	191482617.21	227464515.04
Fab production cost	Mean	1152547200	1150486800	1147841200
	Standard deviation	9210057.91	9040926.79	9363411.35
AT production cost	Mean	52147964.87	51912760.43	51811305.47
	Standard deviation	459039.38	415802.30	472947.49
Fab to AT shipping costs	Mean	14993944.67	14914956	14923571.17
	Standard deviation	101520.93	141995.05	110653.89
AT to customer shipping costs	Mean	10530492.97	9965048.83	9805480.73
	Standard deviation	144375.41	144085.64	125000.62
Total cost	Mean	1230219602.50	1227279565.27	1224381557.37
	Standard deviation	9779730.05	9567257.08	9932489.23
Profit	Mean	**13414510671.43**	12702543738.28	12418668058.04
	Standard deviation	255727242.29	189349021.18	222055473.36

4 Conclusion

Order allocation in the semiconductor supply chain is a multiple-constraint optimization problem. In tackling the problem, the study developed a new constraint handling operators suitable for population-based metaheuristics. The repair operator repairs the solution by adjusting the allocated orders across relevant facilities. Penalty is imposed on the solution only for stubborn constraints upon unsuccessful repair. The experiment result demonstrated that the developed model offered more feasible solutions than conventional penalty constraint handling operator. RP-based GA outperformed others with higher profit and factory utilization, followed by OCRO and DE. OCRO is relatively more consistent in performance. Future works are in two directions. The exploration of artificial intelligence-driven constraint handling operators could lead to more adaptive and efficient solutions. Enhancing the relationships and interactions between constraints in the semiconductor supply chain model could improve the realism and accuracy of the optimization.

Acknowledgements This work was supported by The Malaysia Ministry of Higher Education (Kementerian Pengajian Tinggi) under the Fundamental Research Grant Scheme (FRGS) grant no: FRGS/1/2021/TK0/USM/02/25.

Disclosure of Interest The authors declare no financial or commercial conflict of interest.

References

1. Mönch, L., Fowler, J.W., Mason, S.J.: Production Planning and Control for Semiconductor Wafer Fabrication Facilities, 1st edn. Springer, New York (2013). https://doi.org/10.1007/978-1-4614-4472-5
2. WSTS Historical Billings Report. https://www.wsts.org/67/Historical-Billings-Report. Accessed 1 Sep. 2024
3. Uzsoy, R., Fowler, J.W., Mönch, L.: A survey of semiconductor supply chain models Part II: demand planning, inventory management, and capacity planning. Int. J. Prod. Res. **56**(13), 4546–4564 (2018). https://doi.org/10.1080/00207543.2018.1424363
4. Mönch, L., Uzsoy, R., Fowler, J.W.: A survey of semiconductor supply chain models Part I: semiconductor supply chains, strategic network design, and supply chain simulation. Int. J. Prod. Res. **56**(13), 4524–4545 (2018). https://doi.org/10.1080/00207543.2017.1401233
5. Odili, J.B.: The dawn of metaheuristic algorithms. Int. J. Softw. Eng. Comput. Syst. **4**(2), 49–61 (2018). https://doi.org/10.15282/ijsecs.4.2.2018.4.0048
6. Abdel-Basset, M., Abdel-Fatah, L., Sangaiah, A.K.: Metaheuristic algorithms: a comprehensive review. In: Sangaiah, A.K., Sheng, M., Zhang, Z. (eds) Computational Intelligence for Multimedia Big Data on the Cloud with Engineering Applications, pp. 185–231. Academic Press (2018). https://doi.org/10.1016/B978-0-12-813314-9.00010-4
7. Forstner, L., Mönch, L.: A heuristic to support make-to-stock, assemble-to-order, and make-to-order decisions in semiconductor supply chains. In: 2013 Winter Simulations Conference (WSC), pp. 3696–3706. IEEE, Washington (2013). https://doi.org/10.1109/WSC.2013.6721730
8. Kang, H.-Y., Pearn, W.L., Chung, I.-P., Lee, A.H.I.: An enhanced model for the integrated production and transportation problem in a multiple vehicles environment. Soft. Comput. **20**(4), 1415–1435 (2016). https://doi.org/10.1007/s00500-015-1595-7

9. Ziarnetzky, T., Mönch, L.: Simulation-based optimization for integrated production planning and capacity expansion decisions. In: 2016 Winter Simulation Conference (WSC), pp. 2992–3003. IEEE, Washington (2016). https://doi.org/10.1109/WSC.2016.7822333
10. Haspecker, A., Mönch, L.: Exact and heuristic algorithms for a bi-criteria order-lot pegging problem in a multi-fab setting. In: 2023 Winter Simulation Conference (WSC), pp. 2254–2265. IEEE, San Antonio (2023). https://doi.org/10.1109/WSC60868.2023.10407942
11. Wang, X., Zhan, L., Zhang, Y., Fei, T., Tseng, M.-L.: Environmental cold chain distribution center location model in the semiconductor supply chain: a hybrid arithmetic whale optimization algorithm. Comput. Ind. Eng. **187**, 109773 (2024). https://doi.org/10.1016/j.cie.2023.109773
12. Kalir, A.A., Grosbard, D.I.: Global optimization of a semiconductor IC supply chain network. J. Serv. Sci. Manag. **10**(03), 338–352 (2017). https://doi.org/10.4236/jssm.2017.103027
13. Mirjalili, S.: Genetic algorithm. In: Evolutionary Algorithms and Neural Networks. Studies in Computational Intelligence, vol. 780, pp. 43–55. Springer, Cham (2019). https://doi.org/10.1007/978-3-319-93025-1_4
14. Nguyen, T., Nguyen, T., Nguyen, B.M., Nguyen, G.: Efficient time-series forecasting using neural network and opposition-based coral reefs optimization. Int. J. Comput. Intell. Syst. **12**(2), 1144–1161 (2019). https://doi.org/10.2991/ijcis.d.190930.003
15. Storn, R., Price, K.: Differential evolution—A simple and efficient heuristic for global optimization over continuous spaces. J. Global Optim. **11**(4), 341–359 (1997). https://doi.org/10.1023/A:1008202821328
16. Coello, C.A.C.: Theoretical and numerical constraint-handling techniques used with evolutionary algorithms: a survey of the state of the art. Comput. Methods Appl. Mech. Eng. **191**(11–12), 1245–1287 (2002). https://doi.org/10.1016/S0045-7825(01)00323-1
17. Krawiec, K.: Metaheuristic design pattern: candidate solution repair. In: GECCO 2014—Companion Publication of the 2014 Genetic and Evolutionary Computation Conference, pp. 1415–1418. Association for Computing Machinery, New York (2014). https://doi.org/10.1145/2598394.2609847
18. Van Thieu, N., Deb Barma, S., Van Lam, T., Kisi, O., Mahesha, A.: Groundwater level modeling using augmented artificial ecosystem optimization. J. Hydrol. **617**, 129034 (2023). https://doi.org/10.1016/j.jhydrol.2022.129034
19. Van Thieu, N., Mirjalili, S.: MEALPY: an open-source library for latest meta-heuristic algorithms in Python. J. Syst. Architect. **139**, 102871 (2023). https://doi.org/10.1016/j.sysarc.2023.102871

Integrated Control Strategy for Enhancing Driving Stability of e-Corner Module Based on Vehicle

Hyunmin Lee, Doyeong Kim, Hyomin Jin, and Chulwoo Moon

Abstract This study addresses the challenges posed by narrow urban roads and limited parking, leading to the development of e-Corner module technology. This system uses independently operating wheel motors, eliminating mechanical connections for enhanced maneuverability. However, maintaining wheel alignment can be difficult, affecting straight-line stability. We propose a control method that accounts for performance differences among motors and variations in initial steering angles, aiming to improve lateral acceleration and overall driving performance.

Keywords Control · e-Corner module · Lateral acceleration · Driving stability

1 Introduction

Automobiles have become essential across various industries, playing a significant role in both daily life and industrial activities. Since the 2000s, vehicle demands have increasingly focused on maneuverability and driving stability. However, in urban environments, narrow roads and limited parking spaces pose challenges to vehicle operation, restricting their usability. To address these issues, the automotive industry has developed e-Corner module technology based on by-wire systems, eliminating the reliance on a shared power source and driver input found in traditional powertrain and steering systems [1].

The e-Corner module, unlike conventional engine systems, generates power directly through in-wheel motors, with each steering module operating independently, controlling functions without mechanical connections. This technology enables groundbreaking changes and improvements in vehicle design and operation.

H. Lee · D. Kim · H. Jin
Korea Automotive Technology Institute, Cheonan, Chungnam, Korea

C. Moon (✉)
Department of Intelligent Mobility, Chonnam National University, Gwangju, Korea
e-mail: cwmoon@jnu.ac.kr

M. Awang et al. (eds.), *Proceedings of the International Conference on Mechanical, Manufacturing and Plant Engineering*, Proceedings in Technology Transfer,
https://doi.org/10.1007/978-981-95-7817-7_13

However, for vehicles equipped with such independent steering systems, the absence of linked mechanical constraints makes it difficult to maintain the alignment of each wheel during extended operation, which can compromise straight-line stability and optimized cornering performance.

Additionally, each in-wheel motor within the e-Corner module has different performance characteristics (Torque-RPM curves). When the same input torque is requested, the time it takes to achieve the desired torque varies based on the motor's characteristics. This discrepancy can cause the vehicle to behave as if it were employing skid steering, leading to more significant changes in initial steering angles and affecting lateral acceleration, which in turn degrades driving stability [2, 3].

Therefore, this study presents a control method based on real-world vehicle test results, implementing simulations under real driving conditions. The proposed control method considers the performance of each motor, minimizing the impact on lateral acceleration and enhancing driving performance. Through this approach, the influence on lateral acceleration and driving performance was significantly reduced.

2 Vehicle TEST Modeling

Figure 1 illustrates the process of applying the vehicle stability logic presented in this paper. In the Vehicle Modeling stage, the geometric characteristics of the actual vehicle and the T-N Curve of the In-Wheel Motor are defined, establishing a foundation for accurate model construction. In the Simulation stage, a Carmaker simulation is conducted based on the modeled vehicle. Since there is no data for e-Corner-based vehicles available in Carmaker, the steering and driving characteristics of the e-Corner vehicle are reflected using the Matlab/Simulink program. The generated data is then com-pared and analyzed against the longitudinal test results of the actual vehicle. Based on this analysis, the logic is reapplied in Carmaker for real-time simulation (Co-Simulation) and evaluation. By utilizing this process, comprehensive assessments of driving stability are conducted, ultimately validating the effectiveness of the proposed logic.

2.1 Vehicle Modeling

Figure 2 depicts the vehicle utilized in the tests. The KAMO-M vehicle is based on the e-Corner module and features independent steering and driving capabilities through the Vehicle Control Unit (VCU).

Table 1 presents the specifications of the KAMO-M vehicle, which serves as the fundamental data for this test. The vehicle testing was conducted under Gross Vehicle Weight (GVW) conditions, with the speed set at 25 km/h.

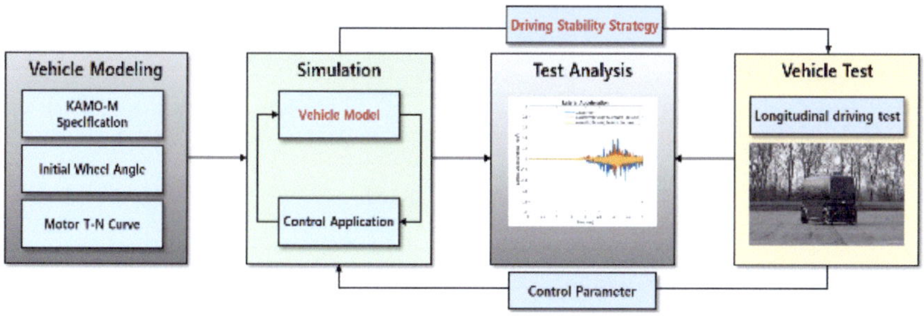

Fig. 1 Vehicle driving stability influence process

Fig. 2 TEST vehicle (KAMO-M)

Table 1 TEST vehicle (KAMO-M) specification

Specification			
Parameter		Measurement	Value
Dimension	Length	mm	5490
	Width		2264
	Height		3080
	Wheel base		3700
Weight	Gross vehicle weight	kg	4000
Battery	Type		Li-ion
	Voltage	V	332
	Capacity	kWh	43.16
Performance	Top speed	km/h	45
	Normal speed		30
Tire	Front	in	245/35R20
	Rear		245/35R20

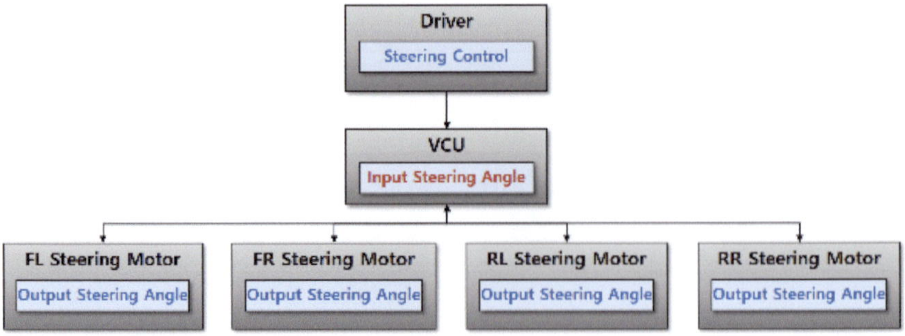

Fig. 3 Steering control diagram

2.1.1 Steering Control Process

Figure 3 illustrates the steering control process. An initial steering value is set, and the resultant value, obtained by combining the driver's steering angle, is transmitted to the steering motor. This enables the adjustment of the wheel angle through the manipulation of the steering motor. However, this control method does not account for changes in the initial steering value of the independent steering sys-tem; thus, it is necessary to apply data based on the proportional relationship between the initial wheel angle and the angular velocity of the steering motor.

2.1.2 T-N Curve

Figure 4 illustrates the torque control process. A motor torque of 450 Nm is applied from the Vehicle Control Unit (VCU) to the In-Wheel Motor. The applied motor torque is output based on the characteristics of each motor.

Figure 5 presents the motor T-N Curve graph based on dynamometer testing. As shown in the graph, there are differences in the T-N curves for each motor. These differences manifest as variations in the torque values of each motor before the output reaches 450 Nm, ultimately causing the vehicle to behave as if it were employing skid steering, which affects driving stability.

2.2 *Driving Stability Strategy*

The torque acting on each wheel, both left and right, is a result of skid steering; therefore, the logic design is based on skid steering. The skid ratio is fundamentally proportional to the steering angle and inversely proportional to the vehicle speed. The relationship of skid application on the left and right wheels can be summarized as follows Definition 1.

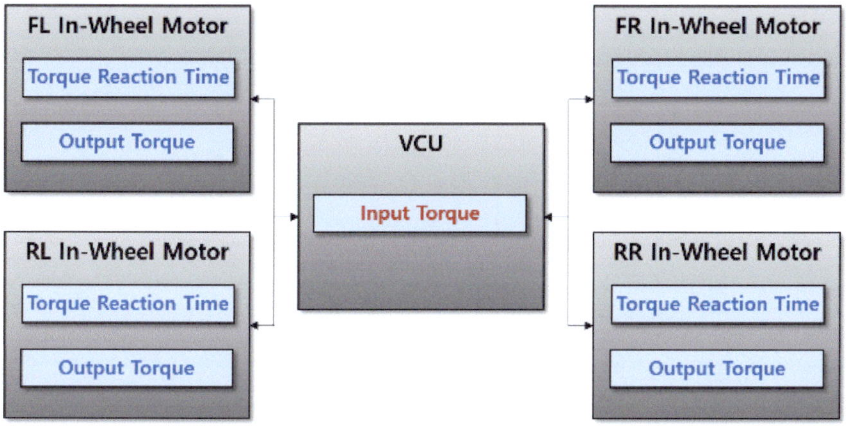

Fig. 4 Torque control diagram

Definition 1: Skid Model formula

$$T_L = K_w \times (V_T + V_S) \tag{1}$$

$$T_R = K_w \times (T_T - V_S) \tag{2}$$

$$V_T = f(A_i) \tag{3}$$

$$V_s = f(\theta_w) \tag{4}$$

Let T_L represent the torque acting on the left wheel and T_R represent the torque acting on the right wheel. V_T is the target speed. The variable V_S, which determines the skid ratio, is proportional to the steering motor angle θ_w and inversely proportional to the current vehicle speed [3].

2.2.1 Steering Control Strategy

Figure 6 presents the overall steering control strategy. The angular velocity of the steering motor is calculated through the driver's steering angular velocity and the change in the Initial Wheel Angle. This relationship is defined in Definition 2.

Definition 2: Output Wheel Angle Calculation

$$\theta_w = K_1 S_i + K_2 \theta_i \tag{5}$$

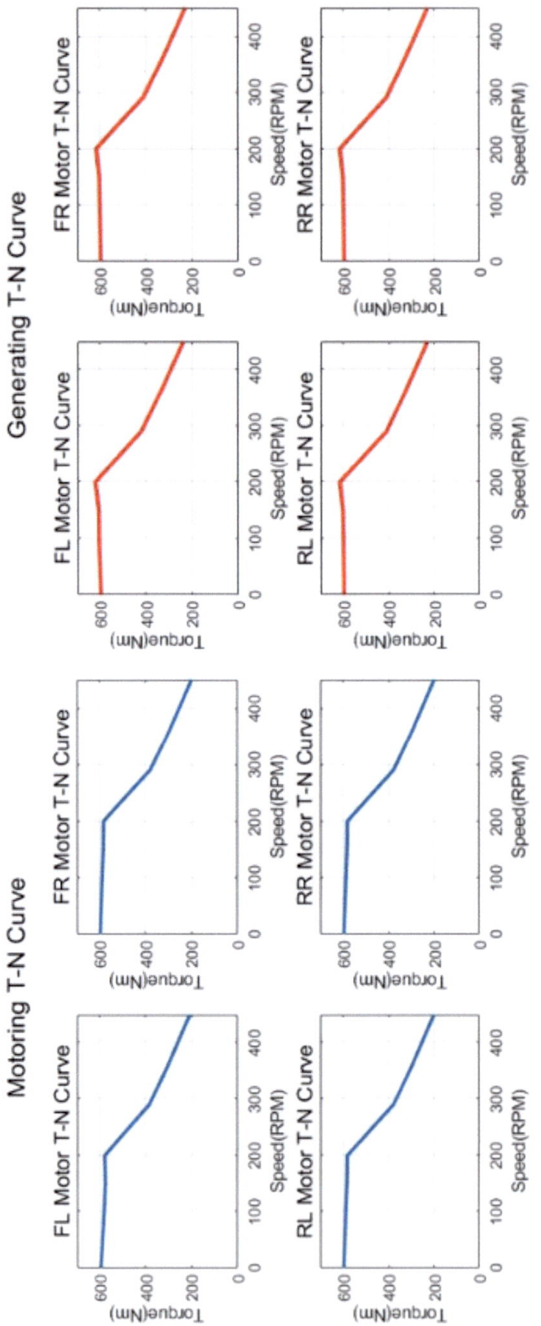

Fig. 5 Motor torque-RPM (T-N) curve graph

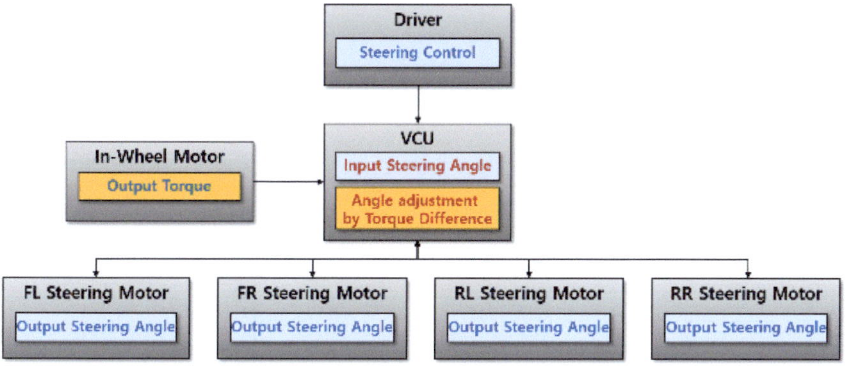

Fig. 6 Steering control strategy diagram

Let θ_w represent the angle of the steering motor, S_i denote the driver's steering angle, and θ_i signify the Initial Wheel Angle.

Figure 7 illustrates the relationship between changes in the Initial Wheel Angle and the steering angular velocity. As the Initial Wheel Angle varies, the driver uses the steering wheel more during longitudinal driving, and this outcome is reflected in the angle of the steering motor. Based on this, the current Initial Wheel Angle can be inferred from changes in the steering angle, allowing for the control of the initial conditions.

Fig. 7 Steering motor angular velocity-initial wheel angle graph

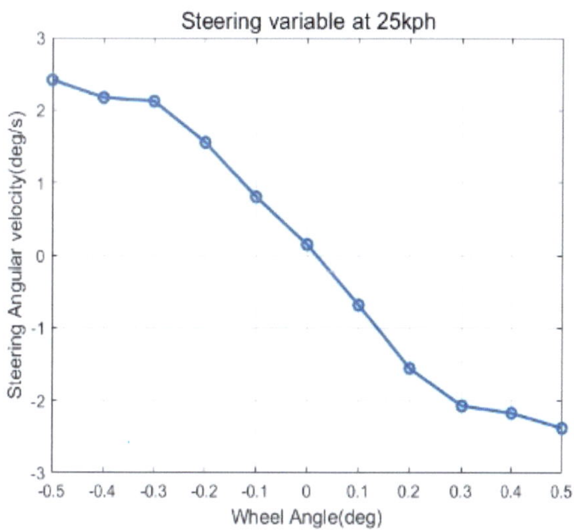

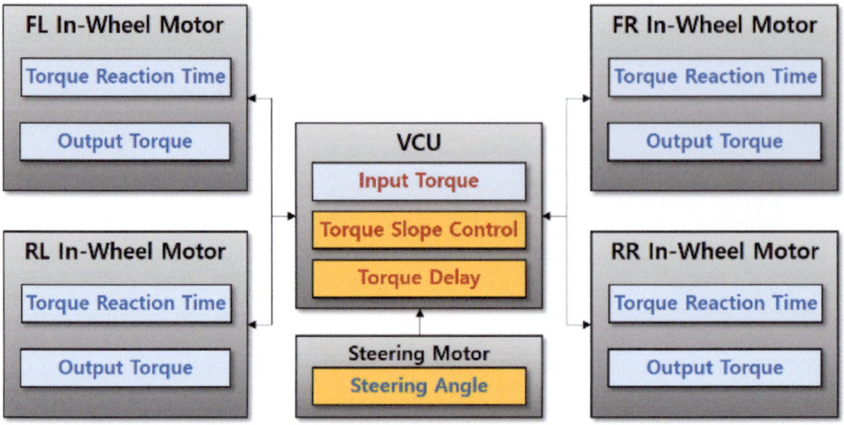

Fig. 8 Torque control strategy diagram

2.2.2 Torque Control Strategy

Figure 8 illustrates the overall torque control strategy. To implement torque control, it is essential to first define the relationship between the initial steering value and torque. Definition 3 describes the relationship between the left and right torques and the initial steering value.

Definition 3: Changing in steering angle due to torque differences.

$$T_L - T_R = 2K_w \times V_S \tag{6}$$

$$f(\theta_i) = \frac{T_L - T_R}{2K_w} \tag{7}$$

By subtracting Eq. (1) from Eq. (2), it can be expressed in the form of Eq. (6). Here, V_s is related to the input steering, and assuming that the driver's steering angle is zero in Eq. (5), we obtain $\theta_w = K_2\theta_i$. This can be represented as shown in Eq. (7), allowing us to derive an expression where the difference in left and right torques indicates the change in the initial steering value. Thus, a control strategy can be established to reduce the θ_i value based on the difference between the input torque and output torque of each motor, as well as the response time.

2.2.3 Driving Stability Strategy

Figure 9 illustrates the proposed Driving Stability Strategy presented in this study, which utilizes a combination of Steering Control and Torque Control. The steering angle of each wheel is adjusted based on the torque differences from the in-wheel

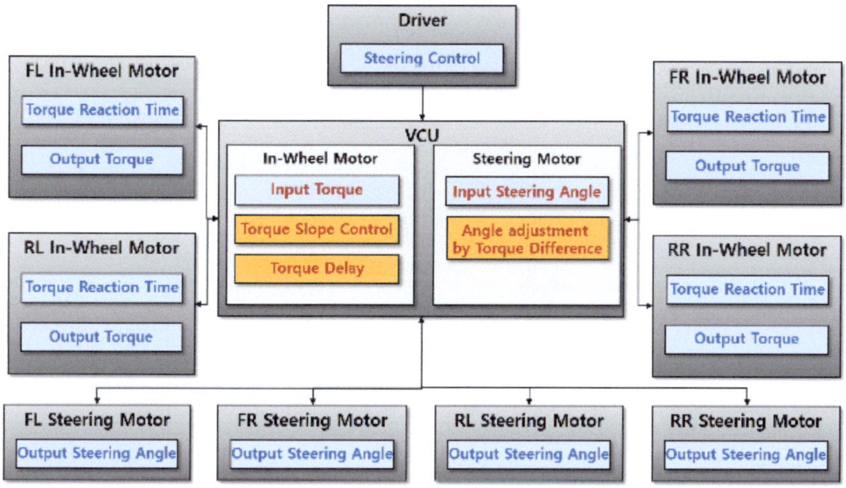

Fig. 9 Driving stability strategy diagram

motors, compensating for the angles of the individual Steering Motors and delivering an optimized Torque Slope for the in-wheel motors. This approach is expected to contribute to the enhancement of the vehicle's driving stability.

3 Performance Validation

Various methods exist for measuring and comparing longitudinal performance; however, this study focused on a straight driving route and conducted experiments by applying an input torque of 450 Nm. The testing method, as outlined in Table 2, followed a Stop and Go procedure. The maximum speed was set to 25 km/h, and longitudinal path accuracy and lateral acceleration were measured by comparing cases with and without the control logic applied. Through this comparison, the effectiveness of the proposed control logic was validated.

3.1 *Vehicle Test Model Validation*

To validate the simulation model, a comparative test was conducted using real vehicle test data. Both the real vehicle test and the simulation applied an input torque of 450 Nm, maintained for 10 s. The validation of the simulation model was performed by assessing the degree of agreement in lateral acceleration.

Figure 10 presents a comparison between the simulation results of the Vehicle Model TEST and the real vehicle longitudinal driving test. Both the simulation and

Table 2 Longitudinal TEST condition

Longitudinal TEST		
Parameter	Measurement	Value
Top speed	km/h	25
Input torque	Nm	450
Clamp force	N	5000
Initial steering angle	deg	0
Driver	Carmaker driver model	
Route	Direct route	
TEST method	Stop&GO(50 s driving, 3 times)	

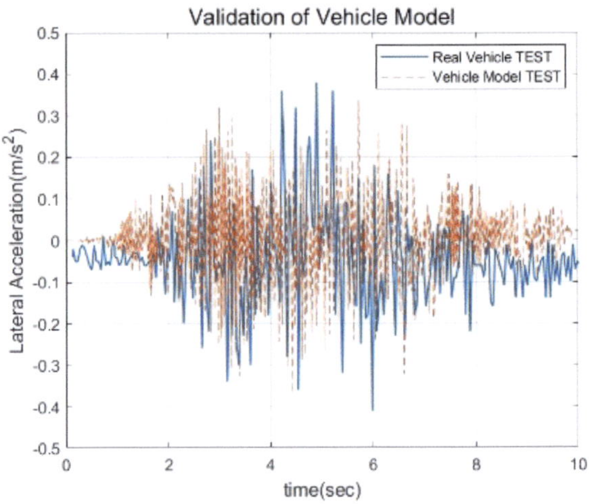

Fig. 10 Validation of vehicle model using lateral acceleration

the real vehicle model demonstrate high accuracy in lateral acceleration. Based on this, Co-simulation is applied to validate the effectiveness of the proposed control logic in this study.

3.2 Vehicle Model Test

In the Vehicle Model Test phase, fundamental data on lateral acceleration and maximum path error were collected. Figures 11 and 12 illustrate the lateral acceleration and driving path of the Vehicle Model TEST. As the driving test progresses, the vehicle gradually deviates more from the intended path, prompting the driver to demand greater steering input for correction. This results in misalignment of the wheel steering angles, leading to the occurrence of path errors and an increasing trend in lateral acceleration.

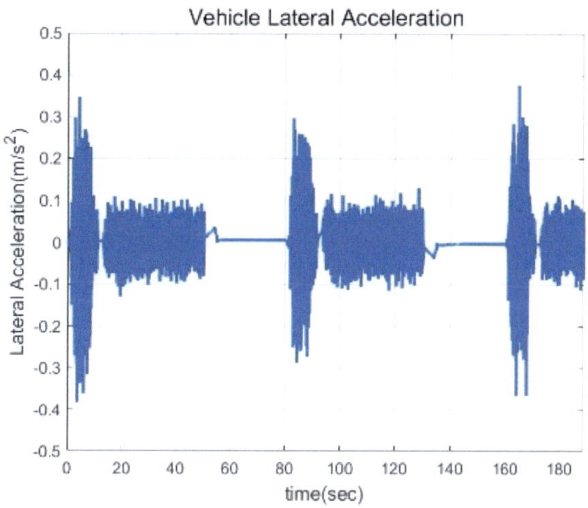

Fig. 11 Basic vehicle model test (Lateral Acceleration)

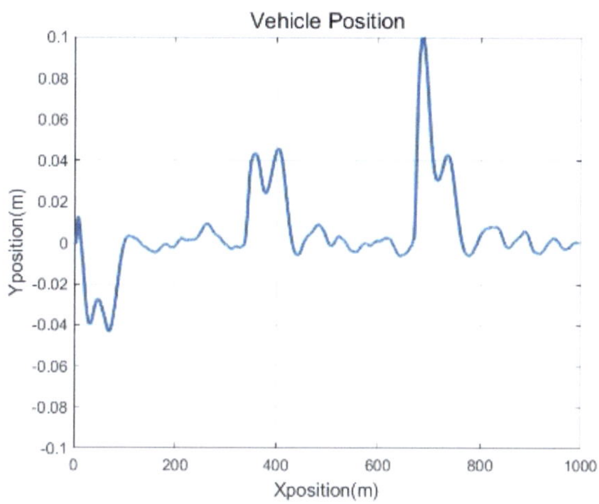

Fig. 12 Basic vehicle model test (Path follow)

3.3 Steering Strategy Test

In the Steering Strategy TEST phase, a logic to correct misaligned steering was introduced, as shown in Fig. 6. As demonstrated in Figs. 13 and 14, the correction of misaligned steering values resulted in a significant reduction in lateral acceleration, allowing the vehicle to maintain a stable trajectory without deviating from the intended path.

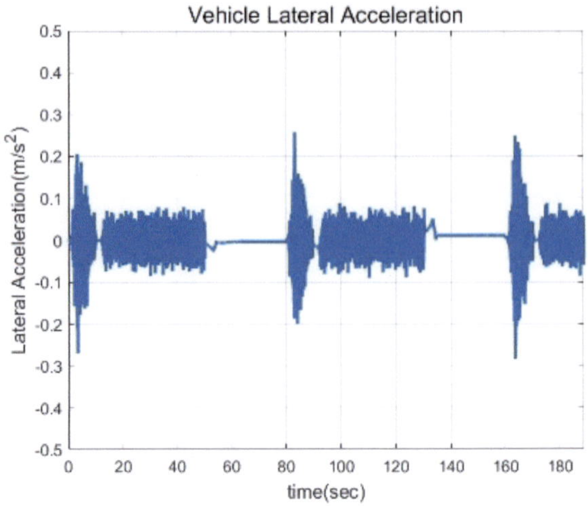

Fig. 13 Steering strategy model test (Lateral Acceleration)

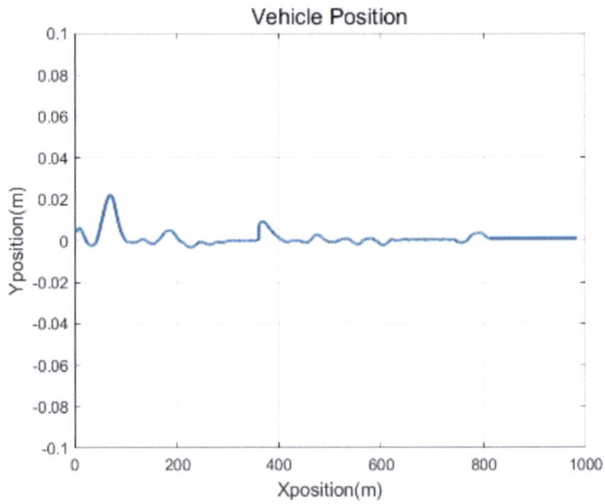

Fig. 14 Steering strategy model test (Path follow)

3.4 *Torque Strategy Test*

The Torque Strategy TEST employs the strategy shown in Fig. 8 to correct the Torque Slope based on the steering values. As illustrated in Figs. 15 and 16, lateral acceleration is reduced to a certain extent, demonstrating the strategy's ability to maintain the intended path. However, the strategy does not address the issue of misalignment in the initial steering, resulting in higher lateral acceleration compared to the Steering TEST, and a corresponding decline in path maintenance performance.

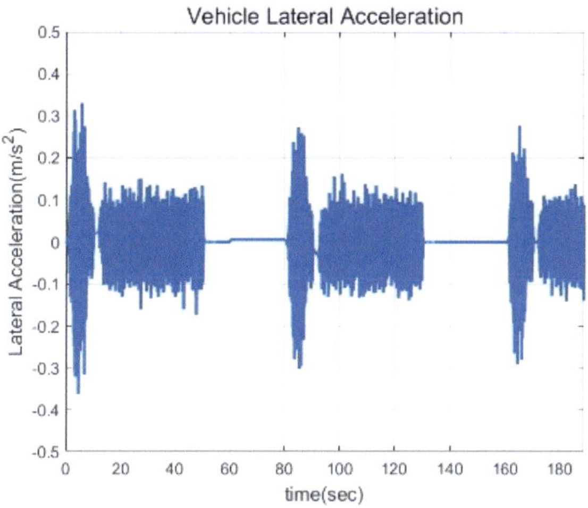

Fig. 15 Torque strategy model test (Lateral Acceleration)

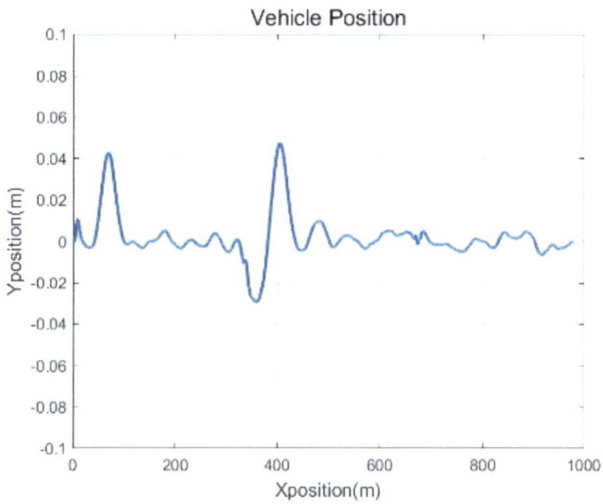

Fig. 16 Torque strategy model test (Path follow)

3.5 Driving Stability Strategy Test

The Driving Stability Strategy TEST simultaneously applies the Steering Strategy and Torque Strategy proposed in this study. As shown in Figs. 17 and 18, the Steering Strategy ensures the steering values, thereby reducing lateral acceleration and maintaining a stable driving path. Subsequently, the Torque Strategy is implemented to adjust the corresponding Torque Slope, resulting in a gradual reduction in lateral acceleration.

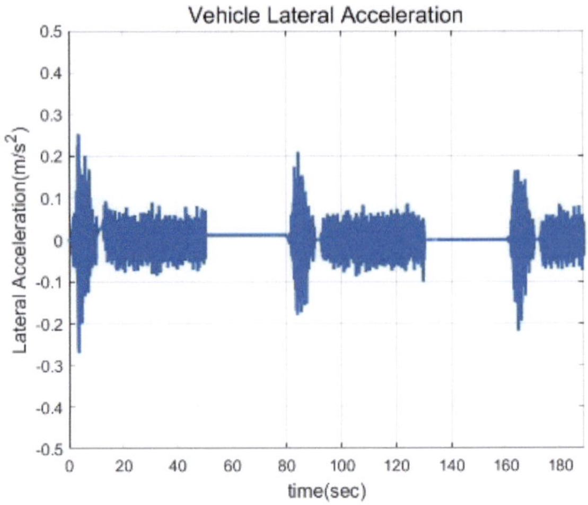

Fig. 17 Driving stability strategy model test (Lateral Acceleration)

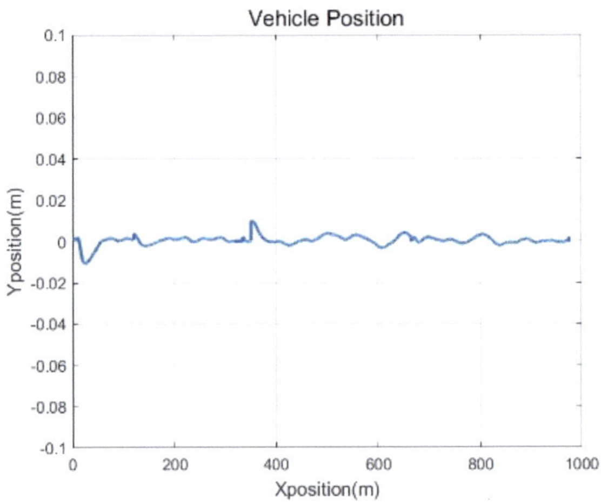

Fig. 18 Driving stability strategy model test (Path follow)

3.6 *Data Analysis*

As shown in Table 3, each strategy successfully reduced lateral acceleration and maximum path error compared to their respective basic models. In particular, the application of the Steering Strategy Model achieved a reduction of 28.27% in lateral acceleration, and ultimately, the Driving Stability Model resulted in a 41.6% decrease. The Torque Strategy Model exhibited a relatively lower reduction rate of 19.2%, which is due to the inability to address the issue of initial steering misalignment and the inherent limitations in adjusting the torque slope and response speed based on motor characteristics, as previously discussed. Nonetheless, the differences in the

Table 3 Lateral acceleration data

TEST	Parameter	
	Lateral acceleration (m/s^2)	Maximum path error (m)
Basic model	0.375	0.10024
Steering strategy model	0.269	0.022143
Torque strategy model	0.303	0.0475
Driving stability model	0.219	0.01073

Torque Strategy allowed for an additional reduction of 13.33% in lateral acceleration, making it an essential strategy in this study.

4 Conclusion

This study proposed a method to enhance vehicle driving stability by integrating steering control and torque control. Various strategies were employed to compare real vehicle tests and simulation models, thereby validating the effectiveness of each strategy. The results demonstrate that the proposed control logic can reduce lateral acceleration and ensure driving stability.

Future work should address not only longitudinal driving stability but also lateral driving stability in e-Corner module vehicles. Future research will involve conducting additional experiments and simulations to more accurately evaluate the vehicle's longitudinal and lateral stability under various driving environments and conditions. Furthermore, integrating multiple sensors and control technologies will be necessary to monitor real-time driving situations and develop improved control algorithms based on this data. Ultimately, applying these research findings to real road driving will be essential for enhancing safety and efficiency.

Acknowledgements This work was supported by the Technology Innovation Program (20014938, Development of autonomous chassis platform for a modular vehicle) and (RS-2024-00468436, Development of implementation and validation technology for 800V EV power train inverter using die embedded power module) funded by the Ministry of Trade, Industry & Energy (MOTIE, KOREA).

References

1. Moon, C., Kim, J.: The future of chassis technology: e-Corner module. Journal of the KSME **62**(12), 32–36 (2022)
2. Kim, W.-S., Hong, J., Lee, D., Yeo, S.: A study on steering performance using various independent steering modes with a steer-by-wire system. In: Proceedings of the Korean Society of Automotive Engineers Fall Conference and Exhibition, Gyeonggi, South Korea (2012)
3. Jeon, D.S.: Study about skid steer of Independent 4wheel drive vehicle. Master's thesis, Korea University, Graduate School of Automotive Engineering, Seoul (2009)

The Application of Vehicle-Mounted Drone—A Review

Yang Yanyong, Mohd Faisal Hushim, and Allyssa Amanda Anak Jihill

Abstract With the continuous development of drone technology, drones are gradually becoming an emerging element in various industries, offering a range of innovative solutions. Their application scenarios include real-time traffic monitoring, outdoor rescue and search operations, geographic surveying, aerial photography for entertainment and tourism, and live broadcasting of large events. Literature analysis reveals that these applications involve key technologies such as remote sensing and data transmission. Furthermore, drones are often transported to outdoor environments by users via vehicles, leading to the emergence of the concept of "vehicle-mounted drones." The vehicle-mounted drone cabin provides a solution for carrying and conveniently using drones with the vehicle, integrating them into the vehicle for easy takeoff and recovery at any time. This paper analyzes the challenges faced by drones in practical applications, such as the convenience of carrying drones with the vehicle, the aerodynamic performance of the cabin, the installation design of the vehicle-mounted drone cabin, the operation of drone takeoff and landing, the transmission and storage of aerial data, and outdoor charging. The aim is to provide valuable references for the application research of vehicle-mounted drones.

Keywords Vehicle-mounted drone · Aerodynamics · Drone cabin · Drone takeoff and landing · Remote sensing technology

Y. Yanyong · M. F. Hushim (✉) · A. A. A. Jihill
Faculty of Engineering Technology, Department of Mechanical Engineering Technology, Universiti Tun Hussein Onn Malaysia, Pagoh, Johor, Malaysia
e-mail: mdfaisal@uthm.edu.my

Y. Yanyong
Faculty of New Energy, Department of New Energy Vehicle Technology, Zhuhai City Polytechnic, Zhuhai, Guangdong, China

M. Awang et al. (eds.), *Proceedings of the International Conference on Mechanical, Manufacturing and Plant Engineering*, Proceedings in Technology Transfer,
https://doi.org/10.1007/978-981-95-7817-7_14

1 Introduction

In recent years, small drones have increasingly been applied across various fields, such as geological disaster monitoring, crop pest and disease diagnostics, and real-time traffic condition monitoring. In the traffic monitoring domain, drones extract traffic parameters through actual traffic observation, providing research data for traffic signal control, traffic emissions, and driving behavior studies [1]. Additionally, vehicle-mounted drones play a crucial role in outdoor rescue operations. When equipped with drones, rescue vehicles can rapidly deploy them at disaster sites to assist in locating trapped individuals, surveying terrain, or transmitting real-time images to command centers, thereby gaining valuable time for rescue operations. This is particularly beneficial in areas that are difficult to access, such as mountainous or forested regions, where vehicle-mounted drones can quickly cover large areas.

As self-driving travel becomes an increasingly popular mode of transportation, the number of items carried has grown, making it essential to find ways to increase vehicle storage capacity due to limited interior space. Installing roof racks is a common solution to this issue. A well-designed roof rack or cargo box becomes a primary means to expand storage space [2].

With the rise of self-driving tourism, more enthusiasts are bringing drones for aerial photography along their journeys. Although the applications do not necessarily require carrying drones, the convenience of loading and unloading drones has become a critical issue in drone usage. As more people opt to bring drones for aerial photography, the challenge of carrying drones with the vehicle has emerged as a technical challenge. During travel, drones need to be frequently used, and operations often require stopping along the route, making the process of retrieving and storing them cumbersome. Therefore, it is necessary to design a platform that facilitates easy transport and operation of drones. The vehicle-mounted drone cabin should include features for storage, charging, and the capability to perform takeoff and landing without requiring a stop during low-speed travel. Such a platform can significantly enhance the efficiency of drone usage.

In 2013, the Ministry of Industry and Information Technology of China issued Announcement No. [2013] 52, promulgating the QC/948–2013 "Automotive Roof Load Device" industry standard, which applies to roof load devices for M1 and G class road vehicles. The regulations permit the installation of load-carrying devices on the roof of vehicles, providing convenience for self-driving travelers. However, since roof load devices alter the vehicle's wind resistance area and drag coefficient, they create air disturbances at high speeds, increasing driving resistance and raising higher design requirements for load-carrying devices [3]. This industry standard provides a basis for the design and application of vehicle-mounted drone cabins, and with the widespread use of vehicle-mounted drones, roof-mounted drone cabins will become one of the auxiliary devices for enhancing drone convenience.

2 Overview of Main Application of Drones

The integration of drones with remote sensing technology is referred to as drone remote sensing. Compared to traditional satellite and manned aircraft remote sensing, drone remote sensing offers advantages such as high timeliness, high spatial and temporal resolution, low-altitude flight capability, and high maneuverability. These characteristics enable drone remote sensing to surpass traditional methods in many aspects. Leveraging these advantages, drone remote sensing has rapidly developed in fields such as ecology and conservation biology. Monitoring tasks based on lightweight drones are generally divided into three phases: preparatory work, data acquisition, and subsequent data processing and analysis [4]. For ecological monitoring and early warning, the timely transmission and storage of data are crucial, with real-time data transmission and storage being key aspects of drone monitoring.

Drones equipped with remote sensing devices are cost-effective, highly maneuverable, and capable of flexible, timely, and high-resolution data collection, making them widely used in disaster emergency response and environmental monitoring in mining areas [5]. Given the extended duration of outdoor operations, the endurance of drones and the convenience of frequent storage are particularly important. Therefore, improving the portability and charging convenience of drones is one of the critical issues that need to be addressed for long-term outdoor applications.

Currently, equipment used for geological disaster monitoring includes crack detectors, landslide extensometers, and high-precision GPS devices. However, the high cost of these specialized devices limits their widespread application. Utilizing drones equipped with various remote sensing sensors for data collection provides a more practical solution for monitoring and early warning of geological disasters [6]. In outdoor geological disaster detection and early warning, staff typically carry drones with their vehicles, but there is currently no highly versatile vehicle-mounted drone cabin to meet such needs. Enhancing the convenience of carrying and charging drones with vehicles will help improve the efficiency of drone applications in geological disaster detection and early warning.

Compared to high-altitude remote sensing, low-altitude remote sensing offers advantages such as lower cost, greater flexibility, and rapid data acquisition, demonstrating unique benefits in crop pest and disease detection. As a key component of low-altitude remote sensing, drone-based agricultural remote sensing has significantly expanded the scope of agricultural monitoring. Small drones equipped with remote sensing devices have become essential tools for detecting crop pests and diseases [7].

Drone-based disaster emergency measurement systems are primarily used for rapid exploration of disaster situations, acquiring spatial information about damaged objects, and providing spatial reference and communication services for emergency coordination. Currently, drones equipped with optical sensors are used internationally for environmental monitoring and disaster investigation [8]. In disaster detection and spatial communication, operators need to frequently take off and land drones, relying on stable power supply devices to recharge drones, thereby enhancing their endurance and the continuity of outdoor operations.

New mobile communication networks are evolving towards diversification, integration, and intelligence. Traditional mobile communication networks mainly depend on infrastructure construction to maintain system deployment and operation, which increases operational costs. In contrast, new mobile self-organizing networks allow user terminals such as ground vehicles, drones, ships, and unmanned sensors to move freely within the network while maintaining communication. These networks are widely used in precision agriculture, network coverage, and target tracking [9]. In complex terrains such as mountainous regions and areas with limited transportation access, drones, with their excellent maneuverability, have become important tools in scenarios where vehicles and ships cannot reach.

3 Technology Related to Vehicle-Mounted Drone

3.1 Drone Technology

Drones, with their high mobility and flexible deployment advantages, offer effective solutions for various commercial and civil applications such as traffic management, remote monitoring, and video streaming. In future Intelligent Transportation Systems (ITS) scenarios, the integration of drones to assist vehicular communication networks demonstrates advantages that traditional ground-based vehicular networks cannot match. Fan Qiying proposed the concept of collaborative communication between drones and ground vehicles, establishing a drone-assisted air-ground integrated vehicular network system. The inclusion of drones not only enhances network throughput and service quality but also introduces higher costs and energy consumption issues, making energy efficiency a key focus in drone application research [10]. The Chinese patent ZL202230803795.5 designs a vehicle-mounted drone cabin that allows drones to be carried and deployed with the vehicle, equipped with an onboard charging device to provide continuous power support and address energy consumption issues.

Figure 1 illustrates the drone compartment within the vehicle and its condition when mounted on the roof. While the vehicle is in motion, the drone remains stored in the compartment. When the drone is prepared for takeoff, the two doors of the compartment open, allowing the drone to launch. The drone can either follow the vehicle during flight or perform its designated mission. Upon completing its mission, the drone returns to the compartment, lands on the platform inside, and the doors are then closed to finalize the operation.

In stable drone flight, gyroscopes, accelerometers, and magnetometers are the primary sensors for obtaining attitude information and are crucial for ensuring smooth flight. However, gyroscopes are prone to temperature drift and integral computation errors that can lead to accumulated inaccuracies; accelerometers can be affected

Fig. 1 Vehicle-mounted drone compartment of Patent Design No.ZL202230803795.5

by vibrations; and magnetometers are susceptible to interference from surrounding magnetic fields, all of which complicate precise attitude calculation. The attitude estimation algorithm based on Mahony-EKF, proposed by Shu Kaibin, effectively suppresses gyroscope drift errors, filters high-frequency noise from accelerometers, and reduces non-gravitational acceleration interference, thus improving the accuracy of attitude estimation [11]. This algorithm provides strong support for drone control stability, enhancing takeoff and landing stability. When the vehicle-mounted drone cabin is activated during vehicle stops or low-speed driving, potential impacts may occur when the drone lands on the cabin platform. This algorithm can improve the stability of drone landings and reduce the impact of landing shocks.

3.2 *Aerodynamics*

The design of aerodynamic kits should minimize air resistance while providing adequate lift. Using Ansys for numerical simulation analysis of 3D models created in Catia can provide reference for optimizing the design of models, thereby improving vehicle aerodynamic performance [12]. When installing a vehicle-mounted drone cabin on the roof, its shape, similar to that of a roof cargo box, may affect the vehicle's aerodynamic performance. Numerical simulation analysis of the 3D model of the vehicle-mounted drone cabin using Ansys can help achieve a reasonable shape design, providing a basis for the aerodynamic performance of the cabin.

To reduce overall vehicle, drag and improve endurance, STAR-CCM+ software can be used for simulations to identify areas of wind resistance in the base vehicle model that can be optimized. Optimization analyses of features such as front nose downforce, rear window side narrowing, rear wing downforce, and overall rear body narrowing lead to a model with reduced drag [13]. These optimized designs offer a technical approach to reducing drag for the vehicle-mounted drone cabin. Adjustments to the cabin's shape using CATIA software and subsequent simulation analysis can yield an optimal design for the vehicle-mounted drone cabin.

Using CATIA software to establish the basic shape of the vehicle body and design aerodynamic kits such as front nose wings, diffusers, and rear wings, and employing FLUENT software for flow field analysis of the complete vehicle model, can optimize vehicle aerodynamic performance. Increasing downforce can enhance overall vehicle stability [14]. Properly adding aerodynamic kits can improve vehicle stability. When installing a vehicle-mounted drone cabin on the roof, its shape may generate lift; therefore, a reasonable design or the addition of aerodynamic kits can improve the cabin's aerodynamic performance and enhance vehicle stability, providing technical support for carrying drones with vehicles.

Wang Shiquan used university formula racing cars as examples, creating 3D models of aerodynamic kit components in CATIA and conducting flow field analysis using the Fluent module of ANSYS software. This significantly improved cornering speed and driving stability of the racing cars while achieving weight reduction under the constraints of safety and reliability [15]. This design method provides a reference for the shape and mounting bracket design of the vehicle-mounted drone cabin.

3.3 Structural Design Analysis Methods for Drone Cabins

Li Siqiang utilized Digital Mock-Up (DMU) in CATIA to simulate the motion of landing gear retraction and extension processes, using sensor collision testing functions to detect interferences among parts during motion and analyzing the minimum gap between the landing gear and cabin door during motion. This helped visualize the swept volume of the landing gear during movement, perform interference checks, and determine the assembly position of the mechanism [16]. Dynamic simulation provided support for the design of similar landing gear assemblies.

Pang Zhanxiang used interference check and physical simulation tools provided by CATIA to perform static and dynamic collision checks on a humanoid robotic arm assembly model, providing a basis for model modifications [17]. The finger movement of the humanoid robotic arm has similarities with the opening and closing drive mechanism of the vehicle-mounted drone cabin door. Based on CATIA/ADAMS modeling and dynamic simulation technology, this can be used to analyze the dynamic characteristics of the drive mechanism for the vehicle-mounted drone cabin door.

Worm gear drives are widely used due to their compact structure, high transmission ratio, and smooth transmission. Wang Min used CATIA software to accurately model the secondary worm gear of a mining valve actuator and used the model for finite element analysis with ANSYS and dynamic analysis with ADAMS. The comparative results verified the accuracy of the modeling method and calculations, providing a reference for product development [18]. The drive mechanism for the vehicle-mounted drone cabin door designed in Chinese patent ZL202230803795.5 can employ worm gear drive for door operation. Using these methods, finite element analysis of the drive mechanism can be conducted to obtain a reasonable structural design for the vehicle-mounted drone cabin.

4 Problems Need to be Solve

4.1 Aerodynamic Performance of the Drone Cabin

Liu Zhendong conducted loading tests on two types of roof cargo carriers and analyzed their aerodynamic drag and lift at different vehicle speeds using ANSYS software. Additionally, tests were performed on a specific SUV model to assess the reliability and potential failure modes of these two devices. The results indicated that a roof cargo basket is more suitable for durability tests of roof cargo carriers, while a vehicle equipped with a cargo box showed decreased safety at high speeds due to increased lift. In contrast, a vehicle equipped with a cargo basket exhibited better handling and stability, and higher safety. Test results also showed that cargo boxes are more prone to damage, indicating lower safety. Therefore, for cases requiring a cargo box to store vehicle items, the design of the cargo box must focus on two key issues: stability at high speeds and structural strength [19]. Aerodynamic performance optimization can enhance stability at high speeds, while structural analysis can improve the cargo box's structural strength.

4.2 Noise Control of the Drone Cabin Mounting Bracket

The noise sources of automotive roof racks mainly consist of dipole sound sources, with quadrupole sound sources as a secondary contributor. When the crossbars of the roof rack are close to the roof, the near wall effect significantly impacts the airflow, and the roof compresses the airflow below the rack, leading to an increase in the tailflow area of the crossbars. Additionally, the presence of the roof rack affects the attachment of airflow to the roof, which impacts the overall vehicle's wake. By establishing a collaborative optimization algorithm mathematical model and calculating the collaborative optimization model under different weight ratios, an optimized design for the roof rack crossbars can be obtained [20]. The mounting bracket for the vehicle-mounted drone cabin is compatible with the crossbars of roof cargo boxes, providing a reference for the installation of the vehicle-mounted drone cabin.

4.3 Stability Control of Drone Takeoff and Landing

During the takeoff and landing process, the vehicle's position can affect the levelness of the drone's landing platform, leading to instability. Zhang Yanzhi designed a three-degree-of-freedom leveling platform that can serve as a leveling device for the vehicle-mounted drone, allowing the platform to adjust its level both statically and dynamically within a 22° tilt angle of the vehicle's orientation, ensuring safe takeoff and landing of the drone under various conditions. This platform expands the application scenarios of three-degree-of-freedom motion platforms, providing structural design and system principle support for the safe takeoff and landing of vehicle-mounted drones [21].

Frequent takeoffs and landings may result in collisions with the ground or landing platform, especially when a drone lands on a cabin platform while the vehicle is moving slowly, which can cause impacts and damage. Deng Xiaolong used CATIA software for 3D modeling of a cup and simulated the free-fall collision process with the ground using ANSYS Workbench software to analyze maximum stress and displacement [22]. This method provides a reference for impact simulation analysis during drone landings, and using CATIA modeling and ANSYS simulation to analyze drone landing processes offers technical support for the design and application of vehicle-mounted drone cabins.

4.4 Transmission and Storage of Drone Imagery Data

In practical applications, small drones face challenges with remote sensing technology, such as data quality degradation due to weather conditions. Drones are suitable for obtaining high-resolution remote sensing data over small areas and can be applied in fields such as urban planning, pollution monitoring, and land resource surveys [23].

Due to weather or other uncontrollable factors, there is a risk of drone communication loss or data loss. The security of drone imagery data is a critical issue in drone applications. The real-time transmission module mounted in the vehicle-mounted drone cabin can store aerial data in the cabin's storage module in real-time, effectively mitigating data loss issues caused by drone communication loss or crashes.

5 Conclusion

This paper aims to investigate the technologies related to vehicle-mounted drones, focusing on the challenges of carrying, takeoff, landing, charging, and data transmission and storage. The paper analyzes advancements in relevant technologies and discusses future development trends for vehicle-mounted drones. The main findings and conclusions are as follows:

1. Convenience of Outdoor Use: The application of drone aerial photography and remote sensing technology in various fields involves the convenience of outdoor carrying. The process of frequently taking out and putting away drones outdoors can be cumbersome, while using a vehicle-mounted drone cabin can simplify this process. The cabin allows for takeoff by simply opening the door and storing the drone by closing the door after landing. The vehicle-mounted drone cabin design outlined in Chinese patent ZL202230803795.5 provides a reference design for this purpose.
2. Charging During Extended Flights: For long-duration flights, a dedicated charging module is required in the vehicle-mounted drone cabin to address charging issues. This module can charge spare batteries while the drone is flying, ensuring a continuous power supply for extended operations.
3. Data Transmission and Storage: The risk of communication loss or crashes with drones underscores the importance of timely transmission and storage of captured data to prevent data loss. Reducing the cost of real-time transmission of large volumes of drone data remains a challenge.
4. Aerodynamic Design: The appearance of the vehicle-mounted drone cabin must meet aerodynamic performance requirements to ensure high-speed driving performance. Relevant studies, such as those by Ma Jian, Rongxin Chen, Quanjun Wu, and Wang Shiquan, provide useful references for this.
5. Design of Takeoff and Landing Mechanisms: The design for vehicle-mounted drone cabin takeoff and landing simplifies the drone usage process but places higher demands on the door opening and closing drive mechanism. This mechanism is similar to the landing gear doors of aircraft, and research by Pang Zhanxiang and Wang Min offers valuable references for the design of these drive mechanisms.
6. Impact During Landing: Drones may experience collisions or impacts with the platform during landing at low vehicle speeds. Deng Xiaolong's method provides a reference for simulating and analyzing impacts between the drone and the ground or landing platform during landing.

References

1. Liu, J., Liu, X., Qiu, J., Yi, Y.: A Lightweight Traffic Parameter Extraction Model for UAV Aerial Photography Videos (2023)
2. Luo, S., Deng, X., Xu, L., Tie, S., Liang, G.: Research on the Design and Development of Car Roof Racks (2021)
3. Industry Standard: Roof Load Carriers for Road Vehicles (2013)
4. Hu, J., Zhang, J.: Advances in UAV remote sensing applications in ecology. Acta Ecol. Sin. **38**(1), 1–10 (2018)
5. Wang, K., et al.: Current status and development trends of UAV remote sensing in the mining industry. Chin. J. Eng. **42**(9), 1085–1095 (2020)

6. Han, W., Ren, Y., Zhao, S.: Major Applications of UAV Remote Sensing in Geological Disaster Response (2011)
7. Lan, Y., Deng, X., Zeng, G.: Research progress on UAV agricultural remote sensing for crop pest and disease diagnosis. Smart Agric. **1**(2), 1–18 (2019)
8. Shen, Y., Liu, J., Wu, L., Li, F., Wang, Z.: Disaster scene reconstruction method based on UAV images and flight control data. Geogr. Geo-Inf. Sci. **27**(6), 13–17 (2011)
9. Dong, C., et al.: A survey on media access control protocols for UAV self-organizing networks and vehicular networks. Dianzi Yu Xinxi Xuebao/J. Electron. Inf. Technol Sci. Press, 790–802 (2022)
10. Xiying, F., et al.: Node access mechanism for interference perception in UAV-assisted vehicular networks. J. Commun. **40**(6), 90–101 (2019)
11. Kaibin, C., Shuang, Z., Chengtao, F.: UAV attitude calculation algorithm based on Mahony-EKF. J. Electron. Meas. Instrum. (2019)
12. Ma, J., Wang, W., Wang, K.: Modeling and flow field analysis of aerodynamic kits for formula student racing cars. Sci. Technol. Inf. **1**(7), 24–25 (2015)
13. Rongxin, C., et al.: Simulation and research on aerodynamic drag reduction of a vehicle surface. Agric. Equip. Veh. Eng. **62**(4), 67–69 (2024)
14. Quanjun, W., et al.: Design and flow field analysis of aerodynamic kits for FSC racing cars. Automot. Appl. Technol. **18**(20), 98–102 (2022)
15. Wang, S., Zhang, Y., Ren, H.: Design of aerodynamic kits for formula student racing cars. Design. **1**(13), 8–10 (2020)
16. Li, S., Zhang, L.: Application of CATIA dynamic simulation in landing gear assembly. Sci. Technol. Vis. **11**(23), 62–64 (2021)
17. Pang, Z., Liu, T., Gong, L., Jiang, N., Jiang, D., Li, L.: Modeling and dynamic simulation of humanoid robotic hands based on CATIA/ADAMS. J. Chang. Univ. Technol. (Nat. Sci. Ed). **34**(3), 264–267 (2013)
18. Wang, M., Wan, C., Wang, W., Du, H., Su, Z.: Study and application of a worm gear modeling method based on CATIA. Mod. Manuf. Eng. **1**(7), 60–64 (2021)
19. Zhendong, L., et al.: Aerodynamic analysis and testing of roof rack loading devices. Automot. Pract. Technol. **1**(5), 133–135 (2020)
20. Lai, C., Jia, H., Fu, H., Ling, Z., Zhou, Y.: Drag reduction and noise reduction synergistic optimization of a certain SUV roof rack. J. Chongqing Univ. Technol. (Nat. Sci.). **35**(10), 23–32 (2012)
21. Zhang, Y.: Research on a three-degree-of-freedom vehicle-mounted UAV lift and leveling device (2021)
22. Deng, X., Feng, G., Li, P.: Dynamic analysis of free-fall collision of a cup based on CATIA modeling and workbench simulation. J. Chengde Pet. Coll. **17**(5), 39–42 (2015)
23. Hong, Y., Gong, J., Hu, S., Huang, M.: Discussion on UAV remote sensing image acquisition and subsequent processing. Remote. Sens. Technol. Appl. **23**(4), 98–102 (2008)

Aqueous-Organic Flow Batteries for a Sustainable European Energy Industry

Barun Kumar Chakrabarti, Andres Parra-Puerto, Claudia Mair-Bauernfeind, Julia Wenger, Aldo Bischi, Kavin Teenakul, Jonas Hereijgers, Serap Hayat Soytas, Torsten Gutmann, Bernhard Marius, J. Rubio-Garcia, Tülay Yilmaz Inan, Mustafa Kemal Bayazit, Ülrich Hirn, and Stefan Spirk

Abstract In recent years, the pursuit of sustainable energy storage solutions has brought aqueous organic redox flow batteries (AORFBs) to the forefront of scientific research and potential industrial application. This perspective critically examines the

---

B. K. Chakrabarti (✉) · S. H. Soytas · T. Y. Inan · M. K. Bayazit
Sabanci University Nanotechnology Research and Application Centre (SUNUM), Tuzla, Istanbul, Turkey
e-mail: barun.chakrabarti@sabanciuniv.edu

A. Parra-Puerto
Department of Physical Chemistry and Electrochemistry Institute, University of Alicante, San Vicente del Raspeig, Spain

C. Mair-Bauernfeind · J. Wenger
Department of Environmental Systems Sciences, University of Graz, Graz, Austria

J. Wenger
Science, Technology and Society Unit, Graz University of Technology, Graz, Austria

A. Bischi
Department of Energy, Systems, Territory and Construction, DESTEC, Universitá di Pisa, Largo Lucio Lazzarino, Pisa, Italy

K. Teenakul · J. Hereijgers
Research Group Applied Electrochemistry & Catalysis, University of Antwerp, Wilrijk, Belgium

T. Gutmann
TU Darmstadt, Eduard-Zintl-Intitute for Inorganic and Physical Chemistry, Darmstadt, Germany

B. Marius · S. Spirk
Ecolyte GmbH, Graz, TU, Austria
e-mail: stefan.spirk@tugraz.at

J. Rubio-Garcia
Department of Energy Storage, Schlumberger Cambridge Research, Cambridge, UK

M. K. Bayazit
Faculty of Engineering and Natural Science, Sabanci University, Istanbul, Turkey

M. Awang et al. (eds.), *Proceedings of the International Conference on Mechanical, Manufacturing and Plant Engineering*, Proceedings in Technology Transfer,
https://doi.org/10.1007/978-981-95-7817-7_15

current literature on AORFBs, with a particular focus on sustainability and performance aspects. We begin by providing an overview of AORFBs and their hybrid (HRFB) technologies, encompassing recent advancements and trends, with a special emphasis on minimizing critical raw materials (redox active species, non-fluorinated membranes, electrodes, etc.) utilized in these systems. A comprehensive discussion on life cycle assessment and circularity of AORFBs and HRFBs with respect to alternative energy storage systems and other RFB configurations is provided. Hereby, an emphasis is placed on the critical reflection of the comparability of different systems, the uncertainties, trade-offs as well as methodological and other challenges. The perspective also summarizes the current state-of-the-art into technoeconomic analyses (TEA), offering insights into cost-effectiveness, scalability potential and their position in the energy storage market. We explore the intersection of technology and policy by discussing the legal and regulatory frameworks shaping the development and deployment of AORFBs. Through this holistic approach, we aim to advance the understanding of AORFBs' role in fostering a sustainable energy future, guiding both academic research, industrial innovation and policy making.

Keywords Aqueous organic redox flow batteries · Hydrogen · Grid-balancing · Circularity · Technoeconomic analysis

1 Introduction

Redox flow batteries (RFBs) can play a critical role for grid-scale energy storage. They have significant potential to minimize grid congestion while supporting renewable energy intermittency [1], providing energy and price security as well as mitigating energy crises in the EU. The sustainable production of redox-active molecules from renewable resources represents a key challenge and opportunity in the development of environmentally friendly biotechnological processes to support the technology to reach the targets set in the SET Plan (the energy storage cost needs to fall below 0.05 €/kWh/cycle by 2030). Amongst these molecules, quinones and their precursors—such as vanillin and structurally related aromatic aldehydes—are of particular interest due to their broad utility in pharmaceuticals, food additives, polymers, and can easily be transformed via biosynthetic approaches into redox active species for aqueous organic redox flow batteries (AORFBs) [2, 3]. Moreover, quinones and their precursors—such as vanillin—are frequently encountered as intermediates in various biobased processes, particularly those involving lignin depolymerization. Such compounds are commonly generated as byproducts in industrial sectors like the pulp and paper industry, where lignin-rich biomass is processed at scale [4].

Ü. Hirn · S. Spirk
Graz University of Technology—Institute of Bioproducts and Paper Technology, Graz, Austria

Despite the potentially lower cost of active materials in AORFBs, their overall system costs remain higher than those of VRFBs, primarily due to their inferior power and energy density characteristics [5, 6]. Thus, the economic viability of AORFBs will critically depend on advancing performance parameters—namely energy density, power density, and long-term durability with respect to both cycling and calendar life—while simultaneously reducing operational and maintenance expenditures [5, 7]. To address these limitations, hybrid system architectures such as regenerative fuel cells (RFCs) or hybrid RFBs (HRFBs) have been proposed. By combining the favorable redox kinetics of AORFBs with the efficient hydrogen-based reactions of RFC configurations, these hybrid approaches aim to enhance energy and power densities, improve round-trip efficiency, and alleviate key barriers that currently constrain the large-scale deployment and scalability of AORFBs. Notably, the first H2–quinone RFC, reported by the Aziz group [8], demonstrated the feasibility of this hybrid concept but also exhibited modest energy efficiency and limited cycling stability, largely due to quinone solubility constraints and crossover-related degradation. These findings highlight the need for improved molecular design and system integration strategies to unlock the full potential of H2–quinone hybrids. In short, it is envisaged that green hydrogen-based HRFBs in combination with quinones will not only advance technological competitiveness but also directly support the objectives of the **European Green Deal** and the **EU Hydrogen Strategy**, thereby contributing to long-term energy security, sustainability, and economic resilience in the Union. Importantly, positioning HRFBs as a next-generation storage technology represents a timely research and innovation opportunity, fully aligned with EU funding priorities aimed at accelerating the transition to a climate-neutral energy system.

2 AORFB Materials and Their Key Performance Indicators (KPIs)

AORFBs, despite their advantages, face challenges in scaling up, with limited large-scale deployment and competition from cheaper, more established Li-ion batteries [9, 10]. While the most mature RFB systems rely on vanadium, a critical raw material with supply and sustainability concerns, AORFBs offer a promising alternative by potentially reducing reliance on such materials [11]. By exploring different redox chemistries and extending the electrochemical window of water, AORFBs can achieve cost reductions and improved efficiency, making them a viable option to reduce the EU's dependence on critical raw materials and contribute to a more sustainable energy storage future.

To assess the effectiveness and scalability of AORFBs, several KPIs are used. These include energy efficiency, cycle life, specific capacity, and power density, which determine the overall performance and longevity of the system. Additionally, energy density, self-discharge rate, and voltage efficiency reflect the storage capability and operational stability of the batteries. Redox reaction reversibility and membrane/electrolyte stability are critical for ensuring minimal degradation over time, while the levelized cost of storage (LCOS) and scalability highlight the economic feasibility and potential for large-scale deployment. By optimizing these KPIs, AORFBs can enhance their competitiveness, offering a promising solution for reducing reliance on critical raw materials and advancing sustainable energy storage.

Quinone electrolytes, including hydroquinone and anthraquinone derivatives, play a critical role in AORFBs due to their reversible redox reactions, high solubility in electrolytes, and electrochemical stability [9, 10]. Various redox-active species, including benzoquinone and anthraquinone derivatives, viologens, TEMPO, and phenazines, have been explored for use in AORFBs (a brief summary of KPIs for some selected systems is highlighted in Table 1). To improve their solubility, stability, electrochemical reversibility, and performance at high charge rates, molecular design strategies have been employed [12]. In addition, ion exchange membranes (IEMs) facilitate the conduction of charge-carrying ions while preventing cross-contamination of redox-active species and potential short circuits. Despite their critical role, IEMs are less studied at the industrial scale compared to redox-active materials (Fig. 1) [13]. Current IEMs, originally designed for applications like fuel cells [14],[14] are not optimized for AORFB conditions. However, IEMs specifically tailored for AORFB chemistries are likely to outperform existing commercial membranes in the near future [9, 10].

3 Sustainability Aspects of AORFBs

Achieving sustainable patterns of consumption and production requires a holistic understanding of potential improvements early in the development of technology lifecycle. Life cycle-based approaches, such as Life Cycle Assessment (LCA), Social Life Cycle Assessment (SLCA), and Life Cycle Costing (LCC), are essential tools for integrating sustainability into product design and evaluating the broader impacts of emerging technologies [19]. These are structured methods that provide valuable insights into the potential environmental, social, and economic consequences of new technologies, helping to identify risks and opportunities for enhancing sustainability from the outset [20].

In the case of AORFBs, sustainability considerations are critical given their potential to serve as an alternative to traditional VRFBs, which rely on scarce and environmentally concerning materials. AORFBs offer a promising solution by reducing dependence on critical raw materials while maintaining high performance. However, assessing their sustainability remains a challenge due to the early stages of development and the inherent uncertainties in projecting their long-term impacts. At these early technological readiness levels (TRLs), data is often sparse, and assessments rely on assumptions or generic databases, making accurate predictions difficult and, at the same time, uncertainty analysis and continuous evaluation key [21]. Furthermore, as AORFBs scale up, energy demands and resource consumption may change,

Table 1 Performance parameters from different reported quinones used in RFC and RFB setups

Ref	Quinone	Potential/V	Operational current density/ mA cm^{-2}	Energy density/ mW cm^{-2}	Cycles	[M]/ mol L^{-1}	Capacity/A h L^{-1}	
							Experimental	Theoretical
[2]	2-methoxyhydroquinone	0.515	1.9	–	250	0.02	0.1	0.107
						1.8		27.1
[15]	1,2-dihydrobenzo-quinone-3,5-disulfonic acid	0.85	100	27.1	200	0.65	13.98	34.84
[16]	3,6-dihydroxy-2,4-dimethylbenzenesulfonic acid	0.652	100	15	30	1	38	53.6
[17]	Tetrasubstituted quinone	0.605	50	17	50	0.25	11.92	13.4
[18]						1		53.6

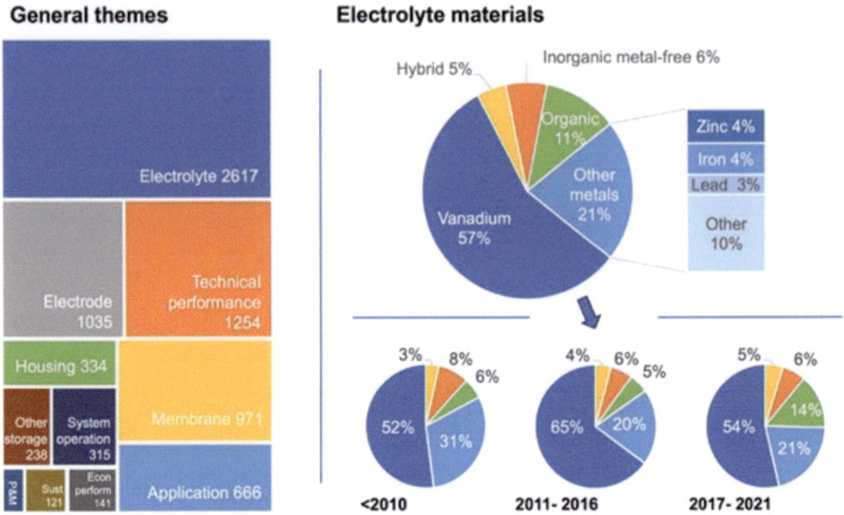

Fig. 1 A clear picture of the thematic distribution in the field of RFBs based on a keyword cluster analysis. (Sust: sustainability; P&M: policy and management). Reproduced with permission from Elsevier [13]

further complicating their sustainability evaluation as assumptions of potential future scenarios must be included, particularly given the potentially long service life of such battery systems [22].

The sustainability impacts of AORFBs span multiple dimensions: environmental, techno-economic, and social. A key aspect of evaluating these impacts is the need to balance trade-offs between different indicators. For instance, while AORFBs could reduce reliance on critical raw materials, such as vanadium, the manufacturing processes for these batteries still require significant energy and resources, which can offset some of the benefits [19]. The production energy demand for AORFBs, which may involve complex electrochemical reactions and materials processing, remains a major challenge for their long-term sustainability [23]. As a result, future improvements in energy efficiency during the production and operation of these batteries are essential to their competitiveness against established technologies, such as lithium-ion batteries, which have already benefited more from technological learning, economies of scale, optimization efforts, and related cost reductions [22].

Incorporating SLCA into the evaluation of AORFBs further highlights the importance of considering societal impacts. For example, raw material extraction, manufacturing practices, and labor conditions in the supply chain could have significant social consequences [24], especially in regions where critical materials are mined or processed under poor working conditions. Moreover, SLCA can help assess how AORFBs could contribute to or mitigate social inequalities by considering who benefits from technological innovations and who may bear the environmental or health costs [22].

An LCC perspective is also crucial for understanding the economic viability of AORFBs. This approach goes beyond capital costs and includes operational, maintenance, and end-of-life disposal costs. It can offer a clearer picture of the total cost over the lifespan of an AORFB system, allowing for more informed decisions about whether these technologies can compete with other energy storage solutions, such as lithium-ion batteries [22].

However, a significant challenge for AORFBs is the lack of standardized metrics and variability in results due to differences in assessment methods and LCA software [21, 25, 26]. Without standardized frameworks for sustainability assessments, comparing the performance of AORFBs against other technologies becomes more difficult, which can slow down their acceptance in both research and industrial contexts [25]. Furthermore, the absence of primary data for key life cycle stages, particularly in early TRLs, makes it harder to make reliable statements about the sustainability of AORFBs and to introduce improvements with confidence [22]. This highlights the need for further research and the development of standardized assessment protocols to enhance the accuracy and reliability of sustainability evaluations for AORFBs.

In the broader context of AORFB sustainability, Fig. 2 provides a visual representation of key sustainability aspects when employing EU-based raw materials, such as wood, for preparing quinone electrolytes [27]. As shown, the sustainability concerns span multiple stages from wood sourcing to AORFB production, including land use change, occupational safety risks, and energy-intensive processes. Each step introduces different challenges—such as regional value chains, multifunctional processes, and the energy demands of production—that must be addressed to ensure AORFBs can meet sustainability goals while scaling up.

In summary, while AORFBs show considerable promise as a sustainable energy storage solution due to their potential to reduce reliance on critical materials, their path to commercialization is not without challenges. High energy demands in production, uncertainties in early-stage assessments, and the need for standardized sustainability metrics must be addressed to fully realize their potential. The future of AORFBs depends on continued technological advancements, improvements in efficiency, a deeper understanding of their environmental, social, and economic impacts, as well as on the socio-technical systems they will be applied in. By focusing on these areas, AORFBs may evolve into a competitive, sustainable alternative in the energy storage market, supporting the transition to a more sustainable energy future.

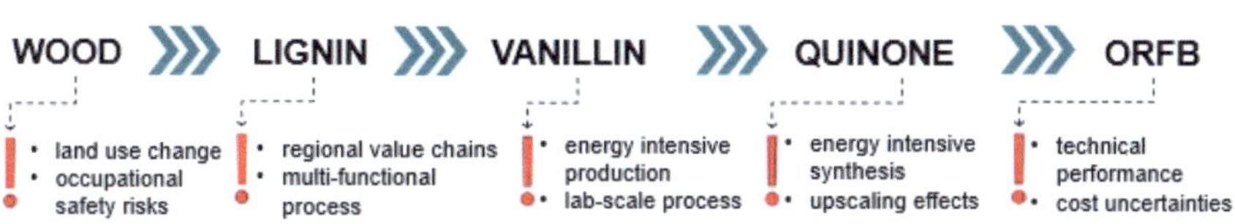

Fig. 2 Sustainability aspects of employing EU based raw materials like wood for preparing quinone electrolytes for AORFBs. Reproduced with permission from Springer Nature [27]

4 Technoeconomic Analysis (TEA)

The technoeconomic performance of AORFBs is evaluated based on key metrics such as LCOS (see Eq. 1) and overall scalability [21, 28]. Unlike other battery technologies, AORFBs may offer advantages in terms of cost-effectiveness and sustainability due to their unique use of organic materials in the electrolyte, which can significantly reduce material costs while providing good performance in large-scale applications.

$$LCOS = -(CC_{NPV} + OM_{PV} + tax_{NPV})\left(\sum_{n=1}^{life_{proj}} \frac{cycles_n}{(1+r)^n}\right)^{-1} + e_{loss} \quad (1)$$

Where $CC_{NPV} ==$ the formula for the net present value of capital investments in the system, $OM_{PV} =$ the present value of O&M (cost of operating and maintaining the system) per kWh of discharged energy in one cycle, $tax_{NPV} =$ an estimate of the net present value of the taxes, $life_{proj} =$ the life of the project from initial operation to the start of decommissioning, $r =$ the nominal discount rate, $n =$ the number of cycles and $e_{loss} =$ the levelized cost of energy loss due to round-trip efficiency [29].

Capital cost plays a critical role in determining the commercial viability of flow battery technologies. For AORFBs, the capital cost is influenced by factors such as the cost of organic electrolyte materials, system design (including stack, tanks, and balance-of-plant components), and the efficiency of the manufacturing process. The specific cost of AORFBs has been reported to be approximately 535 €/kWh for a system with a 4-h discharge time, as shown in Fig. 3 [5]. This places AORFBs at a cost-competitive level compared to other flow battery technologies like VRFBs, which have a reported cost of 438 €/kWh for the same discharge time. Despite the higher initial cost, AORFBs are considered to offer significant potential for cost reduction as manufacturing scales up and optimization of materials and processes takes place.

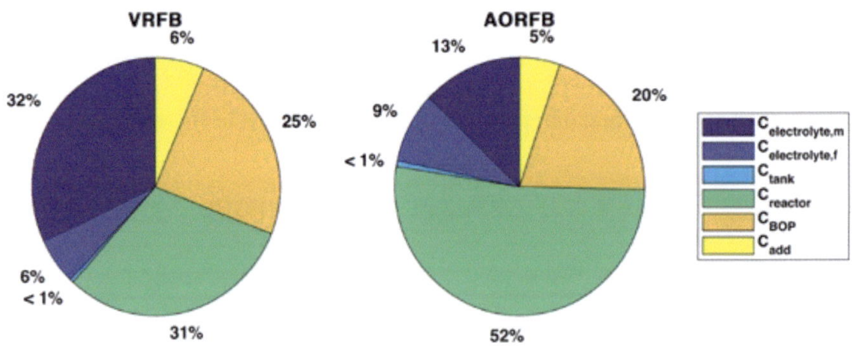

Fig. 3 Breakdown of the capital cost for a VRFB (total specific cost of 438 €/kWh) and an average generic AORFB (total specific cost of 535 €/kWh), for 4 h of discharge time. Reproduced with permission from Elsevier [5]

The LCOS is a key metric in technoeconomic analyses, representing the cost of storage as net present value per delivered-MWh over the system's lifetime, factoring in both capital and operational costs. In the case of AORFBs, LCOS is influenced by the efficiency of energy conversion, capacity fading, and electrolyte maintenance. While AORFBs may initially have higher operational costs due to more frequent maintenance and electrolyte replacement, their overall cost profile is expected to improve with advancements in material science and system design. Moreover, research has shown that optimization of organic materials—such as using more abundant and less expensive compounds—can dramatically reduce the operational costs, thereby improving the overall LCOS of AORFBs over time [5].

A significant challenge for AORFBs, as with other redox flow battery technologies, is efficiency fluctuations over time, which arise due to factors such as electrode degradation, electrolyte degradation, and system losses. These efficiency losses are particularly impactful in long-duration applications, where maintaining high efficiency over extended discharge periods is crucial. However, compared to traditional VRFBs, AORFBs have the potential to mitigate some of these losses through the use of organic electrolytes that are less prone to degradation and scaling. This could reduce the extent of capacity fade and improve the long-term performance of AORFBs [5].

One of the key innovations that could address these concerns is the development of solid booster materials or techniques that can enhance the performance and longevity of AORFB systems. By optimizing the electrolyte composition and improving charge/discharge efficiency, solid boosters have the potential to lower the cost per cycle, thus improving the system's overall economic performance. Research is ongoing to develop these technologies, and their incorporation into AORFBs may represent a crucial step toward making this technology competitive with other storage options [30].

Scalability is another important consideration for the commercialization of AORFBs. Given their modular nature, flow batteries are well-suited for large-scale applications, such as grid energy storage, where they can provide a stable and cost-effective solution for balancing supply and demand. AORFBs, with their reduced reliance on scarce and expensive materials such as vanadium (used in VRFBs), are particularly appealing for large-scale deployment in a variety of industries. As the market for AORFBs grows, economies of scale will likely drive further reductions in both capital and operational costs. Advances in organic materials, improved electrolyte management, and more efficient manufacturing processes are expected to bring down the total system cost, making AORFBs an increasingly attractive option for energy storage [31]. The ability to incorporate renewable and sustainable materials into AORFB systems also positions them as a highly attractive technology in the context of green energy and sustainability goals [2].

The technoeconomic analysis of AORFBs demonstrates that, while these systems may currently have a higher upfront cost compared to other flow batteries like VRFBs, their potential for cost reduction, scalability, supply chain security, and environmental sustainability makes them a promising candidate for large-scale energy storage.

Continued research and innovation, particularly in the optimization of organic materials and system efficiency, will likely enable AORFBs to achieve a cost-competitive position, with lower LCOS and better overall economic performance in the future.

5 Regenerative Fuel Cell (RFC) or Hybrid Redox Flow Battery (HRFB)

HRFBs are a promising solution for advanced energy storage systems that combine both liquid and gas half-cells, optimizing energy storage/conversion efficiency. By incorporating hydrogen into HRFB designs, these systems take advantage of the hydrogen evolution reaction (HER) and hydrogen oxidation reaction (HOR), which provide fast and reversible processes for both charging and discharging [15, 32]. This configuration results in a system that delivers high energy density (theoretically up to 3660 Wh kg^{-1}), exceptional charge/discharge efficiency (over 70%), and long operational lifespans, with lifetimes reaching up to 10,000 h. Furthermore, HRFBs are lightweight, environmentally friendly, and use abundant aqueous molecules as a fuel carrier, making them sustainable and cost-effective [33, 34].

As seen in Fig. 4, different configurations, such as hydrogen/vanadium (RHVFC) [35], hydrogen/manganese (RHMnFC) [36], and polysulfide/air (S-Air) [34], offer various advantages in terms of performance and cost reduction [37]. This approach not only enhances the flexibility of standard RFBs but also improves their overall performance. These improvements can make HRFBs a more viable option for large-scale energy storage systems within the EU. The integration of hydrogen into these systems allows for better energy density and faster response times, which are crucial for meeting the EU's evolving energy demands [15, 32].

Hydrogen-based HRFBs, which have higher energy density than traditional AORFBs, offer a compact and scalable storage solution [34, 35]. This is particularly important for large-scale applications in the EU, where energy storage must be adaptable to both industrial and residential needs. The modular nature of HRFBs means they can be easily scaled to meet the growing demand for clean energy storage solutions. Additionally, the use of hydrogen aligns well with the EU's sustainability goals, reducing reliance on fossil fuels and contributing to decarbonization efforts [15, 34].

Beyond sustainability, HRFBs offer several key benefits. For example, they can enhance grid stability by enabling the storage of excess renewable energy and its release during peak demand. This reduces the need for non-renewable backup sources, improving the overall efficiency and reliability of the grid. The long-term storage capability of HRFBs is also crucial for maintaining energy availability during times of low renewable generation, such as during calm weather or in winter months when sunlight is limited. However, some barriers to the widespread deployment of the technology remains and are associated with the high costs associated with the

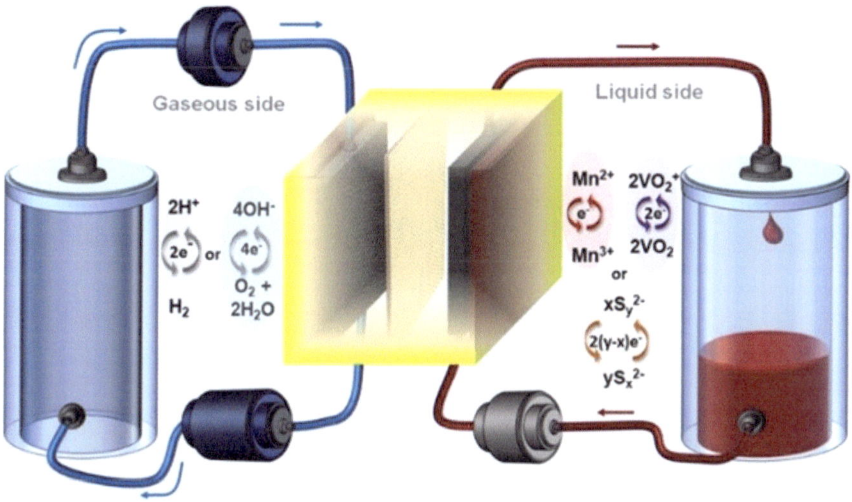

Fig. 4 Three types of HRFBs, namely, hydrogen/vanadium (RHVFC), hydrogen/manganese (RHMnFC), and polysulfide/air (S-Air) developed in recent times. Reproduced with permission from ACS [37]

anodic catalysts (usually platinum group of metals), hydrogen safety infrastructure, dependence on fluorinated Nafion-based IEMs and other factors that are discussed in a recent chapter [33].

6 Summary and Future Perspectives

In conclusion, this chapter has provided a detailed exploration of the technoeconomic landscape of AORFBs, highlighting their current status, challenges, and the potential pathways for future advancements. As we look ahead, the integration of AORFBs with green hydrogen presents a compelling avenue for the future of energy storage and conversion systems (Fig. 5). By leveraging the complementary strengths of both technologies, AORFBs could play a significant role in facilitating the transition to a low-carbon, sustainable energy future. Furthermore, the development of AORFBs, particularly within the context of the European Union's energy framework, holds significant promise. The EU's push toward achieving energy security and sustainability through its Green Deal and hydrogen strategies offers a fertile ground for the adoption and scaling of AORFB-based systems. The synergy between AORFBs and green hydrogen could help bridge the gap between renewable energy generation and storage, ensuring a reliable, low-emission energy supply for the future.

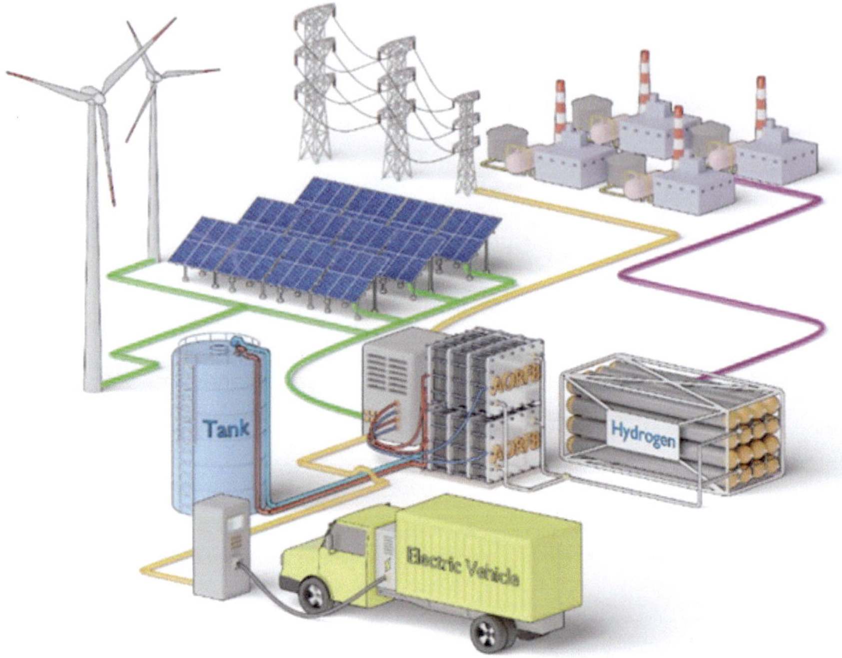

Fig. 5 Schematic predicting how green hydrogen can be integrated with a sustainable AORFB to efficiently store and release energy in an arbitrary EU-based country

Funding This work was supported by the European Innovation Council (EIC) under grant agreement No. 101115293 (VanillaFlow). This research was also funded by the CETPartnership, under the Joint Call 2023, through the project "HydrogEn oRganiC Regenerative FUeL CEll for Stationary Energy Storage (HERCULES)". The CETPartnership's research projects are co-funded by the European Commission (Grant Agreement No. 101069750) and the national TUBITAK 1071 via the project code: 224 N167.

References

1. Brandon, N., et al.: UK research needs in grid scale energy storage technologies. Energy Storage Res. Netw. (2016)
2. Schlemmer, W., et al.: Angew. Chem. Int. Ed. **59**, 22943–22946 (2020)
3. Kim, J., et al.: Nat. Rev. Mater. **8**, 54–70 (2023)
4. Wenger, J., et al.: Curr. For. Rep. **6**, 294–308 (2020)
5. Cremoncini, D., et al.: Appl. Energy. **360**, 122738 (2024)
6. Raihan, M.A., Dyker, C.A.: J. Energy Chem. **100**, 125–143 (2025)
7. Rodby, K.E., et al.: J. Power Sources. **506**, 230085 (2021)
8. Huskinson, B., et al.: ECS Trans. **53**, 101 (2013)
9. Amienghemhen, O.D., et al.: Chem. Pap. **78**, 7335–7374 (2024)
10. Tsehaye, M.T., et al.: Renew. Sust. Energ. Rev. **173**, 113059 (2023)

11. E. Commission, I. Directorate-General for Internal Market, Entrepreneurship, SMEs, B. G. Survey, B. d. R. G. e. Minières, D. Sustainability and TNO: Study on the Review of the List of Critical Raw Materials—Final Report. Publications Office (2017)
12. Li, D., et al.: Chem. Asian J. **20**, e202500020 (2025)
13. Yavorskaya, V., et al.: J. Energy Storage. **99**, 113093 (2024)
14. Seden, M.G., et al.: Int. J. Hydrog. Energy. **48**, 33297–33314 (2023)
15. Rubio-Garcia, J., et al.: J. Mater. Chem. A. **8**, 3933–3941 (2020)
16. Hoober-Burkhardt, L., et al.: J. Electrochem. Soc. **164**, A600 (2017)
17. Gerken, J.B., et al.: Adv. Energy Mater. **10**, 2000340 (2020)
18. Preger, Y., et al.: Joule. **2**, 2722–2731 (2018)
19. Sala, S., et al.: Int. J. Life Cycle Assess. **18**, 1653–1672 (2013)
20. Titirici, M., et al.: J. Phys.: Energy. **6**, 041502 (2024)
21. Emmel, D., et al.: Nat. Commun. **14**, 6672 (2023)
22. Scrucca, F., et al.: J. Clean. Prod. **268**, 122304 (2020)
23. Sacchi, R., et al.: Renew. Sust. Energ. Rev. **160**, 112311 (2022)
24. Mair-Bauernfeind, C., et al.: Int. J. Life Cycle Assess. **30**, 1281–1305 (2025)
25. Dieterle, M., et al.: Sustain. Energy Techn. Assess. **53**, 102457 (2022)
26. Wolf, C., et al.: Energy Sustain. Soc. **15**, 10 (2025)
27. Minke, C., Turek, T.: J. Power Sources. **376**, 66–81 (2018)
28. Hill Balliet, W., et al.: Energy Econ. **142**, 108174 (2025)
29. Gentil, S., et al.: Curr. Opin. Electrochem. **21**, 7–13 (2020)
30. Zhu, F., et al.: Green Energy Environ. **9**, 1641–1649 (2024)
31. Rubio-Garcia, J., et al.: Energy Storage Mater. **31**, 1–10 (2020)
32. Rubio-Garcia, J., et al.: In: Maniam, S. (ed.) Chapter 5: Batteries: The Future of Energy Storage, pp. 217–291. Jenny Stanford Publishing, Singapore (2024) https://doi.org/10.1201/9781003512882
33. Chakrabarti, B.K., et al.: ACS Appl. Nano Mater. **4**, 10754 (2021)
34. Chakrabarti, B., et al.: Int. J. Hydrog. Energy. **44**, 30093–30107 (2019)
35. Rubio-Garcia, J., et al.: Batteries. **9**, 108 (2023)
36. Chakrabarti, B.K., et al.: ACS Appl. Mater. Interf. **12**, 53869–53878 (2020)
37. Ebner, S., et al: ChemSusChem **16**, e202201818 (2023)

Electrophoretic Deposition of Nanomaterial Catalysts on Carbon Cloth Electrodes for Aqueous-Organic Flow Batteries

Reza Afshar Ghotli, Züleyha Kudaş, Shamik Chaudhuri, Süleyman Çelik, Tülay Yılmaz İnan, Serap Hayat Soytaş, Zehra Çobandede, Mustafa Kemal Bayazıt, Andres Parra Puerto, Ülrich Hirn, Stefan Spirk, and Barun Kumar Chakrabarti

Abstract The rising cost and volatility of vanadium have motivated the development of redox flow batteries (RFBs) based on sustainable, widespread availability and lower cost compared to metal-based alternatives. Organic molecules have emerged as attractive alternatives due to their structural tunability, redox properties and solubility, availability, and potential cost advantages. Quinone derivatives, in particular, exhibit fast, reversible redox kinetics in acidic media and are among the most extensively studied organic redox couples. Benzoquinone disulfonic acid (BQDS) has shown promising electrochemical stability and high aqueous solubility, making it a suitable positive electrolyte for hybrid hydrogen/quinone flow battery configurations. However, the electrochemical performance of such systems is strongly dependent on the electrode microstructure, surface area, and catalytic activity. In this study, carbon cloth (CC) electrodes were modified with pristine carbon nanotubes (CNTs), aligned

R. Afshar Ghotli (✉) · Z. Kudaş · S. Chaudhuri · S. Çelik · T. Y. İnan · S. H. Soytaş · Z. Çobandede · M. K. Bayazıt · B. K. Chakrabarti (✉)
Sabanci University Nanotechnology Research and Application Centre (SUNUM), Tuzla, İstanbul, Turkey
e-mail: reza.afsharghotli@sabanciuniv.edu

B. K. Chakrabarti
e-mail: barun.chakrabarti@sabanciuniv.edu

M. K. Bayazıt
Faculty of Engineering and Natural Science, Sabancı University, İstanbul, Turkey

Ü. Hirn · S. Spirk
Institute of Bioproducts and Paper Technology, Graz University of Technology, Graz, Austria

A. P. Puerto
Department of Physical Chemistry and Electrochemistry Institute, University of Alicante, San Vicente del Raspeig, Alicante, Spain

S. Spirk
Ecolyte GmbH, TU Graz Campus Inffeldgasse, Graz, Austria

M. Awang et al. (eds.), *Proceedings of the International Conference on Mechanical, Manufacturing and Plant Engineering*, Proceedings in Technology Transfer,
https://doi.org/10.1007/978-981-95-7817-7_16

CNTs (A-CNTs), and home-synthesized N-doped CNTs through a facile electrophoretic deposition (EPD) process. The resulting nanostructured electrodes exhibited improved surface coverage and enhanced electroactive surface area compared to unmodified CC, as confirmed by cyclic voltammetry. These findings highlight the potential of CNT-modified CC electrodes to improve charge transfer kinetics, increase electrolyte utilization, and enhance round-trip efficiency in aqueous–organic RFBs.

Keywords Electrophoretic deposition · Carbon nanotube · Aligned CNT · Nitrogen doped CNT · Quinone redox couple

1 Introduction

Redox flow batteries (RFBs) are considered one of the most promising technologies for large-scale electrochemical energy storage due to favorable qualities, such as independent power-energy sizing, high efficiency, long charge–discharge cycling and power rating [1, 2]. Among commercialized systems, the all-vanadium redox flow battery (VRFB) remains the most mature; however, the high cost and price fluctuations of vanadium precursors cause limited widespread RFB adoption into the storage market necessitating further research activities [1, 3]. Consequently, attention has shifted toward alternative redox chemistries based on earth-abundant, low-cost, and environmentally benign active species. Organic redox molecules (ORMs) have emerged as a particularly attractive class of active materials because of their molecular diversity, wide natural abundance, and tunable electrochemical properties [4]. Generally, organic active species offer multiple molecular configurations by functionalization with electron-donating or withdrawing groups which enable good redox behavior and solubility [5]. Several classes of organics, including quinones, viologens, nitroxide radicals, and ferrocene derivatives have been investigated [6]. Among these, quinones stand out due to their fast, reversible redox kinetics, which enable high current densities and energy efficiencies. Quinone compounds can generally be defined as a class of cyclic organic compounds including benzoquinones, naphthoquinones and anthraquinones. Benzoquinone disulfonic acid (BQDS) is a water-soluble, sulfonated quinone derivative that offers high redox reversibility and good chemical stability under strongly acidic conditions, making it a suitable positive electrolyte candidate for hybrid hydrogen/organic RFBs [7, 8]. However, conventional carbon felt, and carbon cloth electrodes often suffer from limited electrochemically active surface area (ECSA) and sub-optimal wettability, restricting charge transfer and reducing overall utilization of the redox-active species. Nano structuring and surface functionalization of electrode substrates have proven effective strategies to overcome these limitations. Carbon nanotubes (CNTs) offer a high aspect ratio, excellent electrical conductivity, and a large surface area, which can improve electron transport pathways and provide abundant active sites for redox reactions [9, 10]. Moreover, doping CNTs with heteroatoms such as nitrogen can tailor their electronic structure, increase defect density, and further enhance electrocatalytic activity [3, 11]. Aligned CNTs (A-CNTs) additionally provide directional conductivity and ordered microchannels that can facilitate mass transport within porous electrodes. Electrophoretic deposition (EPD) has emerged as a scalable and facile method for depositing nanomaterials on conductive substrates, enabling uniform coatings with controlled thickness [9, 12, 13]. In this work, a horizontal EPD reactor was employed to functionalize carbon cloth electrodes with pristine CNTs, aligned CNTs, and home-synthesized N-doped CNTs. The electrochemical performance of these nanostructured electrodes was evaluated using cyclic voltammetry in BQDS-based electrolytes to assess improvements in ECSA and charge transfer kinetics.

2 Experimental Materials and Methods

2.1 Material

All reagents utilized in this study were of analytical grade and employed without any further purification. Sulfuric acid (95–98%), 4,5-dihydroxy-1,3-benzenedisulfonic acid disodium salt monohydrate (BQDS, 97%) and Polyvinylpyrrolidone (PVP) were sourced form Sigma-Aldrich. Pristine and aligned multi-walled carbon nanotubes (CNTs) were obtained from Nanografi and used as received. Solvents and reagents such as N,N′-dimethylformamide (DMF), and hydrochloric acid (HCl) were purchased from Merck. Deionized water with a resistivity exceeding 18.2 MΩ·cm was used for solution dilution. Carbon cloth (CC, without microporous layer) was sourced from Fuel Cell Store [10, 14].

2.2 N-Doped CNT Synthesis

Nitrogen doping of CNTs was carried out via thermal treatment, following a procedure adapted from Béjar et al. [15] with minor modifications to optimize dispersion and nitrogen incorporation. 40 mg of pristine and A-CNTs without surface–engineering modifications and 4 mg of PVP were dispersed in 40 mL ethanol. PVP was used as a steric stabilizer to form stable and homogeneous CNT suspensions and preventing agglomeration to provide uniform exposure of CNT surfaces to the nitrogen precursor. Subsequently, suitable amounts of urea was added as the nitrogen source at a mass ratio of 1:20 (MWCNT: Urea) to achieve sufficient nitrogen content for doping. The mixture was placed in an ultrasonic bath for 2 h to ensure complete dispersion and precursor adsorption onto the CNT surface. Thereafter, the suspension was transferred to an oil bath and gently heated at 70 °C for 24 h to facilitate slow solvent evaporation and homogeneous precursor deposition. The resulting dry composite was lightly ground to obtain a fine powder and then subjected to thermal treatment in a tube-furnace. The powder was heated under a constant nitrogen flow (100 mL min^{-1}) from room temperature to 800 °C at a controlled heating rate of 10 °C min^{-1}, followed by an isothermal hold at 800 °C for 1 h. The N-doped CNT powders were collected and stored for further characterization.

2.3 *Electrophoretic Deposition (EPD)*

A horizontal cathodic EPD system was assembled to fabricate uniform, adherent coatings of nanomaterials on conductive substrates. The custom cylindrical reactor, constructed from chemically inert polytetrafluoroethylene (PTFE), had an internal diameter of 40 mm. The horizontal configuration (Fig. 1) was chosen to enhance deposit uniformity and film adhesion compared to traditional vertical EPD setups [10].

A graphite plate (Poco Pyro, Entegris Inc.) served as the anode, while CC functioned as the cathodic deposition surfaces. The electrodes were separated by a 15 mm gap, ensured by a circular glass spacer. During each deposition, 25 mL of colloidal suspensions of CNTs in DMF (20 mg L^{-1}) were prepared via ultrasonic treatment for 2 h. These formulations were based on previously established protocols [11]. Prior to deposition, the CC substrates were cleansed by rinsing with distilled water, followed by acetone, and then dried in an oven at 60 °C to eliminate residual contaminants. For suspensions prepared in DMF, a voltage of 200 V was applied for 30 min using a TT TECHNI-C DC-regulated power supply, following optimized protocols from Chakrabarti et al. for organic-based systems [16, 17].

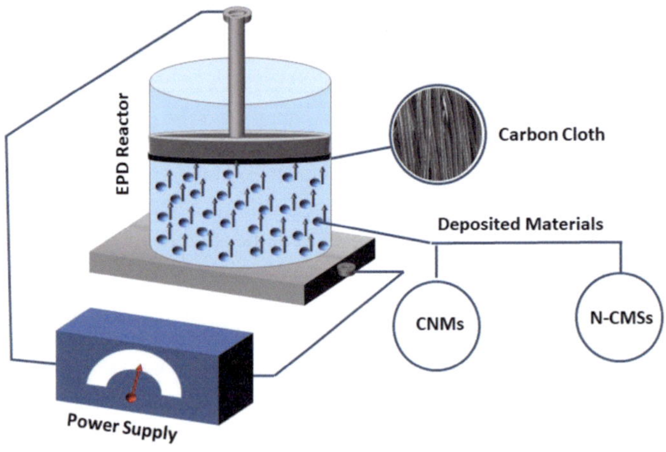

Fig. 1 The design and configuration of the horizontal EPD reactor was inspired from one of our previous investigations [10]

2.4 Morphological Characterization

Fourier transform infrared (FTIR) analyses were performed in a FTIR spectrophotometer (IRAffinity-1S, Shimadzu, Kyoto, Japan) with ATR to identify the bonds of different functional groups. The samples were scanned over the wavenumber range of 4000–600 cm^{-1} with spectral resolution of 4 cm^{-1} and 16 scans for each spectrum. X-ray diffraction (XRD) analyses were conducted in an X-ray diffractometer (D8 Advance, Bruker, Karlsruhe, Germany) with CuKα radiation $\lambda = 0.15405$ nm. Diffraction intensities were measured in the 2θ range 5 to 60° on a step scan mode with 0.02°/step. Raman spectroscopy was executed in a Raman spectrometer (inVia Raman Microscope, Renishaw, Wotton-under-Edge, UK) with a laser emitting at 520 nm for Raman excitation. Scanning electron microscopy (SEM) images were captured via a scanning electron microscope (JSM-6010LV, Jeol, Massachusetts, USA).

2.5 Electrochemical Analysis

Electrochemical analyses were carried out using a Gamry Potentiostat 3000 in a three-electrode electrochemical cell system. In this setup, a platinum wire and an Ag/AgCl (3 M) electrode were used as the counter and reference electrodes, respectively. The working electrodes consisted of bare and modified CCs. The electrolyte solution was prepared by dissolving 1 mM BQDS in 1 M H_2SO_4 using ultrapure water. Cyclic voltammetry (CV), and electrochemical impedance spectroscopy (EIS) experiments were conducted. CV measurements were performed at different scan rates ranging from 2 to 100 mV s^{-1}. EIS measurements were conducted at open circuit voltage (OCV) in the frequency range of 0.1 Hz to 10 kHz using an amplitude of 10 mV.

3 Results and Discussions

3.1 Morphological Characterization

Figure 2a presents the XRD patterns of CC deposited with CNT (EPD-CC-CNT), N-doped CNT (EPD-CC-N-CNT), aligned CNT (EPD-CC-A-CNT), and N-doped A-CNT (EPD-CC-N-A-CNT). All samples exhibit characteristic diffraction peaks at approximately 26 and 43°, corresponding to the (002) and (100) crystallographic planes of graphitic carbon, respectively [18, 19]. Among the samples, CC-CNT shows the highest peak intensity, reflecting improved graphitic character and better stacking order of CNTs on the CC substrate. CC-N-CNTs display slightly broader (002) peaks, which can be attributed to lattice distortions and defect generation caused by nitrogen incorporation into the carbon framework. Similarly, CC-A-CNT and CC-N-A-CNT

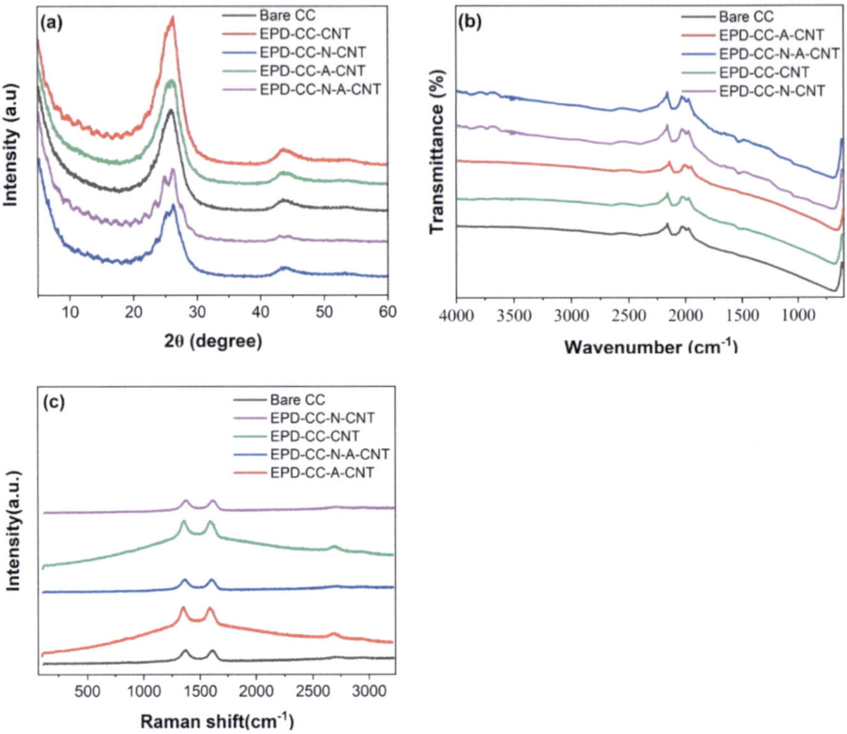

Fig. 2 The comparison of **a** XRD, **b** FT-IR, **c** Raman analysis for CC based electrodes modified with CNT, N-CNT, A-CNT and N-A- CNT

exhibit comparable peak positions with minor differences in intensity, indicating that CNT alignment mainly affects peak sharpness but does not significantly shift the graphitic plane positions.

The FT-IR spectra of CC, CC-CNT, CC-A-CNT, CC-N-CNT and CC-N-A-CNT were acquired between 4000 and 600 cm^{-1} and presented in Fig. 2b. Bare CC, CC-CNT, and CC-A-CNT are dominated by broad C = C and O–H absorptions, making it difficult to distinguish the coating contribution from the carbon substrate, particularly at low coating loadings. For the N-doped powders, distinct spectral features attributable to nitrogen incorporation were observed. The presence of C–N and N–CH_3 vibrational modes is typically reported near 1250 and 1372 cm^{-1}, respectively, corresponding to nitrogen atoms intercalated between graphitic layers or bonded at the inner walls of CNTs [20]. However, N-doped coatings (CC-N-CNT and CC-N-A-CNT) exhibit small stretching bands at ~1530 cm^{-1} which can be because of the shifting or stretching of these peaks. The synthesized N-doped CNT powders exhibited bands consistent with these positions, confirming successful nitrogen doping. However, CC-N-CNT and CC-N-A-CNT electrodes displayed weak, stretching bands near ~1530 cm^{-1}, which may be assigned to N-H bending or C-N stretching modes.

The Raman analysis of CC-based modified electrodes are illustrated in Fig. 2c. Bare CC shows a disorder induced D band at 1353 cm^{-1} and the graphitic G band at 1593 cm^{-1}. Upon deposition of CNTs, both CC-CNT and CC-A-CNT display a considerable enhancement in Raman scattering intensity, indicating a higher concentration of sp^2-hybridized carbon domains. The D and G band positions slightly shifted to 1346 and 1586 cm^{-1}, which are typically associated with structural disorder in graphitic materials and tangential vibrations of carbon atoms in the lattice. These results are consistent with previous reports on CNT-modified carbon substrates [15, 21, 22]. The I_G/I_D ratio for both CC-N-CNT and CC-N-A-CNT are 1.01 and 0.99, respectively, which confirm the presence of defect-rich multi-walled CNTs (MWCNTs) while suggesting that N-doping does not drastically disrupt the graphitic structure.

The SEM images of the electrodes in addition to elemental analysis of CNT powders are presented in Fig. 3. Bare CC displays woven carbon fibers with smooth surfaces and large inter-fiber voids. Deposition of CNTs results in a dense coating of nanotubes bridging the fibers, which significantly enhances surface area and modifies texture. Aligned CNT coatings (CC-A-CNT) form vertically oriented tubular domains on the fiber surfaces, though the strong carbon background makes it difficult to distinguish them at lower magnifications. Nitrogen-doped CNT and A-CNT coatings (CC-N-CNT and CC-N-A-CNT) exhibit more irregular and rough morphologies, consistent with heteroatom-induced defects and functionalization. The elemental analysis of the CNT-based powders, reveals that carbon was the predominant element, accompanied by small amounts of oxygen, aluminum, silicon, and iron for both pristine and aligned CNT powders, which confirms their highly carbonaceous nature with minimal heteroatom incorporation. On the other hand, the N-CNT and N-A-CNT powders exhibit remarkably substantial nitrogen and oxygen contents alongside carbon. These observations verify that nitrogen doping was successful in both CNT and A-CNT powders. In the case of deposited CC based electrodes, it is difficult to distinguish the coating contribution from the carbon substrate, particularly at low coating loadings.

3.2 Cyclic Voltammetry (CV)

The CV results presented in Fig. 4 demonstrate distinct electrochemical behaviors of the modified and unmodified CC electrodes in 1 M H_2SO_4 containing 1 mM BQDS. Deposition of CNTs onto the CC substrate significantly enhanced the conductivity and a number of active sites. The bare CC electrode showed the lowest current response and broad redox features, consistent with its relatively poor conductivity, hydrophobicity and limited active sites. In contrast, incorporation of CNTs significantly increased the current density, reflecting an enlarged ECSA and improved conductivity. Further modification with nitrogen-doped CNTs (EPD-CC-N-CNT) led to even higher current densities and better-defined redox peaks. The use of aligned CNTs (EPD-CC-A-CNT) showed about similar performance to the randomly

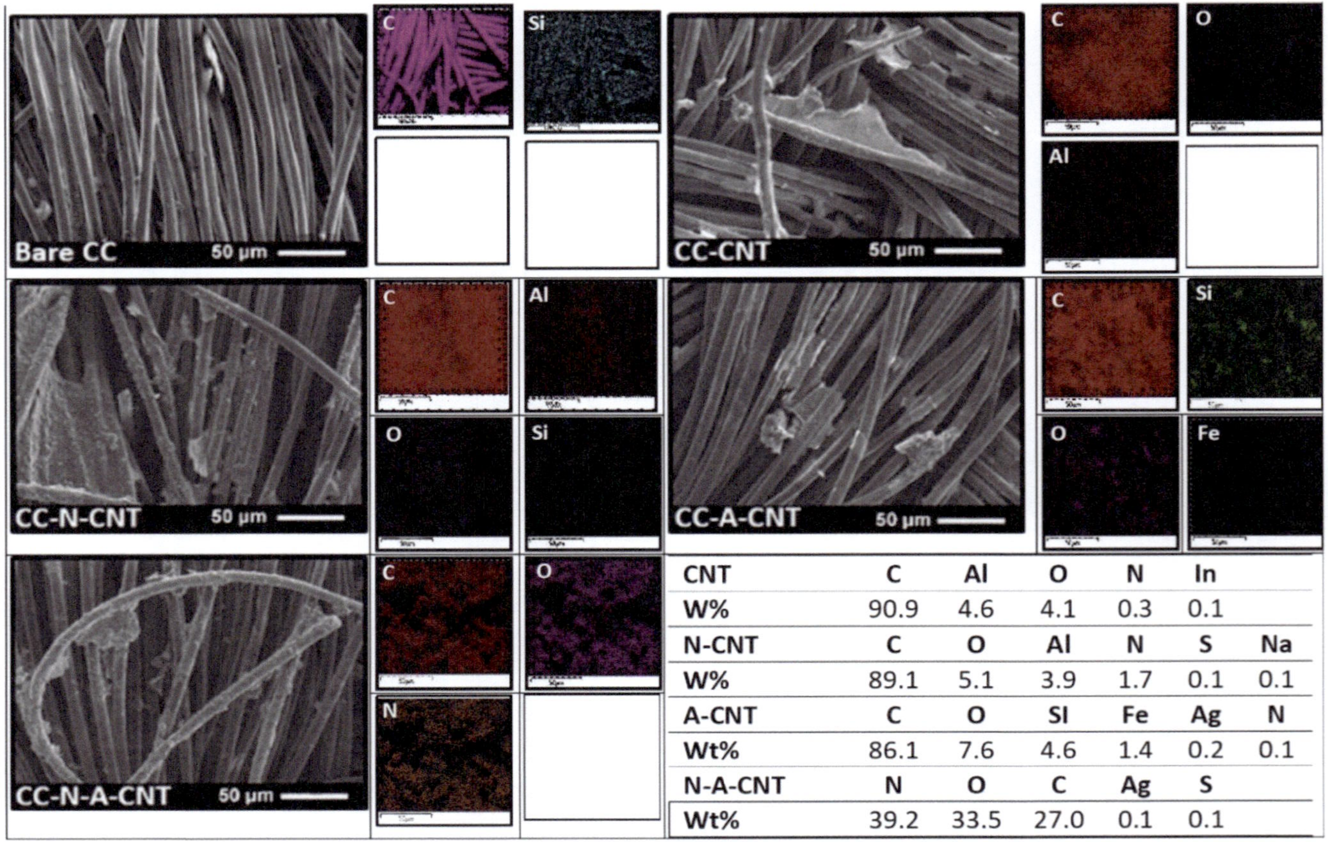

CNT	C	Al	O	N	In	
W%	90.9	4.6	4.1	0.3	0.1	
N-CNT	C	O	Al	N	S	Na
W%	89.1	5.1	3.9	1.7	0.1	0.1
A-CNT	C	O	SI	Fe	Ag	N
Wt%	86.1	7.6	4.6	1.4	0.2	0.1
N-A-CNT	N	O	C	Ag	S	
Wt%	39.2	33.5	27.0	0.1	0.1	

Fig. 3 The SEM and EDS analyses for CC based electrodes modified with CNT, N-CNT, Aligned-CNT and N-Aligned-CNT. The table provides elemental analysis for the CNT powders

oriented CNTs while EPD-CC-N-A-CNT delivered the best performance among the modified electrodes, with the highest current densities and sharpest peaks. This suggests that the structural benefits of alignment and the electrochemical benefits of nitrogen doping by increasing active sites lead to maximizing electrochemical efficiency. Therefore, the results proved that all surface modifications improved the electrochemical performance compared to bare CC.

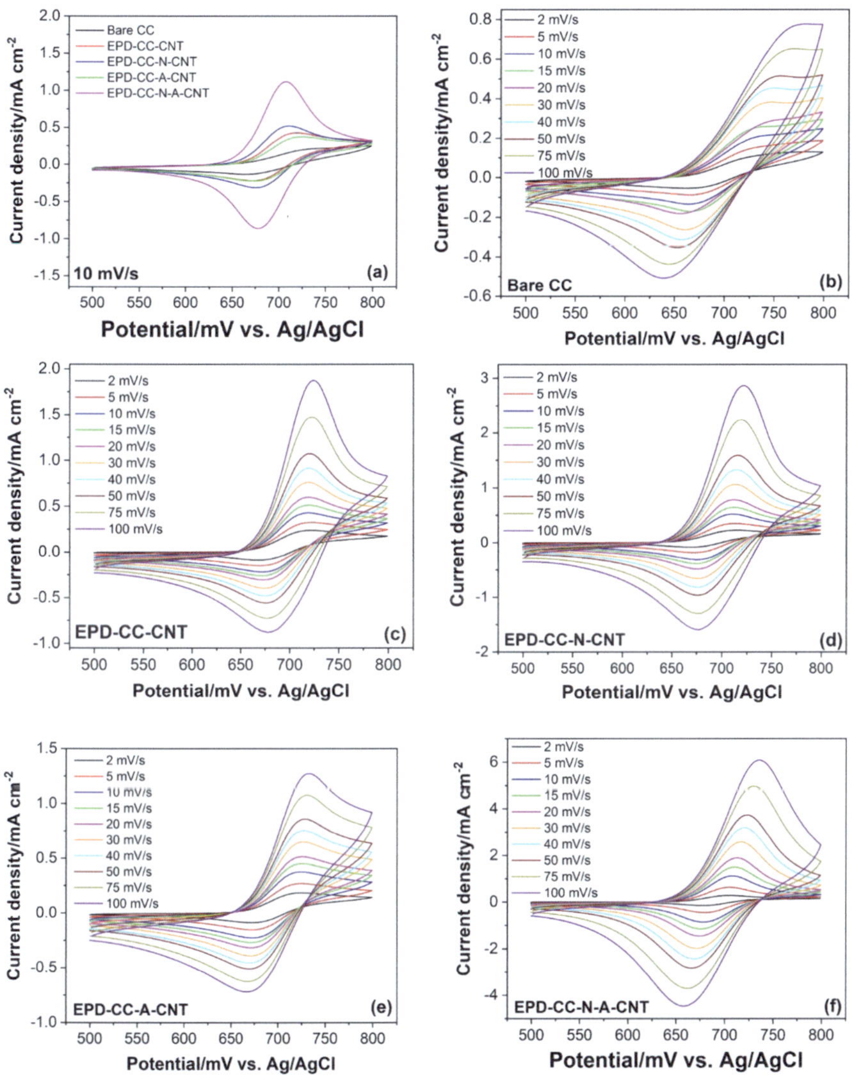

Fig. 4 **a** Comparison of CV at a scan rate of 10 mV s^{-1}, **b–f** Cyclic voltammograms at different scan rates in 1 M H_2SO_4 containing 1 mM BQDS for bare CC, EPD-CC-CNT, EPD-CC-N-CNT, EPD-CC-A-CNT and EPD-CC-N-A-CNT electrodes

The diffusion coefficients of 1 mM BQDS in 1 M H_2SO_4 aqueous solution were estimated using the Randles–Sevcik equation, based on the slope of the linear relationship between peak current (i_p) and the square root of the scan rate ($v^{1/2}$) obtained from CV at various scan rates (Fig. 5). The diffusion coefficients were calculated separately for oxidation and reduction processes for each electrode configuration, as presented in Table 1. The calculated diffusion coefficients ranged from approximately 10^{-6} to 10^{-3} cm^2 s^{-1}, depending on the electrode surface modification. The bare CC electrode exhibited the lowest diffusion coefficients (~9.92×10^{-6} cm^2 s^{-1} for oxidation and 4.98×10^{-6} cm^2 s^{-1} for reduction), which are consistent with typical values reported in the literature for unmodified electrodes in aqueous quinone systems [5].

In contrast, electrodes modified with carbon nanomaterials demonstrated significantly enhanced diffusion behavior. For example, enhancement trends were observed for the EPD-CC-N-A-CNT sample, which yielded diffusion coefficients in the order of 10^{-4} cm^2 s^{-1}, indicating the synergistic effect of nitrogen doping and functionalization of CNTs on electrochemical performance.

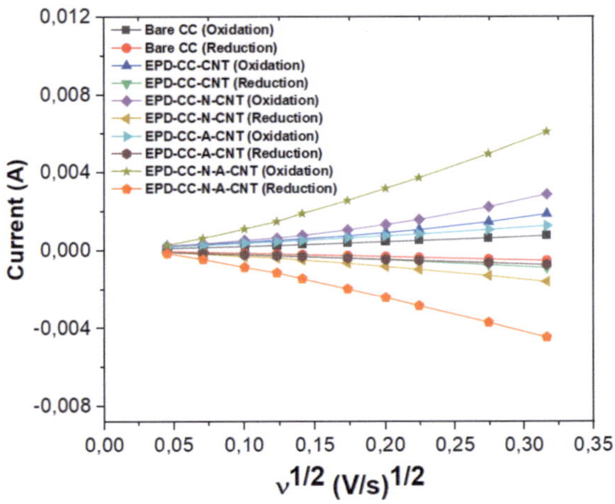

Fig. 5 Linear dependence of peak current (I_p) versus the square root of scan rate ($v^{1/2}$) for 1 mM BQDS in 1 M H_2SO_4 electrolyte solution

Table 1 Diffusion coefficient values calculated from the Randles-Sevcík equation

Sample	$Slope_{Oxidation}/Slope_{Reduction}$ (A (V/s)$^{-1/2}$)	Diffusion $coefficient_{Oxidation}$/Diffusion $coefficient_{Reduction}$ (cm^2 s^{-1})
Bare CC	y = 0.0024/ y = −0.0017	$9.92 \times 10^{-6}/4.98 \times 10^{-6}$
EPD-CC-CNT	y = 0.0059/ y = −0.0029	$6.00 \times 10^{-5}/1.45 \times 10^{-5}$
EPD-CC-N-CNT	y = 0.0079/ y = −0.0056	$1.08 \times 10^{-4}/5.40 \times 10^{-5}$
EPD-CC-A-CNT	y = 0.0040/ y = −0.0023	$2.76 \times 10^{-5}/9.11 \times 10^{-6}$
EPD-CC-N-A-CNT	y = 0.0216/ y = −0.0161	$8.04 \times 10^{-4}/4.47 \times 10^{-4}$

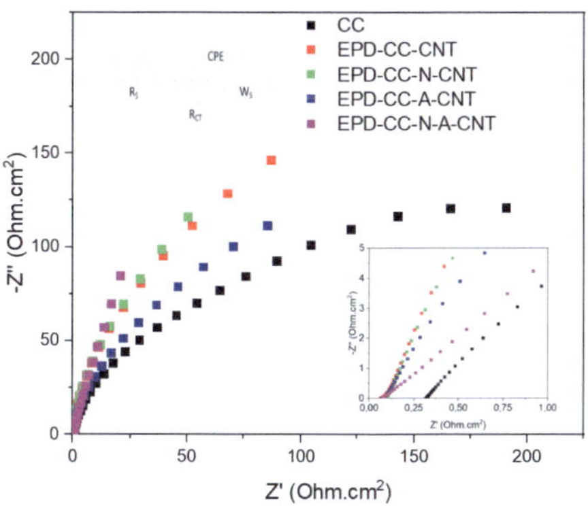

Sample	R_s (Ω.cm^2)	R_{ct} (Ω.cm^2)
Bare CC	0.27	169.0
EPD-CC-CNT	0.16	153.8
EPD-CC-N-CNT	0.08	150.1
EPD-CC-A-CNT	0.15	156.0
EPD-CC-N-A-CNT	0.07	135.2

Fig. 6 Nyquist plots obtained from EIS measurements using bare CC and modified electrodes in a 1 mM BQDS solution containing 1 M H_2SO_4 and equivalent electrical circuit model used for fitting the EIS data [16, 17]. The inset shows the plot at the high frequency region. The table summarizes experimental results from the EIS analysis

3.3 *Electrochemical Impedance Spectroscopy (EIS)*

The EIS analysis provides valuable insights into the interfacial charge transfer and ion transport characteristics of the different modified CC-based electrodes. The Nyquist plots (Fig. 6) for all modified electrodes, except bare CC, demonstrate considerable reduction in R_s values, reflecting improved conductivity and electrolyte accessibility. Bare CC presents the highest R_s (0.27 Ω.cm^2), verify the poor intrinsic conductivity. By contrast, CNT-modified electrodes exhibit lower R_s values confirming the beneficial effect of nanomaterial deposition. Unfortunately, the charge transfer resistances for CC and its variants in the presence of BQDS are too high and displays the fact that EPD requires significant tuning and iterations before it can be of use for preparing electrode materials that function effectively for quinone-based electrolytes. In addition, it is clear that the hydrophobic CC is not a suitable electrode for quinone-based RFBs and requires plasma or thermal treatments prior to its modification with nitrogen doped CNTs. These improvements will be the subject of future investigations.

4 Conclusion

This study demonstrates the modification of carbon cloth (CC) electrodes with pristine CNTs, aligned CNTs and nitrogen doped CNTs via electrophoretic deposition (EPD). This preliminary investigation shows that EPD may slightly improve electrode performances for aqueous–organic flow batteries but the EPD process itself requires significant improvements before any serious studies with real RFBs can become a practical reality. The morphological characterization (XRD, FTIR,

Raman, SEM, and EDS) confirms successful deposition of CNTs on CC. Electrochemical analyses in an organic electrolyte consisting of BQDS and H_2SO_4 reveals that CNT modification slightly enhances electroactive surface area, as shown by increased redox peak currents in cyclic voltammograms. Nitrogen doping further boosts current densities and sharpens redox peaks, indicating potentially faster charge transfer kinetics. The synergistic effect of alignment and N-doping leads to the best overall performance among the CNT-based electrodes, with improved diffusion coefficients (in the order of 10^{-4} cm^2 s^{-1}) compared to bare CC. These results confirm that modified electrodes with CNT coatings, N-doping, and alignment could offer a pathway to optimize CC electrodes for organic redox flow batteries. Future work could focus on optimizing CNT loading, N-doping concentration, and EPD parameters to further minimize R_{ct} and maximize catalytic activity, potentially bringing organic redox flow batteries closer to practical commercialization.

Funding This work was supported by the European Innovation Council (EIC) under grant agreement No. 101115293 (VanillaFlow). This research was also funded by the CETPartnership, under the Joint Call 2023, through the project "HydrogEn oRganiC Regenerative FUeL CEll for Stationary Energy Storage (HERCULES)". The CETPartnership's research projects are co-funded by the European Commission (Grant Agreement No. 101069750) and the national TUBITAK 1071 via the project code: 224N167. R.A.G. is grateful to Türkiye Bilimsel ve Teknolojik Araştırma Kurumu (TÜBİTAK) for the 1001 project for funding his research in SUNUM (223M269).

References

1. Chakrabarti, B.K., et al.: Sustain. Energy Fuels **4**(11), 5433–5468 (2020)
2. Alotto, P., Guarnieri, M., Moro, F.: Renew. Sustain. Energy Rev. **29**, 325–335 (2014)
3. Chakrabarti, B.K., et al.: Mater. Res. Bull. **182**, 113140 (2025)
4. Shoaib, M., et al.: Adv. Energy Mater. **14**(32), 2400721 (2024)
5. Rubio-Garcia, J., et al.: J. Mater. Chem. A **8**(7), 3933–3941 (2020)
6. Zhu, F., Chen, Q., Fu, Y.: Green Energy Environ. **9**(11), 1641–1649 (2024)
7. Hasan, F., Mahanta, V., Abdelazeez, A.A.A.: Adv. Mater. Interfaces **10**(24), 2300268 (2023)
8. Symons, P.: Curr. Opin. Electrochem. **29**, 100759 (2021)
9. Chakrabarti, B.K., et al.: Int. J. Energy Res. **46**(10), 13205–13250 (2022)
10. Chakrabarti, B., et al.: Int. J. Hydrogen Energy **44**(57), 30093–30107 (2019)
11. Chakrabarti, B.K., et al.: ACS Appl. Mater. Interfaces. **12**(48), 53869–53878 (2020)
12. Atiq Ur Rehman, M., et al.: Challenges. Int. Mater. Rev. **66**(8), 533–562 (2021)
13. Chakrabarti, B.K., John Low, C.T.: RSC Adv. **11**(34), 20641–20650 (2021)
14. Chakrabarti, B.K., et al.: ACS Appl. Nano Mater. **4**(10), 10754–10763 (2021)
15. Béjar, J., et al.: Carbon **204**, 411–426 (2023)
16. Chakrabarti, B., et al.: ChemElectroChem **4**(1), 194–200 (2017)
17. Chakrabarti, B., et al.: Batteries **6**(3), 38 (2020)
18. Ahirrao, D.J., et al.: J. Mater. Sci. Technol. **88**, 168–182 (2021)
19. Yu, T.T., et al.: **7**(82), 51807–51813 (2017)
20. Misra, A., et al.: Diam. Relat. Mater. **15**(2), 385–388 (2006)
21. Zhang, X., et al.: Analyst **141**(19), 5527–5534 (2016)
22. Afshar Ghotli, R., et al.: ChemCatChem **18**(2), e01280 (2026)

Transition Metal Chalcogenides as Promising Electrode Materials for High-Performance Supercapacitors

Sezgin Yaşa and Esra Yilmaz Mertsoy

Abstract Metal sulfides are important members of the transition metal chalcogenides family and have attracted great attention as promising electrode materials for supercapacitor applications due to their low cost, abundant redox-active sites, and high specific capacitance. These materials exhibit excellent pseudocapacitive behavior, which arises from their multiple oxidation states and reversible faradaic reactions, enabling rapid charge–discharge processes. In addition, their structural flexibility allows for the accommodation of ion diffusion during cycling, further enhancing electrochemical performance. Despite these advantages, challenges such as limited long-term stability, partial dissolution in electrolytes, and relatively poor electrical conductivity still restrict their practical applications. This study focuses on recent advances in the synthesis strategies, structural tuning, and electrochemical properties of nickel sulfide, cobalt sulfide, and manganese sulfide-based electrodes. Furthermore, the development of high-performance supercapacitors using transition metal sulfides (TMS) and strategies to overcome current limitations are also discussed.

Keywords Transition metal sulfides · Supercapacitor · Pseudocapacitance

1 Introduction

In recent years, effective energy storage techniques have gained significant attention [1]. Supercapacitors are electrochemical energy conversion and storage devices with great promise. Supercapacitors exhibit exceptional power performance, reversibility, long-term stability, ease of use, rapid charging and discharging rates, and high integration viability in electronic devices, such as memory backup systems, consumer

S. Yaşa (✉)
Department of Mechanical Engineering, Çankırı Karatekin University, Cankiri, Türkiye
e-mail: sezginyasa@karatekin.edu.tr

E. Y. Mertsoy
Department of Chemical Engineering, Çankırı Karatekin University, Cankiri, Türkiye

M. Awang et al. (eds.), *Proceedings of the International Conference on Mechanical, Manufacturing and Plant Engineering*, Proceedings in Technology Transfer,
https://doi.org/10.1007/978-981-95-7817-7_17

electronics, and industrial power and energy management systems. Supercapacitors' low energy density, however, restricts their usefulness in comparison to rechargeable batteries. Advanced electrode materials with logical designs have a lot of promise to solve the aforementioned problem. Power density, energy density, operating voltage, and cell capacitance of the supercapacitors are all influenced by the electrode materials. Due to their exceptional energy density, specific capacitance, and rapid charging/discharging rates, as well as their ease of preparation in a high yield from inexpensive and readily available resources, transition metal-based electrode materials are among the most promising electrodes for supercapacitors. Because of their low cost, superior redox reversibility, and electronic conductivity, transition metal sulfides (TMSs) have emerged as a viable material for hybrid supercapacitors in recent years [2, 3]. Transition metal sulfides, as a novel form of electrode material, it has lately shown considerable promise in the field of supercapacitors [4].

In recent years, many different supercapacitor applications have been reported in the literature using nickel sülfide [5], cobalt sülfide [6], and manganese sulfide [7], and they hold an important position as supercapacitor electrode materials. This study will focus on the production methods of metal sulfides, their supercapacitor applications, limitations and applications to enhance the supercapacitive performance of these materials.

Figure 1a–c show the number of research studies conducted between 2018 and 2024, based on searches conducted on the Web of Science using the keywords "Nickel sulfide, Supercapacitor," "Cobalt sulfide, Supercapacitor," and "Manganese sulfide, Supercapacitor." The results in Fig. 1 indicate that supercapacitor applications using these materials have continued in recent years.

2 Synthesis and Supercapacitor Applications of Metal Sulfides

The preparation of chalcogenide nanomaterials has been achieved through a variety of facile and cost-effective methods, each offering specific advantages in terms of morphology control, crystallinity, and scalability. A variety of synthetic techniques have been utilized for the preparation of TMCs, including hydrothermal synthesis, electrochemical deposition, chemical vapor deposition (CVD), and microwave assisted. Among these, hydrothermal or solvothermal methods are most widely adopted due to their simplicity of operation and scalability. In the literature, these structures are generally synthesized by hydrothermal and CVD methods and are mostly investigated for their supercapacitor properties.

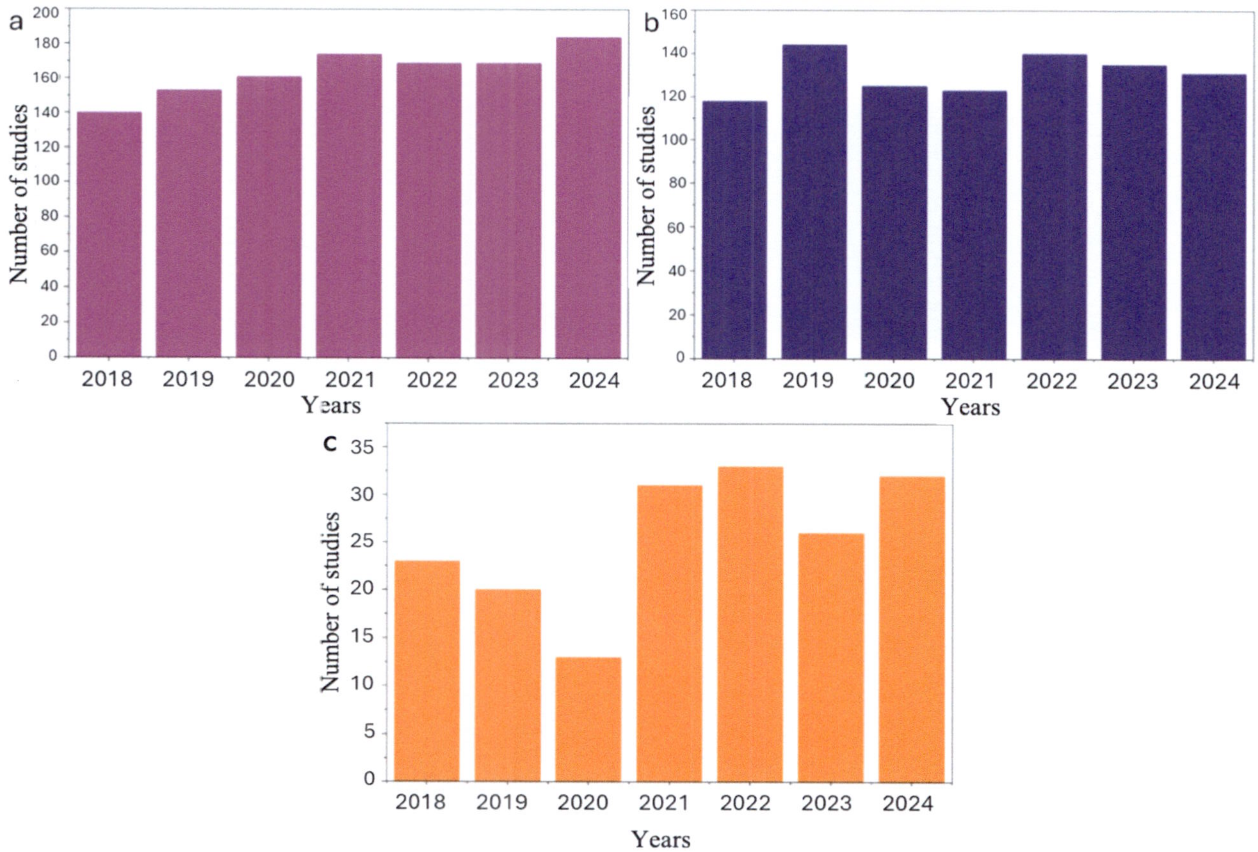

Fig. 1 The number of research studies conducted between 2018 and 2024, based on searches conducted on the Web of Science using the keywords **a** "Nickel sulfide, Supercapacitor," **b** "Cobalt sulfide, Supercapacitor," and **c** "Manganese sulfide, Supercapacitor."

2.1 Hydrothermal Synthesis

Compared with other synthesis techniques, the hydrothermal method offers a simpler route for producing nanoscale materials. The resulting nanostructures generally exhibit uniform distribution in both shape and size, along with a large specific surface area. Moreover, by tuning the hydrothermal reaction parameters, the morphology and dimensions of micro-/nanomaterials can be precisely controlled [8].

Trivedi et al. reported the facile hydrothermal synthesis of a ReS_2/CoS heterostructure catalyst. The nano-interface engineering approach significantly facilitated charge transfer, leading to enhanced performance in both hydrogen evolution reaction (HER) and supercapacitor applications. The heterostructure delivered a high specific capacitance of 663.2 F g^{-1} at 2 A g^{-1} and 643.1 F g^{-1} at a scan rate of 5 mV s^{-1}. These results highlight the promising bifunctional potential of ReS_2/CoS heterostructures for next-generation energy storage and electrocatalytic systems [9].

2.2 Chemical Vapor Deposition (CVD)

Synthesizing nanostructures of transition metal sulfides is a common technique for CVD. This process involves introducing volatile metal precursors and chalcogen sources into a reaction chamber. At elevated temperatures, they decompose or react to form a solid film on the substrate surface. The CVD method is ideal for making large-area, uniform, and atomically thin layers because it allows for precise control over film thickness, composition, and crystallinity [10].

2.3 Chemical Bath Deposition (CBD)

CBD is a simple, economical approach for fabricating thin films of chalcogenides over large areas without sophisticated instrumentation. Parameters such as bath composition, deposition time, temperature, pH, and complexing agents strongly influence film quality [11]. Abdullah et al. were synthesized copper sulfide (CuS) electrodes on nickel foam (NF) via a simple and low-cost chemical bath deposition (CBD) method using citric acid as a complexing agent. Uniform ultrathin CuS nanospheres with a hexagonal crystal structure were obtained, and deposition times of 30–120 min were tested to optimize performance. Electrochemical analysis revealed that the CuS/NF electrode fabricated at 60 min achieved the best results, delivering a specific capacitance of 615 F g^{-1} at 1 A g^{-1}. These results demonstrate that CBD-derived CuS electrodes are promising candidates for high-performance supercapacitors [12].

2.4 Electrodeposition Method

Electrodeposition involves the transfer of electrons from an electrode to metal ions in solution, resulting in the formation of a thin metallic layer on a substrate to tailor its surface properties. As a versatile bottom-up fabrication approach, it can be applied across a wide range of materials and devices. Manyepedza et al. (2024) synthesized $MoS_2/MoSe_2$ and MoS_2/WS_2 heterolayers by electrodeposition and demonstrated that the overlayer strongly affected stability in acidic media: MoS_2 overlayers remained stable, while $MoSe_2$ degraded upon cycling [13]. Zhang et al. synthesized a CuS/CoS heterojunction by CV electrodeposition, where the internal electric field promoted charge separation and enhanced redox kinetics. The electrode exhibited excellent capacitance and cycling stability, and the assembled ASC device achieved an energy density of 1.04 mWh cm^{-2} at a power density of 5.91 mW cm^{-2}, demonstrating its strong potential as a high-performance electrode material [14].

2.5 Microwave-Assisted Synthesis

Microwave heating provides homogeneous energy distribution, ensuring uniform nucleation and highly crystalline nanostructures. Compared with conventional heating, it offers short reaction times, high yield, and minimal side reactions due to the combined electric and magnetic field effects [15]. Konwar et al. reported the microwave-assisted synthesis of ZnxCo1–xS/reduced graphene oxide (rGO) nanocomposite and evaluated its electrochemical performance for supercapacitor applications. The synergistic interaction between ZnxCo1–xS and rGO, together with the conductive framework of rGO, facilitated efficient ion transport and improved charge storage capability. The Zn:Co:rGO stoichiometric ratio was found to play a crucial role in determining the supercapacitor behavior. The nanocomposite delivered a high specific capacitance of 1094 F g^{-1} at 1 A g^{-1} in a three-electrode system, with charge stored through both diffusion-controlled and capacitive processes. An asymmetric supercapacitor fabricated with acetylene black as the negative electrode and ZnxCo1–xS/rGO as the positive electrode exhibited a stable potential window of 0–1.6 V, delivering an energy density of 32.71 Wh kg^{-1} and a power density of 431.95 W kg^{-1}, along with 86% capacitance retention after 10,000 cycles. These results demonstrate that the material possesses superior electrochemical stability and reversibility, making it a promising electrode for high-performance supercapacitors [16].

3 Supercapacitor Application of Ni-, Co- and Mn-Based Sulfides

Among transition metal sulfides, nickel sulfide and cobalt sulfide have attracted considerable attention due to their low cost, facile synthesis, high electrical conductivity, and low toxicity [17, 18].

Moradi et al. synthesized zinc doped nickel metal organic framework (MOF) composed with graphene oxide (GO) hydrothermally [19]. After sulfurization procedure was applied to obtain NiS which are highly favorable for supercapacitor applications due to their redox activity and conductivity. The optimized composite containing a higher amount of GO delivered a specific capacitance of 442.66 F g^{-1} at 1 A g^{-1}. Moreover, the assembled asymmetric supercapacitor (ASC) device exhibited a power density of 774.99 W kg^{-1} and an energy density of 2.142 Wh kg^{-1}, while maintaining 96% capacitance retention after 10,000 cycles, confirming the excellent stability of NiS-derived electrodes.

Recently, Yadav et al. investigated the structural and electronic properties of $NiCo_2X_4$ (X = S, Se, Te) n using DFT calculations. $NiCo_2Se_4$ exhibited superior capacitive performance with a high quantum capacitance (3418 F g^{-1}) and surface charge density (1902 C g^{-1}). Furthermore, its high Young's modulus (189 GPa) highlights the potential application of $NiCo_2Se_4$ in flexible electronic devices. The calculated formation energies were −255.54, −319.23, and −293.47 eV/f.u. (electronVolts per formula unit) for $NiCo_2S_4$, $NiCo_2Se_4$, and $NiCo_2Te_4$ based structures, respectively. Among them, the $NiCo_2Se_4$ structure exhibits the most negative formation energy, indicating that it is the most thermodynamically stable and energetically favorable phase. The stability order of the structures can thus be expressed as $NiCo_2Se_4 > NiCo_2Te_4 > NiCo_2S_4$. Since sulfur (S) is more electronegative than selenium (Se) and tellurium (Te), the Ni–S and Co–S bonds exhibit a stronger ionic character [20].

Ni_3S_4/NiS_2/ACNTs nanocomposite was synthesized by Ouyang et al. The NiS/ACNTs hybrid electrode exhibits electrochemical performance, of specific capacitance of 1266 F g^{-1} at 1.0 A g^{-1}. Furthermore, the NiS/ACNTs//AC asymmetric supercapacitor (ASC), achieves a favorable energy density of 36.0 Wh kg^{-1} at a power density of 806 W kg^{-1} and maintains 83% capacitance retention after 2000 cycles at 2.0 A g^{-1} [21]. These results highlight that employing carbonaceous materials as conductive scaffolds and designing hybrid active structures is an effective strategy to mitigate particle agglomeration and advance the development of next-generation functional electrodes for ASC applications.

Xiong et al. Synthesized hollow Co_xNi_yS spheres through a facile and efficient one-pot solvothermal method, employing cobalt nitrate and nickel nitrate as the metal precursors and thioacetamide as the sulfur source. The resulting hollow spheres are composed of delicately interlaced thin nanosheets. The Co_1Ni_4S material delivers a high specific capacity of 277.2 mAh g^{-1} (2217 F g^{-1}) at 1 A g^{-1}, along with excellent cycling stability, maintaining nearly 100% of its capacity after 5000 cycles, and remarkable rate capability, retaining 84.8% at 20 A g^{-1}. Moreover, the assembled

Co_1Ni_4S//AC device achieves an impressive energy density of 86.9 Wh kg^{-1} at a power density of 800.1 W kg^{-1} and demonstrates outstanding long-term durability, with 97.1% retention over 30,000 cycles, highlighting its strong potential for practical applications [22].

Wang et al. reported an innovative synthesis strategy that combined hydrothermal and constant-pressure deposition techniques to fabricate nanostructured MnO_2/MnS rod electrodes on nickel foam. The formation of a MnO_2/MnS heterostructure provided a significantly enlarged specific surface area, improved conductivity, and strong synergistic effects between the two components, which collectively enhanced the electrochemical performance. The optimized electrode delivered a remarkable specific capacitance of 977.6 F g^{-1} at 1 A g^{-1}, along with an extended cycling lifespan and a high energy density of 122.23 Wh kg^{-1} at a power density of 898.75 W kg^{-1}. Furthermore, the electrode maintained 91.55% of its initial capacity after 5000 cycles at 10 A g^{-1}, confirming its outstanding cycling stability and highlighting the promising potential of MnO_2/MnS heterostructures for advanced high-performance supercapacitor applications [23].

Momeni et al. developed Mn–Ni sulfides with varying Mn/Ni ratios on titania substrates through a co-electrodeposition strategy. Among the fabricated electrodes, MnNiS-3/TN exhibited the highest photocurrent and photovoltage response. Remarkably, its area-specific capacitance increased from 388.3 to 723.3 mF cm^{-2} under illumination, confirming the strong photo-responsive behavior. Using MnNiS-3/TN as the optimal photoelectrode, five asymmetric solid-state photo-supercapacitors (ASSPS) were assembled, demonstrating a significant capacitance gain of 155.7% at 2.0 mA cm^{-2} under irradiation and excellent cycling stability with 97.9% retention after 10,000 cycles [24]. These results clearly demonstrate the advantages of incorporating two transition metals, where the synergistic interaction between Mn and Ni enhances both the electrochemical and photoresponsive properties, thereby motivating further research on bimetallic sulfide systems for advanced supercapacitor applications.

4 Challenges and Performance Enhancement Strategies

TMS-based electrode materials confront several scientific obstacles despite their enormous promise to enhance electrochemical performance. One of the primary constraints on the development of supercapacitors is their energy density. Studies that highlight the limits of their weak conductivity may be found in the literature. Furthermore, it may be noted that managing the volume expansions, slow reaction kinetics, and the potential for some simultaneous side reactions are problems. Two factors may be addressed to enhance their electrochemical performance even further. First, vast surface areas and porous structures may be produced by controlled nanomaterial production, increasing the efficiency of ion transport. Second, the synergy between various materials may be fully used by mixing transition metal sulfides with other energy storage materials, such as metal oxides and carbon compounds.

The creation of a carbon–TMSs composite material may improve electric conductivity, reduce volume fluctuations, and prevent exfoliated TMSs from aggregating. Controlling the volume expansion, surface response, and charge trap will be made easier by customizing various morphologies, such as core–shell and porous-fibre structures [4, 25].

5 Conclusions

TMSs are one of the most important electrode materials used in supercapacitor applications, and research on their use as electrode materials is ongoing. They are produced using various methods, including hydrothermal, solvothermal, electrodeposition, microwave-assisted, CVD, and CBD. Nickel–cobalt and manganese are among the most important TMSs used in supercapacitors. Recent studies have focused on composites of these metal sulfides and their compounds with other metals. Controlled production, particularly of high surface area and porous structures, is an important approach to improving the performance of TMSs.

References

1. Khan, M.S., Shariq, M., Bouzgarrou, S.M., Azooz, R.E., kashif Ali, S., Ghaly, W.A., Hassan, K.F.: Recent advancements in supercapacitors and their charge storage mechanism and progress in transition metal sulfide-based electrodes. Phys. Scr. **99**, 062001 (2024)
2. Rehman, J., Eid, K., Ali, R., Fan, X., Murtaza, G., Faizan, M., Laref, A., Zheng, W., Varma, R.S.: Engineering of transition metal sulfide nanostructures as efficient electrodes for high-performance supercapacitors. ACS Appl. Energy Mater. **5**, 6481–6498 (2022)
3. Zhu, M., Wang, D., Ge, Z., Pan, L., Chen, Y., Wang, W., Mitsuzaki, N., Jia, S., Chen, Z.: Recent advances in transition metal sulfide-based electrode materials for supercapacitors. Chem. Commun. **61**, 8314–8326 (2025)
4. Yan, J., Lu, J., Sheng, Y., Sun, Y., Zhang, D.: Research progress in the preparation of transition metal sulfide materials and their supercapacitor performance. Micromachines (Basel) **15**, 849 (2024)
5. Godlaveeti, S.K., El-marghany, A., Nagireddy, R.R., Gedi, S., Chintaparty, R.: Comparative study of electrochemical supercapacitor performance among various nickel phases: hydroxide, oxide, and sulfide. Ceram. Int. **51**, 20620–20627 (2025)
6. Miya, L.A., Ghosh, S.K., Kumari, P., Morema, C.N.M., Mallick, K.: Eco-friendly and sustainable supercapacitor design: Cobalt sulfide nanoparticles embedded on carbon cloth as an electrode material for asymmetric devices. Chem. Pap. (2025)
7. Rosli, M.I., Omar, F.S., Awang, R., Saidi, N.M.: Optimization of heating temperature on the growth of manganese sulfide nanosheets binder-free electrode for supercapattery. Ionics (Kiel) **30**, 407–420 (2024)
8. Kang, S., Wang, C., Chen, J., Meng, T.: Progress on solvo/hydrothermal synthesis and optimization of the cathode materials of lithium-ion battery. J Energy Storage **67**, 107515 (2023)
9. Trivedi, N.A., Sharma, P.J., Joshi, K.K., Patel, V., Sumesh, C.K., Pataniya, P.M.: ReS2/CoS heterostructure catalysts for electrochemical application in hydrogen evolution and supercapacitor. Int. J. Hydrogen Energy **61**, 1212–1219 (2024)

10. Bahri, M., Gebre, S.H., Elaguech, M.A., Dajan, F.T., Sendeku, M.G., Tlili, C., Wang, D.: Recent advances in chemical vapour deposition techniques for graphene-based nanoarchitectures: from synthesis to contemporary applications. Coord. Chem. Rev. **475**, 214910 (2023)
11. Baviskar, P.K., Karade, S.S.: Chemical bath deposition: thin films with assorted morphologies, simple chemical methods for thin film deposition: synthesis and applications. 27–95 (2023)
12. Abdullah, H., Sundararaj, M., Selvanathan, V., Naim, N.M., Akhtaruzzaman, M., Othman, M.H.D., Yap, W.F., Mohmad, A.R., Azman, N.J., Ahmad, M.F., Sampe, J., Yuliarto, B.: The effect of deposition time on the morphology of CuS electrodes fabricated by chemical bath deposition for supercapacitor applications. Appl. Phys. A Mater. Sci. Process. **130**, 1–9 (2024)
13. Manyepedza, T., Auvray, T., Friščić, T., Rees, N.V.: Electrochemically deposited transition metal dichalcogenide heterostructures as electrocatalysts: accelerated kinetics for the hydrogen evolution reaction. Electrochem. Commun. **160**, 107678 (2024)
14. Zhang, P., Ji, Y., Zhang, B., Zhang, S., Wang, S., Zhou, P.: CuS/CoS heterojunction with enhanced redox kinetics via electrodeposition for high-performance supercapacitors. FlatChem **53**, 100921 (2025)
15. Salvatore, K.L., Fang, J., Tang, C.R., Takeuchi, E.S., Marschilok, A.C., Takeuchi, K.J., Wong, S.S.: Microwave-assisted fabrication of high energy density binary metal sulfides for enhanced performance in battery applications. Nanomaterials **13**, 1599 (2023)
16. Konwar, M., Singh, R.S.K., Borthakur, L.J.: Zinc cobalt sulfide/reduced graphene oxide nanocomposite for high-performance asymmetric supercapacitor: a microwave-assisted synthesis. Energ. Technol. **12**, 2301524 (2024)
17. Sowbakkiyavathi, E.S., Arunachala Kumar, S.P., Maurya, D.K., Balakrishnan, B., Guo, J.Z., Subramania, A.: Research progress in the development of transition metal chalcogenides and their composite-based electrode materials for supercapacitors. Adv. Compos. Hybrid Mater. **7**(4), 1–66
18. Sambathkumar, C., Nallamuthu, N., Kumar, M.K., Sudhahar, S., Devendran, P.: Electrochemical exploration of cobalt sulfide nanoparticles synthesis using cobalt diethyldithiocarbamate as single source precursor for hybrid supercapacitor device. J. Alloys Compd. **920**, 165839 (2022)
19. Moradi, M., Zolfaghari, S., Pooriraj, M., Babamoradi, M., Hajati, S.: One-step synthesis of Zn-doped nickel sulfide/graphene derived from Ni-MOF for supercapacitor application. Mater. Chem. Phys. **329**, 130068 (2025)
20. Yadav, M.S., Kour, S., Sharma, A.L.: Unveiling the potential of bimetallic chalcogenides Ni Co2X4 (X = S, Se, Te): a computational exploration of electronic structure, quantum capacitance, and mechanical properties for advanced energy storage. J. Phys. Chem. Solids **205**, 112830 (2025)
21. Ouyang, Y., Chen, Y., Peng, J., Yang, J., Wu, C., Chang, B., Guo, X., Chen, G., Luo, Z., Wang, X.: Nickel sulfide/activated carbon nanotubes nanocomposites as advanced electrode of high-performance aqueous asymmetric supercapacitors. J. Alloys Compd. **885**, 160979 (2021)
22. Xiong, X., Jin, J., Xie, M., Chen, J., Zhang, Y., Wan, L., Du, C.: Cobalt-nickel sulfide hierarchical hollow microspheres to boost electrochemical activity for supercapacitors. J. Alloys Compd. **1010**, 177730 (2025)
23. Wang, Z., Wang, H., Pei, D., Wan, S., Wang, Y., Yu, M., Lu, H.: Hydrothermal-deposition synthesis of MnO2/MnS nanorods as high-performance asymmetric supercapacitor positive electrode materials. Prog. Nat. Sci. Mater. Int. **34**, 1057–1065 (2024)
24. Momeni, M.M., Najafi, M., Naderi, A., Aydisheh, H.M., Lee, B.K., Farrokhpour, H.: Electrochemical synthesis of Mn–Ni–S on titania nanotubes as new dual-function electrodes for photo-assisted asymmetric supercapacitors. Int. J. Hydrogen Energy **88**, 1085–1097 (2024)
25. Barik, R., Ingole, P.P.: Challenges and prospects of metal sulfide materials for supercapacitors. Curr Opin Electrochem **21**, 327–334 (2020)

Electrochemically Exfoliated Graphene Modified Carbon Paper Electrodes for Use with Vanadium and Quinone Flow Battery Electrolytes

Farouq Sabri Mjalli, Züleyha Kudaş, Reza Afshar Ghotli, Shamik Chaudhuri, Amina Ben Ayed, Mina Namvari, Koray Bahadır Dönmez, Zehra Çobandede, Tülay Yılmaz İnan, Mustafa Kemal Bayazıt, Chee Tong John Low, Andres Parra-Puerto, Stefan Spirk, and Barun Kumar Chakrabarti

Abstract Few layered graphene powders were prepared via electrochemical exfoliation of graphite in a deep eutectic solvent (DES) and with aqueous electrolytes to function as electrode catalysts on commercially sourced carbon papers (CPs) for the positive electrolyte of the all-vanadium redox flow battery (VRFB). Both graphene powders were decorated on the CP by means of a horizontal binder-free electrophoretic deposition (EPD) technique (using dimethylformamide as the colloidal

F. S. Mjalli
Department of Petroleum and Chemical Engineering, Sultan Qaboos University, Muscat, Oman

Z. Kudaş · R. A. Ghotli · S. Chaudhuri (✉) · A. B. Ayed · M. Namvari · K. B. Dönmez · Z. Çobandede · T. Y. İnan · M. K. Bayazıt · B. K. Chakrabarti (✉)
Sabanci University Nanotechnology Research and Application Centre (SUNUM), Tuzla, İstanbul, Turkey
e-mail: shamik.chaudhuri@sabanciuniv.edu

B. K. Chakrabarti
e-mail: barun.chakrabarti@sabanciuniv.edu

M. K. Bayazıt
Faculty of Engineering and Natural Science, Sabancı University, İstanbul, Turkey

C. T. J. Low
WMG, Warwick Electrochemical Engineering Group, Energy Innovation Centre, University of Warwick, Coventry, UK

A. Parra-Puerto
Department of Physical Chemistry and Electrochemistry Institute, University of Alicante, San Vicente del Raspeig (Alicante), Spain

S. Spirk
Graz University of Technology - Institute of Bioproducts and Paper Technology, Graz, Austria

Ecolyte GmbH, TU Graz Campus Inffeldgasse, Graz, Austria

M. Awang et al. (eds.), *Proceedings of the International Conference on Mechanical, Manufacturing and Plant Engineering*, Proceedings in Technology Transfer,
https://doi.org/10.1007/978-981-95-7817-7_18

solvent). The cyclic voltammetry and electrochemical impedance spectroscopy evaluations of the VO^{2+}/VO_2^+ redox couple showed that the graphene prepared electrochemically in an aqueous media provided better performances. Similarly, an attempt was made to electrochemically etch a titanium-based MAX phase to $Ti_3C_2T_X$-MXene using a DES prepared from choline chloride and methane sulfonic acid. However, the aluminum from the MAX phase could not be etched. Instead, MXene prepared via chemical etching with HCl and LiF was of high quality and could be dispersed easily in water to prepare a colloidal suspension for EPD. After a binder free EPD on the CP substrate, the electrochemical results for MXene modified CP were not as good as that of the graphene modified CP owing to insufficient deposition of the former in the non-optimal EPD process. Moreover, 1,2-dihydrobenzoquinone-3,5-disulfonic acid (BQDS), a high-solubility, organic quinone redox alternative dissolved at 10 mM in 1 M sulfuric acid was tested for comparative assessments of cyclic voltammetry. While BQDS was primarily examined in organic flow batteries, its inclusion here offers an insight into the electrode behavior beyond traditional VO^{2+}/VO_2^+ systems.

Keywords Electrochemical exfoliation · Graphene · Electrophoretic deposition · Vanadium · Quinone

1 Introduction

The growing demand for efficient, scalable energy storage solutions has positioned vanadium redox flow batteries (VRFBs) as one of the candidates for grid applications. Their unique architecture decouples energy and power, enables long cycle life and flexible scaling. However, limitations in electrode performance, particularly in facilitating the sluggish VO^{2+}/VO_2^+ redox kinetics in the positive electrolyte, remain a major challenge.

Carbon paper (CP) is widely used as an electrode substrate due to its conductivity and stability, yet its relatively low electrochemical activity necessitates modification with advanced catalysts. Graphene, with its high surface area and conductivity, has been extensively investigated for this purpose. MXenes, a family of two-dimensional carbides and nitrides, also offer promising metallic-like conductivity and hydrophilicity for VRFB applications [1].

Considering previous investigations, this chapter focuses on graphene and $Ti_3C_2T_x$ MXene as CP modifiers, synthesized via electrochemical and chemical approaches, respectively. For the performance comparison, both of them were decorated on the CP electrode via electrophoretic deposition (EPD) and evaluated by cyclic voltammetry (CV) and electrochemical impedance spectroscopy (EIS). Additionally, the electrode behavior was also investigated using the organic redox couple 1,2-dihydrobenzoquinone-3,5-disulfonic acid (BQDS) in 1 M H_2SO_4 [2].

2 Graphene Preparation and EPD

2.1 *Electrochemical Exfoliation of Graphene*

Earlier investigations examined multiple DESs for electrochemical graphene exfoliation [3]. In the present work, few-layered graphene powders were prepared using one of these DES electrolytes consisting of choline chloride (ChCl) as hydrogen bond acceptor (HBA) and methanesulfonic acid (MSA) as hydrogen bond donor (HBD). DES-derived graphene exhibited moderate defect densities and stable yields. For comparison, graphene was also prepared electrochemically in aqueous electrolytes, a route known to produce highly conductive nanosheets with favorable surface functionalization [4].

2.2 *Mechanistic Insights into DES Chemistry*

Constituents and interactions

As mentioned, DESs typically comprise a HBA, (e.g., choline chloride) and a HBD (e.g., urea, ethylene glycol, organic acids). Strong H-bonding and ionic interactions depress the melting point of DESs relative to that of the individual components, yielding a liquid at or near room temperature. In electrochemical exfoliation, DESs provide: (i) high ionic strength for efficient current flow; (ii) wide electrochemical windows in many cases; and (iii) co-intercalation species (e.g., Cl^-, organic anions) that can insert between graphite layers under anodic bias [5, 6].

Role during exfoliation

Under anodic polarization of graphite anions from the DES intercalate, gas evolution (e.g., CO_2, O_2) and electrochemical functionalization create local pressure and ultimately (3) mechanical delamination yields few-layered sheets. The HBD/HBA ratio, viscosity, and water activity control the physicochemical properties of the produced graphene such as defect density, flake size, and surface terminations (e.g., epoxides, hydroxyls). DESs with lower viscosity and appropriate anion size foster faster intercalation and larger lateral flakes. However, it can also introduce oxygenated functionalities that influence VRFB kinetics. In contrast, aqueous electrolytes often yield graphene with higher conductivity and fewer insulating residues, consistent with the superior VO^{2+}/VO_2^+ kinetics as observed here [4].

Implications for VRFB electrodes

DES-derived graphene may present more oxygen-containing groups, improve wettability but increase charge-transfer resistance relative to aqueous-derived graphene. Optimizing DES composition (choice of HBD, water content, chloride versus sulfonate anions) could shift this balance toward higher conductivity while retaining its wetting advantages [3].

2.3 Binder-Free EPD on CP

Both graphene powders, DES-derived and aqueous-derived, were deposited onto CP substrates (Freudenberg H23 as employed in a previous work [7]) by a horizontal, binder-free EPD process [7, 8]. Here, dimethylformamide (DMF) served as the colloidal solvent, providing stable dispersions (0.1 g L^{-1}) and enabling uniform coatings while eliminating the necessity of polymer binders. This ensured direct electrical contact between the nanocarbon coating and the CP fibers.

2.4 Mechanistic Insights into EPD

Particle charging and stability

In DMF, graphene and MXene flakes develop surface charges via dissociation of surface groups such as edge carboxylates/hydroxyls and –O, –OH, –F terminations, respectively. Zeta potential governs dispersion stability and drift under an electric field. Adjusting solids loading and pH-modifiers can tune the deposition rate.

Transport and deposition

Under an applied field, charged colloids migrate toward the oppositely charged CP working electrode (200 V for 30 min for graphene dispersed in DMF). The mass of deposit *m* commonly scales with the Hamaker's law as shown in Eq. (1) [9]:

$$m \propto C\,\mu\,E\,t \tag{1}$$

where C is the colloidal solids concentration, μ electrophoretic mobility, E field strength, and t time. In a horizontal EPD process, minimizing buoyancy and sedimentation can improve uniformity of the deposition.

Adhesion and microstructure

The particles form a percolated network upon arrival near the surface of the substrate. With binder-free EPD, van der Waals forces and mechanical interlocking with the CP fibers provide adhesion that enhances by post-deposition drying. In case of MXene, stacked lamellae produce conductive pathways and high hydrophilicity whereas in graphene, overlapping platelets form in-plane highways but can suffer from restacking issues that can be mitigated by controlled solids loadings and brief sonication prior to EPD. Figure 1 shows the schematic of EPD of graphene (DMF-based suspension) and MXene (aqueous-based suspension).

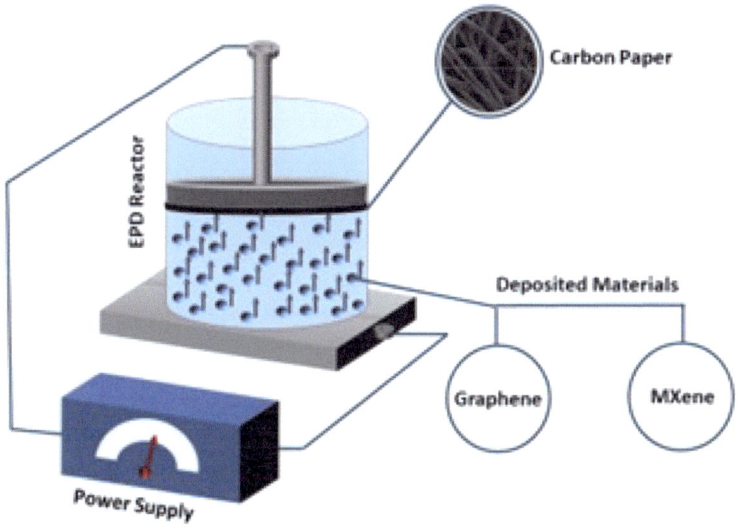

Fig. 1 Schematic of EPD for graphene and MXene deposition on CP

3 Electrochemical Characterization of Graphene-modified CP

3.1 Cyclic Voltammetry

CV of the VO^{2+}/VO_2^+ couple in Fig. 2a showed that CP electrodes modified with aqueous-exfoliated graphene display sharper redox peaks, lower peak-to-peak separation (ΔE_P), and higher currents than that of the DES-derived graphene. These properties indicate faster electron-transfer kinetics and improved reversibility for aqueous-derived graphene, consistent with its higher conductivity and cleaner surfaces [4].

3.2 Electrochemical Impedance Spectroscopy

EIS in Fig. 2b corroborates the CV trend as aqueous-graphene coatings (Fig. 2a) yield smaller semicircles in Nyquist plots manifesting lower charge-transfer resistance (R_{ct}) relative to that of the DES-graphene. The solution resistance (Rs) remained comparable across electrodes. Such enhancement stemmed predominantly from the interfacial kinetics and the percolation within the coating. The corresponding equivalent circuit is shown in the inset of Fig. 2b similar to a previous study on the EPD of reduced graphene oxide on CP [8, 10]. According to this simple circuit, R_S is the bulk resistance mainly associated with the ionic electrolyte and electronic carbon

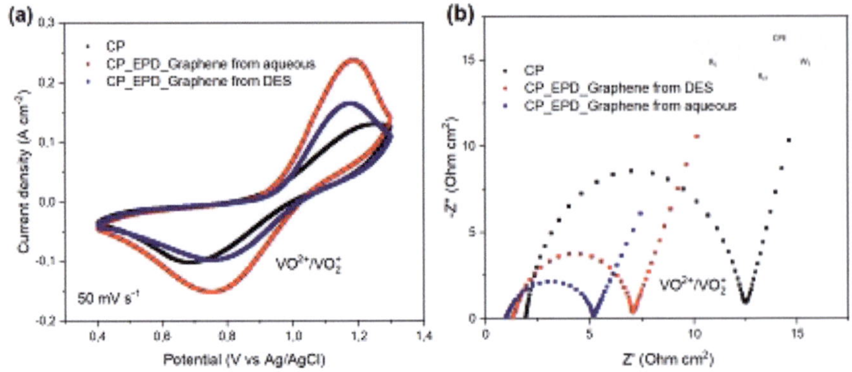

Fig. 2 **a** CV comparison of DES-versus aqueous-exfoliated graphene deposited on CP. **b** Nyquist plots: DES versus aqueous graphene-modified CP (inset shows the equivalent circuit [8])

paper conductivities, R_{CT} is the diameter of the semi-depressed circle and corresponds to the charge-transfer process at the CP/electrolyte interface (corresponding to the VO^{2+}/VO_2^+ reaction), W_S is the generalized Warburg impedance for finite diffusion of the reacting species, and CPE is the constant-phase element characterizing a double-layer capacitance effect of non-ideal porous media such as for the CP [8].

4 MXene: Synthesis and Deposition on CP

4.1 *Electrochemical Etching in DES*

In this study the electrochemical etching of Ti_3AlC_2 to $Ti_3C_2T_x$ in a choline-chloride/methane-sulfonic-acid DES [11] was explored. However, this method produced insufficient MXene for electrode fabrication, presumably, due to slow Al extraction kinetics and surface passivation.

4.2 *Chemical Etching*

Chemical etching of Ti_3AlC_2 using HCl and LiF yielded high-quality $Ti_3C_2T_x$ MXene exhibiting characteristic interlayer expansion and abundant surface terminations such as –O/–OH/–F. The resulting material dispersed readily in water, forming a stable colloidal suspension compatible for EPD.

The surface morphologies of MAX phase and MXene are displayed in Fig. 3a and b. SEM images of the MAX phase exhibit an irregular morphology resembling rock,

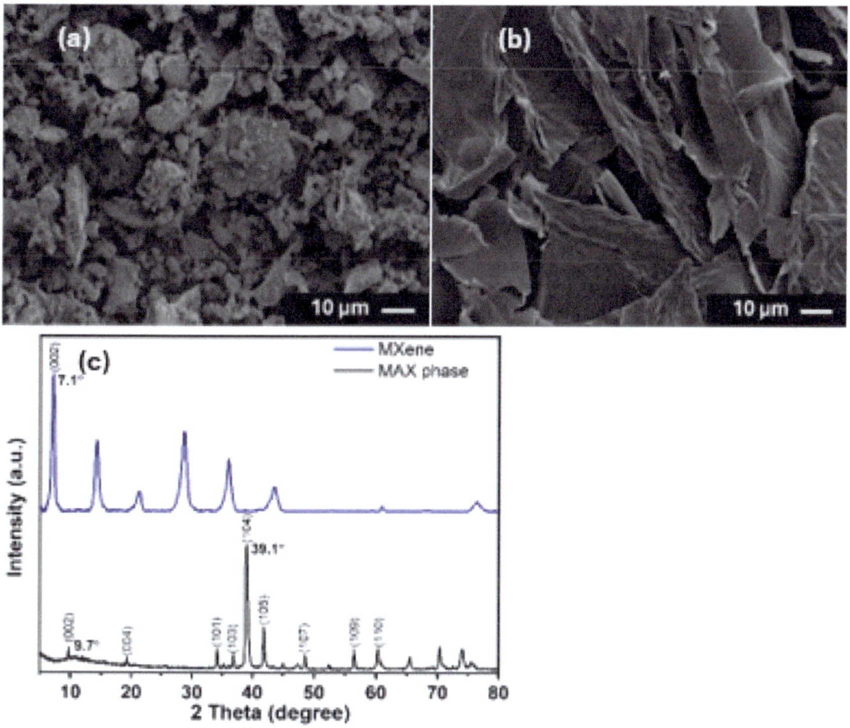

Fig. 3 SEM micrograph of **a** MAX phase, **b** MXene and **c** XRD spetra of MAX phase and MXene

characterized by a broad size distribution [12]. After treatment with hydrofluoric acid, the MXene sheets depicted in Fig. 3b exhibited crumpled shapes and fewer layers, signifying the successful synthesis of thin 2D MXene nanosheets through the removal of aluminum during the etching and exfoliation process.

The XRD spectra of MAX phase and synthesized MXene sample are plotted in Fig. 3c. Sharp peaks at 2θ values of 9.7° (002) and 39.1° (104) are visible in the MAX phase diffraction pattern, suggesting that the MAX phase has a clearly defined crystalline structure. Following the etching process, the (002) peak shifted to a lower value of $2\theta = 7.1°$ and the (104) peak disappeared, suggesting the total removal of Al. Additionally, using the Bragg equation, the interlayer spacing decreased from 0.91 nm for the MAX phase to 1.24 nm for the MXene. All of these results show that MXene was successfully synthesized [13].

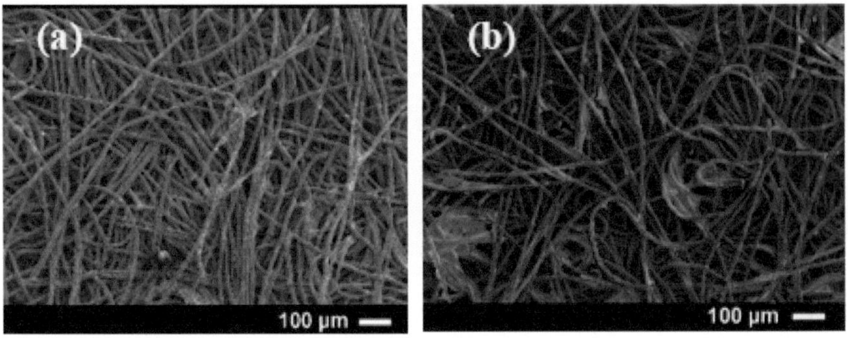

Fig. 4 SEM micrograph of **a** graphene and **b** MXene deposited on the CP

4.3 EPD of MXene on CP

EPD of MXene nanostructures in aqueous dispersion was adopted from a recent review on the subject (2 g L^{-1} colloidal concentration, 30 min sonication, 5 min EPD at 5 V) [14]. Binder-free EPD yielded uniform lamellar $Ti_3C_2T_x$ MXene coatings on CP. The metallic-like conductivity and intrinsic hydrophilicity of $Ti_3C_2T_x$ facilitated efficient charge transport and enhanced electrolyte accessibility to active sites. Figure 4b displays the SEM micrograph of MXene-coated CP, which is compared with the SEM image for a graphene coated CP sample (Fig. 4a). However, the amount of MXene deposited is not optimized and this will be the subject of further work (especially a focused attention should be paid to colloidal concentration, zeta potential, deposition time and EPD voltage).

4.4 VO^{2+}/VO_2^+ Redox Activity on CP

CV and EIS in Fig. 5a, b demonstrated that aqueous-exfoliated graphene outperformed MXene, delivering higher VO^{2+}/VO_2^+ peak currents and lower R_{ct}. The lower peak currents of MXene than that of the aqueous-exfoliated graphene (Fig. 2a) may presumably be due to the non-optimal process involving its EPD on the CP substrate.

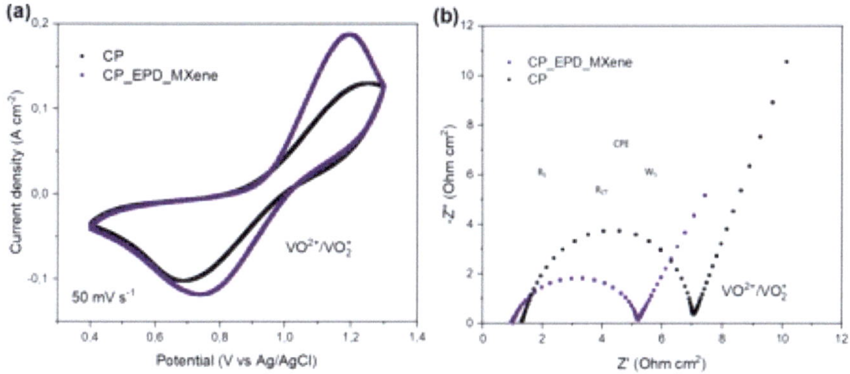

Fig. 5 **a** CV comparison of bare CP versus chemically etched MXene deposited on CP. **b** Nyquist plots of bare CP versus chemically etched MXene deposited on CP (inset shows the equivalent circuit [8])

5 Electrochemical Performance of CP_Graphene and CP_MXene with BQDS Electrolyte

In this work, 10 mM BQDS in 1 M sulfuric acid was used as the electrolyte as reported elsewhere [2]. The CVs of CP showed quasi-reversible redox behavior with well-defined peaks in Fig. 6a, b. Aqueous-exfoliated graphene slightly narrows ΔEp relative to bare CP, reflecting accelerated kinetics. MXene also increases peak currents when compared to the bare CP and reduce ΔEp, consistent with efficient electron transfer pathways and hydrophilic interfaces that accommodate proton-coupled electron transfer typical for quinones. [2].

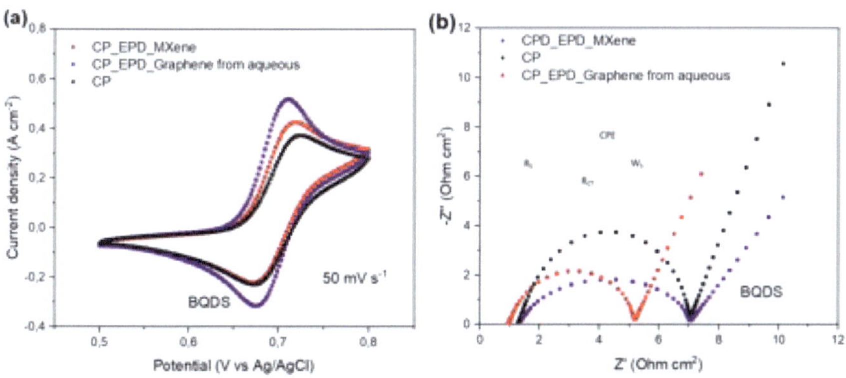

Fig. 6 **a** CV of BQDS on pristine CP, aqueous exfoliated graphene-CP, and MXene-CP. **b** Nyquist plots of bare CP versus chemically etched MXene deposited on CP versus graphene exfoliated from an aqueous electrolyte (inset shows the equivalent circuit [8])

6 Comparative Analysis: Vanadium and Organic Quinone Electrolytes (BQDS)

6.1 Thermodynamics and Kinetics

VO^{2+}/VO_2^+ (acidic)

Inner-sphere, multi-step mechanism with oxygenated vanadium species. The reaction kinetics are sensitive to the surface functional groups and catalytic sites. Typically, higher overpotentials and stronger dependence on electrode activation were the characteristics of this kind of system.

BQDS/H_2SO_4

Both outer-/inner-sphere features with proton-coupled electron transfer (PCET) are the predominant mechanism. This reaction showed generally faster apparent kinetics on hydrophilic, conductive surfaces. The redox potential was tunable via substitution as $-SO_3H$ increases solubility and acid compatibility.

6.2 Solubility, Stability, and Crossover

Vanadium salts ($VOSO_4$)

They have high stability in strongly acidic media and well-established crossover management via ion-exchange membranes in VRFBs [15].

BQDS

It possesses high aqueous solubility at 10 mM in 1 M H_2SO_4 [2]. Chemical stability depends on potential window and pH. Further, the crossover and possible membrane interactions differ from the vanadium systems.

6.3 Electrode Interface Requirements

VO^{2+}/VO_2^+

The catalytic sites such as oxygen functionalities, defects are advantageous as they facilitate inner-sphere electron transfer. Graphene with fewer residues and MXene with abundant terminations both aid kinetics. Further, MXene's metallic nature lowers R_{ct}.

BQDS

It prefers hydrophilic, accessible surfaces for PCET. MXene's –OH/–O terminations and open lamellae are highly advantageous while graphene improves the conductivity with moderate wetting.

Practical implications for CP-based electrodes

- CP modified with aqueous exfoliated graphene offers the most universal enhancement across both electrolytes whereas MXene-CP is the second-best candidate, owing to a non-optimal EPD process.
- Binder-free EPD in DMF (graphene) and water (MXene) enables uniform films without blocking pores. Therefore, tuning solids loading and the deposition time can balance coverage with mass-transport access.

7 Conclusions and Outlook

This focused study on CP substrates shows: (1) Aqueous-exfoliated graphene outperforms DES-derived graphene for VO^{2+}/VO_2^+ due to higher conductivity and cleaner surfaces [4]. (2) MXene (chemically etched, HCl + LiF) delivers a moderate electrochemical response among the tested modifiers, while graphene derived from electrochemical exfoliation from an aqueous medium delivers the lowest R_{ct} and highest peak currents for VO^{2+}/VO_2^+. (3) With BQDS (10 mM in 1 M H_2SO_4) as an organic benchmark, both modifiers enhance the kinetics relative to bare CP, with graphene again showing the most significant performance gains [2].

As further work, a rational DES design is necessary to tune graphene terminations and conductivity [3]. Further, scaling electrochemical routes to MXene in DES [11], hybrid MXene–graphene architectures to combine lamellar conduction with in-plane graphene pathways, and systematic long-term stability studies in both vanadium and quinone electrolytes need be pursued.

Funding This work was supported by the European Innovation Council (EIC) under grant agreement No. 101115293 (VanillaFlow). This research was also funded by the CETPartnership, under the Joint Call 2023, through the project "HydrogEn oRganiC Regenerative FUeL CEll for Stationary Energy Storage (HERCULES)". The CETPartnership's research projects are co-funded by the European Commission (Grant Agreement No. 101069750) and the national TUBITAK 1071 via the project code: 224N167. R.A.G. is grateful to Türkiye Bilimsel ve Teknolojik Araştırma Kurumu (TÜBİTAK) for the 1001 project for funding his research in SUNUM (223M269).

References

1. Botling, E., et al.: RSC Adv. **15**, 13744–13752 (2025)
2. Rubio-Garcia, J., et al.: J. Mater. Chem. A **8**, 3933–3941 (2020)
3. Chakrabarti, M.H., et al.: Chem. Eng. J. **274**, 213–223 (2015)
4. Parvez, K., et al.: J. Am. Chem. Soc. **136**, 6083–6091 (2014)

5. Chakrabarti, B.K., et al.: Batteries **6**, 38 (2020)
6. Chakrabarti, M.H., et al.: J. Solution Chem. **42**, 2329–2341 (2013)
7. Chakrabarti, B.K., et al.: Int. J. Hydrog. Energy **44**, 30093–30107 (2019)
8. Chakrabarti, B.K., et al.: ChemElectroChem **4**, 194–200 (2017)
9. Chakrabarti, B.K., et al.: Int. J. Energy Res. **46**(10), 13205–13250 (2022)
10. Chakrabarti, B.K., et al.: ACS Appl. Mater. Interfaces **12**(48), 53869–53878 (2020)
11. Mishra, S., et al.: J. Mol. Liq. **415**, 126338 (2024)
12. Sheng, X., et al.: J. Mater. Sci. **56**, 4212–4224 (2021)
13. Zahrabi, M., et al.: Biofabrication **17**, 045011 (2025)
14. Namvari, M., Chakrabarti, B.K.: Adv. Colloids Interface Sci. **331**, 103208 (2024)
15. Ghasemiestahbanati, E., et al.: ACS Appl. Energy Mater. **5**(2), 1505–1515 (2022)

Electrophoretic Processing of $LiCoO_2$ Electrodes for Integrated Lithium-Ion Battery Manufacturing

Baidaa Alkhateab, Sanaz Mohammedzadeh Sarabi, Serdar Bozoglu, Koray Bahadir Donmez, Serap Hayat Soytas, Barun Kumar Chakrabarti, and Chee Tong John Low

Abstract Electrophoretic deposition (EPD) has evolved into a scalable and adaptable technique for the fabrication of composite electrodes in lithium-ion batteries, building on its industrial roots in automotive and ceramic coating applications. In this study, EPD is used to fabricate $LiCoO_2$ (LCO) cathodes for Lithium-ion Batteries (LIBs) by co-depositing LCO particles with Super P carbon black and polytetrafluoroethylene (PTFE) onto aluminium foil current collectors under high DC electric fields in an organic-based colloidal solution. The cathodes prepared via EPD are compared with their counterparts prepared via the industrially approved slurry-cast process. Structural, morphological, and electrochemical performance of EPD-based and slurry-cast LCO electrodes are compared via SEM, cyclic voltammetry (CV), electrochemical impedance spectroscopy (EIS), and galvanostatic charge–discharge (GCD) cycling. In most cases, EPD-based half-cells have delivered similar performances to their slurry-cast counterparts, except for the case of successive cycling

B. Alkhateab · S. Hayat Soytas (✉) · B. K. Chakrabarti (✉)
Sabancı University Nanotechnology Research and Application Center (SUNUM), Tuzla, Istanbul, Turkey
e-mail: seraphayat@sabanciuniv.edu

B. K. Chakrabarti
e-mail: barun.chakrabarti@sabanciuniv.edu

S. M. Sarabi
Department of Nanoscience and Nanoengineering, Gebze Technical University, Gebze, Kocaeli, Turkey

S. Bozoglu
Faculty of Science, Department of Chemistry, Istanbul Technical University, Maslak, Istanbul, Turkey

K. B. Donmez
TÜBİTAK National Metrology Institute (TÜBİTAK UME), Gebze, Kocaeli, Turkey

C. T. J. Low (✉)
WMG, Warwick Electrochemical Engineering Group, Energy Innovation Centre, University of Warwick, Coventry, UK
e-mail: C.T.J.Low@warwick.ac.uk

M. Awang et al. (eds.), *Proceedings of the International Conference on Mechanical, Manufacturing and Plant Engineering*, Proceedings in Technology Transfer,
https://doi.org/10.1007/978-981-95-7817-7_19

that clearly displayed the superiority of EPD for preparing highly durable cathodes for energy storage applications.

Keywords Electrophoretic deposition · LCO · Lithium-ion battery · Coin cell

1 Introduction

Lithium-ion Batteries (LIBs) remain among the leading energy storage technologies due to their high-power density and relatively light weight. One of the widely used commercial cathodes is Lithium cobalt oxide ($LiCoO_2$, or LCO) which has high volumetric density, high energy density, and good cyclability [1]. LCO has a theoretical capacity of 270 mAh/cm, however, in practice only half of the lithium is reversibly deintercalated, resulting in a practical capacity in the range 140–160 mAh/g.

In conventional electrode fabrication, the slurry-casting is the most common. This method consists of mixing the active material, conductive carbon and a binder, dissolving the mixture in a usually high boiling point organic solvent to form a consistent slurry texture, and then coating the slurry on a current collector prior to drying. Several problems arise from this method such as inhomogeneity and binder segregation as well as difficulty in controlling the loading of the active materials.

An alternative fabrication method is electrophoretic deposition (EPD) [2]. In this technique the electrical field forces the migration of charged particles in a colloidal suspension toward a charged electrode thereby forming a homogenous deposited layer [3]. In EPD, one can tailor the thickness, and composition of the deposited layer by varying the relevant parameters.

In this work, EPD is used in a horizontal configuration as reported for preparing redox flow battery electrodes in an earlier study [4, 5]. However, since LCO can be negatively charged anodic EPD is therefore used (i.e., particles carry negative charges and deposit toward a positively biased electrode). Additionally, the horizontal EPD process (i.e., the two electrodes are placed within a certain distance on top of each other) uses gravity to settle the agglomerated and unwanted particles on the bottom counter electrode whilst facilitating useful active material deposition on the working electrode held on top owing to the electrophoresis forces, which enhance the uniformity of the coating.

In this research, anodic horizontal EPD was utilized to coat LCO on Al foil, which was then tested as cathode materials in LIBs. Structural properties of EPD cathodes were characterized by SEM. Additionally, half-cell battery tests were carried out via CV, EIS and GCD at 0.5C for 50 cycles for both LCO cathodes prepared by EPD and by slurry-casting.

2 Materials

All reagents utilized in this study were of analytical grade and employed without any further purification. The main chemicals were Lithium cobalt oxide (LCO, 85%, Sigma-Aldrich), ethyl cellulose (EC), Magnesium chloride ($MgCl_2$), poly (vinylidene fluoride) (PVDF), N-methyl-2-pyrrolidone (NMP), and Lithium hexafluorophosphate (1.2 M $LiPF_6$ in EC/EMC = 30/70 (w/w), battery grade) ($LiPF_6$). All chemicals are purchased from Sigma-Aldrich except the Super P carbon (SuperP, 100–200 nm, 75 m^2/g, Timcal).

3 Experimental

To prepare the LCO cathode by EPD, LCO (85%), Super P (10%), and EC (5%) were dispersed in isopropanol (1 g/L). $MgCl_2$ (5 wt.% of solids) was added as a charging agent, and the suspension was sonicated for 90 min. Electrophoretic deposition was performed with aluminium foil (15 μm) as the working electrode and a platinum-coated titanium mesh as the counter electrode, separated by 10 mm. A DC power supply (90–120 V) was applied with stirring at 100 rpm. After 10 min, deposition was stopped, and electrodes were withdrawn and dried in air. When required, coatings were densified to ~70% of their initial thickness using a Gelon LIB calendaring system at 50 °C.

On the other hand, slurry casting of LCO on Al foil was done as follows: the electrode slurry was formulated by combining LCO, Super P carbon, and PVDF in a mass ratio of 80:10:10, respectively. The components were dispersed in NMP and thoroughly homogenized until a viscous, honey-like consistency was obtained. The resultant slurry was uniformly cast onto the current collector using the doctor blade technique. The coated electrodes were subsequently dried in a vacuum oven at 80 °C overnight to ensure complete solvent evaporation and to stabilize the electrode structure prior to electrochemical testing.

4 Characterization

Scanning electron microscopy (SEM) was performed using Leo Supra35VP field emission SEM to observe the surface morphologies of LCO electrodes. The SEM images were produced using an electron power of 3 kV, with a working distance of 8 mm. To find the structure of the cellular units (d-spacing) for LCO coated by EPD, X-ray diffraction (XRD) was carried out. A Bruker D8 ADVANCE diffractometer with Cu Kα radiation (λ = 1.54060 Å) was used. LCO electrodes were placed on a silicon wafer and measured with time lapse of 1 s and increments of 0.02.

LIB devices were assembled with LCO on Al foil as working electrodes (i.e., LCO by EPD and LCO by Slurry-cast), Lithium chip as a reference electrode and cellulose acetate as the separator. The weight of the coated material was calculated by subtracting the weight of bare Aluminium foil, and the separator was soaked in the electrolyte $LiPF_6$ until properly saturated. The components of LIB (separator, electrodes, spring, spacers) were then sealed in a Swagelok-type cell. Assembled cells have been maintained overnight at room temperature before the electrochemical measurements to allow dispersion of electrolyte into LCO electrodes uniformly (Fig. 1).

The main techniques employed to characterize LIB were Cyclic Voltammetry (CV), Electrochemical Impedance Spectroscopy (EIS) and Galvanostatic Charge–Discharge (GCD) for 50 cycles at 0.5 C. Prior to the GCD test, each of the cells were activated by applying 2 cycles charging-discharging using Constant current-constant potential (CC-CV) to achieve a full lithiation and delithiation [6]. Initially a constant current corresponding to 0.1 mA h/g till 4.3 V was reached to allow for a uniform Li^+ insertion, after which the potential was held for 2 h to allow the reaction

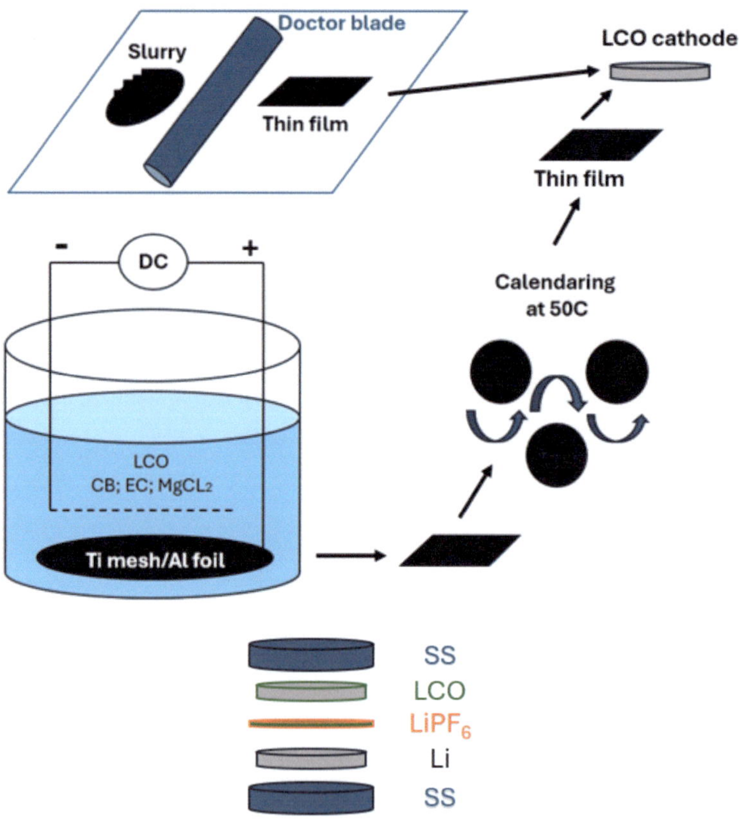

Fig. 1 Scheme of LCO deposition by slurry casting and by EPD and the composition of LIB cell

to reach equilibrium and thus minimizing polarization effect, then -0.1 mA h/g was applied till the potential decreased. Experiments were carried out by the AMATEK Scientific Instruments PARSTAT Multichannel Potentiostat (PMC-1000) device. The EIS measurements were performed with an AC voltage amplitude of 10 mV over a frequency range from 100 mHz to 100 MHz at the open circuit voltage.

The specific capacitance (mA h/g) can be determined from the GCD tests using Eq. (1):

$$C_{SP} = \frac{I \cdot \Delta t}{m} \tag{1}$$

where C_{SP} is the specific capacitance (mAh/g), I represent the discharge current (mA), $\Delta\, t$ is the discharge time (h) and m denotes the mass of the active material (g). Data that are obtained from the EIS spectrums have been fitted on Zview software (Scribner Associates) and a circuit is used to represent the electrochemical elements of the battery system.

5 Results and Discussions

Figure 2a shows the micrograph of pristine LCO. These particles are in the range of 2–5 μm with aggregation into larger secondary particles. Mainly the particles are polyhedral or plate-like which is expected for layered crystal structures. Upon mixing with super P carbon black (CB) and coating LCO on Al foil by EPD some of the surface of LCO is covered due to presence of CB and $MgCl_2$ but the layered particle features remain visible (Fig. 2b). This effect is more prominent in the slurry-cast method where Fig. 2c shows more evidence of binder "wrapping" leading to partial coverage of particle surface. Additionally, there are cracks due to drying and calendaring processes along with localized densification (flat top surface) induced by mechanical pressing [7].

Figure 3 showcases a more detailed comparison of changes in the slurry-cast electrode after cycling, where there are signs for structural degradation and crack propagation which increase the roughness and porosity. There are bundles of grains separated by deep grooves indicating that delithiation process happens on the surface and bulk of the electrode. The corresponding SEM images are not available for EPD samples albeit the electrodes appeared visually the same after 50 cycles at 0.5 C.

Figure 4a presents CV of LCO by EPD in a half-cell configuration. The positions of lithiation and delithiation peaks (at 3.7 V and 4.1 V vs. Li/Li^+, respectively) remain stable with increasing scanning rate from 0.1 to 0.2 mV/s, indicating robust charge transfer kinetics with minimal diffusion limitation. On the other hand, lithiation and delithiation peaks are located at 3.8 and 4.1 V versus Li/Li^+, respectively for the slurry cast sample (Fig. 4b). The increase in scan rate increases peak separation which indicates stronger diffusion control or polarization losses of the slurry cast sample when compared to the EPD electrode. To conclude the narrower peak separation and

Fig. 2 SEM of **a** pristine LCO, **b** pre-cycled LCO deposited on Al foil by EPD, and **c** by slurry-casting

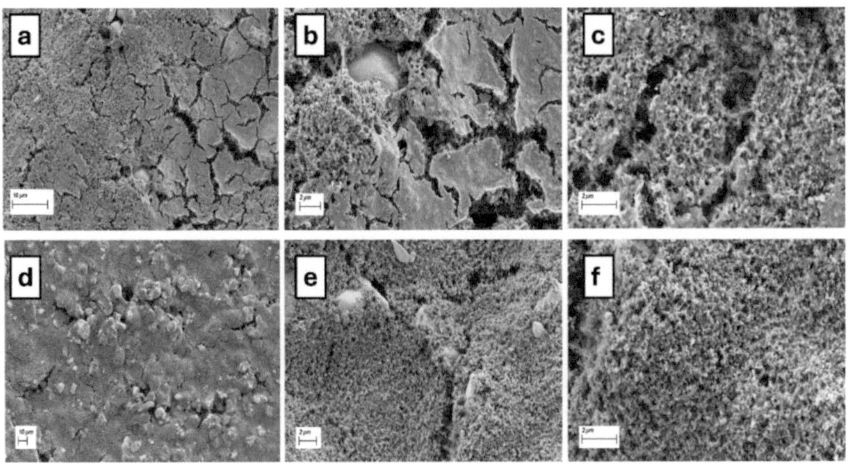

Fig. 3 SEM of LCO deposited on Al foil by slurry-casting, **a–c** pre-cycling, and **d–f** after 50 Cycles

absence of peak shifts at higher scan rates suggests superior kinetics of EPD prepared electrodes compared to the slurry-cast ones [8].

Bearing in mind that the coated layer contains additional materials (i.e., EC and $MgCl_2$ in EPD coating and PVDF in slurry-cast coating) which might complicate the EIS interpretation (Fig. 5), from the Nyquist plot, it seems that the slurry-cast LCO has a smaller semicircle diameter indicating lower charge transfer resistance, but a larger Warburg tail (i.e., correlated to diffusion impedance).

EPD electrode has a larger semicircle but lower Warburg tail which is preferred at high mass loading. Overall, the impedance behavior suggests that while slurry casting gives good ionic/electronic contact initially, EPD provides a superior ion transport for thick-film conditions, a fact that is supported from the literature [8].

LIB half cells were tested for 50 cycles at 0.5 C (estimated based on practical capacitance = 150 mA h/g) and results of specific capacitance (Csp) was calculated using the mass of LCO in the coated layer. The successive cycling for EPD electrodes

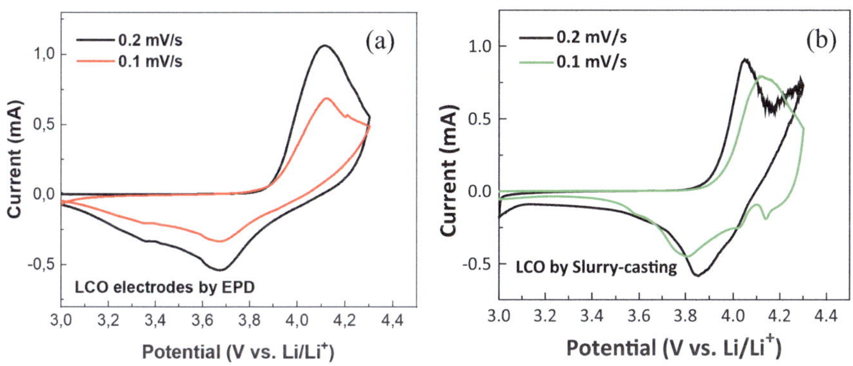

Fig. 4 CV at 0.1 and 0.2 mV/s for LCO deposited on Al foil by **a** EPD and **b** slurry-casting

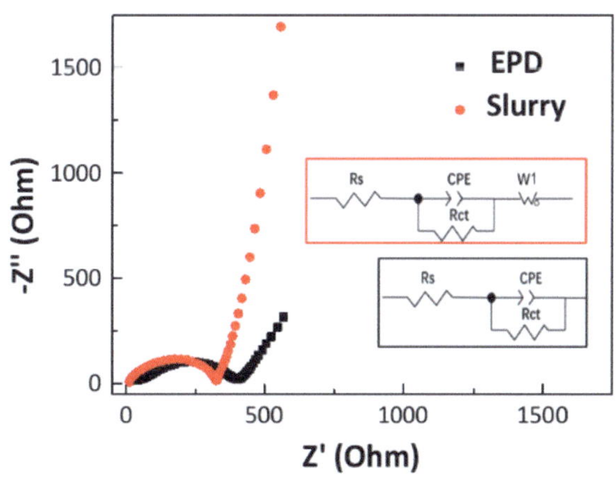

Fig. 5 Nyquist plot of LCO deposited on Al foil by EPD and by slurry-casting

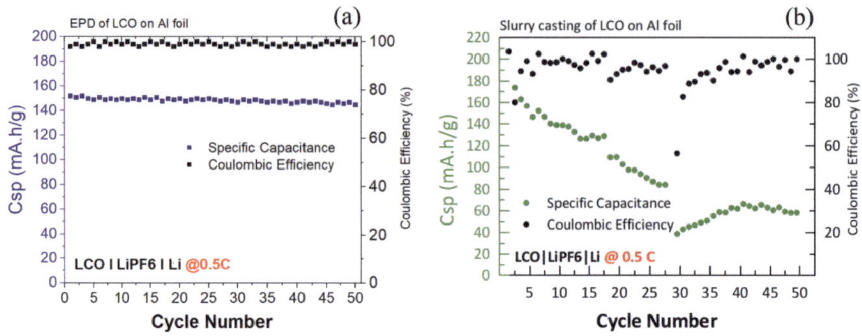

Fig. 6 Specific discharge capacitance of LCO in ||LCO|$LiPF_6$|Li|| half-cell, coated by **a** EPD and **b** slurry casting

is presented in Fig. 6a. The cell performed within the expected specific capacitance for slurry-coated LCO, which was 140–150 mA h/g (Fig. 6b). The coulombic efficiency remained around 100% throughout the test with minimal capacity fading but the EPD electrode maintained far better stability with less polarization over cycles.

6 Conclusion

Electrophoretic deposition (EPD) has emerged as a versatile and scalable approach for the fabrication of composite electrodes in lithium-ion batteries (LIBs), extending its well-established industrial applications in automotive coatings and ceramics to the field of energy storage. In this study, EPD was employed to produce $LiCoO_2$ (LCO) cathodes by co-depositing LCO particles with conductive carbon black (Super P) and polytetrafluoroethylene (PTFE) binder onto aluminum foil current collectors under high direct current (DC) electric fields in an organic-based colloidal suspension. These EPD-fabricated cathodes were systematically compared to conventionally prepared slurry-cast LCO electrodes, which represent the current industrial benchmark for LIB production.

The findings demonstrate that anodic horizontal EPD produces a uniform LCO electrode with strong adhesion to the aluminum substrate. The resulting microstructure retains the layered configuration of the active material, with carbon black and binder homogeneously interspersed throughout the coating. This configuration exhibits fewer defects and improved structural integrity compared to slurry-cast electrodes, which often suffer from inhomogeneous binder distribution and localized delamination. Electrochemical characterization further confirms that EPD-derived electrodes deliver competitive discharge capacity, excellent coulombic efficiency, and stable long-term cycling behavior.

In addition to these performance advantages, EPD substantially reduces the reliance on large quantities of binder and solvent, offering a more straightforward and environmentally friendly fabrication route compared to slurry-casting. This simplification, combined with the method's compatibility with high mass loadings, highlights EPD's scalability and its potential for industrial adoption in the commercialization of next-generation LIBs.

Electrochemical impedance spectroscopy (EIS) analyses further elucidate the interfacial and transport properties of the electrodes. Considering that the coated layers produced by each technique contain additional components, such as ethyl cellulose (EC) and $MgCl_2$ in the EPD coatings, and polyvinylidene fluoride (PVDF) in the slurry-cast coatings, care must be taken when interpreting impedance spectra. From the Nyquist plots, the slurry-cast LCO electrodes exhibit a smaller semicircle, suggesting lower charge-transfer resistance, yet display a longer Warburg tail indicative of higher diffusion impedance. Conversely, EPD electrodes show a slightly larger semicircle but a significantly shorter Warburg tail, which is advantageous under high mass loading conditions. These observations imply that while slurry-cast electrodes may initially ensure strong electronic and ionic contact, EPD electrodes promote superior ion transport in thicker electrode films, an interpretation consistent with existing literature.

Overall, this study establishes that EPD can yield high-performance LCO cathodes that rival, and in some respects surpass, those fabricated by slurry casting. By combining structural uniformity, mechanical robustness, reduced material consumption, and enhanced electrochemical performance, EPD represents a promising pathway for the development of sustainable, scalable, and high-energy-density lithium-ion battery electrodes. The evidence presented herein reinforces EPD's potential as a practical and industrially viable alternative for next-generation electrode manufacturing.

Acknowledgements Prof. Low acknowledges support from the Innovate UK project 'CONTACT: Coated Current Collector for Battery Performance Improvement' (Ref: 10041084), which translated EPD technology innovation into scalable, industry-relevant lithium-ion battery applications. Authors also acknowledge the internal project funded by SUNUM's management (2D nanomaterial electrodes via electro-deposition and electrospinning for electrochemical energy storage applications) that partially supported this work (I.A.SN-23-00110).

References

1. Lyu, Y., et al.: An overview on the advances of $LiCoO_2$ cathodes for lithium-ion batteries (2021) (Wiley-VCH Verlag). https://doi.org/10.1002/aenm.202000982
2. Hajizadeh, A., et al.: Electrophoretic deposition as a fabrication method for Li-ion battery electrodes and separators—A review (2022) (Elsevier B.V.). https://doi.org/10.1016/j.jpowsour.2022.231448
3. Chakrabarti, B.K., John Low, C.T.: Practical aspects of electrophoretic deposition to produce commercially viable supercapacitor energy storage electrodes. RSC Adv. **11**(34), 20641–20650 (2021). https://doi.org/10.1039/d0ra09197a

4. Chakrabarti, B., Nir, D., Yufit, V., Tariq, F., Rubio-Garcia, J., Maher, R., … & Brandon, N. (2017). Performance enhancement of reduced graphene oxide-modified carbon electrodes for vanadium redox-flow systems. *ChemElectroChem, 4*(1), 194–200. https://doi.org/10.1002/CELC.201600402
5. Chakrabarti, B., Rubio-Garcia, J., Kalamaras, E., Yufit, V., Tariq, F., Low, C. T. J., … & Brandon, N. (2020). Evaluation of a non-aqueous vanadium redox flow battery using a deep eutectic solvent and graphene-modified carbon electrodes via electrophoretic deposition. *Batteries, 6*(3), 1–20. https://doi.org/10.3390/BATTERIES6030038
6. Rinkel, B.L.D., Vivek, J.P., Garcia-Araez, N., Grey, C.P.: Two electrolyte decomposition pathways at nickel-rich cathode surfaces in lithium-ion batteries. Energy Environ. Sci. **15**(8), 3416–3438 (2022). https://doi.org/10.1039/d1ee04053g
7. Ludwig, B., Zheng, Z., Shou, W., Wang, Y., Pan, H.: Solvent-free manufacturing of electrodes for lithium-ion batteries. Sci. Rep. **6** (2016). https://doi.org/10.1038/srep23150
8. Bree, G., Tong John Low, C.: Full cell lithium-ion battery manufacture by electrophoretic deposition. Batter Supercaps **6**(2) (2023). https://doi.org/10.1002/batt.202200441

Electrophoretic Deposition of MoS_2 Electrodes for Supercapacitor Applications

Chee Tong John Low, Barun Kumar Chakrabarti, Koray Bahadır Dönmez, and Kaan Demirel

Abstract The synthesis of high-performance MoS_2 nanosheets for energy storage applications has traditionally relied on two main strategies: liquid-phase exfoliation of bulk MoS_2 or chemical intercalation using organolithium compounds. While the former yields semiconducting 2H-MoS_2 with limited conductivity and small flake sizes, the latter (organolithium intercalation) has historically been regarded as the most direct way to obtain metallic 1 T-MoS_2. This method is relevant because the 1 T phase exhibits much higher conductivity and enhanced electrochemical activity, making it particularly attractive for supercapacitor electrodes. However, despite these advantages, organolithium intercalation remains impractical for scalable production due to multi-day reaction times and the use of hazardous, air-sensitive reagents. More recently, an electrochemical route has been introduced as a more accessible method to synthesize metallic-phase MoS_2. In this approach, lithium ions from a lithium salt ($LiClO_4$) are electrochemically intercalated into MoS_2 using a platinum counter electrode, followed by rapid exfoliation upon exposure to water. This bench-top method produces high-quality 1 T-MoS_2 in under two hours, significantly lowering the barrier to producing conductive nanosheets for supercapacitor applications. Equally important is the choice of integration technique, which directly determines the electrochemical performance of the nanosheets. Literature reports show that conventional slurry-cast electrodes based on 1 T-MoS_2–graphene composites can deliver gravimetric capacitances above 70 F g^{-1} with excellent cycling stability. Notably, when the same composite materials were assembled using electrophoretic deposition (EPD), an additional performance enhancement of approximately 10% was observed. This key outcome underscores the potential of EPD to create uniform,

C. T. J. Low · K. B. Dönmez
Warwick Electrochemical Engineering Group, Energy Innovation Centre, WMG, University of Warwick, Coventry, UK

B. K. Chakrabarti (✉) · K. Demirel (✉)
Sabanci University Nanotechnology Research and Application Centre (SUNUM), Tuzla, İstanbul, Turkey
e-mail: barun.chakrabarti@sabanciuniv.edu

K. Demirel
e-mail: kaan.demirel@sabanciuniv.edu

M. Awang et al. (eds.), *Proceedings of the International Conference on Mechanical, Manufacturing and Plant Engineering*, Proceedings in Technology Transfer,
https://doi.org/10.1007/978-981-95-7817-7_20

binder-free, and densely packed electrode architectures, which translate into higher capacitance and long-term stability. These findings position EPD as a compelling and scalable integration strategy for high-performance energy storage devices.

Keywords 1 T-MoS_2 · Electrophoretic deposition · Supercapacitors · Electrochemical intercalation · Energy storage materials

1 Introduction

The global energy transition is driving the need for advanced storage technologies that can bridge the gap between high-energy batteries and high-power capacitors. Supercapacitors have emerged as a vital solution because they combine rapid charge–discharge capability, high power density, and long cycle life. These characteristics make them indispensable for applications such as regenerative braking, backup power systems, and portable electronics. Importantly, as modern devices continue to shrink in size, volumetric energy and power densities are becoming more critical than gravimetric performance, underscoring the need for improved electrode materials and architectures [1, 2].

The choice of electrode material is central to advancing supercapacitor technology. Commercial devices primarily rely on activated carbon due to its abundance, cost-effectiveness, and stability, but its performance is limited to electric double-layer capacitance, resulting in modest volumetric energy density. To overcome these limitations, research has increasingly focused on pseudocapacitive materials—including transition metal oxides, conducting polymers, and transition metal dichalcogenides (TMDs)—which contribute faradaic redox processes in addition to double-layer storage. Among these, molybdenum disulfide (MoS_2) stands out because of its layered structure, intercalation chemistry, and tunable electronic phases. The semiconducting 2H polymorph of MoS_2 has been widely investigated but suffers from poor conductivity, limiting its rate capability. By contrast, the metallic 1 T polymorph provides significantly higher conductivity, an increased density of states, and a stronger pseudocapacitive response.

The differences between the two main polymorphs of MoS_2 are rooted in their atomic coordination and electronic properties. The 2H phase adopts a trigonal prismatic coordination, is thermodynamically stable, and exhibits semiconducting behavior with an indirect bandgap of ~ 1.2 eV in bulk form (1.8 eV for monolayers). By contrast, the 1 T phase exhibits octahedral coordination, is metastable, and displays metallic conductivity that is three to four orders of magnitude higher than the 2H phase. This contrast in conductivity and density of states underpins their distinct electrochemical behaviors and explains why the 1 T phase is particularly desirable for supercapacitor electrodes [3].

Table 1 Comparison of MoS_2 synthesis routes

Route	Dominant phase	Time	Flake size	Safety
Liquid exfoliation [4]	2H	Hours	< 1 μm	Benign
Organo-Li intercalation [5]	1 T	2–3 d	Up to several μm	Hazardous
Electrochemical [5]	1 T	< 2 h	0.5–5 μm	Safe

2H MoS_2: indirect bandgap (~1.2 eV bulk, 1.8 eV monolayer), poor conductivity
1 T MoS_2: metallic, improved conductivity (~103–104 times higher than 2H), increased density of states

Different synthetic routes have been developed to access these polymorphs, each with advantages and drawbacks. Liquid-phase exfoliation, while scalable and relatively benign, typically produces small nanosheets of 2H-MoS_2 with poor conductivity. Organolithium intercalation has long been considered the most direct route to obtain 1 T-MoS_2, but it involves multi-day reactions with hazardous pyrophoric reagents, limiting its practicality for large-scale applications. More recently, electrochemical intercalation has been demonstrated as a safer and faster alternative. In this approach, lithium ions from inert lithium salts are electrochemically intercalated into MoS_2, followed by rapid exfoliation in water, yielding metallic 1 T-MoS_2 nanosheets within a few hours. This advance not only simplifies access to the conductive phase but also makes it feasible to integrate these nanosheets into electrode architectures at larger scales.

Table 1 summarizes the advantages and limitations of these synthesis routes, highlighting the trade-offs between scalability, safety, and product quality. Such comparisons are critical for guiding future efforts in the scalable production of conductive MoS_2 nanosheets for energy storage.

2 Electrochemical Intercalation and Exfoliation

2.1 Materials and Reagents

MoS_2 powder (99%, ~ 6 μm), lithium perchlorate (99.99%), dimethyl carbonate (≥99%), and ethylene carbonate (99%) were obtained from Sigma-Aldrich. Natural MoS_2 crystals were sourced from Manchester Nanomaterials Ltd. (UK). PTFE (0.2 μm) and PVDF (0.1 μm) membranes were used for filtration, and ultrapure water (18.2 MΩ·cm) was produced using a Milli-Q system.

2.2 Electrochemical Intercalation of Li^+ into MoS_2

Electrochemical intercalation was performed using MoS_2 pellets (pressed at 2.5 tons, 12 mm diameter) or natural crystals as the working electrode, with Pt mesh and Ag wire serving as counter and reference electrodes, respectively. The electrolyte was 1 M $LiClO_4$ in a 1:1 mixture of dimethyl carbonate and ethylene carbonate, de-aerated with N_2. Chronoamperometry at − 4.5 V (vs. Ag) for 2 h enabled Li^+ insertion following the procedure of Dryfe and co-workers [5].

After intercalation, samples were rinsed with acetone, exfoliated in deoxygenated water under sonication, and purified by sequential washing with n-hexane, extensive deionized water rinsing, and PTFE membrane filtration. The final dispersion was re-dispersed, sonicated, and centrifuged to remove residual bulk material, yielding a stable aqueous suspension of exfoliated MoS_2.

2.3 Formulation of 1 T-MoS_2/Graphene Composite

A stock dispersion of 1 T-MoS_2 was first diluted in isopropanol (IPA) to a concentration of 0.05 mg mL$^-$1. Separately, solution-exfoliated graphene, obtained through an aqueous electrochemical exfoliation process, [6] was prepared at the same concentration (0.05 mg mL$^-$1). Equal volumes (50 mL each) of the diluted 1 T-MoS_2 and SEG dispersions were then mixed and sonicated for 20 min to yield a uniform 1 T-MoS_2/graphene composite (50% graphene).

2.4 Electrode Preparation for Supercapacitor Testing

1 T-MoS_2/graphene composite electrodes were fabricated by slurry casting on standard aluminum foils. The resulting electrode films were air-dried at room temperature. Symmetrical CR2032 coin cells were assembled by stacking two identical composite electrodes face-to-face with the active layers in contact with the current collectors along with a PVDF-based separator. A small volume of 1.0 M tetraethylammonium tetrafluoroborate ([TEA][BF_4]) in anhydrous acetonitrile was introduced as the electrolyte, and the cells were sealed using a hydraulic crimping machine (MSK-160D). All assembly was carried out in an argon-filled glovebox. All coin cells were tested in a temperature controlled chamber as displayed in Fig. 1.

Fig. 1 EPD coin cells undergoing electrochemical testing in a temperature-controlled chamber

3 Electrophoretic Deposition (EPD)

3.1 Fundamentals

Electrophoretic deposition (EPD), first adopted in the automotive industry, is valued for its automation, low environmental impact, and uniform coatings [7]. The tech nique enables precise control over film architecture but remains underutilized in energy storage devices [8–10]. Common studies typically emphasize ultrathin coatings (<1 μm, ~ 1 mg cm^{-2}), which highlight theoretical performance but are impractical for commercial use [11–13]. Realistic applications demand thicker films (50–200 μm) and higher mass loadings (5–20 mg cm^{-2}) to achieve usable capacities and power delivery as demonstrated for activated carbon recently [14].

3.2 Electrode Preparation for Supercapacitor Testing

1 T-MoS_2/graphene composite electrodes were deposited onto Al foil by EPD from a suspension containing the composite, carbon black, PVDF binder, and iodine as a charging agent in a similar manner to our previous work on activated carbon [10]. Deposition was carried out at 70 V and 45 °C for 20 min using Al working electrodes and Pt/Ti counter electrodes. The films were subsequently washed with ethanol and water to remove residual iodine, dried, vacuum-treated, and hot-pressed to 70% of their original thickness to improve electrical contact while maintaining porosity. Then the electrodes were assembled into CR2032 coin cells as described in Sect. 2.4 above.

4 Experimental Comparison: Slurry Casting vs EPD

EPD enables direct assembly of 1 T-MoS_2/graphene composites onto Al foils from a stable colloidal suspension. Unlike slurry casting, which relies on NMP-based viscous pastes containing high binder content, EPD promotes uniform particle distribution, intimate contact with the current collector, and controlled film thickness without excessive inactive components. This leads to lower interfacial resistance, enhanced charge transfer kinetics, and improved accessibility of active sites.

Additionally, EPD minimizes structural defects such as binder-induced blocking of pores and agglomeration of nanosheets, while also preserving the porous architecture necessary for ion transport. The resulting electrodes therefore combine high electrical conductivity from graphene with efficient ion diffusion pathways through MoS_2, yielding superior capacitance, rate capability, and cycling stability compared to their slurry-cast counterparts as displayed in Fig. 2.

5 Perspective for Scale-Up Consideration

5.1 Roll-To-Roll Deposition

EPD can be adapted for roll-to-roll (R2R) processing, offering a pathway to large-scale electrode manufacturing (Fig. 3). In contrast to slurry casting, EPD allows uniform films with tunable thickness and minimal binder content, making it well suited for continuous deposition on flexible substrates. Recent studies have shown that EPD can deliver consistent performance even at large electrode areas, as demonstrated with activated carbon electrodes on nickel foam where capacitance per mass was maintained across different scales. Similar work with lithium-ion cathodes such as NMC has highlighted the ability of EPD to produce thick films with interconnected porosity, a feature difficult to achieve using conventional slurry casting. Although

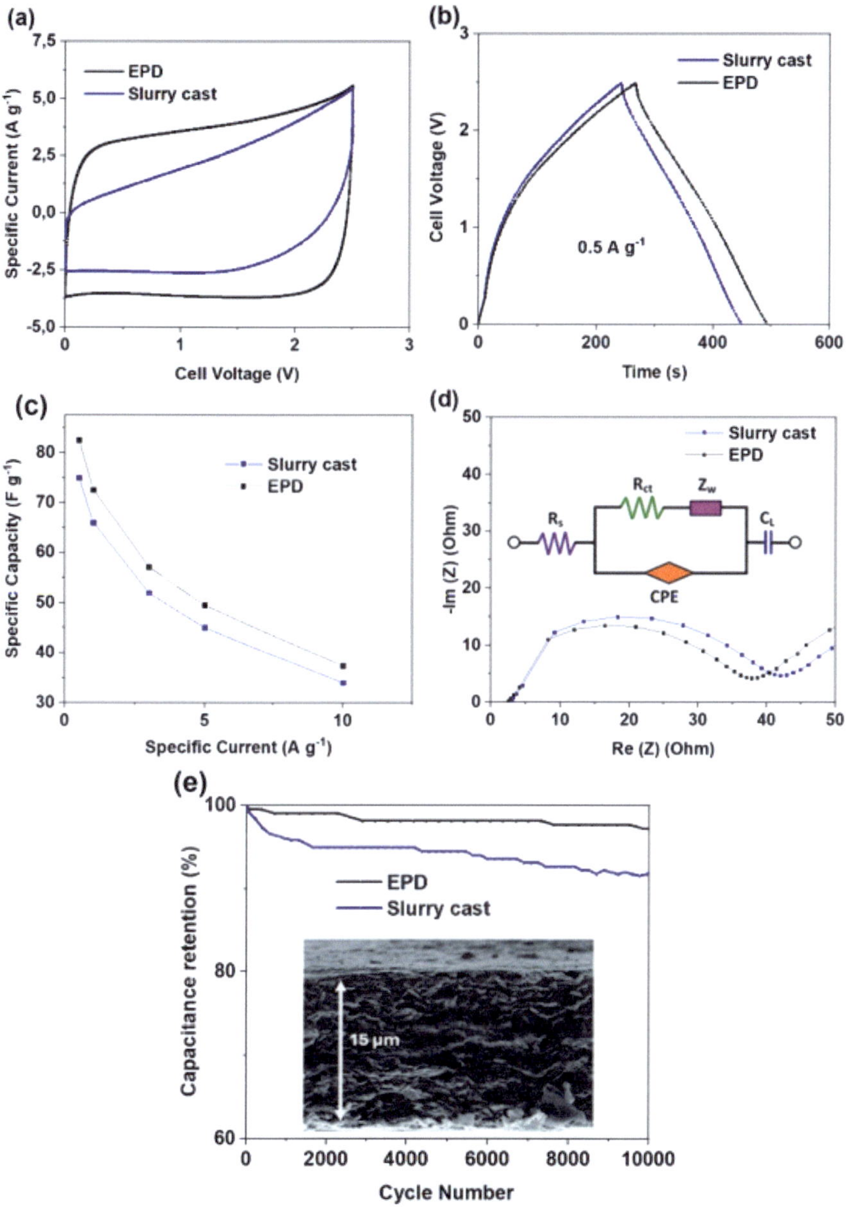

Fig. 2 **a** Cyclic voltammograms recorded at 100 mV s^{-1} in 1.0 M [TEA][BF_4] from 0 to 2.5 V. **b** GCD curves. **c** Rate capability plots. **d** Nyquist plots from impedance (reproduced from Springer Nature) [15]. **e** Successive cycling at = 0.5 A g^{-1}. Electrochemical testing was performed using symmetrical coin cells constructed from 1 T-MoS_2/SEG electrodes. The mass ratio for these devices was 2 (positive mass was 2 × the negative). The inset shows a cross sectional image of the graphene/MoS_2 electrode prepared via EPD with a higher thickness than that reported in a paper published by RSC [5]

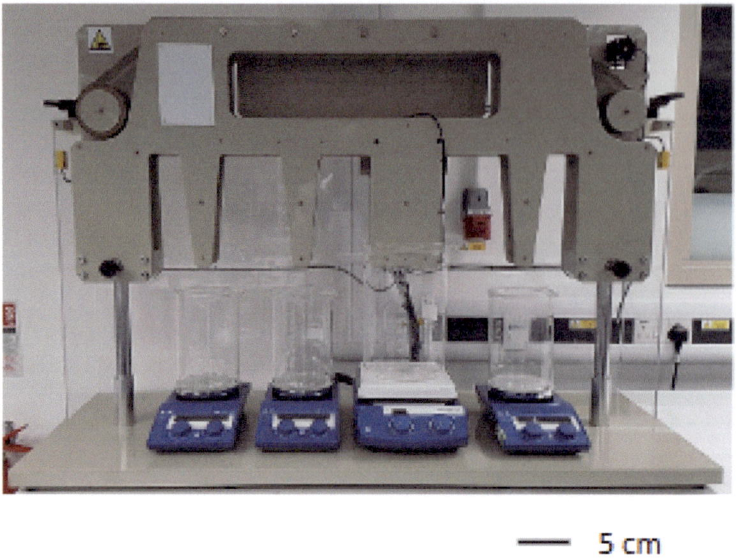

Fig. 3 EPD setup in WMG for R2R manufacturing of electrodes for energy storage applications

full R2R-EPD integration remains limited, advances in continuous electrode coating for batteries and fuel cells suggest strong potential for transferring EPD into scalable manufacturing lines. Challenges remain in maintaining suspension stability and electric field uniformity over wide rolls, but the method is attractive for high-throughput, material-efficient electrode production.

5.2 Integration with Device Architectures

Beyond planar foils, EPD can be directly applied to 3D and porous current collectors, significantly enhancing volumetric utilization. By depositing into the interconnected pores of foams or meshes, active materials penetrate the full thickness of the substrate, improving electrical contact and shortening ion diffusion pathways (Fig. 4). In supercapacitors, EPD of activated carbon onto nickel foam enabled higher areal capacitance compared with slurry-cast films on flat foils, owing to greater electrolyte accessibility. 1 Similar benefits have been reported for NiO/Ni foam composites, where the porous framework provided strong adhesion and increased active surface area. While uniform penetration into deep pores remains a technical challenge, the ability of EPD to conformally coat complex architectures underscores its potential for producing thick, high-performance electrodes that are otherwise difficult to fabricate by slurry casting.

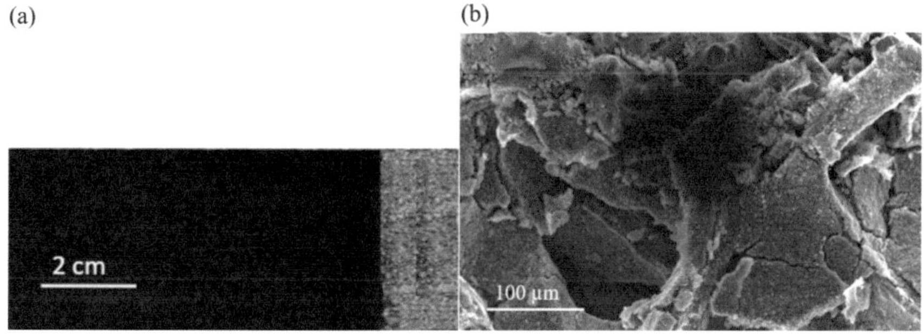

Fig. 4 **a** Digital photo of 50% graphene composite deposited on a 10 cm^2 Ni foam. **b** FESEM of the coated Ni foam (50% graphene composite)

6 Conclusions and Perspectives

Electrochemical intercalation provides a safe and scalable pathway to produce 1 T-MoS_2 nanosheets, while electrophoretic deposition (EPD) enables binder-free, uniform electrodes with superior performance compared to slurry casting. Together, these methods outline a practical route toward compact, high-power supercapacitors that bridge the gap between laboratory innovation and real-world deployment. Future progress will hinge on translating these advances into hybrid architectures that combine 1 T-MoS_2 with graphene, carbon nanotubes, or conductive polymers, thereby enhancing durability and conductivity at device-relevant scales. The use of in-situ and operando tools to monitor phase stability, along with the adoption of aqueous and solid-state electrolytes, will be critical to ensure both reliability and versatility across applications. At the same time, machine learning can accelerate optimization of EPD parameters, reducing development time and cost, while sustainable, green-solvent processing will align these technologies with industrial-scale environmental standards. Collectively, these directions position 1 T-MoS_2/graphene composites not only as a next-generation material for energy storage but as a foundation for future manufacturing strategies that integrate scalability, performance, and sustainability.

Acknowledgements Prof. Low acknowledges support from the Innovate UK project 'CONTACT: Coated Current Collector for Battery Performance Improvement' (Ref: 10041084), which translated EPD technology innovation into scalable, industry-relevant lithium-ion battery applications. In addition, this research was funded by EPSRC ISCF Wave 1: 3D electrodes from 2D materials (EP/R023034/1). Authors also acknowledge the internal project funded by SUNUM's management (2D nanomaterial electrodes via electro-deposition and electrospinning for electrochemical energy storage applications) that partially supported this work (I.A.SN-23-00110).

References

1. Chakrabarti, B.K., Dönmez, K.B., Çobandede, Z., Low, C.T.J.: ChemElectroChem **11**, e202400244 (2024)
2. Wang, Q., Yan, J., Fan, Z.: Energy Environ. Sci. **9**, 729–762 (2016)
3. Das, S., Swain, G., Parida, K.: Mater. Chem. Front. **5**, 2143–2172 (2021)
4. Chavalekvirat, P., Hirunpinyopas, W., Deshsorn, K., Jitapunkul, K., Iamprasertkun, P.: Precis. Chem. **2**, 300–329 (2024)
5. Ejigu, A., Kinloch, I.A., Prestat, E., Dryfe, R.A.W.: J. Mater. Chem. A **5**, 11316–11330 (2017)
6. Parvez, K., Wu, Z.-S., Li, R., Liu, X., Graf, R., Feng, X., Müllen, K.: J. Am. Chem. Soc. **136**, 6083–6091 (2014)
7. Chakrabarti, B.K., Gençten, M., Bree, G., Dao, A.H., Mandler, D., Low, C.T.J.: Int. J. Energy Res. **46**, 13205–13250 (2022)
8. Chakrabarti, B., Nir, D., Yufit, V., Tariq, F., Rubio-Garcia, J., Maher, R., Kucernak, A., Aravind, P.V., Brandon, N.: ChemElectroChem **4**, 194–200 (2017)
9. Chakrabarti, B., Yufit, V., Kavei, A., Xia, Y., Stevenson, G., Kalamaras, E., Luo, H., Feng, J., Tariq, F., Taiwo, O., Titirici, M.-M., Brandon, N.: Int. J. Hydrogen Energy **44**, 30093–30107 (2019)
10. Chakrabarti, B.K., Low, C.T.J.: Next Energy **4**, 100137 (2024)
11. Chakrabarti, B., Rubio-Garcia, J., Kalamaras, E., Yufit, V., Tariq, F., Low, C.T., Kucernak, A., Brandon, N.: Batteries **6,** 38 (2020)
12. Chakrabarti, B.K., Feng, J., Kalamaras, E., Rubio-Garcia, J., George, C., Luo, H., Xia, Y., Yufit, V., Titirici, M.-M., Low, C.T.J., Kucernak, A., Brandon, N.P.: ACS Appl. Mater. Interfaces **12**, 53869–53878 (2020)
13. Namvari, M., Chakrabarti, B.K.: Adv. Coll. Interface Sci. **331**, 103208 (2024)
14. Chakrabarti, B.K., John Low, C.T.: RSC Adv. **11**, 20641–20650 (2021)
15. Lu, Y., Liu, X., Wang, W., Cheng, J., Yan, H., Tang, C., Kim, J.-K., Luo, Y.: Sci. Rep. **5**, 16584 (2015)

Short Review in HER/HOR Electrocatalysts as Anodes for Regenerative Fuel Cells

Jose Luis Pérez-Martínez, Dominik Wickenhauser, Bernhard Marius, Barun Chakrabarti, Tülay Ínan, Stefan Spirk, and Andrés Parra-Puerto

Abstract The utilization of H_2 as energy carrier to generate electricity via electrochemical systems such as polymer electrolyte-membrane fuel cells (PEMFC) requires Pt as electrocatalysts. Pt has a very high activity of hydrogen evolution (HER) and hydrogen oxidation reactions (HOR) in acidic media, with low overpotential, allowing loadings < 0.03 mg Pt $cm^2{}_{geo}$ (in the range of the PEMFC cost demands of < 30$ kW^{-1}). With these characteristics there are not many alternatives to Pt. WO_x based electrocatalysts are the only ones with activity for HOR, while other catalysts do not quite display similar activity for this reaction as for HER (which have more options to Pt with good performance). In alkaline media, HOR kinetics are reduced by at least two times on Pt surfaces but open the possibility to

J. L. Pérez-Martínez · A. Parra-Puerto (✉)
Department of Physical Chemistry and Electrochemistry Institute, Universidad de Alicante, 03690 San Vicente del Raspeig, Spain
e-mail: andres.parra@ua.es

J. L. Pérez-Martínez
e-mail: joseluis.perezm@ua.es

D. Wickenhauser · S. Spirk
Institute of Bioproducts and Paper Technology, Graz University of Technology, 8010 Graz, Austria
e-mail: dominik.wickenhauser@ecolyte.at

S. Spirk
e-mail: stefan.spirk@tugraz.at

D. Wickenhauser · B. Marius · S. Spirk
Ecolyte GmbH, 8020 Graz, Austria
e-mail: bernhard.marius@ecolyte.at

B. Chakrabarti · T. Ínan
Nanotechnology Research and Application Centre (SUNUM), Sabanci University, 34956 Tuzla, İstanbul, Turkey
e-mail: barun.chakrabarti@sabanciuniv.edu

T. Ínan
e-mail: tulay.inan@sabanciuniv.edu

M. Awang et al. (eds.), *Proceedings of the International Conference on Mechanical, Manufacturing and Plant Engineering*, Proceedings in Technology Transfer,
https://doi.org/10.1007/978-981-95-7817-7_21

use transition metals as catalysts. In this media, Nickel shows the best performance for HOR amongst all of them (albeit not being at par with Pt). The fast kinetics and low overpotentials, with low Pt loadings, for H_2 reactions in acidic media can mimic the reversible reactions of the redox flow batteries (RFBs), allowing them to be implemented in energy storage devices called regenerative fuel cells (RFCs). RFCs fuse the fast kinetics of the H_2 reactions in PEMFCs with the fast kinetics of the redox active species in RFBs. In this short review, we overview the fundamentals of H_2 reactions in acidic and alkaline media, review some of the most commonly used electrocatalysts and discuss different RFC chemistries developed at present.

Keywords HOR/HER · Electrocatalysts · Fuel cells · Regenerative fuel cells

1 Introduction

The large deployment of wind and solar energy generation in the EU to be energy independent and carbon neutral needs to go together with large electrochemical energy storage devices (EES) and green H_2 generation. Renewable energies, face challenges such as intermittency and low energy density compared to traditional fuels. For instance, solar energy only produces electricity during daylight, and wind power depends on wind conditions [1, 2]. To mitigate these limitations, excess renewable energy requires to be stored using energy storage systems such as batteries for short and long periods of time [3]. An interesting electrochemical energy storage device is the redox flow battery (RFB), which can meet the growing demand of large-scale, low-cost, safe and sustainable energy storage [4]. In RFBs the storage capacity is decoupled from the stack size, giving the possibility to scale up the capacity or the power output independently [5–7] in comparison with other secondary batteries such as lead acid or lithium-ion.

Another option is the production of green hydrogen, making it a promising complement to renewable integration [8]. Hydrogen (H_2) is a versatile energy carrier with exceptional properties: it has a gravimetric energy density of 142 MJ kg^{-1}, nearly three times higher than that of common fossil fuels like diesel or natural gas. In addition, H_2 can be used as fuel in polymer electrolyte membrane fuel cells or proton exchange membrane fuel cells (PEMFCs) to generate electricity, [9] feedstock, or energy storage medium, while producing zero carbon emissions at the point of use [10]. These features make H_2 a key element in the energy transition and a central component of the European Green Deal, whose goal is to achieve climate neutrality by 2050 through sustainable energy systems [11].

In order to obtain a system with high energy density, low self-discharge, long durability, and low flammability, the combination of RFBs and PEMFCs can be a solution. At the anode side we have the H_2 reactions from a PEMFC, and on the cathode the reversible redox couples from RFBs can be implemented. This type of system is called a regenerative fuel cell (RFC) or Hybrid Redox Flow Battery (HRFB) [12].

2 Electrochemical Hydrogen Reactions

In the scope of energy conversion using electrochemical devices such as electrolyzes and fuel cells the H_2 electrocatalysts play a key role. The utilization of H_2 as fuel in PEMFCs is well known, where it is electrooxidized at the anode following this reaction:

Acid	Alkaline
$2H^+ + 2e^- \rightleftarrows H_2$	$2H_2O + 2e^- \rightleftarrows H_2 + 2OH^-$

This reaction has a very low overpotential and high performance for Platinum (Pt) electrocatalysts requiring very low loadings < 0.03 $mg_{Pt}cm^2_{geo}$,, without compromising the PEMFC performance, which is in the range of the PEMFC cost demands of < 30$ kW^{-1}.[13] The mechanism in acid and alkaline media over Pt adsorption sites (Pt*) are the following [14, 15]:

Mechanism in acid		Mechanism in alkaline	
Tafel:	$H_2 + 2Pt^* \rightleftarrows 2H_{ad}$		
Volmer:	$H_2 + Pt^* \rightleftarrows 2H_{ad} + H^+ + e^-$	Volmer:	$H_2 + OH^- + Pt^* \rightleftarrows H_{ad} + H_2O + e^-$
Heyrovsky:	$2H_{ad} \rightleftarrows H^+ + e^-$	Heyrovsky:	$H_{ad} + OH^- \rightleftarrows Pt^* + H_2O + e^-$

In acid media, due to the high concentration of H^+ near the Pt surface, the hydrogen evolution reaction (HER) overpotential is very low. On the other hand, the hydrogen oxidation reaction (HOR) has, as well, very low overpotential but it is limited by the H_2 diffusion in the electrolyte [16], leading to a potential independent limiting current. In this media the reaction occurs via Tafel-Volmer or Heyrovsky-Volmer mechanisms, depending on the H_2 dissociation (rate limiting step) if it is chemical (Tafel) or electrochemical (Heyrovsky) [17, 18]. Sheng et al. [16] investigated the HOR/HER at different pHs and they observed that when the pH increases > 6 the kinetics of the HOR decreases, reaching up to two orders of magnitude slower in alkaline media [16, 19]. This fast kinetics and low overpotentials in acid media, allow the reaction to be kind of "reversible" allowing to use this reaction in energy storage devices such as RFCs among others.

3 Precious Metal Catalysts

The electrocatalysts, with the highest HOR/HER activity, are in the Pt group metals (PGMs) such as Pt, Ir, Rh and Pd. The baseline parameters for Pt is well agreed that the exchange current density i_0 value is around 1 mA cm_{Pt}^{-2}, with 30 mV dec^{-1} for the anodic and cathodic Tafel slopes [14]. The exchange current density is an important parameter that shows the intrinsic activity of the electrocatalysts under equilibrium conditions [20]. Dust et al. [14] did an incredible work obtaining different parameters of the above-mentioned PGMs, going from the electrochemical surface area (ECSA) obtained from the hydrogen underpotential deposition (H-UDP) and CO stripping using a rotating disk electrode (RDE) and a PEMFC, to j_0 from the Butler-Volmer fittings, Tafel slopes (anodic and cathodic), activation energy (E_a), etc. This work did a very good fundamental study of the different PGMs using different techniques under high mass transport regimes. A summary of some of the important parameters for the PGMs are collected in Table 1.

4 Non-Precious Metal Catalysts

Non-PGMs electrocatalysts for HOR/HER are quite rare in acid, as most of the transition metals are not stable in this media. Only some of them can stand in this media, such as Tungsten (W) oxides; many studies mentioned the stability of WO_x in acid media. WO_x have abundant active sites for H_2 adsorption and oxidation, plus high electrical conductivity and large surface areas [21–23]. One of the approaches done by Meng et al. [24] was the preparation of a WC/WO_2-WO_3 heterostructure by in situ electrochemical reconstruction with high activity for the HOR, proposing an interesting mechanism (Fig. 1).

On the other hand, the majority of the efforts to use transition metal catalysts for HOR/HER have been done for alkaline media, as they are more stable. For HER there are many different catalysts, as they are used in alkaline electrolyzes for green H_2 generation. Those catalysts go from W, Ni, Co, Mo, etc., [25–29] with different supports, or their respective dichalcogenide with activities getting closer to Pt, in the case of MoS_2 and WS_2 [30–32].

However, still for HOR there are not many that can give a good performance to be used in alkaline based PEMFCs [33]. At the moment Ni is the catalyst used as anode for the HOR even though the performance is still far from the Pt based catalysts. Davydova et al. [34] did a very interesting study about Ni-based electrocatalysts for HOR comparing different synthetic methods and Ni alloys with Fe, Co and Cu (Fig. 2).

They observed that depending on the synthetic process the CV of the Ni or Ni alloy catalyst changes due to the nature of the obtained catalysts. In the CR process the catalysts obtained are more a mechanical mix with poor crystallinity and in contrast, SR gives better crystallinity and the formation of Ni-M alloys. The SR catalyst required an activation step to remove the surface hydroxides to reach the peak for the HOR, having a fully reduced surface at 0 V vs. RHE. On the other hand, for the CR catalyst the activation process was observed to have a negative effect on the HOR activity.

Other alternatives for the reduction or replacement of Pt in the anode can be the single-atom catalysts (SACs) type. SACs typically consist of a single metal atom coordinated by four electronegative atoms, all supported on a carbon matrix [35]. The metal content on SACs is quite low keeping an interesting catalytic activity

Table 1 Summary of different parameters from the standard PGMs used in PEMFC (for more detailed information for the acid media check the reference).[14]

Metal	H_{UPD} charge/μC cm^{-2}	$ECSA_{CO\text{-}Stripping}$ PEMFC/$m^2_M g^{-1}_M$	j_0/mA cm^{-2}_M		Tafel slopes/mV dec^{-1} @ 313 K	
			Butler-Volmer	Micropolarization	b_{anodic}	$b_{cathodic}$
Pt	210 ($420_{CO\text{-}st}$)	110 ± 10	135 ± 40	135 ± 40	124 ± 15	124 ± 5
Ir	210 (420 $_{CO\text{-}st}$)	60 ± 10	40 ± 4	40 ± 10	124 ± 5	124 ± 5
Rh	205 (410 $_{CO\text{-}st}$)	90 ± 10	2.7 ± 0.8	2.7 ± 0.8	258 ± 23	127 ± 8
Pd	220 (440 $_{CO\text{-}st}$)	150 ± 40	4.9 ± 0.7	4.9 ± 0.7	180 ± 8	95 ± 3

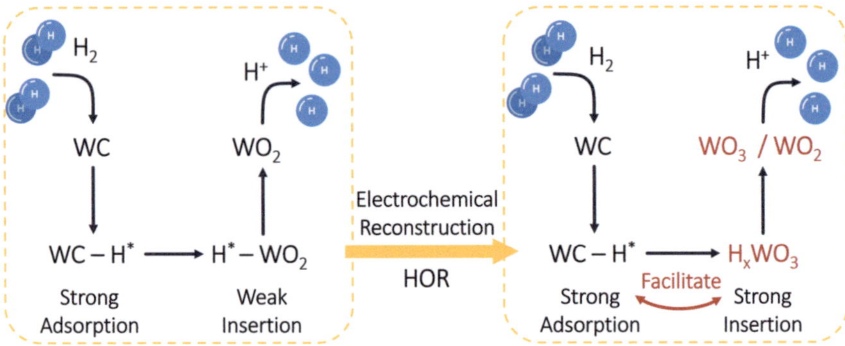

Fig. 1 Scheme adapted from the mechanism proposed by Meng et al. [24]

allowing to reduce the required amount of noble metals or even fully replace them with transition metals. These catalysts have been explored for HER [36] and O2 reactions (reduction [ORR] and evolution [OER] [24, 29, 37, 38].

5 Regenerative Fuel Cells

Regenerative fuel cells (RFC) [12] have been studied quite extensively over the last few years. This technology fuses the best parts of the PEMFCs, the fast kinetics of the HER/HOR, with the fast kinetics of the redox active species in RFBs (scheme in Fig. 3).

This configuration drastically lowers this technology's cost compared with a standard PEMFC as we can completely remove the high loading of PGM catalysts in the cathode for the ORR. On the other hand, this architecture reduces the electrolyte cost, thereby minimising the irreversible crossover of the active species, which is a major concern in the RFBs. A very well-known example of this system is the H_2/Bromine RFC. This system has bromine crossover issues, which poison and corrode the noble metal H_2 catalyst and is a poisonous gas [39, 40]. In addition, the crossover of the Br_2 can dissolve the Pt in the anode, dropping the performance and requiring a system with charge/discharge cycles without a lag between them. This does not allow us to store the energy from solar or wind power for later use. Other chemistries used with high interest are Vanadium and Manganese. Yufit et al. [41] proved the first Vanadium RFC. This configuration lowers the capital cost of the RFB, and reduces the self-discharge problem due to the electrolyte cross-contamination since the electrolyte that passes to the gas side is easily returned to the positive electrolyte reservoir [42, 43]. Other types of chemistries have been developed such as H_2/Ce, [44] H_2/Mn, [45, 46] H_2/V (HCl), [47] H_2/BQ (Benzoquinone), [48] and H_2/Fe [49].

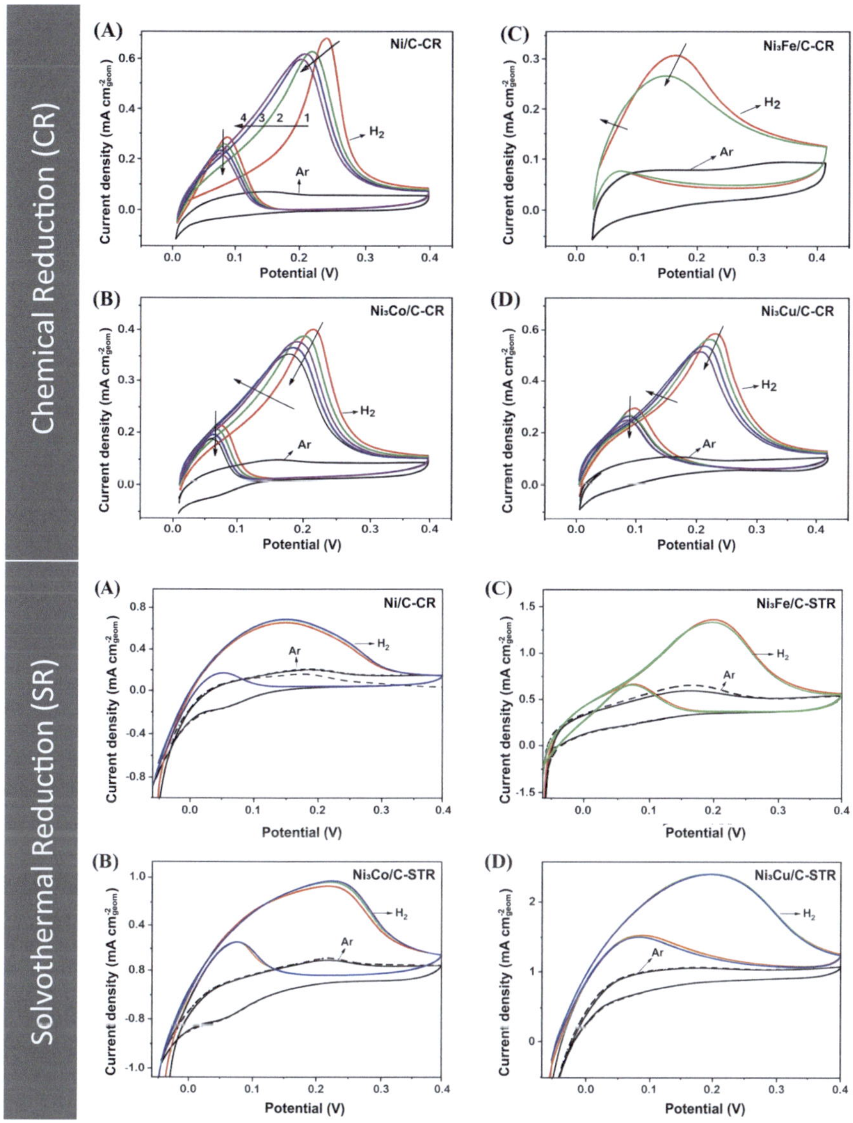

Fig. 2 HOR polarization curves and cyclic voltammograms of the catalysts synthesized by chemical (right column) and solvothermal (left column) reduction. Ar- or H_2-saturated (as indicated) 0.1 M KOH, 25 °C, rotation rate 1600 rpm, sweep rate 1 mV s^{-1}. **a** Ni/C, **b** Ni_3Co/C, **c** Ni_3Fe/C, and **d** Ni_3Cu/C. Reproduced with permission [34]. Copyright 2021, American Chemical Society

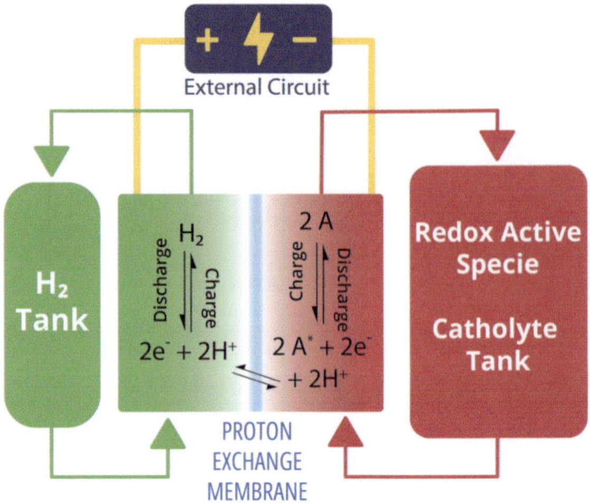

Fig. 3 Scheme of how an RFC works

6 Conclusions

The transition to a sustainable energy system requires the development of efficient, durable, and affordable electrocatalysts for hydrogen-based technologies to remove fossil fuel dependence. Platinum electrocatalysts have the best performance for HOR/HER allowing them to reduce the metal loading to reach the PEMFC cost demands of < 30$ kW^{-1}, especially in acid media. Other alternatives in acid media are the W based electrocatalysts that show interesting performance and stability in this media but far from the performance given by the Pt based electrocatalysts.

On the other hand, in alkaline media Pt kinetics are reduced at least two times with respect to acidic media due to the strong bonding of the Pt–H. This can be weakened by alloying Ru, boosting the HOR/HER performance in this media. In this media, transition metals can be used as they are stable, however, for HOR only nickel have proven to be effective in alkaline media, where they exhibit high stability and good catalytic behavior. Nickel's abundance and versatility make it an attractive option for large-scale applications, especially where the goal is to move away from scarce and expensive noble metals.

In line with the high performance of the HOR/HER with very low overpotential and low loading required in the anodes using Pt, this opens the possibility to be implemented in electrochemical energy storage systems such as Regenerative fuel cells (RFCs). This technology combines the strengths of conventional PEMFCs and RFBs, presenting another alternative to overcome the intermittency of renewables, reducing the costs of the device in comparison with the PEMFC (as they remove the high loading of PGM in the cathode) and the RFBs (they use only one electrolyte tank and can achieve higher potential). The development of alternatives to Pt in acid media needs to be explored to reduce the utilization of this critical raw material and reduce the cost of this technology further.

Acknowledgements A.P.P. is grateful to the Ministry of Universities and the University of Alicante for the María Zambrano fellowship.

Funding This research was funded by the CETPartnership (CETP), under the Joint Call 2023, and the project "HydrogEn oRganiC Regenerative FUeL CEll for Stationary Energy Storage (HERCULES)". The CETP' s research projects are co-funded by the European Commission (Grant Agreement No. 101069750) and the national funding organizations listed on the CETP website (AEI: PCI2024-155046–2/ MICIU/AEI/10.13039/501100011033/Unión Europea).

References

1. Fonseca, J.D., Camargo, M., Commenge, J.M., et al.: Trends in design of distributed energy systems using hydrogen as energy vector: a systematic literature review. Int. J. Hydrogen Energy 9486–9504 (2019). https://doi.org/10.1016/j.ijhydene.2018.09.177
2. Peláez-Peláez, S., Colmenar-Santos, A., Pérez-Molina, C., et al.: Techno-economic analysis of a heat and power combination system based on hybrid photovoltaic-fuel cell systems using hydrogen as an energy vector. Energy **224** (2021). https://doi.org/10.1016/j.energy.2021.120110
3. Njema, G.G., Ouma, R.B.O., Kibet, J.K.: A review on the recent advances in battery development and energy storage technologies. J. Renew. Energy **2024**, 1–35 (2024). https://doi.org/10.1155/2024/2329261
4. Xie, F., Zhang, X., Pan, Z.: Electrochemical systems for renewable energy conversion and storage: focus on flow batteries and regenerative fuel cells. Curr. Opin. Electrochem. **48** (2024)
5. Kebede, A.A., Kalogiannis, T., Van Mierlo, J., Berecibar, M.: A comprehensive review of stationary energy storage devices for large scale renewable energy sources grid integration. Renew. Sustain. Energy Rev. **159**, 112213 (2022). https://doi.org/10.1016/J.RSER.2022.112213
6. Jafari, M., Botterud, A., Sakti, A.: Decarbonizing power systems: a critical review of the role of energy storage. Renew. Sustain. Energy Rev. **158**, 112077 (2022). https://doi.org/10.1016/J.RSER.2022.112077
7. Zhang, H., Sun, C.: Cost-effective iron-based aqueous redox flow batteries for large-scale energy storage application: a review. J. Power. Sources **493**, 229445 (2021). https://doi.org/10.1016/J.JPOWSOUR.2020.229445
8. Hosseini, S.E., Wahid, M.A.: Hydrogen from solar energy, a clean energy carrier from a sustainable source of energy. Int. J. Energy Res. **44**, 4110–4131 (2020)
9. Halder, P., Babaie, M., Salek, F., et al.: Performance, emissions and economic analyses of hydrogen fuel cell vehicles. Renew. Sustain. Energy Rev. **199** (2024)
10. Rasul, M.G., Hazrat, M.A., Sattar, M.A., et al.: The future of hydrogen: challenges on production, storage and applications. Energy Convers. Manag. **272** (2022)
11. The European Green Deal—European Commission. https://commission.europa.eu/strategy-and-policy/priorities-2019-2024/european-green-deal_en. Accessed 3 Oct 2025
12. Brandon, N., Yufit, V., Kucernak, A., Rubio-Garcia, J.: Regenerative fuel cells. (2017)
13. Moschkowitsch, W., Lori, O., Elbaz, L.: Recent progress and viability of PGM-free catalysts for hydrogen evolution reaction and hydrogen oxidation reaction. ACS Catal. **12**, 1082–1089 (2022). https://doi.org/10.1021/ACSCATAL.1C04948
14. Durst, J., Simon, C., Hasche, F., Gasteiger, H.A.: Hydrogen oxidation and evolution reaction kinetics on carbon supported Pt, Ir, Rh, and Pd electrocatalysts in acidic media. J. Electrochem. Soc. **162**, F190–F203 (2014). https://doi.org/10.1149/2.0981501jes
15. Sheng, W., Gasteiger, H.A., Shao-Horn, Y.: Hydrogen oxidation and evolution reaction kinetics on platinum: acid vs alkaline electrolytes. J. Electrochem. Soc. **157**, B1529 (2010). https://doi.org/10.1149/1.3483106

16. Sheng, W., Zhuang, Z., Gao, M., et al.: Correlating hydrogen oxidation and evolution activity on platinum at different pH with measured hydrogen binding energy. Nat. Commun. **1**(6), 1–6 (2015). https://doi.org/10.1038/ncomms6848
17. Kucernak, A.R., Zalitis, C.: General models for the electrochemical hydrogen oxidation and hydrogen evolution reactions: theoretical derivation and experimental results under near mass-transport free conditions. J. Phys. Chem. C **120**, 10721–10745 (2016). https://doi.org/10.1021/acs.jpcc.6b00011
18. Chen, S., Kucernak, A.: Electrocatalysis under conditions of high mass transport: investigation of hydrogen oxidation on single submicron Pt particles supported on carbon. J. Phys. Chem. B **108**, 13984–13994 (2004). https://doi.org/10.1021/JP048641U
19. Zeng, Y., Wang, X., Hu, Y., et al.: Research progress of electrocatalysts for hydrogen oxidation reaction in alkaline media. Carbon Neutral. **3**, 710–736 (2024). https://doi.org/10.1002/CNL2.152
20. Zhou, Q., Liao, L., Zhou, H., et al.: Innovative strategies in design of transition metal-based catalysts for large-current-density alkaline water/seawater electrolysis. Mater. Today Phys. **26**, 100727 (2022). https://doi.org/10.1016/J.MTPHYS.2022.100727
21. Guo, X., Hao, A., Li, S., et al.: Selective hydrogen oxidation catalyst for PEM fuel cells: tungsten cluster-tuned platinum single atoms. Small **21**, 2410101 (2025). https://doi.org/10.1002/SMLL.202410101
22. Xu, H., Xu, Z., Wang, K., et al.: Tungsten oxide-based electrocatalysts for energy conversion. Chem. Commun. **60**, 13507–13517 (2024). https://doi.org/10.1039/D4CC04767B
23. Tian, H., Cui, X., Shi, J.: Emerging electrocatalysts for PEMFCs applications: tungsten oxide as an example. Chem. Eng. J. **421**, 129430 (2021). https://doi.org/10.1016/J.CEJ.2021.129430
24. Meng, G., Yao, H., Tian, H., et al.: An electrochemically reconstructed WC/WO2–WO3 heterostructure as a highly efficient hydrogen oxidation electrocatalyst. J. Mater. Chem. A Mater. **10**, 622–631 (2022). https://doi.org/10.1039/D1TA08872F
25. Gaikwad, M.A., Burungale, V.V., Malavekar, D.B., et al.: Self-supported Fe-based nanostructured electrocatalysts for water splitting and selective oxidation reactions: past, present, and future. Adv. Energy Mater. **14**, 2303730 (2024). https://doi.org/10.1002/AENM.202303730
26. Zhu, Y., Li, L., Cheng, H., Ma, J.: Alkaline hydrogen evolution reaction electrocatalysts for anion exchange membrane water electrolyzers: progress and perspective. JACS Au **4**, 4639–4654 (2024). https://doi.org/10.1021/JACSAU.4C00898
27. Feidenhans'l, A.A., Regmi, Y.N., Wei, C., et al.: Precious metal free hydrogen evolution catalyst design and application. Chem. Rev. **124**, 5617–5667 (2024). https://doi.org/10.1021/ACS.CHEMREV.3C00712
28. Bhatt, M.D., Lee, J.Y.: Advancement of platinum (Pt)-Free (Non-Pt Precious Metals) and/or metal-free (non-precious-metals) electrocatalysts in energy applications: a review and perspectives. Energy Fuels **34**, 6634–6695 (2020). https://doi.org/10.1021/ACS.ENERGYFUELS.0C00953
29. Xiong, B., Zhao, W., Chen, L., Shi, J.: One-step synthesis of W2C@N,P-C nanocatalysts for efficient hydrogen electrooxidation across the whole pH range. Adv. Funct. Mater. **29**, 1902505 (2019). https://doi.org/10.1002/ADFM.201902505;REQUESTEDJOURNAL:JOURNAL:16163028;WGROUP:STRING:PUBLICATION
30. Liu, L., Zhou, X., Wang, X., et al.: H2 regulates the sulfur vacancy of Mo-doped WS2 for electrochemical hydrogen evolution reaction. Energy Fuels **38**, 23847–23856 (2024). https://doi.org/10.1021/ACS.ENERGYFUELS.4C04913
31. Xie, L., Wang, L., Zhao, W., et al.: WS2 moiré superlattices derived from mechanical flexibility for hydrogen evolution reaction. Nat. Commun. **12**, 1–9 (2021). https://doi.org/10.1038/S41467-021-25381-1;TECHMETA
32. Van Nguyen, T., Tekalgne, M., Nguyen, T.P., et al.: Electrocatalysts based on MoS2 and WS2 for hydrogen evolution reaction: an overview. Battery Energy **2**, 20220057 (2023). https://doi.org/10.1002/BTE2.20220057
33. Cong, Y., Yi, B., Song, Y.: Hydrogen oxidation reaction in alkaline media: from mechanism to recent electrocatalysts. Nano Energy **44**, 288–303 (2018). https://doi.org/10.1016/J.NANOEN.2017.12.008

34. Davydova, E.S., Manikandan, M., Dekel, D.R., Sunde, S.: Effect of the synthetic method on the properties of Ni-based hydrogen oxidation catalysts. ACS Appl. Energy Mater. **4**, 3404–3423 (2021). https://doi.org/10.1021/ACSAEM.0C03157
35. Mehmood, A., Gong, M., Jaouen, F., et al.: High loading of single atomic iron sites in Fe–NC oxygen reduction catalysts for proton exchange membrane fuel cells. Nat. Catal. **4**(5), 311–323 (2022). https://doi.org/10.1038/s41929-022-00772-9
36. Song, W., Xiao, C., Ding, J., et al.: Review of carbon support coordination environments for single metal atom electrocatalysts (SACS). Adv. Mater. **36**, 2301477 (2024). https://doi.org/10.1002/ADMA.202301477
37. Rukmani Krishnan, S., Verstraete, D., Aguey-Zinsou, F.: Performance of non-precious metal electrocatalysts in proton-exchange membrane fuel cells: a review. ChemElectroChem **11**, e202400299 (2024). https://doi.org/10.1002/CELC.202400299
38. Wang, L., Xu, Z., Kuo, C.H., et al.: Stabilizing low-valence single atoms by constructing metalloid tungsten carbide supports for efficient hydrogen oxidation and evolution. Angew. Chem. Int. Ed. **62**, e202311937 (2023). https://doi.org/10.1002/ANIE.202311937;PAGE:STRING:ARTICLE/CHAPTER
39. Yeo, R.S.: A hydrogen-bromine cell for energy storage applications. J. Electrochem. Soc. **127**, 549 (1980). https://doi.org/10.1149/1.2129710
40. Lin, G., Chong, P.Y., Yarlagadda, V., et al.: Advanced hydrogen-bromine flow batteries with improved efficiency, durability and cost. J. Electrochem. Soc. **163**, A5049–A5056 (2016). https://doi.org/10.1149/2.0071601jes
41. Yufit, V., Hale, B., Matian, M., et al.: Development of a regenerative hydrogen-vanadium fuel cell for energy storage applications. J. Electrochem. Soc. **160**, A856–A861 (2013). https://doi.org/10.1149/2.086306JES/XML
42. Muñoz, C.A.P., Dewage, H.H., Yufit, V., Brandon, N.P.: A unit cell model of a regenerative hydrogen-vanadium fuel cell. J. Electrochem. Soc. **164**, F1717–F1732 (2017). https://doi.org/10.1149/2.1431714jes
43. Chakrabarti, B.K., Kalamaras, E., Ouyang, M., et al.: Trichome-like carbon-metal fabrics made of carbon microfibers, carbon nanotubes, and Fe-based nanoparticles as electrodes for regenerative hydrogen/vanadium flow cells. ACS Appl. Nano Mater **4**, 10754–10763 (2021). https://doi.org/10.1021/ACSANM.1C02195/ASSET/IMAGES/LARGE/AN1C02195_0005.JPEG
44. Hewa Dewage, H., Wu, B., Tsoi, A., et al.: A novel regenerative hydrogen cerium fuel cell for energy storage applications. J. Mater. Chem. A Mater. **3**, 9446–9450 (2015). https://doi.org/10.1039/C5TA00571J
45. Rubio-Garcia, J., Kucernak, A., Zhao, D., et al.: Hydrogen/manganese hybrid redox flow battery. J. Phys. Energy **1**, 015006 (2018). https://doi.org/10.1088/2515-7655/AAEE17
46. Chen, W., Li, G., Pei, A., et al.: A manganese-hydrogen battery with potential for grid-scale energy storage. Nat. Energy **3**, 428–435 (2018). https://doi.org/10.1038/s41560-018-0147-7
47. Rubio-Garcia, J., Cui, J., Parra-Puerto, A., Kucernak, A.: Hydrogen/vanadium hybrid redox flow battery with enhanced electrolyte concentration. Energy Storage Mater. **31**, 1–10 (2020). https://doi.org/10.1016/j.ensm.2020.05.031
48. Rubio-Garcia, J., Kucernak, A., Parra-Puerto, A., et al.: Hydrogen/functionalized benzoquinone for a high-performance regenerative fuel cell as a potential large-scale energy storage platform. J. Mater. Chem. A Mater. **8**, 3933–3941 (2020). https://doi.org/10.1039/C9TA12396B
49. Zhu, Z., Meng, Y., Wang, M., et al.: A high-performance aqueous iron–hydrogen gas battery. Mater. Today Energy **19**, 100603 (2021). https://doi.org/10.1016/J.MTENER.2020.100603

Development of DES-Processed Graphene-Based Electrocatalysts for Quinone Redox Flow Battery Applications

Shamik Chaudhuri, Züleyha Kudaş, Reza Afshar Ghotli, Nadir Deniz Akgül, Mustafa Kemal Bayazıt, Serap Hayat Soytaş, Tülay Yılmaz İnan, Farouq Sabri Mjalli, Andres Parra-Puerto, Stefan Spirk, and Barun Kumar Chakrabarti

Abstract Deep eutectic solvents (DESs) are emerging as tunable and sustainable electrolytic media for electrochemical applications due to their low volatility, high ionic conductivity, and environmental compatibility. In this study, a series of acidic DESs was systematically evaluated through physicochemical characterizations (e.g., viscosity, density, conductivity) and electrochemical techniques (e.g., cyclic voltammetry) to determine their electrochemical potential windows and stability. Among the formulations tested, a selection of high-performing DESs was screened using ferrocene as a model redox couple to assess their compatibility and redox behavior. The most promising DESs were employed for the exfoliation of activated carbon derived from sustainable biomass sources. This green exfoliation approach was used to synthesize two-dimensional carbon nanomaterials: graphene (EG) and nitrogen-doped graphene (N-EG). In parallel, nitrogen doped carbon nanotubes (CNTs) were prepared via a chemical process to compare their performances with N-EG. These

S. Chaudhuri (✉) · Z. Kudaş · R. Afshar Ghotli · N. D. Akgül · M. K. Bayazıt · S. H. Soytaş · T. Y. İnan · B. K. Chakrabarti (✉)
Sabanci University Nanotechnology Research and Application Centre (SUNUM), Orta Mahalle Üniversite Caddesi, Tuzla, İstanbul, Turkey
e-mail: shamik.chaudhuri@sabanciuniv.edu

B. K. Chakrabarti
e-mail: barun.chakrabarti@sabanciuniv.edu

F. S. Mjalli
Department of Petroleum and Chemical Engineering, Sultan Qaboos University, Muscat, Oman

A. Parra-Puerto
Department of Physical Chemistry and Electrochemistry Institute, University of Alicante, San Vicente del Raspeig (Alicante), Spain

S. Spirk
Graz University of Technology—Institute of Bioproducts and Paper Technology, Graz, Austria

Ecolyte GmbH, TU Graz Campus Inffeldgasse, Graz, Austria

M. Awang et al. (eds.), *Proceedings of the International Conference on Mechanical, Manufacturing and Plant Engineering*, Proceedings in Technology Transfer,
https://doi.org/10.1007/978-981-95-7817-7_22

modifiers were deposited onto carbon cloth substrates via cathodic electrophoretic deposition (EPD), forming high-surface-area, conductive electrodes. Further, the coated electrodes were evaluated in a three-electrode cell configuration using the positive electrolyte of the all-vanadium redox flow battery (RFB) as well as a functionalized benzoquinone-based electrolyte, a promising class of redox-active organic molecules for aqueous RFB systems. Key electrochemical performance metrics, such as cyclic voltammetry and electrochemical impedance spectroscopy were used to optimize the most effective electrode–electrolyte combinations. This work established a sustainable, DES-enabled pathway for producing advanced carbon-based electrocatalysts. The best-performing electrode materials from this study will be further investigated as catalytic electrodes in full-cell hydrogen/quinone regenerative fuel cells aiming to enhance the efficiency and scalability of next-generation organic energy storage systems.

Keywords Deep eutectic solvents · Graphene · Redox flow batteries · Quinone electrolytes · Electrophoretic deposition

1 Introduction

Electrochemical energy storage is rapidly evolving towards systems that combine low cost, high cyclability, and environmentally benign materials. Organic redox flow batteries (RFBs) that employ quinone-based molecules are attractive candidates because many quinones are derived from abundant feedstocks and can be chemically tuned to the desired redox potential and solubility [1]. However, the performance of quinone electrolytes critically depends on the electrode surface chemistry and architecture, active surface area, conductivity, catalytic sites for fast electron transfer, wettability, and mass-transport pathways.

In parallel, deep eutectic solvents (DESs) are quite promising as green, tunable media for materials processing. They can dissolve a range of precursors, stabilize ionic species, and, in specific formulations, assist exfoliation of layered carbons. Using DESs for exfoliating biomass-derived activated carbon into graphene-like sheets and nitrogen-doped graphene offers a sustainable feedstock pathway (bio-waste to advanced carbon), the ability to introduce heteroatoms via DES chemistry or secondary treatments, and a processing route compatible with electrophoretic deposition (EPD) for scalable electrode fabrication [2]. Based on this understanding, this chapter projects a research plan that integrates DES selection and characterization, green exfoliation, nitrogen-doping strategies, EPD electrode fabrication, and systematic electrochemical testing against quinone electrolytes.

2 Objectives of the Study

The main objectives of this investigation involved:

1. Screening and identification of various acidic DES formulations that offer a suitable electrochemical window, reasonable conductivity, and capacity to exfoliate biomass-derived activated carbon into 2D graphene-like sheets.
2. Synthesizing exfoliated graphene (EG) and nitrogen-doped exfoliated graphene (N-EG) from biomass-derived activated carbon using the selected DES(s) and prepare nitrogen-doped carbon nanotubes (N-CNTs) for comparison.
3. Deposition of as-synthesized carbon nanomaterials onto carbon-cloth substrates via cathodic EPD to form conductive, high-surface-area electrodes [3].
4. Evaluation of the electrode/electrolyte combinations using benzoquinone-derived redox-active organic molecules in three-electrode cells and optimize the best-performing electrode formulations for follow-up full-cell hydrogen/quinone regenerative fuel cell tests [1].

3 Conceptual Insight-Rationale

3.1 DES

DESs combine a hydrogen-bond donor (HBD) and a hydrogen-bond acceptor (HBA) to form low-melting mixtures with desirable solvent properties such as low vapor pressure, wide tunability, and the possibility to dissolve a broad range of materials [4]. In acidic DES formulations, proton activity and hydrogen-bonding networks can become effective promoters of carbon delamination and heteroatom incorporation when combined with heat, sonication or mechanical shear. Importantly, many DES components are low-cost and biodegradable in comparison to certain ionic liquids, aligning with sustainability objectives.

3.2 Biomass-Derived Activated Carbon as Precursor

Biomass-derived activated carbons offer a low-cost precursor that contain a hierarchical porosity and a network of defect sites. Such defect-rich carbons are often more readily exfoliated than pristine graphite, and after controlled exfoliation and doping the resulting materials can have high surface area and abundant edge sites, traits that are beneficial for catalysis and capacitor-like behavior at the electrode–electrolyte interfaces.

3.3 Nitrogen Doping

Nitrogen functionalization of carbon materials modifies electronic structure, introduces active sites for redox catalysis, and can improve wettability. N-doping typically enhances the heterogeneous electron-transfer kinetics for organic redox couples. N-CNTs serve as a complementary morphology to 2D N-EG to elucidate the role of dimensionality and curvature on quinone reactions and transport [5].

3.4 EPD

EPD provides a facile, scalable route to form uniform, adherent coatings on conductive substrates [6]. Cathodic EPD is especially attractive when suspensions contain negatively charged carbon sheets or when surface functionalization can impart electrophoretic mobility. EPD enables control overcoating thickness and microstructure through deposition parameters (voltage, time, suspension composition), which is crucial for tailoring electrode mass loading and pore accessibility.

4 Experimental

4.1 DES Preparation and Characterization

Preparation. Table 1 presents various low toxicity HBDs (e.g., citric, lactic, p-toluene sulfonic acid, and others) that can combine with HBAs like choline chloride forming a tailor-made DES for certain applications. With a certain molar ratio, as enlisted in the table, HBDs and HBAs were mixed in a glass beaker via magnetic stirring at 60–80 °C as reported by Mjalli et al. [4]. Prior drying was important for choline chloride due to its highly hygroscopic nature that can alter properties like viscosity.

Physicochemical characterization. Physico-chemical properties of DESs such as density, viscosity, ionic conductivity pH and electrochemical window were determined as reported earlier [7]. Conductivity of the DESs was measured using a DZS-708 multi-parameter analyzer calibrated with 0.001 M KCl at different temperatures. Viscosity was determined with a Brookfield DV-II + Pro EXTRA viscometer as the average of 3–5 replicates, with an uncertainty below ± 1%. Densities were measured at various temperatures using a DMA 4100 density meter (Anton Paar, Austria) with triplicate measurements (±0.00008 g/cm^3). Instrument accuracy was verified with distilled, degassed water at 298 K, showing a deviation of ± 0.00005 g/cm^3 from reference data. pH values were obtained using a calibrated Metrohm pH meter with standard buffer solutions.

Table 1 DES (made with choline chloride salt as the hydrogen bond acceptor) properties measured inside a glovebox

Hydrogen bond donor	Molar ratio	Density (g/cm^3)	Viscosity (mPa s)	Conductivity (mS/ cm)	pH	Electrochemical potential window (V)
Ethylene glycol	1:2	1.2	66	7.5	7.3	3.2
Urea	1:2	1.22	450	2.3	9	3.5
Oxalic acid	1:1	1.3	460	2	3	2.1
Acetic acid	1:2	1.11	45	4.6	0.8	N/A
Propionic acid	1:2	1.08	56.2	1.5	1	2.11
Citric acid	1:1	2.9	350	0	0.5	N/A
Lactic acid	1:1	1.1	309	0.04	0.4	N/A
p-toluene sulfonic acid (PTSA)	1:2	1.08	206	5.6	0.5	2
Methane sulfonic acid (MSA)	1:2	1.23	35	6480	1	2.5

Redox Compatibility. Cyclic voltammetry (CV) measurements were conducted on a platinum (Pt) electrode in a DES solution. The CVs were recorded within a potential window of 0.1 V to 0.7 V at different scan rates ranging from 10 to 100 mV/s both with and without 1 mM ferrocene.

4.2 *Biomass Derived Activated Carbon*

Biomass-derived activated carbon (BAC) was synthesized from locally sourced hazelnut shells, using NaCl as the activating agent at a precursor-to-salt ratio of 1:3. The shells were first cleansed with deionized (DI) water, then oven-dried at 100 °C for 24 h. Subsequently, they were carbonized under nitrogen at 700 °C for 2 h to obtain hydrochar, which was further activated at 750 °C for 2 h in a CO_2 atmosphere, following a previously reported procedure [8]. The activated material was treated with 1 M HCl to eliminate residual salts, washed repeatedly with DI water until neutral, and finally dried in a vacuum oven at 120 °C overnight.

4.3 *DES-Assisted Preparation of EG and N-EG*

Exfoliation concept. BAC was dispersed in a selected DES and sonication, or shear-assisted milling was applied to promote delamination into graphene-like sheets (EG or exfoliated graphene). EG was obtained by dispersing BAC in a certain DES (in different concentrations) and subsequent sonication, or shear-assisted milling was performed. In some cases, quenching the DES suspension with a controlled anti-solvent was required. Further, the suspension was subjected to filtration and then washed to remove residual DES and recover the exfoliated materials. The materials were collected under mild conditions to preserve surface functional groups. N-EG was prepared either by introducing a nitrogen source in the DES composition (e.g., urea) or by incorporating a nitrogen precursor during or after exfoliation followed by a mild thermal treatment to fix N into the lattice.

4.4 EPD onto Carbon Cloth

A horizontal cathodic EPD system was assembled to fabricate uniform, adherent coatings of nanomaterials on conductive substrates as reported elsewhere [3, 6]. The custom-made cylindrical reactor, constructed from chemically inert polytetrafluoroethylene (PTFE), had an internal diameter of 40 mm. The horizontal configuration was chosen to enhance deposit uniformity and film adhesion compared to vertical EPD [3]. A graphite plate (Poco Pyro, Entegris Inc.) served as the anode, while carbon cloth (CC) functioned as the cathodic deposition surfaces. The electrodes were separated by a 15 mm gap, ensured by a circular glass spacer. Prior to deposition, the CC substrates were cleansed by rinsing with distilled water, followed by acetone, and then dried in an oven at 60 °C to eliminate residual contaminants. EG, and N-EG, powders in an EPD-compatible medium (N,N-Dimethylformamide or DMF) were dispersed with any necessary charge-stabilizing agents.

4.5 Electrochemical Evaluation with Vanadium Electrolytes

Electrochemical analyses were carried out using a Gamry Potentiostat 3000 in a three-electrode electrochemical cell system. In this setup, a platinum wire and an Ag/AgCl (3 M) electrode were used as the counter and reference electrodes, respectively. The working electrodes consisted of bare CC and modified CC. A 0.5 M $VOSO_4$ solution in 3 M H_2SO_4 was employed as the aqueous electrolyte. All the electrodes were soaked in the electrolyte for 30 min prior to investigation. To determine the electrochemical activity, cyclic voltammetry (CV) and electrochemical Impedance spectrometry (EIS) experiments were performed. CV was conducted between 0.4 and 1.0 V at a scan rate of 50 mV/s in a similar manner to that reported previously [3, 9]. CV assesses the peak separation and reversibility while EIS assesses charge transfer resistances.

4.6 *Electrochemical Evaluation with Quinone Electrolytes*

Similarly, the solution for quinone electrolyte was prepared by dissolving 10 mM 4,5-dihydroxy-1,3-benzenedisulfonic acid disodium salt monohydrate (BQDS, 97%, Sigma-Aldrich) in 1 M H_2SO_4 using ultrapure water. CV measurements were performed at a scan rate of 50 mV/s. EIS measurements were conducted at open circuit voltage (OCV) in the frequency range of 0.1 Hz to 10 kHz using an amplitude of 10 mV.

5 Results and Discussion

The electrochemical potential window was measured in a three-electrode cell using glassy carbon working, platinum gauze counter and a silver wire quasi-reference electrode. All measurements reported in Table 1 were performed inside a glovebox. The ChCl: acetic acid medium exhibits acidic characteristics and is rich in complexing species, particularly chloride ions as reported elsewhere [10]. These conditions prevent the use of conventional redox couples, such as $Fe(CN)_64^-/Fe(CN)_63^-$ or ferrocene/ferrocenium, as reliable electrochemical standards. From Table 1, it was clear that MSA-ChCl DES (MSA DES) was the best one for electrochemical testing.

CV analyses were performed in the MSA DES, both with and without ferrocene for the first time [11]. The influence of the DES on the electrochemical behavior of ferrocene was evaluated by comparing the voltammograms obtained in MSA-ChCl in the absence and presence of ferrocene, depicted in Figs. 1a and 2b, respectively.

Similarly, the investigation was carried out with urea-ChCl DES (reline) as a comparison. All the conditions including ferrocene were kept identical as in the previous experiment with MSA-ChCl. Figure 2a and b illustrated the CV plots for urea-ChCl without and with ferrocene, respectively. From Figs. 1 and 2, it was evident that MSA DES performed better than reline as also confirmed from Table 1. *As a*

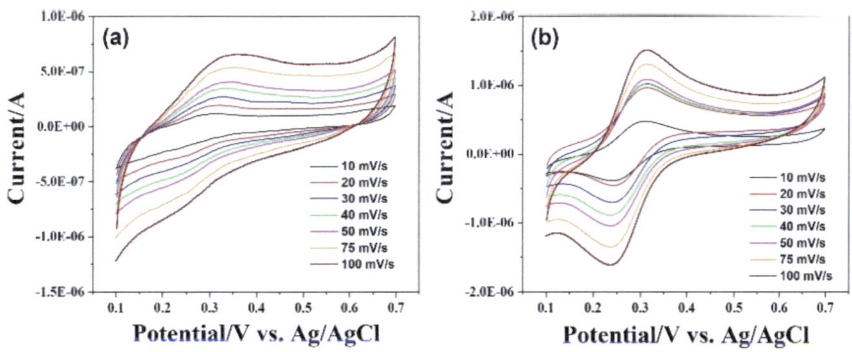

Fig. 1 CV in MSA-ChCl **a** without 1 mM ferrocene and **b** with ferrocene at different scan rates

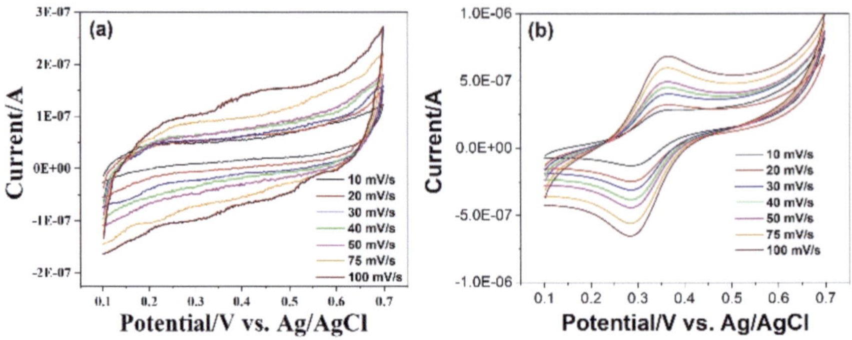

Fig. 2 CV in urea-ChCl **a** without 1 mM ferrocene and **b** with ferrocene at different scan rates

consequence, MSA DES was selected for the exfoliation of BAC to prepare EG and N-EG (a ternary DES of ChCl:MSA:urea was prepared at a ratio of 1:2:2 for this purpose).

Structure and morphology. Structural and morphological characterizations of bare CC were carried out via XRD, FTIR, Raman spectra and SEM microscopy. XRD of bare CC in Fig. 3a exhibited peaks near 26° and 43°**,** consistent with the (002) and (100) planes [12–14]. Figure 3b showed from the FTIR curves that C = C was predominating. Figure 3c depicted the Raman spectra of CC (D band at 1353 cm^{-1} and G band at 1593 cm^{-1}). The I_G/I_D ratio was obtained as 1.00. Figure 3d displayed a SEM image of bare CC containing woven carbon fibers with smooth surface and prominent inter-fiber voids.

Electrochemical Evaluation with Vanadium Electrolytes. Figure 4a and b illustrated the CV (50 mV/s scan rate) and Nyquist plot of bare CC, CC_CNT, CC_N-CNT, CC_EG and CC_N-EG samples in the positive electrolyte of the all vanadium RFB, respectively. It is clearly seen that CC_N-CNT has significantly enhanced peak current densities compared to all other samples. Nyquist plot also confirmed that CC_N-CNT exhibited the lowest semi-circle corresponding to the lowest charge-transfer resistance (R_{ct}). Although solution resistance of all the samples is comparable, CC_N-CNT again showed the lowest value. CC_CNT showed the second-best performance.

Electrochemical evaluation with quinone electrolytes. Figure 5a and b illustrated the CV (50 mV/s scan rate) and Nyquist plot of bare CC, CC_CNT, CC_N-CNT, CC_EG and CC_N-EG samples in a 10 mM BQDS system, respectively. CC_N-CNT displays a significantly enhanced peak current density compared to all other samples once again. Also compared to the VRFB electrolyte this enhancement is almost 2.5-fold. However, the second-best performance was exhibited by CC_N-EG in this system. The corresponding Nyquist plot also displayed a similar trend.

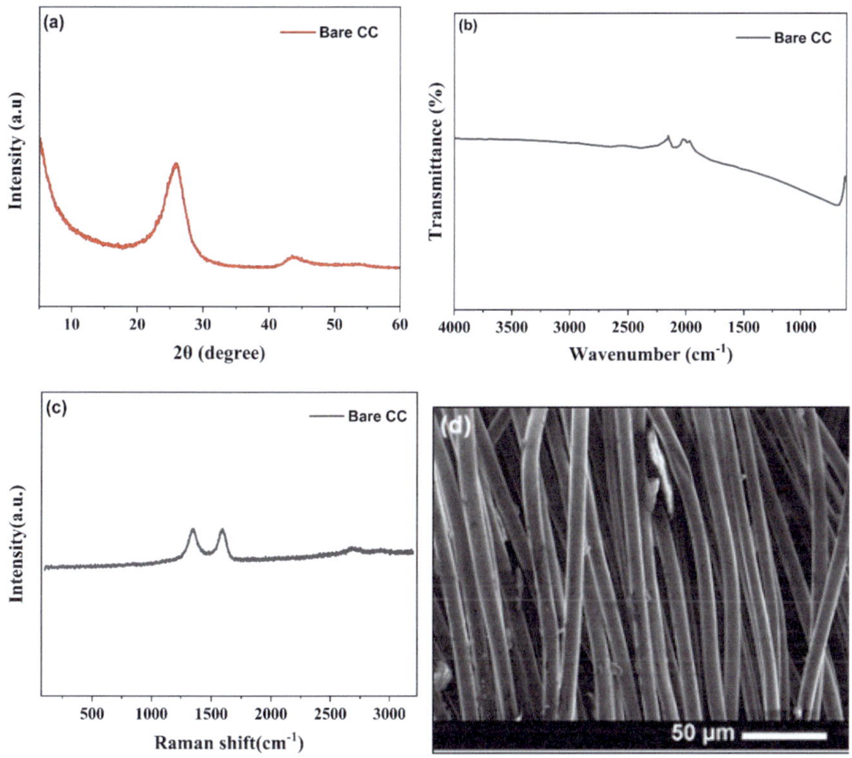

Fig. 3 Structure and morphology of bare CC **a** XRD **b** FTIR **c** Raman spectra **d** SEM

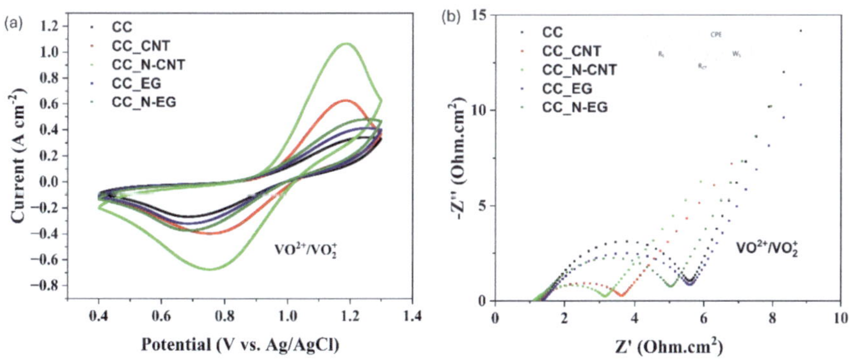

Fig. 4 Comparison of CNT, N-doped CNT, EG and N-doped EG modified CC with bare CC in the positive electrolyte of all vanadium RFB **a** CV **b** Nyquist plot with the equivalent circuit displayed in the inset

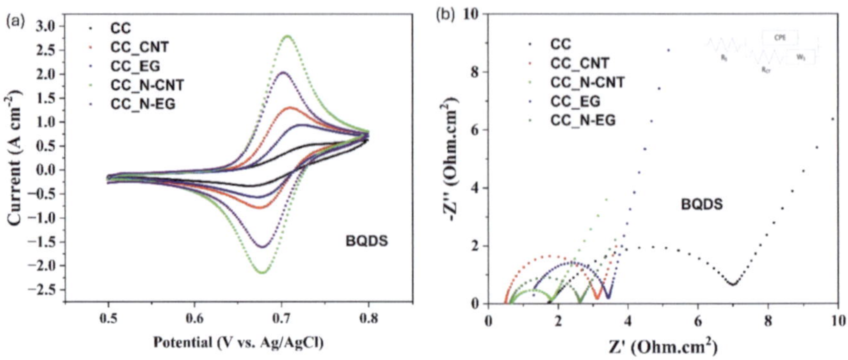

Fig. 5 Comparison of CNT, N-doped CNT, EG and N-doped EG modified CC with bare CC in 10 mM BQDS **a** CV **b** Nyquist plot

6 Conclusions and Outlook

In this study, a broad range of DESs was examined to enhance the exfoliation efficiency of biomass-derived activated carbon, with and without ferrocene as an electrolyte additive. Acidic DES formulations demonstrated particular promises for both exfoliation and nitrogen doping. The electrochemical performance of bare and surface-modified carbon cloth electrodes was further evaluated in vanadium and quinone systems. Among the tested materials, N-CNT–modified carbon cloth exhibited the highest activity in both the VO_2^+/VO_2^+ and BQDS redox couples. CNT modification ranked second in the vanadium system, whereas N-EG modification proved more favorable for the quinone system. These findings highlight the versatility of DESs in electrochemical applications and underscore the importance of rational DES design to control graphene surface terminations, conductivity, and related properties. Future work should focus on tailoring DESs to broaden electrochemical stability windows and on identifying the most effective electrode–electrolyte combinations to advance next-generation organic energy storage systems.

Acknowledgements This work was supported by the European Innovation Council (EIC) under grant agreement No. 101115293 (VanillaFlow). This research was also funded by the CETPartnership, under the Joint Call 2023, through the project "HydrogEn oRganiC Regenerative FUeL CEll for Stationary Energy Storage (HERCULES)". The CETPartnership's research projects are co-funded by the European Commission (Grant Agreement No. 101069750) and the national TUBITAK 1071 via the project code: 224N167. Authors also acknowledge the internal project funded by SUNUM's management (2D nanomaterial electrodes via electro-deposition and electrospinning for electrochemical energy storage applications) that supported this work (I.A.SN-23-00110). R.A.G. is grateful to Türkiye Bilimsel ve Teknolojik Araştırma Kurumu (TÜBİTAK) for the 1001 project for funding his research in SUNUM (223M269).

References

1. Garcia, J.R., Kucernak, A., Puerto, A.P., Liu, R., Chakrabarti, B.: J. Mater. Chem. A **8**, 3933–3941 (2020)
2. Chakrabarti, B., Garcia, J.R., Kalamaras, E., Yufit, V., Tariq, F., Low, C.T.J., Kucernak, A., Brandon, N.: Batteries **6**, 38 (2020)
3. Chakrabarti, B., Nir, D., Yufit, V., Tariq, F., Garcia, J.R., Maher, R., Kucernak, A., Arvind, P.V., Brandon, N.: Chem. Electro. Chem. **4**, 194–200 (2017)
4. Chakrabarti, M.H., Brandon, N.P., Mjalli, F.S., Bahadori, L., Nashef, I.M.A., Hashim, M.A., Hussain, M.A., Low, C.T.J., Yufit, V.: J. Solution Chem. **42**, 2329–2341 (2013)
5. Chakrabarti, B.K., Feng, J., Kalamaras, E., Garcia, J.R., George, C., Luo, H., Xia, Y., Yufit, V., Titirici, M.M., Low, C.T.J., Kucernak, A., Brandon, N.P.: ACS Appl. Mater. Interfaces **12**, 53869–53878 (2020)
6. Chakrabarti, B., Yufit, V., Kavei, A., Xia, Y., Stevenson, G., Kalamaras, E., Luo, H., Feng, J., Tariq, F., Taiwo, O., Titirici, M.M., Brandon, N.: Int. J. Hydrogen Energy **44**, 30093–30107 (2019)
7. Bahadori, L., Chakrabarti, M.H., Hashim, M.A., Manan, N.S.A., Mjalli, F.S., Nashef, I.M.A., Brandon, N.P.: J. Electrochem. Soc. **162**, H617 (2015)
8. Gharibi, A.S.R.A., Mjalli, F.S., Tarboush, B.A., Saadi, M.A.H.A.: Mater. Sci. Forum **1059**, 145–155 (2022)
9. Ghasemiestahbanati, E., Shaibani, M., Konstas, K., Chakrabarti, B.K., Low, C.T.J., Majumder, M., Hill, M.R.: ACS Appl. Energy Mater. **5**, 1505–1515 (2022)
10. Polomski, D., Chotkowski, M.: J. Solid State Electrochem. **28**, 1463–1474 (2024)
11. Mishra, S., Jadhao, P.R., Pandey, V., Pant, K.K., Harbottle, D.: J. Mol. Liq. A **415**, 126338 (2024)
12. Bai, R., Zhao, Y., Lu, C., Meng, Y., Gao, W., Wang, Y., Dang, R., Mu, M., Wang, J., Jiao, Y.: J. Chem. Res. **47**, 17475198231158744 (2023)
13. Zhang, X., Zhang, J., Quan, J., Wang, N., Zhu, Y.: Analyst **141**, 5527–5534 (2016)
14. Haruna, A., Dönmez, K. B., Hooshmand, S., Avcı, E., Qamar, M., Zaidi, S. A., Shahzad, F., Miller, T. S., Chakrabarti, B. K., Howard, C. A., Bayazıt, M. K. (2024). Harmony of nanosystems: Graphitic carbon nitride/carbon nanomaterial hybrid architectures for energy storage in supercapacitors and batteries. Carbon **226**, 119177. https://doi.org/10.1016/j.carbon.2024.119177

Single-Cell Mathematical Modeling of an Iron–Manganese Redox Flow Battery for Enhanced Electrochemical Performance

Mohammad Jakir Hossain Khan, Audrius Senulis, Zilvinas Kryzevicius, Arturas Tadzijevas, Jochen Uebe, Reza Afshar Ghotli, Barun Kumar Chakrabarti, and Yashar Hajimolana

Abstract Redox flow batteries (RFBs) offer a promising pathway for scalable and long-duration energy storage. Among emerging chemistries, the iron–manganese (Fe/Mn) system stands out due to its low cost, environmental compatibility, and favorable redox potentials. Mathematical models for Iron–Manganese style redox flow batteries (RFBs) employ the same multi-scale electrochemical, transport and fluid-dynamic frameworks used across RFB research, but the existing literature shows few dedicated iron–manganese based RFBs studies. Major trends include pore-scale and continuum multi-physics, reduced-order stack models, and emerging physics-informed ML and topology-optimization tools. This chapter provides an extensive investigation of mathematical modeling approaches pertinent to iron–manganese (Fe-Mn) redox flow batteries, with a highlighting on the detailed explanation of electrochemical performance via computational design and optimization strategies. We explore foundational electrochemical principles, governing equations, and modeling methodologies that encompass the fluid dynamics, state of charge (SoC), impact of pressure, equations essential for measuring performance and multidimensional battery electrode simulations. The chapter also addresses critical challenges inherent in Fe-Mn systems, encompassing iron-manganese speciation stability, precipitation kinetics, and transport phenomena, while accentuating recent advancements in computational techniques and their applications for performance enhancement. Practical case studies illustrate the manner in which mathematical modeling informs

M. J. H. Khan (✉) · A. Senulis · Z. Kryzevicius · A. Tadzijevas · J. Uebe
Engineering Department, Faculty of Marine Technology and Natural Sciences, Klaipėda University, H. Manto 84, 92294 Klaipėda, Lithuania
e-mail: khan.mohammad-jakir-hossain@ku.lt

R. A. Ghotli · B. K. Chakrabarti
Sabanci University Nanotechnology Research and Application Centre (SUNUM), Orta Mahalle Üniversite Caddesi No:27, 34956 Tuzla, İstanbul, Turkey

Y. Hajimolana
HyEnSys Consultancy, Enschede, The Netherlands

M. Awang et al. (eds.), *Proceedings of the International Conference on Mechanical, Manufacturing and Plant Engineering*, Proceedings in Technology Transfer,
https://doi.org/10.1007/978-981-95-7817-7_23

electrode design, electrolyte formulation, and the selection of operational conditions to optimize energy efficiency and extend cycle life.

Keywords Iron–manganese redox flow battery · Mathematical modeling · Single · Cell simulation · Battery management · Electrochemical performance · State-of-charge estimation · Energy storage · System optimization

1 Introduction

Redox flow batteries (RFBs) represent a highly advantageous technology for large-scale energy storage systems, as they facilitate the independent scaling of both power and energy capacity, in addition to offering an extended cycle life and improved safety characteristics relative to conventional battery technologies [1]. Among the various redox systems, the adaptation of iron–manganese (Fe-Mn) system has drawn substantial research interest on account of their abundance, cost-effectiveness, and possible relevance in high energy density applications. Nonetheless, the intricate electrochemical dynamics associated with manganese species, particularly the instability of $Mn3^{+}$ ions and their less-studied reaction kinetics, introduce distinctive challenges that necessitate advanced mathematical modeling methodologies to comprehend and address effectively [2]. Single-cell mathematical modeling functions as an essential instrument for comprehending the fundamental transport, kinetic, and thermodynamic processes that dictate the performance of Fe-Mn redox flow batteries (RFB). These models grant researchers and engineers the capability to forecast cell-level behavior over a spectrum of operational conditions, adjust design parameters, and ascertain performance-limiting elements without the necessity of thorough experimental investigations. The intricate nature of Fe-Mn systems, which encompasses multiple redox couples, side reactions, and potential solid phase formation, necessitates the employment of multi-physics modeling methodologies that integrate electrochemical kinetics with mass transport, fluid dynamics, and thermodynamic equilibria [3].

2 Chapter Scope and Objectives

This chapter endeavors to deliver an exhaustive examination of mathematical modeling at the single-cell level for iron-manganese redox flow batteries (Fe-Mn RFBs), with particular emphasis on:

i. Fundamental electrochemical principles governing Fe-Mn redox reactions and their mathematical representation.
ii. Modeling methodologies from simplified zero-dimensional approaches to detailed three-dimensional simulations.

iii. Key performance parameters and their optimization through computational design.
iv. Recent advances in computational methods and their application to Fe-Mn systems.
v. Practical applications and case studies demonstrating model validation and design optimization.
vi. Future challenges and research directions in Fe-Mn RFB modeling.

3 Iron-Manganese Chemistry Overview

Iron-manganese redox flow batteries (RFBs) are capable of functioning within various electrolyte systems, each of which poses distinct modeling complexities. Two fundamental configurations have been verified through experimental demonstration.

Alkaline All-Soluble System.

This configuration employs iron-oxo complexes in conjunction with manganese oxides in redox couples, culminating in a model cell voltage of roughly 1.43 V. Although this system exhibits elevated Coulombic efficiency, it is adversely affected by escalating voltage losses attributed to internal resistance throughout cycling, necessitating meticulous modeling of resistance evolution and mechanisms of capacity degradation [1].

Acidic Chloride System.

In this arrangement, the couples of Fe^{2+}/Fe^{3+} are associated with Mn^{2+}/Mn^{3+} redox processes within chloride media. The Mn^{3+} species is rendered stable in the form of $[MnCl_4(H_2O)_2]^-$ complexes to inhibit the precipitation of MnO_2, thereby facilitating operation at voltages up to 0.7 V under conditions of elevated current densities. This system necessitates advanced speciation modeling to accurately elucidate the impact of chloride complexation on electrochemical behavior [2].

4 Fundamental Electrochemical Principles

Redox Reactions and Thermodynamics

The electrochemical characteristics of Fe-Mn redox flow batteries (RFBs) are determined by the intrinsic redox processes that transpire at the anode and cathode. A comprehensive understanding of these reactions and their thermodynamic attributes is imperative for the formulation of precise mathematical models. In acidic media, the iron redox couple operates according to

$$Fe^{3+} + e^- \rightleftharpoons Fe^{2+} E^\circ = +0.77V \text{ vs. SHE} \tag{1}$$

The standard electrode potential serves as the foundational reference for conducting thermodynamic assessments; however, the actual electrochemical cell potentials are contingent upon the concentrations of species present, the temperature, and the composition of the electrolyte. In alkaline environments, iron manifests in multiple oxo-complexes, which profoundly modifies the redox potential and necessitates the development of comprehensive speciation models. Consequently, during operation within the Redox Flow Battery (RFB) system, the Mn^{3+}/Mn^{2+} and Fe^{3+}/Fe^{2+} redox couples function as the catholyte and anolyte, respectively. Given that the electrochemical conversion of Fe/Fe^{2+} (0.35 V vs SHE) necessitates functioning under a controlled atmosphere to enhance activity on the anolyte side, it is recommended that Fe^{3+}/Fe^{2+} (0.77 V vs SHE) be employed as the redox couple in the anode half-cell reaction, albeit with a minor compromise on cell potential. By coupling these redox species, an overall cell voltage of 0.79 V can be achieved, with the corresponding redox reactions outlined below.

Half-cell reaction at cathode:

$$Mn^{3+} + e^- \leftrightarrow Mn^{2+}(1.56V \text{ vs SHE}) \tag{2}$$

Half-cell reaction at anode:

$$Fe^{3+} + e^- \leftrightarrow Fe^{2+}(0.77V \text{ vs SHE}) \tag{3}$$

Overall cell reaction:

$$Fe^{3+} + Mn^{2+} \leftrightarrow Fe^{2} + Mn^{3+} \text{ Ecell } = 0.79V \tag{4}$$

In this context, the Fe^{3+}/Fe^{2+} pair functions as the anolyte during the redox process, notwithstanding the pronounced negative electrode potential. Conversely, the Mn^{3+}/Mn^{2+} redox couple is widely recognized for its elevated cathodic potential. Consequently, the amalgamation of the Fe^{3+}/Fe^{2+} and Mn^{3+}/Mn^{2+} systems yields a net practical cell voltage amounting to 0.79 V. Although this cell voltage is comparatively lower than that of other noble metal systems, the solubility of both metal precursors is notably elevated in a commonly utilized supporting electrolyte; thus, it facilitates the adjustment of pH levels, which is critical for maintaining the equilibrium of the redox reaction throughout the experimental procedure.

5 Manganese Redox Chemistry

The chemistry of manganese exhibits a significantly higher level of complexity attributable to the presence of various accessible oxidation states and pronounced propensities for disproportionation reactions. Within acidic chloride environments, the principal redox reaction is:

$$Mn^{3+} + e^{-} \rightleftharpoons Mn^{2+} E^{\circ} = +1.51V \text{ vs. SHE} \quad (5)$$

However, the Mn^{3+} species is highly unstable and undergoes disproportionation:

$$2Mn^{3+} + 2H_2O \rightarrow Mn^{2+} + MnO_2 + 4H^{+} \quad (6)$$

This reaction results in a reduction of capacity and the degradation of the electrode, rendering it an essential factor in the formulation of mathematical models. The process of chloride complexation facilitates the stabilization of Mn^{3+} via the synthesis of $[MnCl_4(H_2O)_2]^{-}$ complexes; however, it is imperative that the equilibrium constants and reaction kinetics associated with these processes are precisely depicted within the models.

In this regard, the employment of methane sulfonic acid as the supporting electrolyte (acidic medium) allows the methane sulfonate to form a complex with the Mn3 + cations, characterized by a substantial molecular volume, thereby mitigating collisions between the cations and ultimately minimizing the capacity fade of the redox flow battery, characterized by a substantial molecular volume, thereby mitigating collisions between the cations and ultimately minimizing the capacity fade of the redox flow battery. In this context, the transient generation of MnO_2 demonstrates reversible characteristics attributable to the acidic composition of the overall electrolyte. Conversely, within an aqueous neutral environment, the deposition of MnO_2 on the electrode surface exhibits limited reversibility. The half-cell reaction mechanism pertaining to the MSA and $MnCl_2{\cdot}4H_2O$ anticipated in the current study is delineated below.

$$MnCl_2 \cdot 4H_2O + 2CH_3SO_3H \leftrightarrow \left[Mn(CH_3SO_3)^2\right] + 4H_2O + 2HCl \quad (7)$$

$$[Mn(CH_3SO_3)2] \leftrightarrow [Mn(CH_3SO_3)] + \ e- \quad (8)$$

6 Overall Reaction Mechanism

$$MnCl_2.4H_2O + 2CH_3SO_3H \leftrightarrow [Mn(CH_3SO_3)] + +4H_2O + 2HCl + e-$$

As the energy density is significantly influenced by the concentration of the electrolyte and its volume, it is feasible to attain a high energy density through the utilization of appropriate electrode materials. Furthermore, the precursors utilized are comparatively low-cost and abundant, especially when juxtaposed with vanadium. To substantiate this assertion, the initial evaluation of the current innovative systems demonstrated enhanced electrochemical performance over extended cycling life, accompanied by satisfactory efficiencies.

A schematic illustration of the Fe-Mn flow cell can see in Fig. 1. In Fig. 1, a lab-scale flow cell setup of Fe-Mn redox flow battery exhibited considerable user-friendly applicability, demonstrating elevated and consistent efficiencies throughout prolonged operational cycles. The single-cell configuration incorporated several fundamental components: the cover plate was constructed from polyvinyl chloride (PVC), gold-coated copper was employed as the current collector, while graphite supporting electrodes and heat-treated graphite felt (HT-GF) with an active area of 5 cm^2 were also implemented. The Nafion-117 cation exchange membrane was found to be compatible with the Fe-Mn electrochemical reaction. The electrolytes were delivered into the system via peristaltic pumps at a predetermined flow rate of 30 ml per minute on both sides.

7 Nernst Equation and Efficiency Measurement:

The Nernst equation plays a crucial role in modeling the electrochemical behavior of Iron-Manganese Redox Flow Batteries (RFBs), particularly in predicting the open circuit voltage (OCV) and understanding the dynamics of charging and discharging processes. This equation is essential for accurately simulating the performance of RFBs, which are pivotal for integrating renewable energy sources into power grids. The application of the Nernst equation in RFB modeling involves several considerations, including the impact of transport delays, activity coefficients, and the state of charge (SOC).

Role of the Nernst Equation in RFB Modeling

Open Circuit Voltage (OCV) Prediction: The Nernst equation is fundamental in determining the OCV of RFBs, which is a key parameter for assessing battery performance. It describes the equilibrium potential based on ion concentrations and temperature. However, simplifications in the equation, such as neglecting activity coefficients, can lead to inaccuracies in OCV predictions, as these coefficients significantly influence the OCV, especially in Iron-Cadmium and All-Vanadium RFBs [5]. The open-circuit voltage (OCV) of Fe-Mn cells is determined by the Nernst equation:

$$\mathrm{E\ cell} = \mathrm{E^\circ\ cell} - (\mathrm{RT/nF})\ln(\mathrm{Q}) \tag{9}$$

where Q represents the reaction, quotient derived from the activities of the species involved. In concentrated electrolyte solutions characteristic of redox flow batteries (RFBs), the activity coefficients exhibit considerable divergence from unity, thereby necessitating a thermodynamically rigorous approach to prevent the erroneous assignment of open circuit voltage (OCV) deviations to kinetic losses during the process of parameter extraction.

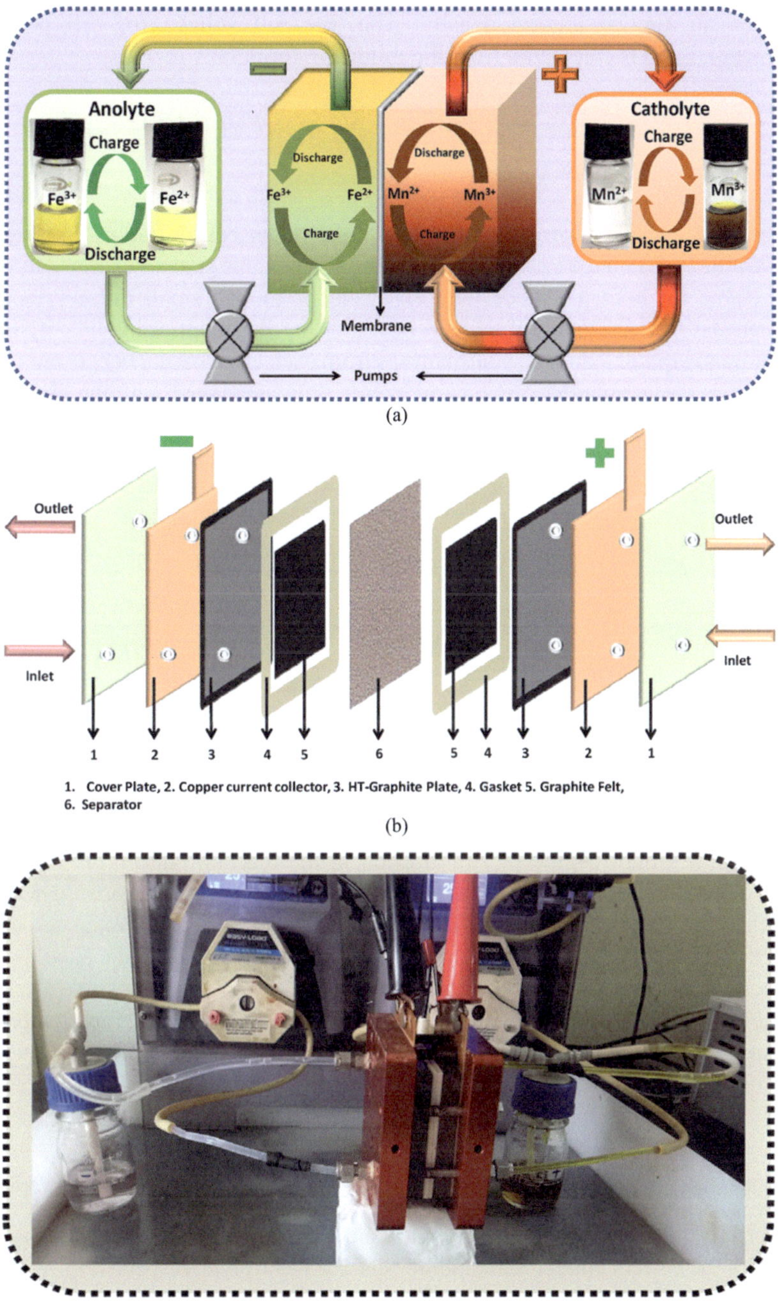

Fig. 1 A schematic illustration depicting the operational framework of the Fe-Mn redox flow battery system [4]

Given an applied current density, japp, the cell voltage (or cell potential difference), Ecell is calculated from the following formula:

$$E_{\text{cell}} = E_{\text{cell}}^{rev} - \sum (\text{IR})\,k - \sum |\eta| = E_{\text{cell}}^{rev} - (IR)_m - (IR)_e - (IR)_c - \eta_a \tag{10}$$

in which E_{cell}^{rev} is the reversible OCV (since the outlet concentrations are used to calculate the OCV in the experimental monitoring cell, the state of charge definition is the same for both the model and experiments, though small differences exist between the true OCV and the OCV measured by the monitoring cell, η_a is the activation overpotential (with contributions from each electrode), $(IR)_m$ is the ohmic drop across the membrane, (IR)e is the ohmic drop associated with the electrolyte and $(IR)_e$ is the ohmic drop associated with the current collector. The OCV can be approximated using Nernst's equation (assuming unit activity coefficients) as follows:

$$\begin{aligned} E_{\text{cell}}^{\text{rev}} &= \left(E_2^0 - E_1^0\right) + \frac{RT}{F}\ln\left(\frac{c_V(II)c_V(II)\left[H^+\right]}{c_V(II)c_V(IV)}\right) \\ &= E^0 + \frac{RT}{F}\ln\left(\frac{c_V(II)c_V(V)}{c_V(III)c_V(IV)}\right) + \frac{4.6RT}{F}\log\left[H^+\right] \end{aligned} \tag{11}$$

where: E_1^0 and E_2^0 (both functions of temperature) are the formal potentials for the reactions at the negative and positive electrodes, respectively and $\left(E_2^0 - E_1^0\right)$; R is the molar gas constant; T is the cell temperature; F is the Faraday constant; and ci is the molar concentration of (spatially distributed) species i. For very flow rates (<1 ml s^{-1}), a high degree of stratification in the reactant concentrations would develop between the inlet and outlet to the cell. For the flow rates considered here, the stratification is expected to be low. The proton activity has been approximated by the concentration (in mol dm^{-3}) of protons, [H^+] in the positive electrode. Note that the final term in Equation x could be written in terms of the pH of the positive electrode solution, pH = −log10[H^+]. The formal potentials depend on temperature. The variations are, however, small and therefore neglected. The ohmic losses associated with the current collector, membrane and the electrolyte can be modelled as follows.

$$(IR)_C = j_{\text{appl}}\frac{W_c}{\sigma_c};\ (IR)_m = j_{\text{appl}}\frac{W_m}{\sigma_m};\ (IR)_e = j_{\text{appl}}\frac{W_e}{\varepsilon^{3/2}\sigma_e}; \tag{12}$$

respectively, where, σ_c, σ_m and σ_e are the conductivities and W_c, W_m and W_e are the widths of the current collector, membrane and the electrolyte, respectively, as shown in Fig. 2.

The coefficients of "excess OCV" (Eex) should be introduced in modeling algorithm, defined as the term in the Nernst equation that accounts for the influence of activity coefficients and stems from the excess chemical potential [7]. In the

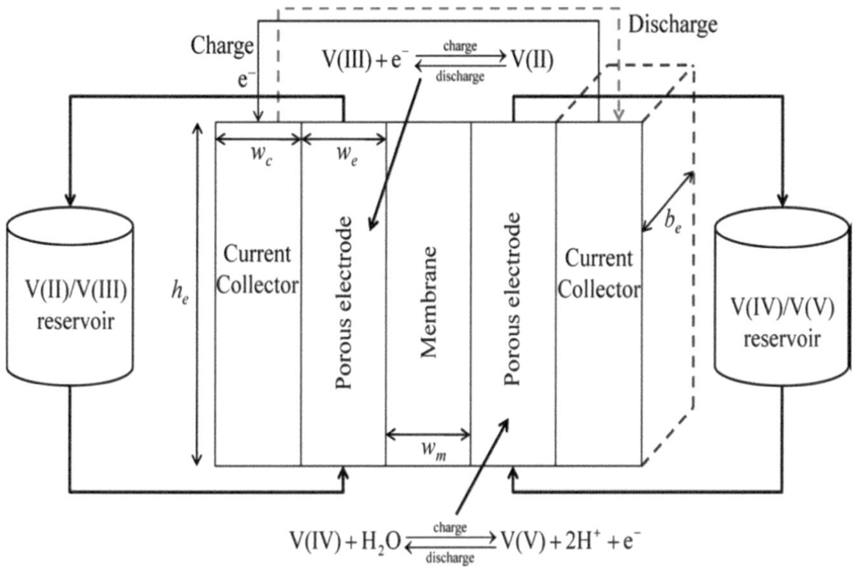

Fig. 2 A diagrammatic representation of the comprehensive vanadium redox flow battery alongside the constituent elements embedded within the theoretical framework [6]

context of iron-based redox flow batteries (for instance, Fe/Cd), the chemical reactions governing the half-cells may be articulated, and mass balances pertaining to the molalities of the species must be formulated under the assumption that the quantity of solvent remains invariant due to the minimal diffusion of water across the membrane [3].

It is imperative to acknowledge that activity coefficients exert a substantial impact on the open-circuit voltage (OCV) of both Iron-Cadmium (Fe/Cd) and All-Vanadium (AVRFB) redox flow batteries (RFBs). This effect is analogous in magnitude to the conventional voltage losses experienced during operational conditions. Consequently, activity coefficients ought to be incorporated into the Nernst equation.

Figure 3 delineates contour plots that depict the simulated Open Circuit Voltage (OCV) for an Iron-Cadmium (Fe/Cd) Redox Flow Battery (RFB) at a State of Charge (SOC) of 0.5. The plots illustrate the OCV's variation as a function of the total molar concentrations of $FeCl_2$ and HCl. The overall molarity of $CdCl_2$ is consistently maintained at fifty percent of that of $FeCl_2$. In the ideal scenario (Fig. 4a) the plot illustrates the OCV under the assumption that activity coefficients are disregarded. It demonstrates the expected behavior of the OCV under idealized conditions. The regions depicted in white within the plot signify areas where simulation data cannot be produced due to an inadequate quantity of protons for charge equalization throughout the total SOC. In the realistic scenario (Fig. 4b), the plot incorporates the OCV with activity coefficients derived from the Pitzer model. This approach offers a more thermodynamically sound representation of the OCV, thereby accounting for the non-ideal characteristics associated with highly concentrated electrolyte solutions. The 'X' marks present in both plots signify the typical overall composition of the electrolyte solution employed in the Fe/Cd RFB featuring chloride salts. The assessment of HCl Molarity demonstrates divergent trends between the ideal and real scenarios. In the ideal scenario, OCV increases with higher HCl molarity; conversely, the real scenario indicates a declining trend, suggesting that neglecting activity coefficients may mislead OCV optimization efforts through electrolyte variation. This underscores the necessity for a thermodynamically rigorous methodology, as reliance on the ideal scenario erroneously implies that reducing acid concentration is advantageous, whereas the reality suggests otherwise. Additionally the results from the Figs. 3 highlight the pivotal significance of activity coefficients in the precise forecasting of the OCV in Fe/Cd RFBs, with their omission potentially resulting in flawed interpretations of optimal electrolyte configurations and an underappreciation of the system's actual performance capabilities.

The Nernst equation functions as the fundamental theoretical framework for the modeling and simulation of iron–manganese flow batteries. It establishes a connection between microscopic electrochemical phenomena and macroscopic battery performance by correlating electrode potential with ionic concentration, temperature, and charge state. The incorporation of the equation-set [5] into simulation frameworks facilitates the accurate forecasting of system behavior, thereby informing design enhancements aimed at optimizing efficiency, stability, and cost-effectiveness in emerging energy storage technologies.

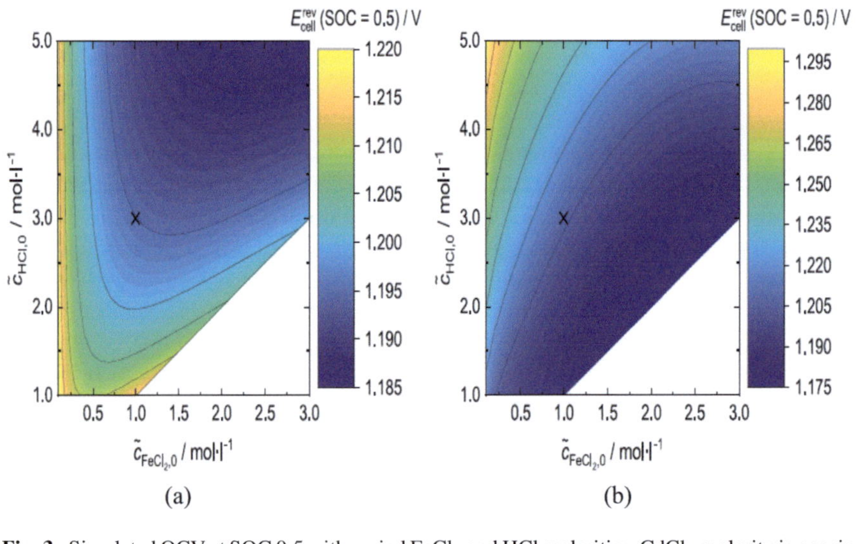

Fig. 3 Simulated OCV at SOC 0.5 with varied $FeCl_2$ and HCl molarities. $CdCl_2$ molarity is consistently half that of $FeCl_2$. The cross represents the typical electrolyte composition for the Fe/Cd RFB with chloride salts. No simulation data is obtainable for the white area due to insufficient protons for charge equalization across the SOC. **a** Ideal scenario (ignoring activity coefficients). **b** Realistic scenario (activity coefficients derived from the Pitzer model) [5]

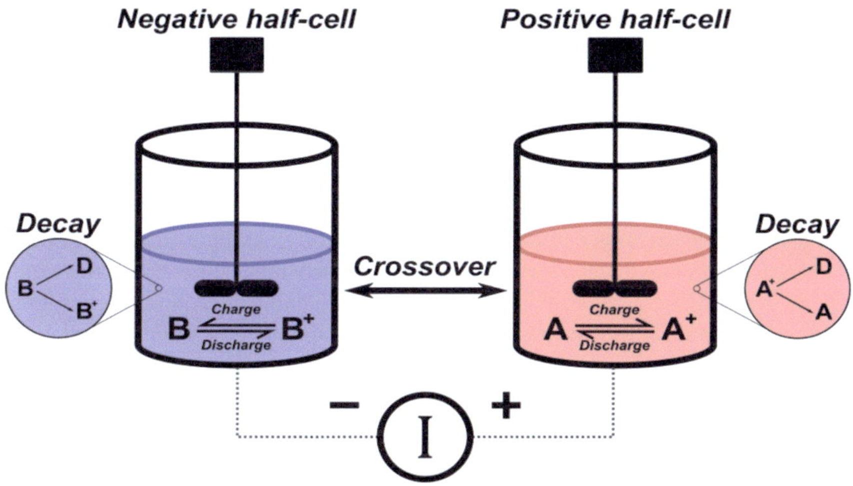

(a)

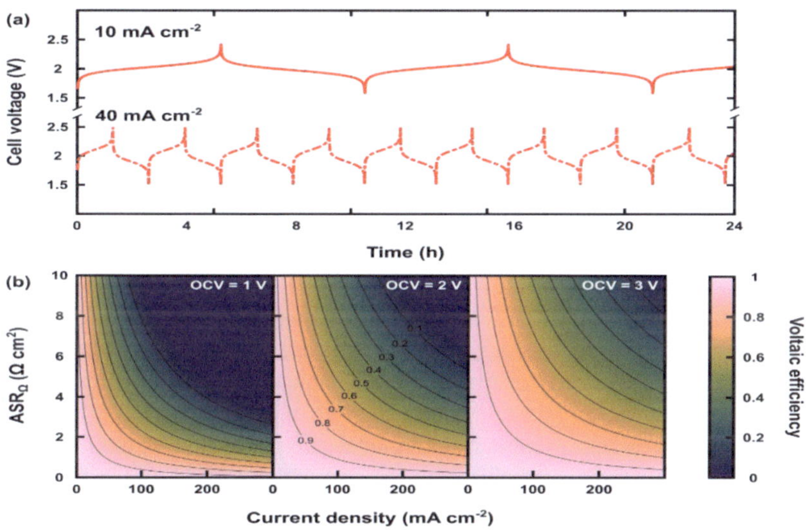

(b)

Fig. 4 **a** Charge/discharge profiles without parasitic losses are presented. **b** Voltaic efficiency (VE) is analyzed concerning current density and area-specific ohmic resistance (ASRΩ) for electrolytes with different open-circuit voltages (OCV), with black lines indicating constant VE values from 0.1 to 0.9

8 Existing Modeling Methodologies Pertinent to the Simulation of Electrochemical Reactions Relevant to Fe-Mn Redox Flow Batteries

Modeling and simulation play a critical role in understanding the coupled electrochemical, transport, and thermal phenomena within these systems. Modeling approaches typically range from zero-dimensional (0D)—which focus on system-level performance and mass balance—to multidimensional (1D–3D) frameworks that resolve spatial gradients, flow distributions, and localized reaction kinetics. Integrating these approaches enables accurate, computationally efficient representations of Fe-Mn RFBs across scales.

Zero-Dimensional Model Overview

The development of a 0-dimensional (0D) model for iron-manganese flow batteries (IMFBs) is a promising approach to enhance the understanding and performance of these energy storage systems. The 0D model simplifies the complex interactions within the battery by focusing on key parameters and processes, allowing for efficient simulation and analysis. This model is particularly useful for evaluating the performance of IMFBs, which are known for their potential in sustainable and cost-effective energy storage. The following sections will delve into the specifics of the 0D model, its applications, and its implications for IMFBs.

The 0D model is a computationally efficient framework that simplifies the simulation of redox flow batteries (RFBs) by focusing on mass balances and cell voltages under galvanostatic cycling conditions. This approach allows for rapid simulation of battery performance, facilitating the exploration of various operating conditions and material properties without the need for extensive computational resources [8, 9]. By employing analytical expressions, the 0D model can simulate hundreds of charge–discharge cycles in seconds, making it an invaluable tool for parametric sweeps and system optimization.

In general, zero-dimensional cycling models are predicated on the resolution of mass balance equations for each species present within the system; in this context, it is advantageous to postulate that both half-cells exhibit homogenous mixing and that the reaction occurs uniformly across the entire volume of the electrolyte. Consequently, each reactor half-cell, electrolyte reservoir, and interconnecting tubing are regarded as a singular continuous domain, analogous to two stirred tank reactors as depicted schematically in Fig. 4a. Here, the stoichiometric coefficients pertaining to the charged species, A + and B, are omitted from the opposing half-cells, as the electrochemical potentials within those specific environments inhibit their formation; consequently, $C_{A+}^{\infty,-} = 0$ and $C_{B}^{\infty,-} = 0$ consistently. All mass balance equations are governed by an initial condition delineated by Eq. (13).

$$C_j^{\infty,h}(t = 0) = C_j^{0,h} \tag{13}$$

Specifically, where, $C_j^{\infty,h}(mol\, m^{-3})$, represents the initial concentration of species j at the commencement of the half-cycle. For instance, the initial concentration at the onset of the discharge phase will correspond to the final concentration observed at the conclusion of the charge phase.

All simulations utilize a flow rate of Q = 10 mL min–1 for both electrolytes, with $D_j^{\infty} = 1 \times 10^{-9} m^2 s^{-1}$ for all species. However, the requisite governing equations are delineated in the research of [9], which additionally enhance the analysis of more subtle relationships between molecular features and cellular performance that may be challenging to extract from empirical research. Such insights are invaluable in the formulation of experimental designs when contemplating appropriate operational parameters and in the interpretation of data when clarifying potential performance constraints. These explicit analytical formulations offer a streamlined approach to investigate the extent to which particular design variables influence overall performance. For instance, they can be employed to evaluate the efficacy of a specific electrolyte composition and cell architecture under ideal conditions, thereby establishing a foundational expectation for performance within a defined set of cycling parameters. As a demonstrative example, the representative cycling characteristics of a redox flow cell subjected to varying current densities are illustrated in Fig. 2a. In this context, the augmentation of current density results in enhanced power density and expedited cycling intervals; however, this advantage is counterbalanced by elevated overpotentials, ultimately leading to diminished voltage efficiency (VE). To further delineate these losses, Fig. 2b elucidates the resultant VE as a function of the imposed current density and area-specific ohmic resistance (ASRΩ = Am (Rc + Rm)) across different open-circuit voltage (OCV) values. In alignment with Fig. 2a, the escalation of current density engenders increased ohmic and mass transport losses, thereby reducing the VE; specifically, ohmic losses are directly proportional to both the ASRΩ and current density, establishing a robust relationship between these variables. Notably, the VE exhibits a significant increase with elevated OCV, indicating that higher voltage redox chemistries, such as those facilitated by nonaqueous electrolytes, possess the ability to accommodate greater ASRΩ while sustaining acceptable cycling efficiencies.

Applications 0-D approach in Iron-Manganese Flow Batteries

IMFBs benefit from the 0D model as it enables the assessment of key performance metrics such as energy efficiency, capacity fade, and power density. This is crucial for understanding the trade-offs between different material properties and operating conditions. The model can be used to evaluate the impact of parasitic losses, such as shunt currents and species crossover, which are significant factors in the performance and durability of IMFBs [8].

Advantages and Challenges

The 0D model's computational efficiency allows for its integration into techno-economic assessments, providing a physics-informed basis for cost predictions and technology comparisons. This is particularly important for the development of cost-effective and scalable IMFB systems.

Despite its advantages, the 0D model must be carefully validated against experimental data to ensure its accuracy and reliability. This involves comparing model predictions with experimental results to refine the model parameters and improve its predictive capabilities [10, 11].

Broader Implications

The use of 0D models in IMFBs represents a significant step forward in the design and optimization of flow batteries. By providing a clear link between material properties and battery performance, these models can guide the development of new materials and configurations that enhance the efficiency and longevity of IMFBs. Additionally, the insights gained from 0D models can inform the design of other manganese-based flow batteries, which face similar challenges in terms of reaction mechanisms and cycling stability.

While the 0D model offers a streamlined approach to simulating IMFB performance, it is important to recognize its limitations. The model's simplifications may overlook certain complex interactions and spatial variations within the battery, which could affect its accuracy in predicting real-world performance. Therefore, it is essential to complement 0D modeling with experimental validation and more detailed models when necessary to ensure comprehensive understanding and optimization of IMFB systems.

One-dimensional (1D) models for IMFBs

A 1D model can provide insights into the electrochemical processes, mass transport, and performance characteristics of these batteries, facilitating their design and operation. This subsection explores the key aspects of developing a 1D model for IMFBs, drawing on insights from related research on redox flow batteries (RFBs).

Model Development and Components

Electrochemical Reactions: The IMFB operates using the Mn^{3+} /Mn^{2+} and Fe^{3+} / Fe^{2+} redox couples. The model must accurately represent these reactions, including the potential for Mn^{3+} disproportionation, which can lead to MnO2 precipitation. Complexation with Cl- ions has been shown to stabilize Mn^{3+}, preventing this issue and ensuring efficient battery operation [12].

Mass Transport: A 1D model should incorporate species conservation equations, including advection and diffusion terms, to describe concentration changes within the porous electrodes. This approach is similar to models developed for vanadium RFBs, which have successfully predicted cell performance across different flow rates [13].

Hydraulic and Ohmic Losses: The model should account for hydraulic losses due to pressure drops in the porous electrodes and ohmic losses across the cell. These factors are critical in determining the overall efficiency and performance of the battery [14].

Performance Optimization

Current Density and Flow Rates: The model can be used to optimize operating conditions such as current density and electrolyte flow rates. High current densities can lead to mass-transfer limitations, while high flow rates may reduce efficiency due to increased pressure drops.

Membrane and Electrode Design: The choice of membrane and electrode materials significantly impacts battery performance. For instance, the use of different substrates for iron deposition and various membrane types can influence cycle life and efficiency [15].

The one-dimensional (1-D) model facilitates (provided in Fig. 5 as an ideal illustration) the integration of a Newman-type cell alongside a resistor network, enabling the assessment of shunt current contributions, which are vital for the operational efficacy and design of the stack.

Furthermore, it incorporates considerations for hydraulic losses and membrane crossover phenomena, thereby enhancing the optimization of system performance in relation to the characteristics of the electrolyte and electrode materials. Within

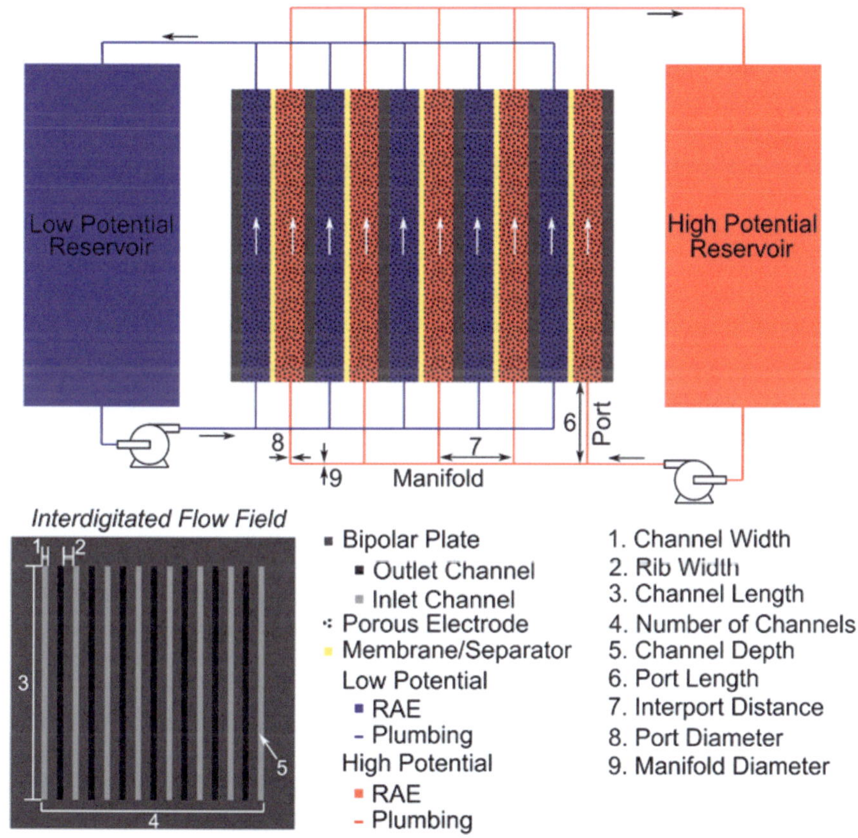

Fig. 5 Elucidating representations of the geometrical parameters of the system that facilitate the expansion beyond the analysis of single cell [14]

the domain of Porous Electrode Modeling, the 1-D methodology integrates asymmetric kinetic behaviors and fluctuating states of charge, thereby quantifying losses attributable to mass transfer and conductivity through a numerical resolution of a polarization model. Hydraulic losses occurring within the porous electrode are accounted for to improve electrochemical efficacy under conditions of elevated flow rates. By employing a one-dimensional approach, the membrane can be modeled as a separator characterized by linear concentration gradients, which fluctuate in accordance with reactor conditions for both passive and active membranes, incorporating a fundamental model for the crossover of active species driven by concentration gradients and electrochemical potential. Tanks are conceptualized as fixed-volume reservoirs that undergo rapid reaction processes, concurrently representing two oxidation states of vanadium. Component models may be constructed through the interconnection of a circuit network to illustrate shunt current pathways, employing a comprehensive resistance network that takes system geometry into account. The model linearizes the relationship between cell voltage and current density by utilizing resistance to facilitate computational efficiency. The modeling processes encompass user inputs, recalibration of parameters, and adaptation to variables that influence input interpretation, thereby creating a matrix for the shunt current network and updating cell parameters over time.

The utilization of the one-dimensional (1-D) methodology for the assessment of single cell performance emphasizes the pivotal significance of mass-transfer coefficients and the inherent trade-offs among current density, flow rate, and overall system efficiency. Furthermore, it implies that although kinetic rates hold considerable importance, alternative factors such as membranes and flow fields may present more substantial opportunities for enhancing performance.

Challenges and Future Directions

Cycling Stability: One of the main challenges for IMFBs is achieving stable cycling performance. The model should help identify optimal conditions to minimize issues such as hydrogen evolution and iron precipitation, which can degrade battery performance.

Scalability and Cost: The model should also address scalability and cost-effectiveness, crucial for the commercial viability of IMFBs. The use of inexpensive materials like iron and manganese is a significant advantage, but the model must ensure that these benefits are realized in large-scale applications [16].

While the development of a 1D model for IMFBs offers numerous advantages, it is essential to consider the limitations and potential improvements. For instance, while 1D models provide valuable insights into the fundamental processes within the battery, they may not capture all the complexities of real-world systems, such as three-dimensional flow patterns and localized phenomena. Therefore, integrating 1D models with higher-dimensional simulations or experimental data could enhance their predictive capabilities and provide a more comprehensive understanding of IMFB performance.

9 2-Dimensional (2D) Model for Iron-Manganese Flow Battery

The development of a 2-dimensional (2D) model for iron-manganese flow batteries is a promising area of research that combines the benefits of both iron and manganese-based systems. These models are crucial for understanding the complex interactions within the battery, optimizing performance, and addressing challenges such as energy density and cycling stability. The integration of 2D modeling techniques allows for a detailed analysis of the electrochemical and transport phenomena occurring within the battery, providing insights into improving efficiency and longevity.

2D Modeling in Flow Batteries.

Physics-Based Models: The use of physics-based models, such as those developed for solid oxide iron-air redox flow batteries, provides a comprehensive framework for simulating the chemical, electrochemical, and transport behaviors within the battery. These models are built on fundamental reaction engineering theories and have shown good agreement with experimental data, highlighting their accuracy and reliability in predicting battery performance [17].

Machine Learning Approaches.

Physics-informed neural networks (PINNs) have been employed to enhance the predictive capabilities of 2D models. By incorporating physics constraints, these models can accurately predict cell voltage and other critical parameters, although challenges remain in correcting potential shifts. The integration of machine learning with 2D models offers a promising avenue for further improving the accuracy and efficiency of flow battery simulations [18]

Iron-Manganese Flow Battery Specifics.

Manganese-Based Systems: Manganese-based flow batteries (MFBs) are recognized for their environmental sustainability and cost-effectiveness. However, they face challenges such as suboptimal energy density and cycling stability. The development of 2D models can help address these issues by providing a detailed understanding of the reaction mechanisms and transport phenomena within the battery [19].

Iron-Manganese Combinations.

Combining iron and manganese in a flow battery system can leverage the strengths of both materials. For instance, the use of Mn^{3+}/Mn^{2+} and Fe^{3+}/Fe^{2+} redox couples has been demonstrated to achieve significant output voltages and current densities, making them viable for large-scale energy storage applications. The 2D modeling of such systems can optimize the interactions between these redox couples, enhancing overall battery performance [12, 20].

Challenges and Future Directions.

Material and Design Innovations: The development of advanced materials, such as 2D material-based membranes, is crucial for improving the performance of redox flow batteries. These materials offer high ion-selective transport and can significantly enhance the efficiency and stability of the battery system. Future research should focus on integrating these materials into 2D models to explore their full potential.

Addressing Heterogeneity: The Real 2D (R2D) model addresses the physico-chemical heterogeneity within battery systems, which is a critical factor affecting performance. By quantifying the impact of multiphysics coupling, these models can provide insights into optimizing battery design and operation under various cycling conditions [21].

While 2D models offer significant advantages in understanding and optimizing flow battery systems, they also present challenges, such as computational complexity and the need for accurate parameterization. Additionally, the integration of machine learning techniques, while promising, requires careful validation to ensure reliability. As research progresses, the combination of advanced modeling techniques and innovative materials will be key to overcoming these challenges and realizing the full potential of iron-manganese flow batteries for sustainable energy storage solutions.

10 3D Modeling of IMFBs

The development of a 3-dimensional (3D) model for iron-manganese flow batteries is a promising approach to enhance the performance and efficiency of these energy storage systems. This model aims to address the challenges associated with mass transfer, flow distribution, and electrochemical reactions within the battery. By leveraging advanced modeling techniques, researchers can optimize the design and operation of iron-manganese flow batteries, potentially leading to improved energy density, cycling stability, and overall efficiency. The following sections explore key aspects of 3D modeling in the context of iron-manganese flow batteries.

Flow and Mass Transfer Optimization

A 3D model can significantly improve the understanding of flow and mass transfer characteristics within the battery. For instance, in an all-iron semi-solid redox flow battery, a 3D model demonstrated that optimizing the flow channel length could enhance the operating efficiency by improving mass transfer efficiency and ion concentration distribution [22].

The use of computational fluid dynamics (CFD) tools in 3D modeling allows for the analysis of flow hydrodynamics, which is crucial for achieving uniform flow distribution and minimizing pressure drops in the battery system [23].

Electrode Design and Polarization Management.

3D models can predict local polarization within electrodes, which is essential for managing gas side reactions that can reduce battery efficiency. A multi-scale model incorporating 3D continuum modeling can accurately predict local polarization and help design electrodes that minimize overpotential and improve uniformity [24].

Advanced 3D electrode designs, such as those incorporating engineered electrolytes, can mitigate issues like hydrogen evolution reactions, thereby enhancing the storage capacity and efficiency of iron-based flow batteries [25].

Structural and Geometric Optimization.

The topology optimization of 3D flow fields can lead to significant performance improvements by minimizing electrical and flow pressure power losses. This approach allows for the design of flow fields that reduce dead zones and enhance reactant distribution within the electrode [26].

Different stack geometries, such as trapezoidal and radial designs, have been explored using 3D hydraulic-electrochemical models to increase electrolyte velocity and improve mass transport, which is critical for enhancing power density in flow batteries [27].

Challenges and Future Directions.

Despite the advancements in 3D modeling, challenges remain in achieving optimal energy density and cycling stability, particularly for manganese-based flow batteries. The contentious reaction mechanisms and suboptimal performance of manganese redox couples need further investigation and refinement [19].

The integration of 3D models with experimental data is crucial for validating and refining these models. For example, the use of deep eutectic solvent electrolytes in 3D models has shown promise in reducing overpotentials and enhancing energy efficiency, highlighting the importance of electrolyte optimization in conjunction with structural design [28].

While 3D modeling offers significant potential for improving iron-manganese flow batteries, it is important to consider the limitations and challenges associated with these models. The complexity of accurately simulating electrochemical reactions and flow dynamics in a 3D space requires sophisticated computational tools and validation against experimental data. Additionally, the economic feasibility and scalability of implementing these advanced designs in commercial applications remain critical considerations for the widespread adoption of 3D-modeled flow batteries. Nonetheless, continued research and development in this area hold promise for advancing the performance and efficiency of iron-manganese flow batteries in energy storage applications.

11 Future Research Priorities

Based on the analysis presented in this chapter, several research priorities emerge for advancing Fe-Mn RFB modeling and technology.

Immediate Priorities (1–3 years):

1. Development of comprehensive thermodynamic databases for Fe-Mn-Cl systems
2. Validation of models against high-quality experimental datasets with in-situ diagnostics
3. Integration of speciation-precipitation models with transport and electrochemical models
4. Development of efficient multi-scale coupling strategies.

Medium-term Goals (3–7 years):

1. Machine learning-enhanced models for rapid design optimization
1. Digital twin approaches for real-time system monitoring and control
2. Advanced manufacturing integration for electrode and membrane optimization
3. System-level integration with renewable energy and grid applications.

Long-term Vision (7–15 years):

1. Fully integrated multi-physics models spanning molecular to system scales
2. Autonomous discovery of new electrolyte and electrode materials
3. Sustainable lifecycle optimization including recycling and environmental impact
4. Commercial deployment of optimized Fe-Mn RFB systems for grid-scale storage.

Broader Implications

The development of sophisticated mathematical models for Fe-Mn RFBs has implications beyond this specific technology:

Methodology Transfer.

Many of the modeling approaches developed for Fe-Mn systems are applicable to other complex electrochemical energy storage technologies, including other RFB chemistries, metal-air batteries, and fuel cells.

Computational Infrastructure.

The multi-scale, multi-physics nature of RFB modeling is driving advances in computational methods, software tools, and high-performance computing approaches that benefit the broader electrochemical community.

Experimental Design.

The insights gained from modeling are informing the development of new experimental techniques and diagnostic methods that advance fundamental understanding of electrochemical systems.

Education and Training.

The complexity of RFB modeling is creating new educational opportunities and training needs, driving curriculum development in electrochemical engineering and computational methods.

12 Final Remarks

Iron–manganese redox flow batteries represent a promising technology for grid-scale energy storage, but their complex chemistry and multi-scale behavior present significant modeling challenges. The mathematical modeling approaches reviewed in this chapter provide powerful tools for understanding, predicting, and optimizing these systems, but continued advances in both modeling methods and experimental validation are needed to fully realize the potential of this technology.

The success of Fe-Mn RFBs will ultimately depend on the continued integration of sophisticated mathematical models with innovative materials development, advanced manufacturing techniques, and system-level optimization. The modeling framework presented in this chapter provides a foundation for these efforts and will continue to evolve as our understanding of these complex systems advances.

As the energy storage landscape continues to evolve, the lessons learned from Fe-Mn RFB modeling will inform the development of next-generation energy storage technologies, contributing to the broader goal of enabling a sustainable, renewable energy future.

References

1. Neyhouse, B.J., Brushett, F.: A spreadsheet-based redox flow battery cell cycling model enabled by closed-form approximations **171**(8), 080518 (2024)
2. Neyhouse, B.J., Fenton, A.M., Brushett, F.: Too much of a good thing? Assessing performance tradeoffs of two-electron compounds for redox flow batteries **168**(5), 050501 (2021)
3. Crawford, A.J.: A Chemistry Neutral Flow Battery Performance Model Development, Validation, and Application: Portland State University (2016)
4. Archana, K.S., Suresh, S., Ragupathy, P., Ulaganathan, M.J.E.A.: Investigations on new Fe-Mn redox couple based aqueous redox flow battery **345**, 136245 (2020)
5. Hayer, N., Kohns, M.: Thermodynamically rigorous description of the open circuit voltage of redox flow batteries **167**(11), 110516 (2020)
6. Shah, A.A., Tangirala, R., Singh, R., Wills, R., Walsh, F.: A dynamic unit cell model for the all-vanadium flow battery **158**(6), A671 (2011)
7. Sum, E., Skyllas-Kazacos, M.: A study of the V (II)/V (III) redox couple for redox flow cell applications **15**(2–3), 179–90 (1985)
8. Price, N.A., Neyhouse, B.J., Brushett, F.R., (Eds).: A computationally-efficient, zero-dimensional stack model for simulating redox flow battery performance. Electrochemical Society Meeting Abstracts 243. The Electrochemical Society, Inc (2023)
9. Neyhouse, B., Lee, J., Brushett, F.: Connecting material properties and redox flow cell cycling performance through analytical zero-dimensional models (2022)
10. Howard, A.A., Yu, T., Wang, W., Tartakovsky, A.: Physics-informed CoKriging model of a redox flow battery **542**, 231668 (2022)
11. Wang, H., Pourmousavi, S.A., Li, Y., Soong, W.L., Zhang, X., Xiong, B.: A new zero-dimensional dynamic model to study the capacity loss mechanism of vanadium redox flow batteries. **603,** 234428 (2024)
12. Wu, J., Yang, J., Leong, Z.Y., Zhang, F., Deng, H., Ouyang, G.F., et al.: A method to inhibit disproportionation of Mn3+ for low-cost Mn–Fe all-flow battery **5**(12):14646–51 (2022)
13. Chen, C., Yeoh, H., Chakrabartia, M., (Eds).: An enhanced one-dimensional stationary model for the all-vanadium redox flow battery. In: Proceedings of the 6th International Conference on Process Systems Engineering (PSE ASIA) (2013)
14. Barton, J.L., Brushett, F.R.J.B.: A one-dimensional stack model for redox flow battery analysis and operation **5**(1), 25 (2019)
15. Noack, J., Wernado, M., Ortner, J., Pinkwart, K., (Eds).: Iron/Iron redox flow battery optimisation studies. electrochemical society meeting abstracts 242. The Electrochemical Society, Inc (2022)
16. Dinesh, A., Olivera, S., Venkatesh, K., Santosh, M.S., Priya, M.G., Inamuddin, N., et al.: Iron-based flow batteries to store renewable energies **16**(3), 683–94 (2018)

17. Guo, M., Zhao, X., White, R.E., Huang, K.: A multi-physics model for solid oxide iron-air redox flow battery: simulation of discharge behavior at high current density **160**(11), A2085 (2013)
18. Chen, W., Fu, Y., Stinis, P.: Physics-informed machine learning of redox flow battery based on a two-dimensional unit cell model **584**, 233548 (2023)
19. Wang, X., Ding, M., Jia, C.: A perspective on manganese-based flow batteries **57**(42):423002 (2024)
20. Shen, X., Kellamis, C., Tam, V., Sinclair, N., Wainright, J., Savinell, R., et al.: An all-soluble Fe/Mn-based alkaline redox flow battery system **16**(15):18686–92 (2024)
21. Sun, Z.-T., Chen, S., Zhao, T., Guo, Y., Xu, Z., Zhou, S., et al.: Real 2D galvanostatic model: encoding physicochemical heterogeneity into a full battery **135**(6):068001 (2025)
22. Zhang, Z., Tian, Y., Li, H.: Exploring the Flow and Mass Transfer Characteristics of All-Iron Semi-Solid Redox Flow Battery
23. Krishnappa, R.B., Subramanya, S.G., Deshpande, A.: Metals, Fuels. Numerical investigation of flow hydrodynamics in a flow field and porous substrate configuration for redox flow battery application 71(12) (2023)
24. Luo, Y., Zheng, M., Wenrui, L., (eds).: Predictable local polarization inside electrodes by three-dimensional multi-scale model in flow batteries. Electrochemical Society Meeting Abstracts 244. The Electrochemical Society, Inc (2023)
25. Jayathilake, B.S., Ivanovskaya, A.N., Overland, A.E., Chandrasekaran, S., Clemens, A.L., (Eds).: Advancing Low-Cost Iron Redox Flow Batteries for Grid Storage. Electrochemical Society Meeting Abstracts 245. The Electrochemical Society, Inc (2024)
26. Lin, T.Y., Baker, S.E., Duoss, E.B., Beck, V.: Topology optimization of 3D flow fields for flow batteries **169**(5), 050540 (2022)
27. Gurieff, N., Menictas, C., Timchenko, V., Skyllas-Kazacos, M., Noack, J., (Eds).: Performance enhancing stack geometry concepts. In: Proceedings of the International Flow Battery Forum, Lyon, France (2019)
28. Xiao, Z., Zhang, R., Lu, M., Ma, Q., Li, Z., Su, H., et al.: Numerical simulation of impact of different redox couples on flow characteristics and electrochemical performance of deep eutectic solvent electrolyte flow batteries **11**(1), 18 (2025)

Electrospray-Deposited Nitrogen-Doped Carbon Nanostructures for Enhanced Electrode Performance in Quinone-Based Redox Flow Batteries

Mahmut Taş, Züleyha Kudaş, Reza Afshar Ghotli, Shamik Chaudhuri, Baidaa Alkhateab, Serap Hayat Soytaş, Tülay Yılmaz İnan, Nadir Deniz Akgül, Mustafa Kemal Bayazıt, Andres Parra-Puerto, Stefan Spirk, and Barun Kumar Chakrabarti

Abstract The development of high-performance electrodes is critical to unlocking the full potential of Redox Flow Batteries (RFBs). In this study, a Carbon Cloth (CC) electrode coated with Nitrogen-doped Reduced Graphene Oxide (N-rGO) nanostructures was successfully fabricated using a controlled Electrospray (ES) technique followed by thermal treatment. The subsequent thermal treatment confirmed the successful conversion of Graphene Oxide (GO) into N-rGO, resulting in a significant improvement in electrical conductivity. Electrochemical performance evaluation demonstrated that the CC/N-doped rGO electrode exhibited superior electrocatalytic activity towards the Benzoquinone-2-sulfonic acid sodium salt (BQDS) redox couple. The most striking result is the dramatic reduction in the electrode's Charge Transfer Resistance (R_{ct}): the value plummeted from 169.0 Ω for the bare CC to a remarkable 1.26 Ω for the CC/N-doped rGO electrode. This signifies a massive acceleration in

M. Taş (✉) · Z. Kudaş · R. A. Ghotli · S. Chaudhuri · B. Alkhateab · S. H. Soytaş · T. Y. İnan · N. D. Akgül · M. K. Bayazıt · B. K. Chakrabarti
Sabanci University Nanotechnology Research and Application Centre (SUNUM), Tuzla, İstanbul, Turkey
e-mail: mahmut.tas@sabanciuniv.edu

B. K. Chakrabarti
e-mail: barun.chakrabarti@sabanciuniv.edu

M. K. Bayazıt
Faculty of Engineering and Natural Science, Sabancı University, Tuzla, İstanbul, Turkey

A. Parra-Puerto
Department of Physical Chemistry and Electrochemistry Institute, University of Alicante, San Vicente del Raspeig (Alicante), Spain

S. Spirk
Institute of Bioproducts and Paper Technology, Graz University of Technology, Graz, Austria

Ecolyte GmbH, TU Graz Campus Inffeldgasse, Graz, Austria

M. Awang et al. (eds.), *Proceedings of the International Conference on Mechanical, Manufacturing and Plant Engineering*, Proceedings in Technology Transfer,
https://doi.org/10.1007/978-981-95-7817-7_24

the reaction kinetics at the electrode surface, leading to potentially much lower overpotentials and higher energy efficiency. A substantial increase in the diffusion coefficient, calculated from the Randles–Sevcík equation, was also observed, confirming faster diffusion-controlled reaction processes. This work proves that electrospray is a scalable and promising method for the production of highly efficient electrodes for RFBs. The significant performance enhancements achieved, particularly in kinetics and conductivity, reinforce the critical role of N-doped carbon nanostructures in realizing the full potential of quinone-based redox flow batteries.

Keywords Graphene · Carbon nanotube · Electrophoretic deposition · Vanadium redox flow battery · Quinone redox couple

1 Introduction

The transition toward sustainable energy systems requires storage technologies that can effectively buffer the variability of renewable sources such as solar and wind. Redox flow batteries (RFBs) are well-suited for this task due to their scalability, long operational life, and decoupling of power and energy. However, conventional RFB chemistries, while technologically mature, are constrained by resource costs and environmental concerns. This has stimulated significant research interest in organic redox-active molecules, which can be synthesized from abundant feedstocks, tailored at the molecular level, and integrated into aqueous or non-aqueous electrolytes [1].

Among organic candidates, quinones—a class of redox-active aromatic molecules with carbonyl groups—stand out for their tuneable electrochemistry, structural diversity, and potential sustainability. Quinones undergo reversible two-electron, two-proton redox reactions, making them highly attractive for aqueous RFB systems [2]. Importantly, their redox potentials can be systematically adjusted through substitution with functional groups, allowing flexible design of cell voltages.

A prominent example for aqueous RFBs is Benzoquinone-2-sulfonic acid sodium salt (BQDS), a highly water-soluble quinone derivative that has attracted attention for its reversible redox chemistry and potential for cost-effective synthesis [3]. BQDS's sulfonate group enhances its hydrophilicity and stability in acidic aqueous environments, making it compatible with carbon-based electrodes. Despite these merits, the electrochemical kinetics of BQDS on conventional carbon cloth (CC) electrodes remain sluggish, leading to increased overpotentials and reduced energy efficiency.

To overcome these limitations, surface engineering of electrodes has become a critical strategy. Among the most promising approaches is the integration of nitrogen-doped carbon nanostructures (such as graphene) [4–7]. Nitrogen doping introduces heteroatomic sites that alter the electronic structure of the carbon lattice, creating catalytically active sites—particularly pyridinic and graphitic nitrogen—that facilitate electron transfer [8]. In addition, the incorporation of nanostructures increases surface area, improves electrical conductivity, and enhances wettability, all of which are crucial for efficient redox reactions with quinone molecules like BQDS.

However, depositing these nanostructures uniformly on porous electrodes presents a challenge. Traditional slurry casting methods, which rely on binders such as PVDF in N-methyl-2-pyrrolidone (NMP), often result in blocked pores, inhomogeneous coatings, and poor adhesion, ultimately reducing electrochemical performance. In contrast, electrospray deposition (ESD) provides a binder-free alternative, enabling conformal coatings of nanostructured carbons onto porous substrates with controlled morphology and thickness. By atomizing a suspension under an electric field, ESD generates uniform films that preserve the porosity of the carbon cloth and ensure intimate contact between the active material and current collector.

This study investigates the electrospray deposition of Nitrogen-doped Reduced Graphene Oxide (N-rGO) onto carbon cloth substrates for use in BQDS-based aqueous RFBs. The research highlights how nanoscale morphology and tailored nitrogen functionalities significantly enhance BQDS's redox kinetics, reduce charge transfer resistance, and improve electrode stability under acidic aqueous conditions. Specifically, the charge transfer resistance (R_{ct}) decreased from 169.0 Ω for the bare CC to a remarkable 1.26 Ω for the CC/N-doped rGO electrode, coupled with a significant increase in the diffusion coefficient calculated from the Randles–Sevcík equation. The findings highlight the potential of combining sustainable organic electrolytes with engineered nanostructured electrodes to advance the performance and sustainability of next-generation flow batteries.

2 Experimental Materials and Methods

2.1 Materials

All chemicals were used as received without further purification. Benzoquinone-2-sulfonic acid sodium salt (BQDS) was employed as the active organic redox species. Aqueous electrolytes were prepared by dissolving BQDS in deionized (DI) water at specific concentrations, with the pH adjusted using concentrated sulfuric acid (H_2SO_4) to ensure the stability of the redox couple. Carbon cloth (CC) substrates (plain weave, thickness ~350 μm) served as the base electrodes. Graphene Oxide (GO) was utilized as the precursor material for the synthesis of the nanostructured coating. Isopropanol (IPA) and deionized water (DI H_2O) were used as solvents for preparing the electrospray suspensions.

2.2 Synthesis of Nitrogen-Doped Reduced Graphene Oxide (N-rGO)

The Nitrogen-doped Reduced Graphene Oxide (N-rGO) nanostructures were synthesized in a two-step procedure. First, Graphene Oxide (GO) was prepared from graphite powder via the modified Hummer's method. Subsequently, the synthesized GO was subjected to a controlled thermal treatment in a nitrogen-containing atmosphere (such as ammonia or a nitrogen precursor) to simultaneously achieve the reduction of GO to rGO and the incorporation of nitrogen heteroatoms. This thermal reduction process generated the final N-rGO material, which is rich in catalytically active nitrogen functionalities, including pyridinic and graphitic sites.

2.3 Preparation of Electrospray Suspensions

A stable colloidal suspension of the N-rGO nanostructures was prepared by dispersing the material in a mixed solvent (typically a blend of isopropanol and deionized water) at a concentration of approximately $\sim$1 mg mL^{-1}. To ensure a highly homogeneous dispersion and prevent significant agglomeration prior to deposition, the suspension was subjected to intensive ultrasonication for one hour while being maintained in an ice-bath. Crucially, no polymer binders or conductive additives were included in the final suspension, resulting in a completely binder-free active layer on the electrode.

2.4 Electrospray Deposition (ESD) of N-rGO Electrodes

Electrospray deposition (ESD) was utilized to achieve a uniform and conformal coating of the N-rGO onto the porous carbon cloth. The custom laboratory-scale ESD setup consisted of a syringe pump for controlled fluid delivery, a stainless-steel capillary nozzle, a high-voltage power supply, and a grounded substrate holder. The carbon cloth substrates were cleaned and dried before deposition. The N-rGO suspension was dispensed at a constant, controlled flow rate (e.g., 0.2 mL h^{-1}). A high DC voltage (e.g., 15–18 kV) was applied between the nozzle and the substrate, maintained at a fixed distance (e.g., 12 cm). The deposition process was sustained for a specific period (e.g., 60 min) to achieve the desired mass loading and uniform coverage while critically preserving the inherent porosity of the CC. Following deposition, the resulting CC/N-doped rGO electrodes were dried under vacuum at 80 °C to ensure the complete removal of residual solvent.

2.5 Physicochemical Characterization

The structural, morphological, and chemical properties of the functionalized electrodes were thoroughly characterized. The uniform deposition and nanostructure morphology of the N-rGO films were examined using Scanning Electron Microscopy (SEM). The chemical composition, specifically the degree of nitrogen doping and the distribution of N-bonding configurations (pyridinic, pyrrolic, and graphitic N), was quantified using FTIR and XRD. Finally, the surface wettability, which is vital for electrolyte access, was assessed through static contact angle measurements using deionized water.

2.6 Electrochemical Testing

All electrochemical measurements were carried out in a three-electrode cell configuration. The CC/N-doped rGO served as the working electrode, a platinum mesh as the counter electrode, and Ag/AgCl (3 M KCl) as the reference electrode. The electrolyte comprised 0.1–0.5 M BQDS dissolved in 0.5 M H_2SO_4. Electrochemical analyses included Cyclic Voltammetry (CV), which was performed at various scan rates ($5 - 100$ mV s^{-1}) to evaluate redox reversibility, peak separations, and most importantly, to calculate the diffusion coefficient of the BQDS species using the Randles–Sevcík equation. Electrochemical Impedance Spectroscopy (EIS) was recorded over a frequency range of 100 kHz to 0.1 Hz with a 10 mV perturbation amplitude to extract the critical Charge Transfer Resistance (R_{ct}).

3 Results and Discussions

Scanning electron microscopy (SEM) revealed that electrospray deposition produced uniform and conformal coatings of nitrogen-doped graphene (N-G) across the carbon cloth substrate. The deposited N-G formed interconnected, sheet-like networks that bridged the woven carbon fibres, increasing the effective surface area while maintaining the open porosity of the cloth (Fig. 1).

At higher magnification, the N-G films displayed wrinkled nanosheets with lateral dimensions of several hundred nanometres, suggesting good dispersion during the electrospray process. The wrinkling and folding likely contribute to increased defect density, which can serve as active sites for electron transfer with Tiron molecules (Fig. 1).

Fourier-transform infrared spectroscopy (FTIR) was employed to confirm the successful chemical transformations on the electrode surface, comparing the Bare Carbon Cloth (CC), the Electrospray-deposited Graphene Oxide coated CC (ES-CC-GO), and the final Nitrogen-doped Reduced Graphene Oxide coated CC (ES-CC-N-rGO) samples. The spectrum of the Bare CC (black line) presents a relatively flat baseline with fundamental peaks characteristic of the carbon cloth material itself (see Fig. 2). The red line, corresponding to the ES-CC-GO sample, clearly shows characteristic peaks associated with oxygen-containing functional groups, such as a broad band attributed to O–H stretching vibrations and peaks corresponding to C–O stretching, confirming the successful deposition and presence of Graphene

Fig. 1 Scanning electron microscopes of bare carbon cloth and N-doped graphene decorated CC

Oxide on the carbon cloth substrate. Crucially, the spectrum of the final ES-CC-N-rGO electrode (blue line) exhibits a noticeable decrease in the intensity of these oxygen-containing functional group peaks compared to the ES-CC-GO spectrum. This significant reduction in oxygen species confirms the successful deoxygenation and partial reduction of GO to rGO during the thermal treatment process, a chemical change vital for the observed improvement in electrical conductivity and electrocatalytic performance (Fig. 2). Furthermore, the resulting spectral profile supports the successful introduction of nitrogen functionalities into the carbon lattice, validating the formation of the intended N-rGO active material.

The structural characteristics of the bare carbon cloth and the CC/N-doped rGO samples were investigated using X-ray Diffraction (XRD). In the XRD analysis, the characteristic (002) diffraction peak of the bare carbon cloth sample was initially

Fig. 2 FTIR of bare carbon cloth and N-doped graphene decorated CC

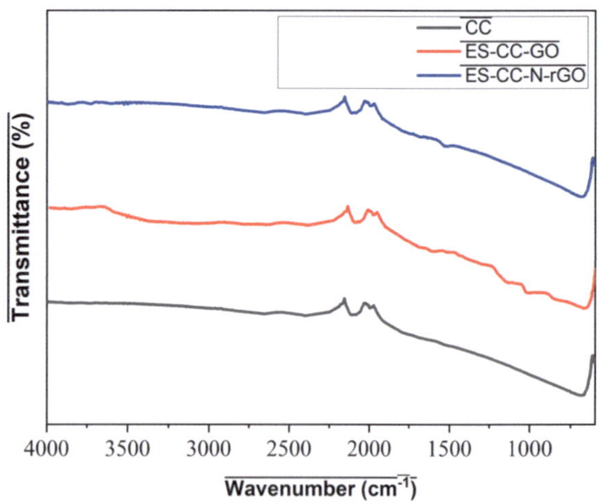

observed at $2\theta = 25.33°$. Upon modification with the N-doped rGO layer, this specific (002) peak exhibited a subtle yet important shift to a higher angle, appearing at $2\theta = 25.54°$ in the CC/N-doped rGO sample (Fig. 3). This observed shift indicates a change in the interlayer spacing of the carbon structure, consistent with the successful incorporation of the nitrogen-doped reduced graphene oxide nanosheets onto the carbon cloth, which confirms the structural modification and successful synthesis of the final active material.

Cyclic Voltammetry (CV) measurements were performed to evaluate the redox behaviour and electrocatalytic activity of the prepared electrodes towards the Benzoquinone-2-sulfonic acid sodium salt (BQDS) redox couple. Measurements were conducted in a 1 mM BQDS solution dissolved in 1 M H_2SO_4 electrolyte at various scan rates (2–100 mV/s). The performance of the bare Carbon Cloth (CC) was directly compared against the functionalized CC/N-doped rGO electrode as shown in Fig. 4.

The CVs obtained for the Bare CC electrode showed low redox peak currents with broad and poorly defined peaks, clearly indicating a weak interaction and sluggish

Fig. 3 XRD of bare carbon cloth and N-doped graphene decorated CC

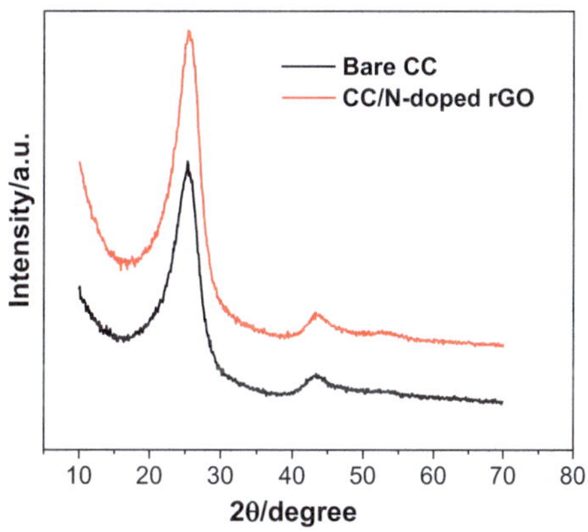

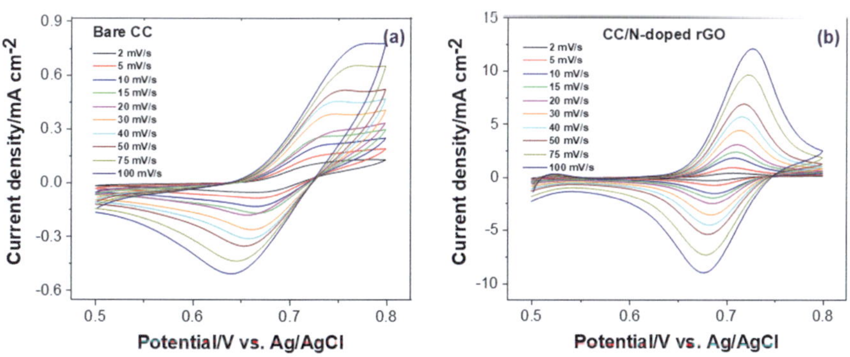

Fig. 4 CV results of bare carbon cloth and N-doped graphene decorated CC

reaction kinetics with the BQDS redox couple. In stark contrast, the CC/N-doped rGO electrode exhibited well-defined, sharper redox peaks and substantially higher current densities (Fig. 4). This pronounced difference demonstrates the significant enhancement in the electrocatalytic activity and a massive acceleration in the reaction kinetics provided by the N-doped rGO coating. Furthermore, analysis of the relationship between the redox peak current and the square root of the scan rate revealed a linear dependence for both electrodes. This linearity confirms that the electrochemical processes involving BQDS are primarily diffusion-controlled, thus allowing the application of the Randles–Sevcík equation to quantitatively determine the diffusion coefficient (D). The calculated diffusion coefficients show a striking improvement upon electrode modification. For the Bare CC electrode, the oxidation and reduction diffusion coefficients were calculated as 9.92×10^{-6} cm^2 s^{-1} and 4.98×10^{-6} cm^2 s^{-1}, respectively. Following the N-doped rGO modification, these values increased dramatically to 3.08×10^{-3} cm^2 s^{-1} (oxidation) and 1.79 $\times$ 10^{-3} cm^2 s^{-1} (reduction) for the CC/N-doped rGO electrode. This substantial increase in the diffusion coefficient confirms that the N-doped rGO coating facilitates faster diffusion-controlled reaction processes, which is key to achieving higher power density in flow battery applications.

The analysis of the Diffusion Coefficient (D) values, calculated from the Randles–Sevcík equation for 1 mM BQDS in 1 M H_2SO_4 electrolyte solution, provides a quantitative measure of the enhanced reaction kinetics at the modified electrode surface. The linear relationship between the redox peak current and the square root of the scan rate supports the use of the Randles–Sevcík equation for these diffusion-controlled redox processes. A comparison of the calculated diffusion coefficient values for the Bare Carbon Cloth (CC) and the CC/N-doped rGO electrodes reveals a striking improvement as shown in Table 1.

The results show that the diffusion coefficients for both the oxidation and reduction of the BQDS species are significantly enhanced following the deposition of the N-doped rGO coating (Table 1). Specifically, the oxidation diffusion coefficient saw a significant increase.

Electrochemical Impedance Spectroscopy (EIS) measurements were conducted to evaluate the charge transfer kinetics and overall internal resistance at the electrode/electrolyte interface, directly comparing the performance of the Bare Carbon Cloth (CC) and the CC/N-doped rGO electrodes. The series resistance (R_s) and the charge transfer resistance (R_{ct}) values were extracted by fitting the Nyquist plots to a suitable equivalent circuit model (shown in the inset of Fig. 5). The results

Table 1 Steps shown for the determination of diffusion coefficients for BQDS in sulphuric acid compared with two different electrodes

Sample	$Slope_{oxidation}/slope_{reduction}$ (A (V/s)$^{-1/2}$)	Diffusion $coefficient_{oxidation}$/diffusion $coefficient_{reduction}$ (cm^2 s^{-1})
Bare CC	y = 0.0024/y = −0.0017	$9.92 \times 10^{-6}/4.98 \times 10^{-6}$
CC/N-doped rGO	y = 0.0422/y = −0.0322	$3.08 \times 10^{-3}/1.79 \times 10^{-3}$

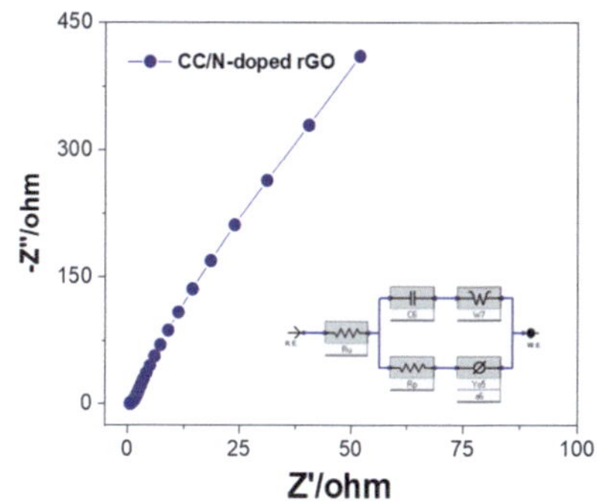

Sample	R_s (ohm)	R_{ct} (ohm)
Bare CC	3.05	169.0
CC/N-doped rGO	0.67	1.26

Fig. 5 EIS results of bare carbon cloth and N-doped graphene decorated CC

clearly demonstrate the transformative effect of the N-doped rGO modification on the system's resistive properties. The Bare CC electrode exhibited a series resistance (R_s) of 3.05 Ω, while this value was significantly reduced to 0.67 Ω for the CC/N-doped rGO electrode (Fig. 5). This decrease in R_s reflects a general reduction in ohmic losses, attributed to the higher electrical conductivity of the N-doped rGO coating and improved electrode–electrolyte contact. Most importantly, the drop in the Charge Transfer Resistance (R_{ct}), which is a direct measure of the electrochemical reaction kinetics, is dramatic. The Bare CC electrode exhibited a high R_{ct} value of 169.0 Ω, whereas this value was remarkably reduced to 1.26 Ω for the CC/N-doped rGO electrode. This represents a greater than 130-fold decrease in R_{ct}, unequivocally indicating a massive acceleration in the reaction kinetics at the electrode surface. This substantial decrease confirms that the catalytic sites introduced by the N-doped rGO significantly lower the kinetic barrier for the BQDS redox couple, leading to substantially reduced overpotentials and a higher potential energy efficiency.

4 Conclusion

This study successfully demonstrated a highly effective strategy for enhancing the performance of quinone-based Redox Flow Batteries (RFBs) through the fabrication of Nitrogen-doped Reduced Graphene Oxide (N-rGO) coated Carbon Cloth (CC) electrodes using the scalable Electrospray Deposition (ESD) technique. The physicochemical analyses confirmed that ESD is a promising, binder-free method capable of producing uniform, conformal, and highly reproducible N-rGO coatings onto the porous CC substrate, as evidenced by FTIR and XRD results. The thermal reduction successfully converted Graphene Oxide (GO) into N-rGO, yielding a material with improved electrical conductivity and the desired N-functionalities (e.g., pyridinic and graphitic nitrogen) which act as catalytic active sites. The electrochemical evaluation revealed an outstanding acceleration in the reaction kinetics toward the Benzoquinone-2-sulfonic acid sodium salt (BQDS) redox couple. The kinetic barrier was dramatically reduced, with the Charge Transfer Resistance (R_{ct}) plummeting from 169.0 Ω for the bare CC to a remarkable 1.26 Ω for the CC/N-doped rGO electrode. This greater than 130-fold reduction signifies a massive enhancement in electrocatalytic activity, promising substantially lower overpotentials and higher energy efficiency for the full battery system. Furthermore, analysis using the Randles–Sevcık equation confirmed a significant increase in the diffusion coefficients (over 300-fold), validating faster diffusion-controlled reaction processes at the modified electrode surface. In summary, this work provides compelling evidence that the combination of sustainable organic electrolytes (BQDS) with engineered nanostructured electrodes (CC/N-rGO) fabricated via a scalable ESD technique is a critical pathway for developing the next generation of high-performance and sustainable RFBs. The demonstrated enhancements in kinetics and conductivity reinforce the crucial role of N-doped carbon nanostructures in unlocking the full potential of quinone-based flow batteries for effective renewable energy integration.

Acknowledgements This work was supported by the European Innovation Council (EIC) under grant agreement No. 101115293 (VanillaFlow). This research was also funded by the CETPartnership, under the Joint Call 2023, through the project "HydrogEn oRganiC Regenerative FUeL CEll for Stationary Energy Storage (HERCULES)". The CETPartnership's research projects are co-funded by the European Commission (Grant Agreement No. 101069750) and the national TUBITAK 1071 via the project code: 224N167. Authors also acknowledge the internal project funded by SUNUM's management (2D nanomaterial electrodes via electro-deposition and electrospinning for electrochemical energy storage applications) that partially supported this work (I.A.SN-23-00110). R.A.G. is grateful to Türkiye Bilimsel ve Teknolojik Araştırma Kurumu (TÜBİTAK) for the 1001 project for funding his research in SUNUM (223M269).

References

1. Wolf, C., et al.: Safe-and-sustainable-by-design redox active molecules for energy storage applications. Energy Sustain. Soc. **15**(1) (2025)
2. Schlemmer, W., et al.: 2-Methoxyhydroquinone from vanillin for aqueous redox-flow batteries. Angew. Chem.-Int. Ed. **59**(51), 22943–22946 (2020)
3. Rubio-Garcia, J., et al.: Hydrogen/functionalized benzoquinone for a high-performance regenerative fuel cell as a potential large-scale energy storage platform. J. Mater. Chem. A **8**(7), 3933–3941 (2020)
4. Chakrabarti, B.K., et al.: Trichome-like carbon-metal fabrics made of carbon microfibers, carbon nanotubes, and Fe-based nanoparticles as electrodes for regenerative hydrogen/vanadium flow cells. ACS Appl. Nano Mater. **4**(10), 10754–10763 (2021)
5. Chakrabarti, B., et al.: Performance enhancement of reduced graphene oxide-modified carbon electrodes for vanadium redox-flow systems. ChemElectroChem **4**(1), 194–200 (2017)
6. Chakrabarti, B., et al.: Evaluation of a non-aqueous vanadium redox flow battery using a deep eutectic solvent and graphene-modified carbon electrodes via electrophoretic deposition. Batteries-Basel **6**(3) (2020)
7. Parra-Puerto, A. et al.: Carbon aerogel based thin electrodes for zero-gap all vanadium redox flow batteries—quantifying the factors leading to optimum performance. ChemElectroChem **9**(5) (2022)

8. Chakrabarti, B.K., et al.: Hybrid redox flow cells with enhanced electrochemical performance via binderless and electrophoretically deposited nitrogen-doped graphene on carbon paper electrodes. ACS Appl. Mater. Interfaces **12**(48), 53869–53878 (2020)

Electrochemical Performance of Electrophoretically Deposited Carbon Nanotube and Graphene Hybrid Electrode in Vanadium- and Quinone-Based Redox Flow Cell Electrolytes

Züleyha Kudaş, Kamal Asadipakdel, Faizah Md. Yasin, Reza Afshar Ghotli, Barun Kumar Chakrabarti, and Mustafa Kemal Bayazıt

Abstract The growing concerns over the environmental impact and limited reserves of fossil fuels have accelerated the global shift toward renewable energy sources such as solar and wind. Despite their sustainability, these energy sources are inherently intermittent and unpredictable due to their dependence on time and weather conditions, which pose significant challenges for stable grid integration. To overcome these challenges, large-scale energy storage systems (ESS) have emerged as a critical solution. Among various ESS technologies, redox flow batteries have gained considerable attention owing to their flexible design, independent scaling of power and energy components, and suitability for large-scale applications. In this study, modified electrodes composed of a carbon nanotube (CNT) and graphene hybrid mixture were fabricated on carbon paper (CP) substrates via electrophoretic deposition. The

Z. Kudaş (✉) · K. Asadipakdel · F. Md. Yasin · R. A. Ghotli · B. K. Chakrabarti (✉) · M. K. Bayazıt (✉)
Nanotechnology Research and Application Centre (SUNUM), Sabanci University, Tuzla, İstanbul, Türkiye
e-mail: zuleyha.kudas@sabanciuniv.edu

B. K. Chakrabarti
e-mail: barun.chakrabarti@sabanciuniv.edu

M. K. Bayazıt
e-mail: mkbayazit@sabanciuniv.edu

F. Md. Yasin
Department of Chemical and Environmental Engineering, Faculty of Engineering, Putra University of Malaysia, Serdang, Malaysia

Institute of Nanoscience and Nanotechnology, ION2, Putra University of Malaysia, Serdang, Malaysia

M. K. Bayazıt
Faculty of Engineering and Natural Science, Sabanci University, Tuzla, İstanbul, Türkiye

M. Awang et al. (eds.), *Proceedings of the International Conference on Mechanical, Manufacturing and Plant Engineering*, Proceedings in Technology Transfer,
https://doi.org/10.1007/978-981-95-7817-7_25

resulting hybrid electrodes were evaluated in both vanadium redox flow battery and quinone-based organic redox flow battery electrolytes. The combination of CNT and graphene significantly enhanced the electrode performance due to their high surface area, excellent electrical conductivity, and electrochemical stability. Electrochemical analyses revealed that the modified electrodes exhibited superior performance compared to bare CP, in terms of both redox kinetics and charge–discharge capacity. This study highlights the potential of hybrid nanocarbons in electrode modification for advanced flow battery systems.

Keywords Graphene · Carbon nanotube · Electrophoretic deposition · Vanadium redox flow battery · Quinone redox couple

1 Introduction

The transition from fossil fuels to clean and renewable energy sources such as solar and wind has become a pressing global necessity due to increasing concerns over climate change and energy security [1, 2]. However, the intermittent and unpredictable nature of renewable energy generation creates significant imbalances between supply and demand, threatening the stability and reliability of electrical grids. To address this challenge, efficient and scalable energy storage solutions are essential for the widespread integration of renewable technologies into power infrastructures [3, 4].

Among the various electrochemical storage technologies, redox flow batteries (RFBs) have emerged as promising candidates for large-scale stationary energy storage, owing to their unique architecture that decouples energy and power capacities. RFBs offer several advantages, including long cycle life, high safety, design flexibility, and the ability to independently scale energy capacity simply by increasing electrolyte volume [5–7]. Within this category, vanadium redox flow batteries (VRFBs) are considered the most advanced, as they employ vanadium ions in multiple oxidation states in both half-cells, thereby minimizing issues such as cross-contamination and electrolyte degradation [8, 9].

Despite these advantages, VRFB commercialization is hindered by high system costs and performance limitations, particularly due to sluggish electrode kinetics on the positive side involving the VO^{2+}/VO_2^+ redox couple. Conventional carbon-based electrodes such as carbon felt and carbon paper (CP) often exhibit poor electrochemical activity and low wettability in acidic electrolytes, which leads to high activation overpotentials and reduced energy efficiency [10, 11]. Enhancing the electrochemical performance of electrode materials is therefore crucial for minimizing voltage losses and improving system efficiency at high current densities.

Several strategies have been explored to improve carbon electrode performance, including thermal and chemical treatments, heteroatom doping (e.g., nitrogen, oxygen), and surface modification with nanomaterials [12, 13]. Among these, the incorporation of carbon nanostructures such as multi-walled carbon nanotubes

(MWCNTs) and graphene (G) derivatives has shown significant promise. These nanomaterials provide large surface areas, excellent electrical conductivity, and enhanced catalytic activity, while also offering superior mechanical strength and robust chemical and electrochemical stability under VRFB operating conditions [14, 15].

Electrophoretic deposition (EPD) has emerged as a versatile and scalable technique for modifying electrode surfaces with nanomaterials. Compared to other deposition methods such as chemical vapor deposition and layer-by-layer assembly, EPD offers several advantages, including low-cost instrumentation, uniform coatings on complex geometries, controllable film thickness, and suitability for industrial-scale applications [16]. Among its variants, cathodic EPD is particularly attractive because it prevents undesired oxidation and allows the co-deposition of functional materials, enabling the fabrication of composite electrodes with enhanced electrochemical properties [17]. Recent studies have demonstrated the successful use of EPD to deposit graphene oxide, reduced graphene oxide (rGO), and carbon nanotubes (CNTs) onto carbon-based electrodes, leading to significantly improved redox kinetics for the VO^{2+}/VO_2^+ couple [17, 18]. However, these performance improvements are often attributed solely to nanomaterials, without considering potential structural or chemical transformations in the carbon substrate during EPD. Moreover, achieving stable adhesion between nanomaterials and electrode surfaces remains a challenge, frequently requiring the use of binders such as Nafion or additional surface treatments [19]. Beyond metal-based electrolytes, there is growing interest in organic redox-active molecules, particularly quinones, as alternatives for next-generation redox flow batteries. Quinones offer several attractive features, including high theoretical capacities, rapid redox kinetics, environmental benignity, and molecular tunability [20]. A representative redox reaction of benzoquinone (BQ) in acidic aqueous media is: $\mathrm{BQ} + 2\mathrm{H}^+ + 2\mathrm{e}^- \rightleftharpoons \mathrm{Hydroquinone}\ E^\circ \approx +0.70$ V versus SHE. This reversible two-electron, two-proton redox couple enables efficient charge storage and offers opportunities for chemical tunability. However, similar to vanadium-based systems, the performance of quinone-based flow batteries is strongly influenced by electrode surface properties such as wettability, conductivity, and redox kinetics.

In this context, the present study investigates the electrochemical behavior of CP electrodes modified with a composite of CNTs and G deposited via EPD. The modified electrodes are systematically evaluated in both vanadium- and quinone-based electrolytes to provide a comparative perspective on their performance across inorganic and organic redox systems. Particular emphasis is placed on the role of nanomaterial coating, EPD processing parameters, and electrode–electrolyte interactions in determining redox kinetics, surface morphology, and overall energy efficiency. The findings are expected to contribute to the development of cost-effective, high-performance electrodes for scalable flow battery technologies.

2 Experimental Materials and Methods

2.1 *Materials*

MWCNTs with diameters of 3–15 nm, lengths of 15–30 μm, and a purity of 97% were procured from a supplier in China. G was synthesized in-house via electrochemical exfoliation of graphite (details provided below). *N*,*N*-Dimethylformamide (DMF, >99.9% purity, Sigma-Aldrich) was used as the dispersing medium owing to its effectiveness in stabilizing carbon-based nanomaterials. Sulfuric acid (H_2SO_4, 95–98%) and 1,2-dihydrobenzoquinone-3,5-disulfonic acid (BQDS, 97%) were obtained from Sigma-Aldrich. Deionized water with a resistivity greater than 18.2 MΩ cm was used for solution preparation and dilution.

2.2 *Preparation of MWCNT/G Dispersion*

A dispersion of the MWCNT and G mixture was prepared in DMF at a concentration of 1 mg/mL using a multi-step sonication protocol to ensure high dispersion quality. The mixture was first ultrasonicated for 2 h to obtain a homogeneous suspension. Following this step, the solution was left undisturbed for 10 min to allow larger agglomerates to sediment. The clear supernatant was carefully collected to minimize the presence of poorly dispersed particles. To further enhance stability and uniformity, the supernatant was subjected to an additional 60 min of ultrasonication. This sequential sonication approach significantly improved the dispersibility and long-term stability of the composite suspension, which are critical for achieving uniform and reproducible coatings during the EPD process.

2.3 *Preparation of G Sheets via Electrochemical Exfoliation of Graphite Flakes*

G sheets were synthesized via electrochemical exfoliation of graphite flakes in an aqueous 0.1 M ammonium sulfate ($(NH_4)_2SO_4$) solution, following a reported method [21]. The process employed a two-electrode setup, with a graphite flake as the working electrode (anode) and a platinum gauze as the counter electrode (cathode), separated by a 2 cm inter-electrode distance. A constant direct current of +10 V was applied until either the desired amount of G was obtained, or the graphite flake was fully consumed. The exfoliated product was collected by vacuum filtration through nylon membrane filter paper and thoroughly rinsed with distilled water to remove residual salts. The resulting solid was then dispersed in DMF and ultrasonicated for 10 min. Since G is soluble in DMF whereas graphite is not, the dispersion was left undisturbed for 48 h to allow sedimentation of unexfoliated graphite. The precipitate was removed, and the supernatant was heated to evaporate the solvent, ultimately yielding few-layer G sheets.

2.4 *Electrophoretic Deposition of CNT/G Composite on CP Electrodes*

A horizontal cathodic EPD technique was employed to fabricate composite electrodes ($CP/(CNT + G)_{mix}$). A graphite plate (Poco Pyro, Entegris Inc.) served as the conductive anode, while CP was used as the cathodic deposition substrate. The inter-electrode distance was fixed at ca. 15 mm using a circular glass spacer to ensure consistent spacing and uniform electric field distribution. The deposition medium consisted of a 25 mL colloidal dispersion of the CNT-G mixture. A constant direct current (DC) voltage of 200 V was applied for 10 min using a TT TECHNI-C regulated DC power supply, with current monitored in real time to ensure process stability. These parameters were selected based on optimized conditions previously reported by Chakrabarti et al. for rGO systems [18]. The EPD process proceeds in two steps: (i) electrophoresis, during which positively charged composite particles migrate toward the negatively charged substrate, and (ii) deposition, where particles adhere to the cathodic surface to form a uniform coating. Effective and homogeneous deposition requires a stable suspension with uniformly dispersed particles smaller than ~30 μm, minimizing sedimentation and agglomeration during processing [22].

Following deposition, the coated CP substrates were dried and subjected to morphological and structural characterization to evaluate coating uniformity, thickness, and composite distribution. The optimized EPD parameters enabled the successful formation of conductive, homogeneous composite films, suitable for application as high-performance electrode materials.

2.5 *Structural and Morphological Characterization*

X-ray diffraction (XRD), Raman spectroscopy, and scanning electron microscopy (SEM) were employed to investigate the structural, molecular, and morphological properties of bare CP and CP modified with the CNT-G mixture ($CP/(CNT + G)_{mix}$).

XRD measurements were carried out using a Bruker D8 Advance diffractometer (Bruker, Karlsruhe, Germany) equipped with Cu Kα radiation ($\lambda = 0.15405$ nm). Diffraction patterns were recorded over a 2θ range of 5–60° in step-scan mode with an increment of 0.02° per step, enabling precise identification of crystalline phases and evaluation of interlayer spacing. Raman spectra were obtained with an inVia Raman Microscope (Renishaw, Wotton-under-Edge, UK) using a 520 nm laser as the excitation source to assess the molecular structure and degree of disorder within the graphitic domains. Surface morphology and microstructural features of the electrode materials were examined by SEM using a JSM-6010LV scanning electron microscope (JEOL, Massachusetts, USA), providing insights into coating uniformity, particle distribution, and composite integrity.

2.6 Electrochemical Measurements

Electrochemical analyses were conducted using a Gamry Potentiostat 3000 in a conventional three-electrode cell configuration. A platinum wire was used as the counter electrode, and an Ag/AgCl (3 M KCl) electrode served as the reference electrode. The working electrodes were either bare CP or CP/$(CNT + G)_{mix}$. Two different electrolyte systems were employed to evaluate electrode performance. In the first system, the electrolyte consisted of 1 mM BQDS dissolved in 1 M H_2SO_4 prepared with ultrapure water. Cyclic voltammetry (CV) measurements were performed in this electrolyte at scan rates ranging from 2 to 100 mV s^{-1} to examine the redox behavior and charge storage characteristics of the electrodes. In the second system, a 0.5 M vanadyl sulfate ($VOSO_4$) solution in 3 M H_2SO_4 was used to investigate the electrochemical properties associated with vanadium-based redox processes. CV experiments were carried out under identical conditions to probe the V^{4+}/V^{5+} redox couple. Additionally, electrochemical impedance spectroscopy (EIS) measurements were performed in the $VOSO_4/H_2SO_4$ electrolyte at open-circuit voltage over the frequency range of 0.1 Hz to 10 kHz, using a sinusoidal perturbation of 10 mV.

3 Results and Discussions

Raman spectroscopy was employed to evaluate both the quality of the deposited CNT/G composite films and the effect of deposition on structural properties. Spectra were collected for bare CP and CP/$(CNT + G)_{mix}$. The characteristic Raman features observed in all samples included the D band (~1350 cm^{-1}), G band (~1580 cm^{-1}), and 2D band (~2700 cm^{-1}) (Fig. 1a) [23–28]. The D band arises from structural defects and disorders in the carbon lattice, such as vacancies, substitutional impurities, dangling bonds, and edge sites [23, 26–28]. The G band corresponds to the in-plane vibrational mode of sp^2-bonded carbon atoms and reflects the degree of graphitic ordering. The 2D band, a second-order overtone of the D band, provides insights into stacking and electronic interactions of graphitic layers. For bare CP, the Raman spectrum exhibited a relatively low I_D/I_G ratio (0.59), consistent with a predominantly graphitic structure containing fewer defects (Table 1). The D peak intensity was markedly lower than the G peak, indicating a more ordered crystalline carbon arrangement. In contrast, after deposition of the CNT/G composite (CP/(CNT + $G)_{mix}$), a significant increase in the D band intensity was observed, leading to an elevated I_D/I_G ratio of 0.99 (Table 1). This increase indicates a higher density of structural defects and disorders [23, 26, 28]. Similar behavior has been reported for rGO-coated systems, where deposition led to greater amorphous carbon content [18]. The broader and more intense D band in CP/(CNT + G)$_{mix}$ suggests the presence of edge sites, functional groups, and disruptions in the sp^2 carbon lattice, typical of hybrid materials containing graphene derivatives and CNTs. The increase in the I_D/I_G ratio implies that the composite film is structurally more disordered compared to bare CP. Such disorder can play a dual role in electrochemical performance: defect sites can act as active centers that enhance charge transfer and catalysis, but excessive disorder may negatively impact conductivity.

Based on the Raman spectroscopy results, which revealed increased disorder and defect density in the composite electrodes, XRD analysis was conducted to further examine the crystalline structure and interlayer spacing of bare CP and CP/(CNT + G)$_{mix}$. The bare CP exhibited a characteristic diffraction peak at $2\theta = 25.76°$, corresponding to an interlayer spacing of approximately 3.45 Å, indicative of a well-ordered graphitic structure (Fig. 1b, Table 1). This reflection is consistent with the typical d-spacing of graphitic carbon. Upon modification with the CNT/G composite, the main diffraction peak shifted to a higher angle at $2\theta = 26.31°$, accompanied by a slight reduction in interlayer spacing to 3.38 Å (Fig. 1b).

The shift to a higher angle indicates a reduction in interlayer spacing, suggesting that the G sheets and CNTs are more tightly packed, although the overall structure becomes more disordered (Table 1) [24]. The slight decrease in crystallite size further implies that the composite coating introduces defects and structural distortions within the graphite lattice. These effects likely arise from the incorporation of CNTs and G sheets, which disrupts the stacking order and reduce crystallinity. The combination of denser packing and increased disorder is consistent with the presence of edge defects and functional groups typically observed in such composites.

The XRD results demonstrate that while bare CP retains a more conventional graphitic structure with larger interlayer spacing and crystallite size, the CP/(CNT + G)$_{mix}$ electrode exhibits tighter stacking and smaller crystallite dimensions, reflecting structural modifications induced by the composite coating. These structural changes are expected to influence the electrochemical behavior of the electrodes, particularly by affecting ion transport pathways and electrical conductivity.

Following the Raman and XRD analyses, the morphological features of the electrodes were examined by SEM to elucidate surface and microstructural differences between bare CP and CP/(CNT + G)$_{mix}$. The SEM images of bare CP revealed its intrinsic layered, plate-like morphology, characteristic of stacked carbon fibers and graphitic domains (Fig. 2a, b). The surface appeared relatively smooth, with large flat facets and sharp edges, reflecting the untreated and unmodified CP structure.

In contrast, the CP/(CNT + G)$_{mix}$ electrode displayed a markedly different morphology (Fig. 3a, b). The surface was densely coated with a homogeneous network of CNTs and G sheets, where CNTs formed entangled fibrous structures and G sheets appeared as wrinkled, overlapping flakes [24, 29]. This composite architecture generated a porous three-dimensional network with enhanced surface roughness, providing an enlarged electroactive area and improved electrolyte accessibility.

Elemental mapping further supported these observations. While bare CP consisted predominantly of carbon (97.37%) with an oxygen content of 2.63% (Fig. 2c–e),

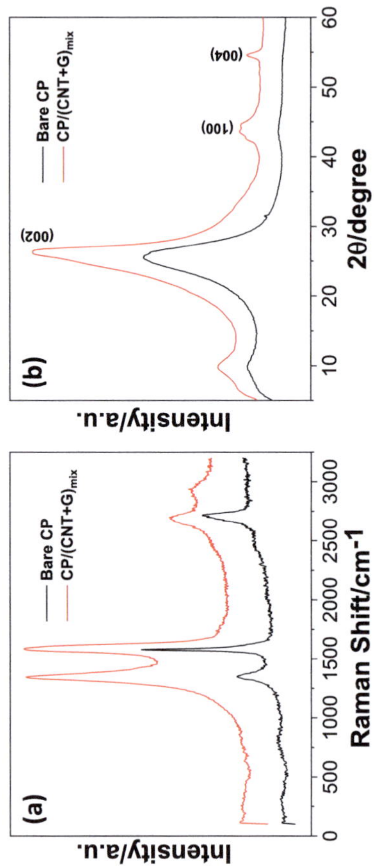

Fig. 1 The comparison of **a** Raman and **b** XRD analysis for bare CP and CP/(CNT + G)$_{mix}$ electrodes

Table 1 The results of Raman and X-ray diffraction (XRD) analysis for bare CP and CP/(CNT + G)$_{mix}$ electrodes

Sample	Peak position		I_D/I_G	2θ (degree)	Crystallite size (Å)
	D band (cm^{-1})	G band (cm^{-1})			
Bare CP	1350.09	1580.54	0.59	25.76	3.45
CP/(CNT + G)$_{mix}$	1345.01	1585.58	0.99	26.31	3.38

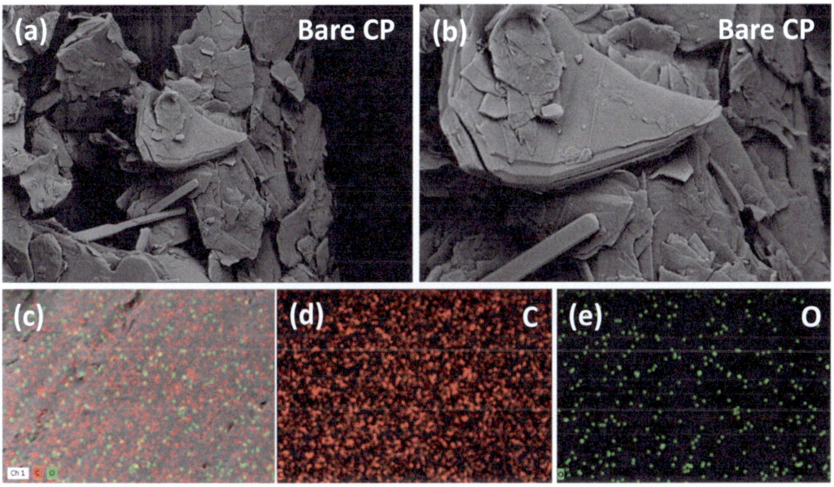

Fig. 2 The SEM images, SEM-EDS elemental mapping of C and O elements, and EDS analyses for the bare CP electrode

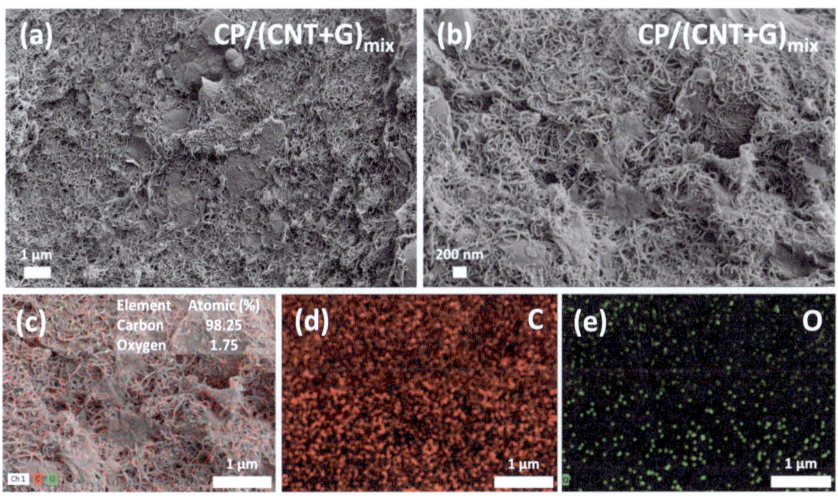

Fig. 3 The SEM images, SEM-EDS elemental mapping of C and O elements, and EDS analyses for the CP/(CNT + G)$_{mix}$ electrode

the CP/(CNT + G)$_{mix}$ electrode exhibited a higher carbon contribution (98.25%) with correspondingly reduced oxygen content (1.75%) (Fig. 3c–e). Such a high carbon content in the CP/(CNT + G)$_{mix}$ electrode is advantageous for redox flow battery (RFB) electrodes, as they promote high conductivity and facilitate rapid charge transfer.

3.1 Cyclic Voltammetry

The CV results in Fig. 4 highlight the distinct electrochemical behaviors of bare CP and CP/(CNT + G)$_{mix}$ electrodes in 1 M H_2SO_4 containing 1 mM BQDS. Bare CP exhibited the lowest current response, consistent with its poor conductivity and limited number of active sites (Fig. 4a). The diffusion coefficients of BQDS in 1 M H_2SO_4 were estimated using the Randles–Sevcík equation, based on the slopes of the linear relationships between peak current (i_p) and the square root of the scan rate ($v^{1/2}$) obtained from CV measurements at various scan rates (Fig. 4c, d). Diffusion coefficients were calculated separately for the oxidation and reduction processes for both electrode configurations.

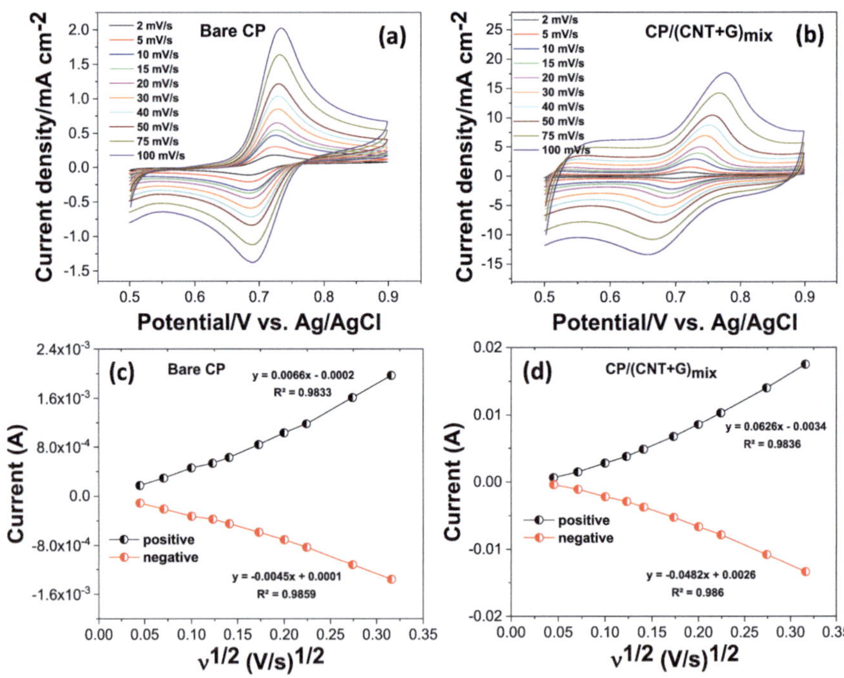

Fig. 4 Cyclic voltammograms obtained at different scan rates in 1 M H_2SO_4 containing 1 mM BQDS on bare CP and CP/(CNT + G)$_{mix}$ electrodes. Linear dependence of I_p versus $v^{1/2}$ for 1 mM BQDS in 1 M H_2SO_4 electrolyte

For bare CP, i_p versus $v^{1/2}$ plots showed an almost linear dependence but with relatively small slopes, indicative of diffusion-controlled processes with limited efficiency. The calculated diffusion coefficients were correspondingly low (7.54 × 10^{-5} cm^2 s^{-1} for oxidation and 3.50 × 10^{-5} cm^2 s^{-1} for reduction), confirming the poor catalytic activity and restricted electroactive surface area of bare CP. This behavior is consistent with its smooth layered morphology and lack of functional sites observed in SEM analysis.

In contrast, the CP/(CNT + G)$_{mix}$ electrode displayed more pronounced and symmetric redox peaks across all scan rates, suggesting enhanced reversibility and faster charge-transfer kinetics of the quinone-based redox reaction (Fig. 4b). The peak current values increased substantially compared with bare CP, reflecting a larger electroactive surface area. Linear fitting of i_p versus $v^{1/2}$ yielded steeper slopes and improved linearity, confirming efficient diffusion-controlled behavior. Importantly, the calculated diffusion coefficients were significantly higher (6.78 × 10^{-3} cm^2 s^{-1} for oxidation and 3.40 × 10^{-3} cm^2 s^{-1} for reduction), nearly two orders of magnitude greater than those for bare CP (Fig. 4d).

This substantial improvement can be attributed to the synergistic effect of CNTs and G. CNTs provide highly conductive pathways for electron transport, while G contributes additional surface area and oxygen-containing functional groups, improving wettability and creating more active sites for redox reactions. The electrochemical results underscore the critical role of electrode surface modification in enhancing the redox kinetics of quinone-based electrolytes. Whereas bare CP exhibits poor electron-transfer efficiency and limited diffusion, the incorporation of CNTs and G transforms the electrode into a highly conductive, porous, and functionalized matrix. This structural advantage facilitates faster ion transport and improves redox reversibility, establishing CP/(CNT + G)$_{mix}$ as a markedly superior electrode material for RFB applications.

After evaluating the electrochemical performance of quinone-based electrolytes, which demonstrated the beneficial effect of CNT and G modification on charge-transfer kinetics and diffusion behavior, further investigations were extended to 0.5 M $VOSO_4$ in 3 M H_2SO_4. This approach enabled a direct comparison of two distinct electrolyte systems, organic quinones and inorganic vanadium salts, on bare CP and CP/(CNT + G)$_{mix}$ electrodes, thereby providing deeper insight into the universality and effectiveness of electrode surface modification strategies for RFB applications.

For the V^{4+}/V^{5+} redox couple, the bare CP electrode exhibited a large peak-to-peak separation of 0.50 V at 20 mV s^{-1}, indicative of sluggish electron-transfer kinetics and poor reversibility (Fig. 5a). The calculated diffusion coefficients were also relatively small (1.78 × 10^{-5} cm^2 s^{-1} for oxidation and 7.92 × 10^{-6} cm^2 s^{-1} for reduction), reflecting limited ion transport and a restricted electroactive surface area.

In contrast, the CP/(CNT + G)$_{mix}$ electrode displayed sharper, more symmetric redox peaks with a markedly reduced peak-to-peak separation of 0.28 V at 20 mV s^{-1}, signifying enhanced reversibility and faster kinetics (Fig. 5b). The diffusion coefficients were correspondingly higher (2.18 × 10^{-5} cm^2 s^{-1} for oxidation and 1.23 ×

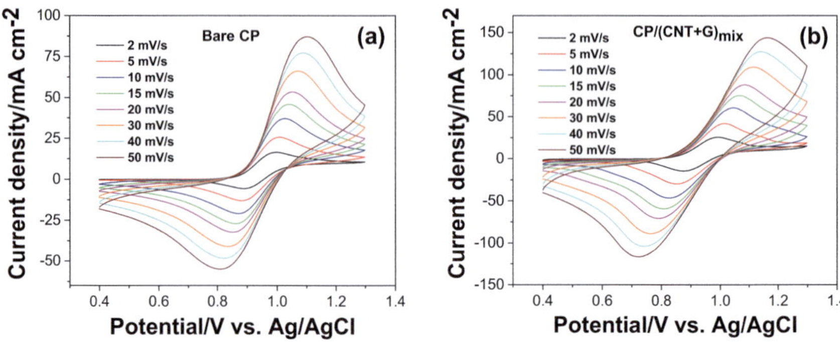

Fig. 5 Cyclic voltammograms obtained at different scan rates for 0.5 M $VOSO_4$ in 3 M H_2SO_4 electrolyte solution

10^{-5} cm^2 s^{-1} for reduction), confirming the beneficial role of CNT and G modification in improving both charge transfer and ion diffusion. These enhancements are attributed to the synergistic effects of CNTs, which provide efficient conductive pathways for rapid electron transport, and G sheets, which increase the accessible surface area and introduce oxygen functional groups that promote electrolyte wettability.

3.2 *Electrochemical Impedance Spectroscopy*

The EIS analysis provided further insight into the interfacial charge-transfer and ion-transport characteristics of bare CP and CP/(CNT + G)$_{mix}$ electrodes. In the Nyquist plot, the bare CP electrode displayed a large semicircle, corresponding to a high charge-transfer resistance (R_{ct} = 3.72 Ω) and a relatively high solution resistance (R_s = 1.10 Ω) (Fig. 6). These values reflect sluggish electron-transfer kinetics, low electrocatalytic activity, and weak electrode–electrolyte interfacial interaction.

In contrast, the CP/(CNT + G)$_{mix}$ electrode exhibited a much smaller semicircle, with a markedly reduced R_{ct} of 1.08 Ω and a lower R_s of 0.97 Ω (Fig. 6). This significant decrease in resistance demonstrates that CNT and G modification improves electrical conductivity, facilitates ion diffusion, and accelerates charge-transfer kinetics at the electrode–electrolyte interface. Specifically, CNTs provide efficient conductive pathways for electron transport, while G contributes a large accessible surface area and oxygen-containing functional groups that enhance wettability and strengthen electrolyte interactions. The synergistic effect of CNTs and G is therefore responsible for the substantial reduction in charge-transfer resistance compared with bare CP.

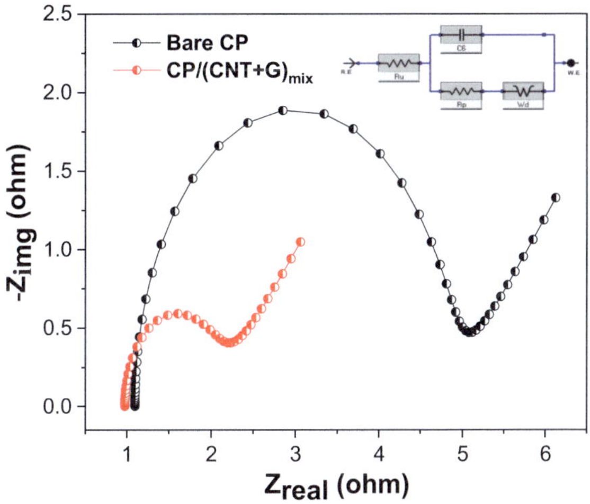

Fig. 6 Nyquist plots obtained from EIS measurements using bare CP and CP/(CNT + G)$_{mix}$ electrodes in a 0.5 M $VOSO_4$ solution containing 3 M H_2SO_4, and an equivalent electrical circuit model used for fitting the EIS data

4 Conclusion

Comprehensive electrochemical and structural characterizations demonstrate that surface modification of bare CP electrodes with CNT and G composite markedly enhances their performance in RFB applications. CV analyses reveal that the CP/(CNT + G)$_{mix}$ electrode exhibits reduced peak-to-peak potential separation and higher diffusion coefficients compared with bare CP, indicating improved electron-transfer kinetics and more efficient redox activity in $VOSO_4$ electrolytes. EIS further confirms this enhancement, as the modified electrode shows a substantially lower R_{ct}, reflecting superior interfacial conductivity and ion-transport properties. The integration of CNTs and G with CP electrodes produces a synergistic effect that enhances electrochemical activity, minimizes interfacial resistance, and optimizes structural features. These findings underscore the promise of hybrid nanocarbon architecture as high-performance electrodes for both vanadium- and quinone-based RFB systems.

Acknowledgements This work was supported by the European Innovation Council (EIC) under grant agreement No. 101115293 (VanillaFlow). This research was also funded by the CETPartnership, under the Joint Call 2023, through the project "HydrogEn oRganiC Regenerative FUeL CEll for Stationary Energy Storage (HERCULES)". The CETPartnership's research projects are co-funded by the European Commission (Grant Agreement No. 101069750) and the national TUBITAK 1071 via the project code: 224N167. Authors also acknowledge the internal project funded by SUNUM's management (2D nanomaterial electrodes via electro-deposition and electrospinning for electrochemical energy storage applications) that partially supported this work (I.A.SN-23-00110). R.A.G. is grateful to Türkiye Bilimsel ve Teknolojik Araştırma Kurumu (TÜBİTAK) for the 1001 project for funding his research in SUNUM (223M269).

References

1. Hassan, Q., Viktor, P., Al-Musawi, T.J., Ali, B.M., Algburi, S., Alzoubi, H.M., Al-Jiboory, A.K., Sameen, A.Z., Salman, H.M., Jaszczur, M.: The renewable energy role in the global energy transformations. Renew. Energy Focus **48**, 100545 (2024)
2. Khare, V., Nema, S., Baredar, P.: Solar–wind hybrid renewable energy system: a review. Renew. Sustain. Energy Rev. **58**, 23–33 (2016)
3. Enasel, E., Dumitrascu, G.: Storage solutions for renewable energy: a review. Energy Nexus **17**, 100391 (2025)
4. Yang, Y., Bremner, S., Menictas, C., Kay, M.: Battery energy storage system size determination in renewable energy systems: a review. Renew. Sustain. Energy Rev. **91**, 109–125 (2018)
5. Naseer, M.U.H., Pan, B., Wang, S., Lyu, Y., Liu, B., Li, L., Qi, J., Du, H.: A comprehensive review of advancements in vanadium electrolyte preparation for Vanadium redox flow batteries. J. Environ. Chem. Eng. **13**(5), 118402 (2025)
6. Yao, Y., Lei, J., Shi, Y., Ai, F., Lu, Y.-C.: Assessment methods and performance metrics for redox flow batteries. Nat. Energy **6**(6), 582–588 (2021)
7. Park, M., Ryu, J., Wang, W., Cho, J.: Material design and engineering of next-generation flow-battery technologies. Nat. Rev. Mater. **2**(1), 16080 (2016)
8. Lee, W., Park, G., Shin, M., Emmel, D., Schröder, D., Kwon, Y.: Challenges and advances in redox flow batteries utilizing sustainable and cost-effective non-vanadium active materials. J. Mater. Chem. A **13**(21), 15491–15516 (2025)
9. Zorrilla, F., Montiel, M., Mustata, R., Losantos, R., Valiño, L.: Impact of electrolyte composition on the mitigation of electrolyte imbalance in a vanadium redox flow battery: a 3D multiphysics model. J. Energy Storage **107**, 114899 (2025)
10. Bayeh, A.W., Kabtamu, D.M., Demeku, A.M., Chen, G.-C., Wang, C.-H.: Perspective on electrocatalysts and performance hindrances at the negative electrode in vanadium redox flow batteries. J. Energy Storage **102**, 114026 (2024)
11. Opar, D.O., Nankya, R., Lee, J., Jung, H.: Three-dimensional mesoporous graphene-modified carbon felt for high-performance vanadium redox flow batteries. Electrochim. Acta **330**, 135276 (2020)
12. Schnucklake, M., Kuecken, S., Fetyan, A., Schmidt, J., Thomas, A., Roth, C.: Salt-templated porous carbon–carbon composite electrodes for application in vanadium redox flow batteries. J. Mater. Chem. A **5**(48), 25193–25199 (2017)
13. Li, W., Zhang, Z., Tang, Y., Bian, H., Ng, T.-W., Zhang, W., Lee, C.-S.: Graphene-nanowall-decorated carbon felt with excellent electrochemical activity toward VO^{2+}/VO_2^+ couple for all vanadium redox flow battery. Adv. Sci. **3**(4), 1500276 (2016)
14. Han, P., Wang, H., Liu, Z., Chen, X., Ma, W., Yao, J., Zhu, Y., Cui, G.: Graphene oxide nanoplatelets as excellent electrochemical active materials for VO^{2+}/VO_2^+ and V^{2+}/V^{3+} redox couples for a vanadium redox flow battery. Carbon **49**(2), 693–700 (2011)
15. Li, W., Liu, J., Yan, C.: Multi-walled carbon nanotubes used as an electrode reaction catalyst for VO^{2+}/VO_2^+ for a vanadium redox flow battery. Carbon **49**(11), 3463–3470 (2011)
16. Gerard, O., Numan, S., Khalid, M., Subramaniam, R., Kasi, R.: A review on the recent advances in binder-free electrodes for electrochemical energy storage application. J. Energy Storage **50**, 104283 (2022)
17. Chakrabarti B., Yufit V., Kavei, A., Xia, Y., Stevenson, G., Kalamaras, E., Luo H., Feng, J., Tariq, F., Taiwo, O., Titirici M.M.: Charge/discharge and cycling performance of flexible carbon paper electrodes in a regenerative hydrogen/vanadium fuel cell. Int. J. Hydrogen En. **44**(57), 30093–30107 (2019)
18. Chakrabarti, B., Nir, D., Yufit, V., Tariq, F., Rubio-Garcia, J., Maher, R., Kucernak, A., Aravind, P.V., Brandon, N.: Performance enhancement of reduced graphene oxide-modified carbon electrodes for vanadium redox-flow systems. ChemElectroChem **4**(1), 194–200 (2017)
19. Reynolds, C.D., Slater, P.R., Hare, S.D., Simmons, M.J.H., Kendrick, E.: A review of metrology in lithium-ion electrode coating processes. Mater. Des. **209**, 109971 (2021)

20. Rubio-Garcia, J., Kucernak, A., Parra-Puerto, A., Liu, R., Chakrabarti, B.: Hydrogen/functionalized benzoquinone for a high-performance regenerative fuel cell as a potential large-scale energy storage platform. J. Mater. Chem. A **8**(7), 3933–3941 (2020)
21. Parvez, K., Wu, Z.-S., Li, R., Liu, X., Graf, R., Feng, X., Müllen, K.: Exfoliation of graphite into graphene in aqueous solutions of inorganic salts. J. Am. Chem. Soc. **136**(16), 6083–6091 (2014)
22. Corni, I., Ryan, M.P., Boccaccini, A.R.: Electrophoretic deposition: from traditional ceramics to nanotechnology. J. Eur. Ceram. Soc. **28**(7), 1353–1367 (2008)
23. Bayazit, M.K., Coleman, K. S.: Probing the selectivity of azomethine imine cycloaddition to single-walled carbon nanotubes by resonance Raman spectroscopy. Chem. –Asian J. **7**(12), 2925–2930 (2012)
24. Bayazit, M.K., Xiong, L., Jiang, C., Moniz, S.J.A., White, E., Shaffer, M.S.P., Tang, J.: Defect-free single-layer graphene by 10 s microwave solid exfoliation and its application for catalytic water splitting. ACS Appl. Mater. Interfaces **13**(24), 28600–28609 (2021)
25. Clancy, A.J., Bayazıt, M.K., Hodge, S.A., Skipper, N.T., Howard, C.A., Shaffer, M.S.P.: Charged carbon nanomaterials: redox chemistries of fullerenes, carbon nanotubes, and graphenes. Chem. Rev. **118**(16), 7363–7374 (2018)
26. Hodge, S.A., Bayazıt, M.K., Coleman, K.S., Shaffer, M.S.P.: Unweaving the rainbow: a review of the relationship between single-walled carbon nanotube molecular structures and their chemical reactivity. Chem. Soc. Rev. **41**(12), 4409–4429 (2012)
27. Hodge, S.A., Bayazit, M.K., Tay, H.H., Shaffer, M.S.P.: Giant cationic polyelectrolytes generated via electrochemical oxidation of single-walled carbon nanotubes. Nat. Commun. **4**(1), 1989 (2013)
28. Yau, H.C., Bayazit, M.K., Steinke, J.H.G., Shaffer, M.S.P.: Sonochemical degradation of N-methylpyrrolidone and its influence on single walled carbon nanotube dispersion. Chem. Commun. **51**(93), 16621–16624 (2015)
29. Bayazit, M.K., Yau, H.C., Leese, H., Lee, W.J., Shaffer, M.S.P.: Mono-acetylenes as new crosslinkers for all-carbon living charge carbon nanotubide organogels. ChemistrySelect **7**(30), e202202469 (2022)

Empowering Sustainability: A Comprehensive Energy Review of a Higher Education Institution as a Catalyst for Energy Management System Development

Lailah A. Cadeliña, Antonio P. Doroliat, and Basil Aloysius P. Bautista

Abstract The Philippines faces rising energy demand, high electricity costs, and increasing environmental impacts, which highlight the urgent need for sustainable practices in higher education institutions (HEIs). This study conducted a comprehensive energy review of Adamson University, a designated establishment under Republic Act No. 11285 due to its annual electricity consumption of 4,338,620 kWh. The Ozanam (OZ) Building being the highest energy-consuming facility, recording nearly 22% of the total, was therefore selected as the focus of the study. The objectives were to identify significant energy use (SEU), establish an energy baseline (EnB), define energy performance indicators (EnPIs), and propose strategies for energy efficiency improvement. Using energy consumption data, statistical analysis, and a Level 1 energy audit, the study confirmed the air-conditioning (AC) system as the dominant SEU, consuming 67% of the building's annual use. EnPIs were established, including 65 kWh/m^2 and 101 kWh/student for the building. The results showed that energy consumption patterns closely followed the academic calendar, with peaks during full semesters and significant drops during breaks. Improvement measures were proposed across technological, operational, and behavioral dimensions. These included replacing non-inverter units with inverters, introducing high-efficiency window-type and split-type ACs, implementing thermal retrofits, enforcing strict runtime controls, and promoting awareness campaigns. Collectively, these interventions could reduce electricity consumption, with the AC system providing the largest savings potential. The study concludes that a systematic, ISO 50001-aligned Energy Management System (EnMS) is essential for HEIs. Beyond cost reduction, such

L. A. Cadeliña (✉) · A. P. Doroliat · B. A. P. Bautista
Adamson University, Manila, Philippines
e-mail: cadelinalailah@gmail.com

A. P. Doroliat
e-mail: antonio.doroliat@adamson.edu.ph

B. A. P. Bautista
e-mail: basil.aloysius.bautista@adamson.edu.ph

M. Awang et al. (eds.), *Proceedings of the International Conference on Mechanical, Manufacturing and Plant Engineering*, Proceedings in Technology Transfer,
https://doi.org/10.1007/978-981-95-7817-7_26

initiative ensures compliance with national energy laws and position universities as sustainability leaders.

Keywords Energy management system · Energy baseline · Significant energy use · Energy efficiency

1 Introduction

The Philippines has one of the highest electricity prices in Asia, ranking second in 2018 behind Japan [1]. Despite these costs, electricity demand continues to rise, with a 6.3% growth recorded in 2019, largely driven by economic expansion [2]. This increasing demand intensifies environmental impacts such as air pollution, climate change, and resource depletion, exemplified by the expected depletion of the Malampaya gas fields [3]. To address these challenges, Republic Act No. 11285, or the Energy Efficiency and Conservation Act, mandates designated establishments including Higher Education Institutions (HEIs) with high annual consumption to adopt Energy Management Systems (EnMS) aligned with ISO 50001 [4]. A Catholic HEI in Manila, Philippines, consuming about 1.4 million kWh annually, falls under this category, requiring compliance with energy audits, performance monitoring, and structured energy planning [5].

However, developing an EnMS in HEIs remains difficult due to the absence of a structured energy plan, lack of comprehensive energy reviews, and limited identification of significant energy use (SEUs). Many institutions do not have energy baselines, defined performance indicators (EnPIs), or systematic data collection methods, which restrict their ability to track improvements and comply with RA 11285.

This study aims to conduct a comprehensive energy review of a HEI as a foundation for developing an EnMS. Specifically, it seeks to: (1) analyze energy use and consumption by identifying SEUs, establishing an energy baseline (EnB), and determining EnPIs; and (2) recommend opportunities for improving energy performance through an energy audit.

The study is significant as it provides the institution with practical insights to optimize energy use, reduce costs, and promote sustainability. Its findings can guide other HEIs in adopting EnMS, assist local government units (LGUs) in energy planning, contribute to environmental protection by reducing electricity-related emissions, and serve as a reference for future researchers in energy management.

The scope of the study is limited to engineering building, the most energy-intensive facility on campus. It evaluates electricity consumption from January to December 2023, covering major loads such as air-conditioning, lighting, laboratory equipment, IT devices, and other plug loads. The study focuses solely on electricity data, excluding fuel use, energy pricing, and long-term ecological impacts.

2 Methodology

This study carried out a comprehensive energy review of the Engineering Building of the HEI, identified as the campus's highest-consuming facility. Monthly electricity consumption data from January to December 2023 were collected from the facility office records to establish a clear profile of energy use. The review focused on identifying significant energy use (SEU), setting an energy baseline (EnB), and developing energy performance indicators (EnPIs) as measurable benchmarks for monitoring and improvement.

In addition to data analysis, a Level 1 energy audit was conducted to identify opportunities for improving energy performance. The audit included a walk-through inspection of major energy systems, review of operational schedules, and interviews with facility personnel to verify equipment usage and maintenance practices. Observations focused on air-conditioning units, laboratory equipment, lighting, and plug loads to pinpoint inefficiencies and potential savings measures.

Data was organized and processed using Microsoft Excel to generate baseline values, SEU profiles, and EnPIs. The methodology was guided by ISO 50001:2018 standards and supported by related literature to ensure consistency with internationally recognized energy management practices. This framework enabled the study to present an evidence-based account of consumption trends and provide practical insights into efficiency improvement.

The figure below shows the step-by-step process applied in this study. It begins with the collection and analysis of electricity consumption data, followed by the identification of the most energy-intensive building. The subsequent steps, analyzing the building's energy use through energy audit, determining Significant Energy Uses (SEUs), assessing SEU data, and establishing both Energy Baseline (EnB) and Energy Performance Indicators (EnPIs). The process concludes with identifying opportunities to improve energy performance (Fig. 1).

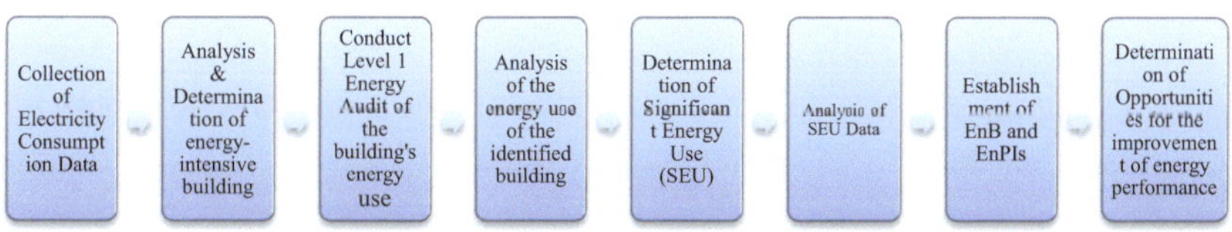

Fig. 1 Energy review methodology framework (ISO, 2018)

3 Results and Discussion

3.1 Campus Energy Consumption

As shown in Fig. 2 and Table 1, the HEI's total electricity consumption for 2023 amounted to 4,338,620 kWh, distributed across its major academic and support buildings. Among these, the Building 3, the College of Engineering Building, recorded the highest annual consumption at 951,600 kWh, accounting for 22% of the campus total. The Building 1 (849,600 kWh; 20%) and Building 2 (821,100 kWh; 19%) followed as the next largest consumers, while smaller facilities such as the Building 5 (89,040 kWh; 2%) contributed minimally.

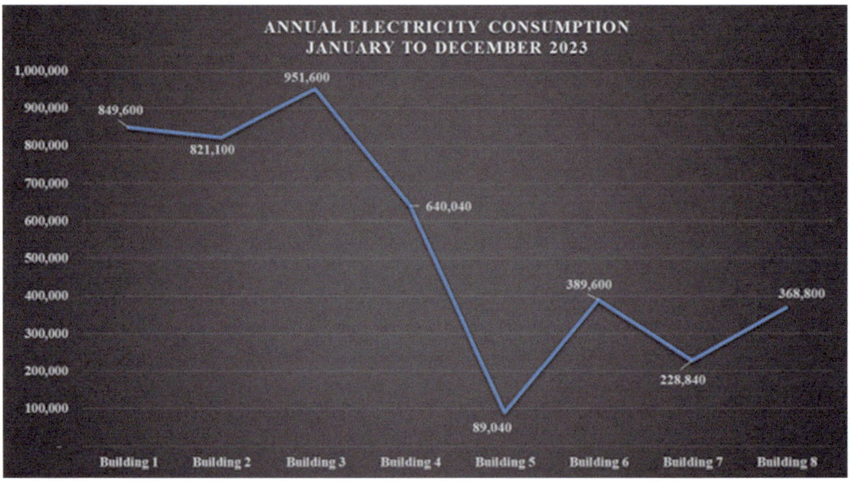

Fig. 2 Campus annual energy consumption trend

Table 1 Campus annual energy consumption data

Building	Consumption (kWh)	% Share of total (%)
Building 3	951,600	22
Building 1	849,600	20
Building 2	821,100	19
Building 4	640,040	15
Building 6	389,600	9
Building 8	368,800	9
Building 7	228,840	5
Building 5	89,040	2
Total	**4,338,620**	100

Given its diverse functions, housing engineering laboratories, classrooms, offices, and support facilities, the College of Engineering Building exhibited the broadest range of end-uses, resulting in its dominant share of energy demand. For this reason, it was selected as the target building of the study, as efficiency interventions here would deliver the most significant impact on overall campus consumption. Concentrating efforts on this facility thus provides a strategic starting point for the university's energy management system development.

3.2 *Significant Energy Use (SEU) of the Engineering Building*

Figure 3 and Table 2 show that the air-conditioning (AC) system dominated electricity consumption, with 633,254 kWh annually, accounting for 67% of the building's total use. Laboratory equipment contributed 12% (117,882 kWh), followed by appliances/plug loads (8%), lighting (6%), and office/IT equipment (3%). The elevator and unaccounted loads made up the remaining share. This profile confirms that cooling requirements are the primary driver of energy demand in the Engineering building.

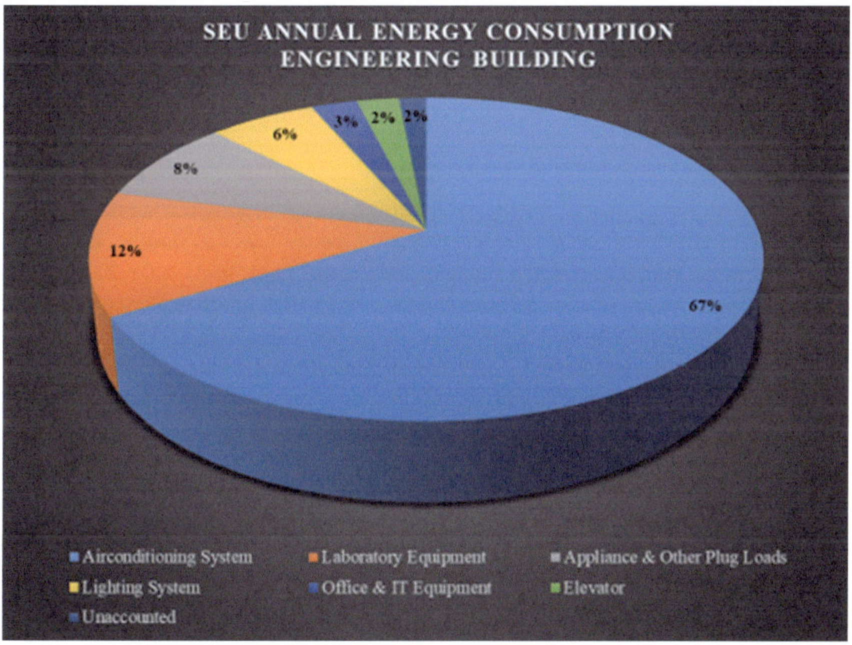

Fig. 3 Significant energy use trend in engineering building

Table 2 Distribution of significant energy use in engineering building

Engineering building SEU	Energy consumption (kWh)
Air conditioning system	633,254
Laboratory equipment	117,882
Appliance and other plug loads	80,075
Lighting system	58,400
Office and IT equipment	24,852
Elevator	22,380
Unaccounted	14,758
Total annual kWh	**936,842**

3.3 Air-Conditioning System Performance

Figure 4 and Table 3 present the monthly energy consumption of the air-conditioning (AC) system in the Engineering Building. It showed significant fluctuations throughout 2023, closely reflecting academic operations and seasonal climate conditions. Consumption peaked in April (66,113 kWh), coinciding with the end of the second semester and the hot dry season, while another high was observed in October (63,316 kWh) during the regular semester. In contrast, the lowest values were recorded in June (39,608 kWh) and December (33,281 kWh), corresponding to the summer break and Christmas holiday period, when occupancy and cooling demand were minimal. Overall, the trend highlights the dominant role of academic scheduling and thermal comfort requirements in driving the building's electricity use, with the AC system consistently accounting for the largest share of monthly demand.

Further analysis of the AC inventory as shown in Fig. 5, showed that non-inverter units consumed 74% of total AC energy, despite being fewer in number compared to inverter units, which accounted for only 26%. This disproportionate share underscores the inefficiency of non-inverter models and their significant contribution to the building's high electricity consumption. Together, these findings confirm the AC system as the primary target for energy-saving interventions, with substantial potential for improvement through equipment upgrades, runtime control, and operational reforms.

3.4 Energy Baseline and Performance Indicators

An energy baseline (EnB) was established for the Engineering Building at 951,600 kWh, with the AC system baseline at 633,254 kWh. Key EnPIs were computed. These indicators provide measurable benchmarks for monitoring future efficiency improvements.

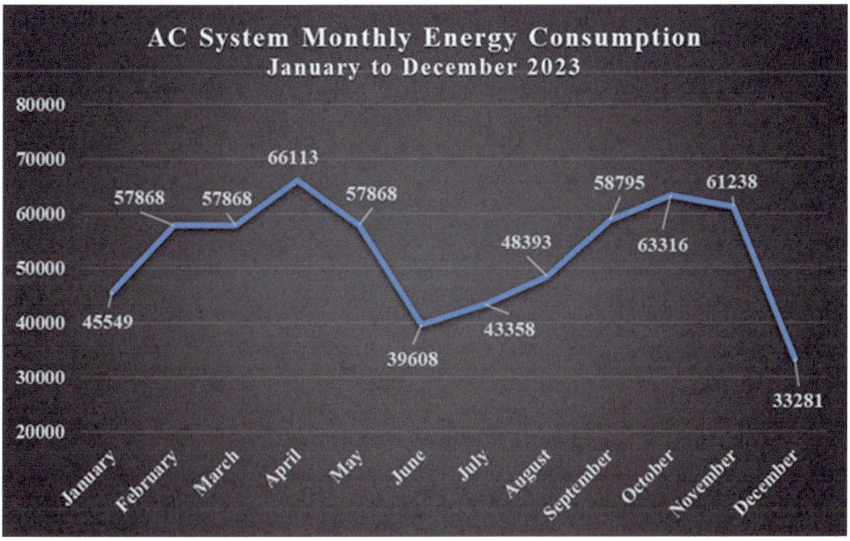

Fig. 4 AC system monthly energy use trend

Table 3 Monthly energy use of AC system

Month	Monthly consumption (kWh)
January	61,200
February	81,600
March	89,400
April	82,200
May	100,800
June	57,000
July	54,600
August	84,600
September	116,400
October	82,200
November	99,000
December	42,600
Total annual kWh	**951,600**

Energy intensity: 65 kWh/m^2 (building), 43 kWh/m^2 (AC system)

Energy per student: 101 kWh/student (building), 67 kWh/student (AC system)

Energy per enrolled unit: 10 kWh/unit (building), 6 kWh/unit (AC system)

Energy per degree-day: 279 kWh/DD (building), 186 kWh/DD (AC system).

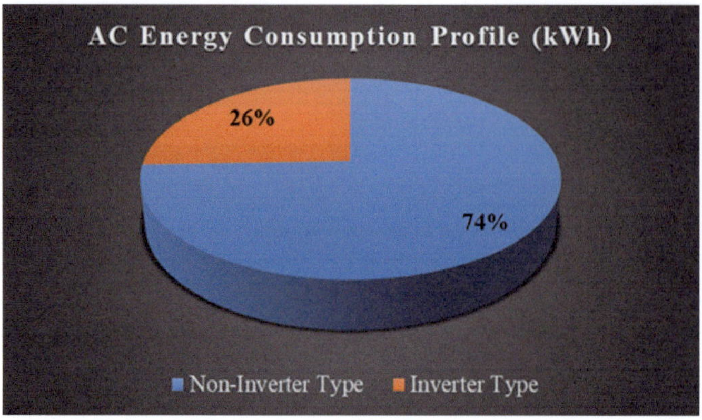

Fig. 5 AC system energy consumption profile

3.5 Opportunities for Energy Performance Improvement

Based on the findings of the energy review and the results of the Level 1 audit, several opportunities for improving energy performance were identified, with emphasis on the air-conditioning (AC) system as the building's dominant Significant Energy Use (SEU). The results show that targeted improvements in the AC system can substantially reduce the Engineering Building's electricity consumption. These opportunities range from no cost and low-cost actions to capital-intensive technological measures, offering a tiered strategy that delivers immediate savings while supporting long-term efficiency goals. Collectively, the findings and audit outcomes provide a practical foundation for implementing an ISO 50001-aligned Energy Management System (EnMS), ensuring both operational cost reductions and sustained environmental benefits.

1. No-Cost and Low-Cost Opportunities

    - Enforce strict shutdown of AC units, lights, and laboratory equipment when not in use.
    - Standardize thermostat set-points between 23 and 25 °C to prevent overcooling.
    - Rotate the use of multiple AC units to balance load and extend equipment lifespan.
    - Promote awareness campaigns and designate energy champions in classrooms and laboratories to encourage responsible usage.
    - Strengthen preventive maintenance practices, including regular cleaning of filters and coils, which improve AC efficiency at minimal cost.

Operational Opportunities

- Optimize AC operation by aligning schedules with actual class and laboratory hours.

- Integrate automated scheduling and controls to reduce unnecessary runtime.
- Develop clear energy policies to guide equipment usage and set priorities for future upgrades.

Technological Opportunities

- Replace inefficient non-inverter AC units with inverter models, reducing cooling energy demand by up to 28%.
- Introduce high-efficiency window-type and split-type units for high-load spaces.
- Implement thermal retrofits such as reflective window films, improved insulation, and sealing of air leaks to reduce heat gain and cooling requirements.
- Install submeters for major energy-consuming areas to improve monitoring and accountability.

4 Conclusion

This section summarizes the key findings of the energy review and provides practical recommendations for improving energy performance of the College of Engineering Building.

4.1 Conclusion

This study conducted a comprehensive energy review of the HEI's Engineering Building, the campus's highest-consuming facility with an annual electricity demand of 951,600 kWh. Results confirmed that the air-conditioning (AC) system accounted for 67% of total use (633,254 kWh), establishing it as the dominant Significant Energy Use (SEU). Consumption patterns closely followed the academic calendar, with peaks during regular semesters and significant reductions during holiday and midyear breaks.

The establishment of an energy baseline and performance indicators (EnPIs), including energy intensity (65 kWh/m^2), per-student consumption (101 kWh/student), and per-degree-day use (279 kWh/DD), provided measurable benchmarks for monitoring efficiency and guiding future improvements.

The Level 1 energy audit reinforced these findings by identifying inefficiencies in equipment configuration. Non-inverter units were found to consume disproportionately more energy than inverter units, while operational practices such as extended runtimes and inconsistent set-point management further increased demand. The audit also revealed opportunities for simple thermal retrofits and better maintenance practices to enhance system performance.

Overall, the integration of consumption analysis and audit findings confirms that targeted interventions in the AC system could deliver significant efficiency gains. These results emphasize the importance of a systematic, ISO 50001-aligned

Energy Management System (EnMS) for higher education institutions, both to reduce operational costs and to advance sustainability commitments.

4.2 Recommendations

To optimize performance and align with Republic Act No. 11285 and ISO 50001:2018, the following actions are recommended:

1. Institutional Measures: Adopt an Energy Management System (EnMS) that institutionalizes policy, targets, and regular reviews of energy performance. Allocate funds for capital investments in efficient technologies and for periodic energy audits.
2. Technological Measures: Prioritize replacement of non-inverter AC units with inverter models, introduce high-efficiency window-type and split-type systems, and implement thermal retrofits such as reflective films and improved sealing.
3. Operational Measures: Enforce stricter runtime controls, integrate automated scheduling, and improve preventive maintenance practices for laboratory and classroom equipment.
4. Behavioral and Awareness Initiatives: Promote energy-conscious practices among faculty, staff, and students, integrate efficiency topics into academic discussions, and designate energy champions within units.
5. Future Directions: Extend the review to other campus buildings, conduct higher-level audits, and explore integration of renewable energy sources such as solar PV to complement demand-side management.

References

1. Lectura, L.: Average Electricity Price in PHL 2nd Highest in Asia–Think Tank | Lenie Lectura. BusinessMirror, 7 Aug. 2018. https://businessmirror.com.ph/2018/08/07/average-electricity-price-in-phl-2nd-highest-in-asia-think-tank/
2. Alves, B. P: Power Production Share by Source 2021 (2022). www.statista.com/statistics. Accessed 20 Oct. 2024
3. International Trade Administration: Philippines Energy Market. www.trade.gov. Accessed 22 Apr. 2020; www.trade.gov/market-intelligence/philippines-energy-market. Accessed 20 June 2024
4. Republic Act No. 11285: Lawphil.net, 12 Apr. 2019, lawphil.net/statutes/repacts/ra2019/ra_11285_2019.html. Accessed 7 Nov. 2023
5. 2019 Power Situation Report. Department of Energy Philippines. Doe.gov.ph (2019). legacy.doe.gov.ph/electric-power/2019-power-situation-report. Accessed 15 May 2024

Enhancement of Classroom Indoor Air Quality in Adamson University Using Room Ventilators with HEPA Filters Installed

Francis Andre Ynota, John Benedict Ilano, Charles Villanueva, Angela May D. Flores, Basil Aloysius P. Bautista, Lester Alfred M. Olasiman, and Lailah C. Bautista

Abstract The abstract should summarize the contents of the paper in short terms, i.e. 150–250 words. The study aimed to improve indoor air quality in academic environments by developing and evaluating a room ventilator prototype equipped with a High-Efficiency Particulate Air (HEPA) filter. The project focused on Room OZ 101B at Adamson University, where poor ventilation was found to contribute to elevated levels of airborne pollutants. Using computational fluid dynamics (CFD) through SolidWorks Flow Simulation 2023, airflow patterns were analyzed to optimize the design. After fabrication, the prototype was tested under real classroom conditions. Air quality was assessed by measuring particulate matter (PM2.5 and PM10) before and after prototype operation. Results showed a significant reduction in airborne particles, indicating improved indoor air quality and more effective air circulation. The study concludes that integrating HEPA-based ventilation systems in classrooms is a practical and cost-efficient solution to reduce pollutant exposure

F. A. Ynota · J. B. Ilano · C. Villanueva · A. M. D. Flores (✉) · B. A. P. Bautista · L. A. M. Olasiman · L. C. Bautista
Adamson University, Manila, Philippines
e-mail: angela.may.flores@adamson.edu.ph

F. A. Ynota
e-mail: francis.andre.ynota@adamson.edu.ph

J. B. Ilano
e-mail: john.benedict.ilano@adamson.edu.ph

C. Villanueva
e-mail: charles.villanueva@adamson.edu.ph

B. A. P. Bautista
e-mail: basil.aloysious.bautista@adamson.edu.ph

L. A. M. Olasiman
e-mail: lester.alfred.olasiman@adamson.edu.ph

L. C. Bautista
e-mail: lailah.bautista@adamson.edu.ph

M. Awang et al. (eds.), *Proceedings of the International Conference on Mechanical, Manufacturing and Plant Engineering*, Proceedings in Technology Transfer,
https://doi.org/10.1007/978-981-95-7817-7_27

and promote healthier learning environments. Further research is recommended to explore broader applications and long-term performance in varied academic settings.

Keywords Indoor air quality (IAQ) · High-efficiency particulate air (HEPA) filter · Ventilator · SolidWorks flow simulation · Computational fluid dynamics (CFD)

1 Introduction

The Maintaining healthy indoor air quality in schools is essential, as it directly affects students' health, comfort, and academic performance. Poor air quality in classrooms has been linked to various health issues, including allergies, asthma, fatigue, and respiratory infections, which can lead to increased absenteeism and reduced productivity. Concerns about indoor pollutants—such as particulate matter, VOCs, and pathogens—have grown over the past decade, especially with the heightened awareness brought by the COVID-19 pandemic.

To address these issues, schools have increasingly considered the use of air purifiers and room ventilators equipped with HEPA filters, activated carbon filters, UV-C lamps, and mechanical ventilation systems. These technologies aim to reduce airborne pollutants and create safer learning environments. However, their effectiveness in real-world classroom settings remains under investigation, as factors like room layout, equipment placement, maintenance, and cost can influence their performance.

Given these variables, there is a strong need for context-specific research that evaluates the practical outcomes of using air-cleaning technologies in educational spaces. Such studies should offer data-driven insights and implementation guidelines to help school administrators and policymakers make informed decisions about indoor air quality improvements. Artificial Ventilation technology plays a critical role in maintaining energy efficiency and ensuring optimal indoor environmental quality, particularly in educational environments where occupants are vulnerable to poor air conditions. Azuma and Ohura [1] conducted a comprehensive review of the diffusion of energy-saving technologies and found that ventilation systems—especially those integrated with energy-efficient features—have gained momentum due to their dual impact on environmental sustainability and occupant well-being. In academic environments, where students and teachers spend extended hours indoors, the quality of indoor air directly affects health, comfort, and cognitive performance. A landmark study by Seppänen et al. [2] established a strong correlation between low ventilation rates and adverse health responses, including headaches, fatigue, and respiratory problems, in commercial and institutional buildings. This was further reinforced by Sundell et al. [3], whose multidisciplinary review concluded that ventilation rates are not merely technical parameters but essential determinants of human health outcomes, reinforcing the demand for properly designed and maintained systems.

The influence of ventilation on academic performance has also been quantitatively investigated. Haverinen-Shaughnessy et al. [4] revealed that students exposed to classrooms with substandard ventilation scored lower on standardized academic assessments, underscoring the need to treat IAQ as an academic resource rather than a background condition. Zhao and Lian [5] contributed to the field by reviewing airflow modeling techniques and emphasized that accurate measurement and simulation of indoor airflow are essential for designing systems that distribute fresh air effectively. Their work laid the foundation for integrating computational tools into indoor environment planning. Zong et al. [6] expanded this discussion through their experimental evaluation of naturally ventilated classrooms, showing that a combination of passive ventilation and mechanical fans significantly improves both air quality and thermal comfort.

Recent innovations in airflow strategies have been shown to enhance air distribution uniformity and pollutant dilution. Galla and Antony [7] demonstrated the effectiveness of vertical airflow ventilation schemes in maintaining homogeneous air quality throughout the classroom. Similarly, Moon et al. [8] compared various ventilation setups and showed that mechanical systems could outperform natural methods in maintaining consistent air quality if appropriately controlled. Park et al. [9] combined experimental setups with CFD simulations to validate the efficiency of diffuser-based ventilation, offering a robust framework for performance-based system design.

Natural ventilation remains a widely studied and applied solution, especially in regions with mild climates. Stabile et al. [10] showed that when natural ventilation strategies are carefully implemented, they can significantly reduce indoor concentrations of CO_2 and other pollutants. Furthermore, decentralized ventilation systems, as studied by Catalina et al. [11], offer a modular and energy-efficient alternative by delivering conditioned air directly to occupied zones without reliance on centralized ductwork. The FRESH intervention study by Rosbach et al. [12] provided longitudinal data confirming that improved ventilation in classrooms leads to lower CO_2 levels and fewer student complaints related to discomfort and drowsiness.

Collectively, these studies reinforce the importance of integrating both mechanical and natural ventilation strategies with advanced monitoring systems in educational environments. Such integration ensures not only compliance with IAQ standards but also supports cognitive function, occupant health, and environmental sustainability—core goals in modern classroom design and air management.

2 Methodology

The successful implementation of this study required a systematic process involving the design, simulation, fabrication, installation, and testing of a HEPA-based room ventilator intended to improve indoor air quality in a classroom setting. The process began with the creation of a three-dimensional model using SolidWorks 2023, which included all key components such as the centrifugal fan, HEPA filter, intake and exhaust vents, and an acrylic casing. This model served as the foundation for visualizing airflow behavior and refining the physical layout of the device.

To optimize airflow distribution and identify areas of stagnant air within the classroom, Computational Fluid Dynamics (CFD) simulations were performed using SolidWorks Flow Simulation. The simulation parameters included inlet velocity, volumetric flow rate, and environmental pressure, replicating realistic classroom conditions. The insights gained from the simulation guided the proper placement of the intake and exhaust openings to ensure maximum circulation of clean air.

Following the design phase, the prototype was fabricated using accessible and cost-efficient materials, including a 4 DC fans, HEPA filter, electrical wiring, and transparent acrylic sheets. The components were assembled according to the dimensions and layout established in the 3D model, with careful attention to airflow alignment and structural stability. The HEPA filter was securely positioned to ensure that all air entering the system passed through the filtration layer before being released into the room.

For performance evaluation, a classroom at Adamson University was selected based on its size, layout, and ventilation characteristics. Before the installation of the device, initial air quality readings were collected using digital sensors capable of measuring particulate matter (PM2.5), (PM10), carbon dioxide (CO_2), total volatile organic compounds (TVOC), and formaldehyde (HCHO). These readings served as baseline data for comparison.

After installation, the system was activated and left to operate under typical classroom conditions. Observations were made to assess airflow behavior and comfort levels. Post-installation air quality readings were then recorded using the same sensors and under similar environmental conditions. These data were analyzed to determine the effectiveness of the HEPA-based ventilation system in improving indoor air quality and reducing harmful pollutants in the classroom environment analysis.

3 Results and Discussion

The results present a thorough summary of the key findings based on the conducted simulations, prototype testing, and indoor air quality (IAQ) measurements inside Room OZ101 at Adamson University. The research was primarily focused on understanding and addressing poor air circulation using a locally designed and fabricated room ventilator equipped with HEPA filters. The work involved using Computational Fluid Dynamics (CFD) via SolidWorks Flow Simulation and hands-on physical air quality testing to validate the performance of the proposed solution.

It started by simulating the natural airflow behavior of the classroom without any mechanical ventilation. The simulation assumed standard classroom conditions with an inlet airflow of 0.08 m^3/s, outlet flow of 0.04 m^3/s, and inlet velocity of 3.175 m/s under standard atmospheric pressure. Results from the simulation showed that airflow was highly uneven across the room. In several areas—particularly corners and between obstacles—air was stagnant. The average velocity measured in these

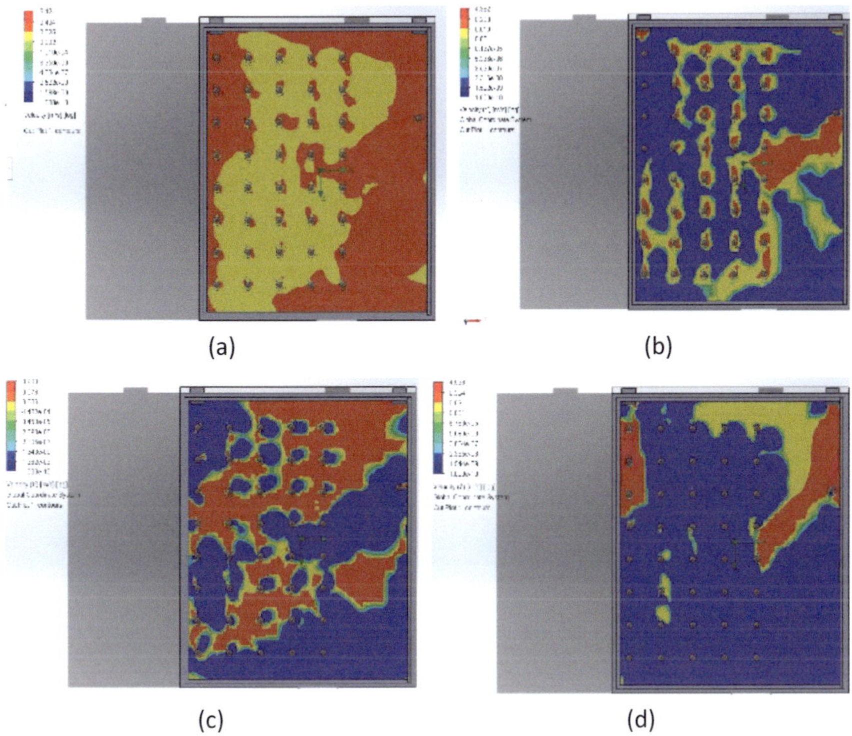

Fig. 1 Cut plot, classroom velocity (**a**), velocity X direction (**b**), velocity Y direction (**c**) velocity Z direction (**d**)

zones was just 0.009 m/s. The data confirmed that natural airflow was not enough to ensure good IAQ under normal classroom use as shown in Fig. 1.

In Fig. 2, the CFD analysis showed that areas with improved airflow in the simulation directly matched the regions were physical sensors recorded pollutant reduction. This strong correlation validated the reliability of using simulation to predict actual performance.

The 3D model presented illustrates in Fig. 3, a conceptual design of a ventilator system. The structure consists of a transparent rectangular housing with internal compartments that guide airflow.

In Table 1, the designed prototype consisting of four axial fans, each rated at 0.189 m^3/s, resulting in a total simulated flow capacity of 0.756 m^3/s. This setup was tested in SolidWorks, and the prototype successfully achieved a bulk outlet velocity of 17.039 m/s with a mass flow rate of –0.2273 kg/s, indicating strong and consistent airflow suitable for pollutant removal.

From the results shown in Fig. 4, it can be observed that the airflow velocity is highest near the fan outlet, shown in the red region, and gradually decreases as it

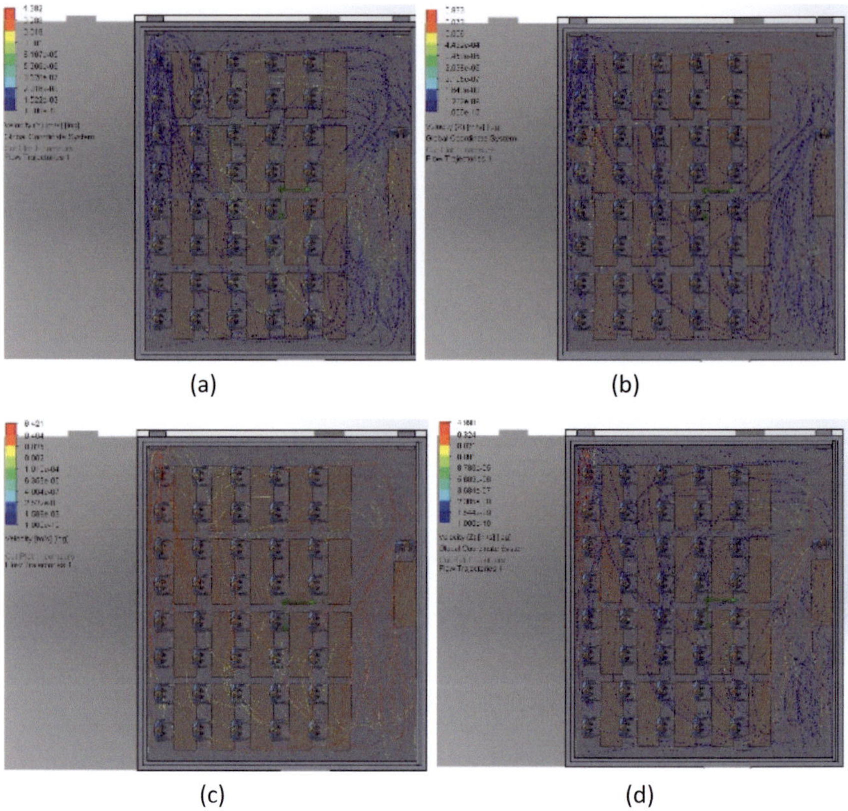

Fig. 2 Flow trajectory, classroom velocity (**a**), X direction (**b**), Y direction (**c**), Z direction (**d**)

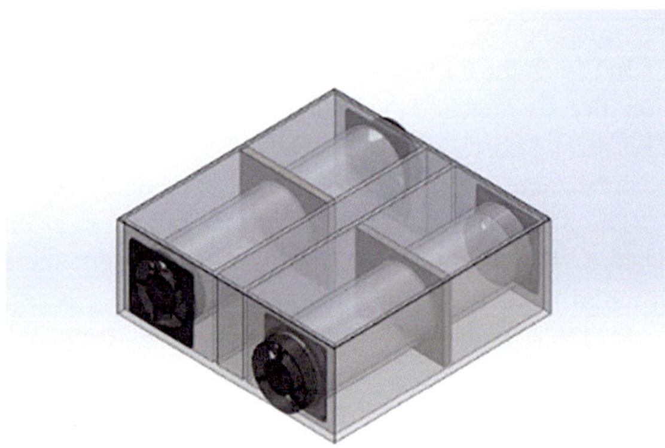

Fig. 3 3D model of ventilator using SolidWorks

Table 1 Surface parameters at fan outlet

Parameter	Value	Unit
Simulated volume flow rate	0.189	m^3/s
Expected system flow rate	0.756	m^3/s
Bulk average velocity	17.039	m/s
Mass flow rate	−0.2273	kg/s

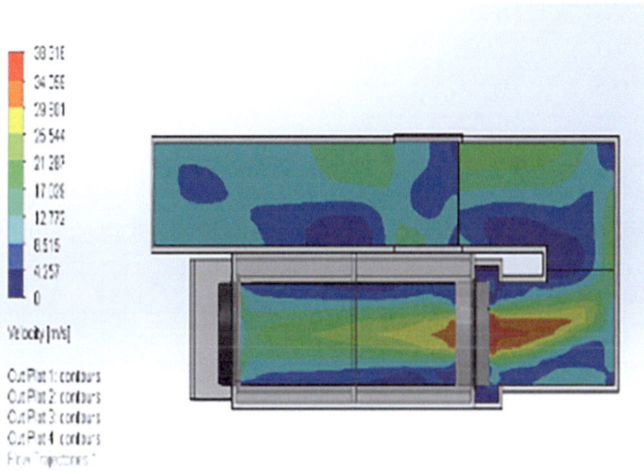

Fig. 4 Flow simulation of 3D modeled ventilator

moves through the ventilator channels. The blue and green regions indicate lower velocities, showing how the air spreads evenly across the compartments.

In Table 2, pollutants such as formaldehyde (HCHO), total volatile organic compounds (TVOC), and carbon monoxide (CO) showed sudden spikes between 5 and 7 PM, with CO peaking at 282 ppm, which could be caused by cooking, cleaning, or other activities. CO_2 levels started high (1362 ppm) but decreased steadily to 735 ppm, suggesting better ventilation in the evening.

Overall pollutant values were lower and more stable shown in Table 3, with only slight increases around 6–7 PM. CO_2 levels ranged between 580 and 713 ppm, which indicates good ventilation when the ventilator has been installed.

The data shown in Fig. 5 also revealed how pollutant levels fluctuated depending on classroom usage. From 3:00 PM to 7:00 PM, pollution levels increased in the absence of the prototype due to human occupancy and closed windows.

With the prototype active, these pollutant spikes were significantly flattened shown in Fig. 6, proving that the system worked dynamically to prevent harmful buildup.

In Table 4, explains the formaldehyde dropped from a peak of 0.923–0.012 mg/m^3, which is a 98.7% reduction. TVOC and CO also decreased by 93.0% and 92.2%, respectively. CO_2 levels, which directly relate to classroom occupancy, dropped from an average of 1048.5–646.5 ppm, proving better ventilation and air exchange.

Table 2 Data of testing a room without prototype

Time	HCHO (mg/m^3)	TVOC (mg/m^3)	PM2.5 (μg/m^3)	PM10 (μg/m^3)	CO (ppm)	CO_2 (ppm)
1–2 PM	0.005	0.011	22	28	2	1362
2–3 PM	0.005	0.014	24	29	3	1204
3–4 PM	0.006	0.012	20	26	2	1054
4–5 PM	0.004	0.010	23	30	1	1030
5–6 PM	0.138	0.905	22	28	37	924
6–7 PM	0.923	1.999	22	28	282	908
7–8 PM	0.005	0.014	23	30	2	740
8–9 PM	0.007	0.012	29	38	2	735

Table 3 Data of testing a room with prototype

Time	HCHO (mg/m^3)	TVOC (mg/m^3)	PM2.5 (μg/m^3)	PM10 (μg/m^3)	CO (ppm)	CO_2 (ppm)
1–2 PM	0.004	0.009	18	22	1	713
2–3 PM	0.004	0.010	19	23	1	624
3–4 PM	0.004	0.009	17	21	1	660
4–5 PM	0.003	0.008	18	22	1	638
5–6 PM	0.008	0.075	20	24	10	629
6–7 PM	0.012	0.140	21	25	22	605
7–8 PM	0.004	0.010	19	23	1	592
8–9 PM	0.005	0.009	20	25	1	580

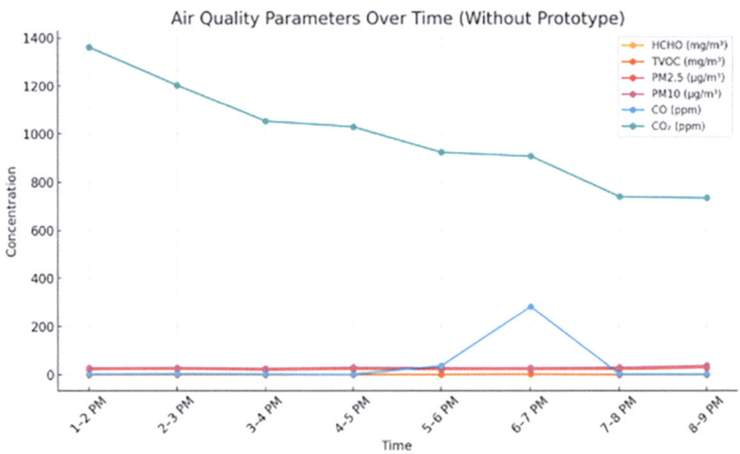

Fig. 5 Air quality parameters (without prototype)

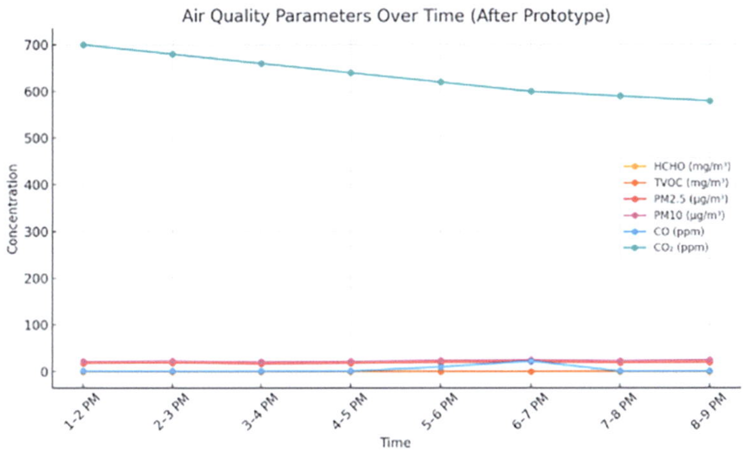

Fig. 6 Air quality parameters (with prototype)

Table 4 Comparison with indoor air quality standards

Pollutant	Before (peak)	After (peak)	% Reduction (%)
HCHO (mg/m^3)	0.923	0.012	98.7
TVOC (mg/m^3)	1.999	0.14	93.0
PM2.5 (μg/m^3)	23	19	17.4
PM10 (μg/m^3)	29	23	20.7
CO (ppm)	282	22	92.2
CO_2 (ppm)	1048.5 (avg)	646.5 (avg)	38.3

In Table 5, when compared to global indoor air quality standards such as those from the WHO and EPA, all post-installation readings fell well below the maximum allowable limits, confirming that our system met international benchmarks.

The results also aligned with global studies from WHO, ASHRAE, and other researchers, reinforcing the broader relevance of the findings. The study contributes

Table 5 Comparison with indoor air quality standards

Pollutant	After (measured)	Standard limit	Compliant?
HCHO (mg/m^3)	0.012	0.1	Yes
TVOC (mg/m^3)	0.14	≤0.3	Yes
PM2.5 (μg/m^3)	19	25	Yes
PM10 (μg/m^3)	23	50	Yes
CO (ppm)	22	35	Yes
CO_2 (ppm)	580–713	≤1000	Yes

practical insight into how a low-cost, locally built ventilation system can significantly improve classroom IAQ, especially in institutions that lack centralized HVAC systems.

4 Conclusion

This study thoroughly evaluated the effectiveness of a ventilation prototype in improving the indoor air quality of a fully occupied mechanical engineering classroom at Adamson University, specifically Room OZ101. By integrating computational fluid dynamics (CFD) simulations, physical prototype testing, and actual air quality monitoring, the research provided a holistic understanding of airflow behavior and pollutant distribution within a classroom setting. The CFD simulations revealed airflow patterns and areas with poor ventilation and stagnation, highlighting the necessity of improved air circulation in such environments. The fan system demonstrated the ability to produce efficient airflow with a bulk average velocity of 17.039 m/s, although localized flow rates were found to be lower than the total expected output.

Physical implementation of the prototype led to notable improvements in key indoor air pollutants. Concentrations of formaldehyde (HCHO), total volatile organic compounds (TVOC), particulate matter (PM2.5 and PM10), carbon monoxide (CO), and carbon dioxide (CO_2) were significantly reduced after deployment. CO and HCHO levels dropped by over 90%, confirming the system's potential to mitigate harmful contaminants in active classroom conditions. Improvements in particulate matter were also evident, albeit to a lesser degree. The comparative analysis between pre- and post-installation conditions further validated the effectiveness of the ventilation system, especially during peak classroom activity hours when pollutant buildup is typically higher.

Despite the positive findings, the study acknowledged several limitations, including the focus on a single classroom environment, limited data collection duration, the lack of temperature and humidity measurements, and inconsistencies in CO sensor readings. Nonetheless, the research concluded that the combined use of simulation tools and real-world testing effectively demonstrated the prototype's practicality and efficiency. Overall, the study supports the use of targeted ventilation systems as a viable solution for enhancing indoor air quality in educational settings, ultimately contributing to a healthier and more conducive learning environment.

References

1. Azuma, D., Ohura, T.: Diffusion of energy-saving technologies: a review. Energy Rep. **5**, 416–429 (2019)

2. Seppänen, O., Fisk, W.J., Mendell, M.J.: Association of ventilation rates and CO_2 concentrations with health and other responses in commercial and institutional buildings. Indoor Air **9**(4), 226–252 (1999)
3. Sundell, J., Levin, H., Nazaroff, W.W.: Ventilation rates and health: multidisciplinary review of the scientific literature. Indoor Air **21**(3), 191–204 (2011)
4. Haverinen-Shaughnessy, U., Moschandreas, D.J., Shaughnessy, R.J.: Association between substandard classroom ventilation rates and students' academic achievement. Indoor Air **21**(2), 121–131 (2011)
5. Zhao, B., Lian, Z.: Review on measurement and simulation of ventilation and airflow in buildings. Indoor Built Environ. **23**(4), 504–515 (2014)
6. Zong, J. et al.: Experimental study on indoor environment quality in a naturally ventilated classroom of a university using natural ventilation and ventilation fan. Build. Environ. **165** (2020)
7. Galla, B., Antony, J.: Enhancement and homogenization of indoor air quality in a classroom using a vertical airflow ventilation scheme. Build. Environ. (2022)
8. Moon, H.J., Sohn, J.R., Hwang, T.: Indoor air quality performance of ventilation systems in classrooms. Build. Environ. (2016)
9. Park, K., Kwon, W.-D., Kwon, H.-M., Yang, J.-H.: Examination of the ventilation function of a combined air-diffuser ventilation system using experiments and CFD. Sustainability **16**(3) (2017)
10. Stabile, L. et al.: The effect of natural ventilation strategy on indoor air quality in schools. Build. Environ. **595** (2017)
11. Catalina, T., Lungu, C.: Influence of a decentralized ventilation system on the indoor air quality of a primary school classroom. IOP Conf. Ser.: Mater. Sci. Eng. **664** (2021)
12. Rosbach, J. et al.: Classroom ventilation and indoor air quality—results from the FRESH intervention study. Indoor Air **26**(4) (2016)

Ultrasound-Assisted Preparation of Chitosan-Based Microsphere for Oil and Water Separation as Sustainable Oil Spill Adsorbent

Ezekiel Alilano, Michelle Rienagrace Basbas, Alyssa Marie Caacbay, Anne Kristie Pearl Manansala, Cherrylen Santos, Cyelle Andrei Tuaña, Rugi Vicente Rubi, and Carlou Eguico

Abstract The rapid growth of industry and economic development for the last few decades has significantly contributed to increasing water pollution, primarily due to factors such as oil spills and leaks, leading to substantial environmental and ecological damage. This study explored the synthesis of ultrasound-assisted preparation of chitosan-based microspheres under varying ultrasonic power (90, 180, 270 W) for marine diesel oil–seawater separation. Scanning Electron Microscope (SEM), Fourier Transform Infrared Spectroscopy (FTIR), X-ray Diffraction (XRD), and

E. Alilano · M. R. Basbas · A. M. Caacbay · A. K. P. Manansala · C. Santos · C. A. Tuaña · R. V. Rubi · C. Eguico (✉)
Chemical Engineering Department, College of Engineering, Adamson University, Manila, Philippines
e-mail: carlou.eguico@adamson.edu.ph

E. Alilano
e-mail: ezekiel.alilano@adamson.edu.ph

M. R. Basbas
e-mail: michelle.rienagrace.basbas@adamson.edu.ph

A. M. Caacbay
e-mail: alyssa.marie.caacbay@adamson.edu.ph

A. K. P. Manansala
e-mail: anne.kristie.pearl.manansala@adamson.edu.ph

C. Santos
e-mail: cherrylen.santos@adamson.edu.ph

C. A. Tuaña
e-mail: cyelle.andrei.tuana@adamson.edu.ph

R. V. Rubi
e-mail: rugi.vicente.rubi@adamson.edu.ph

R. V. Rubi · C. Eguico
Adamson University Laboratory of Biomass, Energy and Nanotechnology (ALBEN), Chemical Engineering Department, College of Engineering, Adamson University, Manila, Philippines

M. Awang et al. (eds.), *Proceedings of the International Conference on Mechanical, Manufacturing and Plant Engineering*, Proceedings in Technology Transfer,
https://doi.org/10.1007/978-981-95-7817-7_28

X-ray Fluorescence (XRF) were used to characterize the microspheres. Batch experiments were conducted to investigate the effect of oil–seawater concentration and contact time on the adsorption performance of the microspheres. Models were used to describe in detail the adsorption kinetics (Pseudo-first order and Pseudo-second order) and isotherms (Langmuir and Freundlich). Maximum removal efficiencies were observed at 99.49% for lower oil–seawater concentration (10 g/L) and 90.54% at longer contact time (4 h). The adsorption process also followed a pseudo-first order kinetic model ($R^2 = 0.991$) and fitted the Langmuir isotherm ($R^2 = 0.9957$), indicating a monolayer adsorption. The results obtained in this study will offer valuable insight into the adsorption mechanism and support future applications in the design of treatment systems for real-world oily wastewater.

Keywords Ultrasound-assisted · Chitosan-based microspheres · Oil–seawater separation · Adsorption kinetics · Langmuir isotherm · Pseudo-first order

1 Introduction

The rapid growth of industry and economy over the last few decades has caused a significant increase in water pollution, resulting from several causes such as oil spillage and leakage, thereby imposing severe environmental and ecological damage. These pollutants may be released through petroleum products and byproducts, oil transportation, oil extraction, and oily wastewater discharge from a variety of sources posing serious threats to water resources and quality of life, marine environments, and their wildlife.

To resolve this problem, various oil spill response techniques are being used like bioremediation or biological methods (e.g., microorganisms), thermal methods (e.g., in-situ burning), mechanical methods (e.g., booms), chemical treatments (e.g., dispersants and emulsifiers), and sorbents [1]. Other techniques include solidification, skimming, and absorption [2]. However, these techniques have disadvantages. Mechanical methods for oil spill clean-up are expensive and labor-intensive. Chemical dispersants, while effective, may be more harmful to marine life than the oil itself, as they persist in the environment and threaten vulnerable species like diving birds and shellfish. In-situ burning causes secondary pollution, and oil skimmers are limited to small areas around deployment vessels. Overall, these methods suffer from high costs, limited efficiency, recontamination risks, and corrosion problems.

In the past few years, natural oil sorbents have gained attention for being non-toxic, low-cost, and abundant. These hydrophobic and oleophilic materials function as absorbents or adsorbents, depending on their interaction with oil. Among them, macroporous microspheres particularly those derived from chitosan are promising for oil–water separation due to their high adsorption capacity, reusability, large surface area, and porous structure [3]. Recent studies have shown that ultrasound treatment can further enhance the structural and functional properties of chitosan-based microspheres, improving their efficiency in adsorption applications [4].

Considering these advantages, this study focuses on the application of ultrasonication to modify chitosan-based microspheres, with the goal of enhancing their structural and functional properties for more efficient oil–water separation.

2 Methodology

2.1 Materials and Methods

Chitosan powder (80% degree of deacetylation) was sourced from a supplier based in Jiangsu, China. Reagents including 2% acetic acid, sodium tripolyphosphate (STP), and distilled water were used in the preparation of ultrasound-assisted chitosan-based microspheres via ionic gelation. The marine diesel oil used as the target pollutant was provided by Valvoline and Frabelle Fishing Corporation. Natural seawater used as the simulated marine environment for the experiments was collected from Cavite City, Philippines.

2.1.1 Synthesis of Ultrasound-Assisted Chitosan-Based Microspheres

The synthesis of ultrasound-assisted chitosan-based microspheres was carried out through ionic gelation followed by sonication to enhance microsphere formation. A chitosan solution was prepared by dissolving 1 g of 80% DD chitosan in 100 mL of 2% (v/v) acetic acid under magnetic stirring until complete dissolution. Separately, a 5% (w/v) sodium tripolyphosphate (STP) solution was prepared by dissolving 5 g of STP in 100 mL of distilled water. The chitosan solution was then added in a dropwise method into the STP solution [5].

2.1.2 Ultrasound Treatment

The resulting suspension was subjected to ultrasonication using an ultrasonic bath at varying power levels (90, 180, and 270 W) for 2 h [5] to enhance crosslinking efficiency and improve the morphology and surface characteristics of the microspheres. Following ultrasonication, some of the collected chitosan microspheres (CSM) were dried using a drying oven to eliminate residual moisture, then ground into a fine powder using a mortar and pestle to ensure uniformity for analysis.

2.2 Material Characterization

Fourier-transform infrared (FTIR) spectroscopy was performed to identify the characteristic functional groups on the microsphere surface, providing insight into chemical bonds and molecular interactions. X-ray fluorescence (XRF) analysis was used to detect and quantify metallic elements within the microsphere, essential for understanding its composition. Scanning electron microscopy (SEM) and X-ray diffraction (XRD) analyses were carried out to further characterize the microspheres. SEM examined surface morphology, texture, and porosity, while XRD determined crystallographic structure and phase composition, offering a comprehensive evaluation of the microsphere's structural and material properties.

2.3 Batch Adsorption Experiment Using Experimental Design

To determine the optimal conditions for oil removal and to identify the ideal ultrasound-assisted chitosan-based microsphere batch adsorption experiment was conducted. A fractional factorial design was applied using Minitab Statistical Software, resulting in 36 experimental runs with varying combinations of chosen parameters: contact time, oil–seawater concentration, and adsorbent dosage. Each parameter was tested at two levels: adsorbent dosage at 0.1 and 0.6 g, oil–seawater concentration at 10 and 50 g/L, and contact time at 1 and 3 h. The percentage oil removal was calculated for each trial to evaluate adsorption performance. The optimal condition and best-performing chitosan variant were identified by comparing the adsorption efficiencies across all treatments. The % removal was calculated based on the following equation [6]:

$$\% \text{ oil removal} = \frac{C_o - C_f}{C_o} \times 100 \tag{1}$$

where C_o is the initial oil–seawater concentration and C_f is the concentration after the adsorption.

2.4 Adsorption Isotherm and Kinetic Studies

The adsorption isotherm and kinetic studies were carried out using simulated oil-contaminated seawater prepared by introducing marine diesel oil into seawater collected from Cavite City, the method is presented in Fig. 1. To investigate the effect of oil–seawater concentration, marine diesel oil was added at varying concentrations of 10, 20, 30, 40, and 50 g/L. A 0.6 g of chitosan microspheres were then

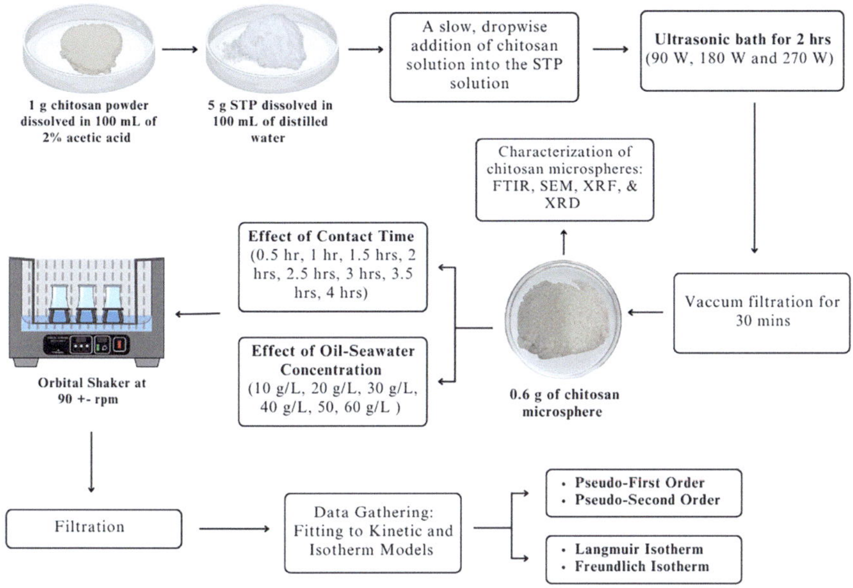

Fig. 1 Schematic diagram of synthesis of chitosan-based microspheres and batch adsorption experiment

introduced into each oil–seawater solution then agitated using an orbital shaker at a constant agitation rate of 90± rpm for a fixed contact time of 3 h.

To study the effect of contact time, a separate set of experiments was conducted by maintaining a fixed oil–seawater concentration of 50 g/L while varying the contact time between 0.5 and 4 h under the same adsorbent dosage and agitation conditions. The resulting data were used to evaluate the adsorption capacity and percentage oil removal. Kinetic behavior was analyzed by fitting the data to the plot of pseudo-first order and pseudo-second order models (Eqs. 2 and 3) [7]. Adsorption equilibrium data were further evaluated using Langmuir and Freundlich isotherm models (Eqs. 4 and 6) to describe the interaction between the oil molecules and the microsphere surface [8].

Pseudo-First Order:

$$\ln(q_e - q_t) = \ln\left(q_e - k_1 t\right) \tag{2}$$

Pseudo-Second Order:

$$\frac{t}{q_t} = \frac{1}{k_2 q_e^2} + \frac{t}{q_e} \tag{3}$$

Langmuir:

$$\frac{C_e}{q_e} = \frac{1}{K_L q_m} + \frac{C_e}{q_m} \tag{4}$$

where:

$$q_e = \frac{(C_o - C_e)V}{M} \tag{5}$$

Freundlich:

$$\ln q_e = \ln b + \frac{1}{n} \ln C_e \tag{6}$$

where:

$$q_e = K_F \times C_e^{1/n} \tag{7}$$

2.5 *Water Capacity Experiment*

The water capacity experiment was conducted by subjecting 0.1 g of chitosan microspheres in a 50 mL of distilled water for 1 h. After soaking, the wet chitosan was then filtered using vacuum filtration for 30 min to allow only free water and not bound to be removed from the microspheres [5]. The water capacity is then computed using the formula:

$$\%\ \text{Water Capacity} = \frac{(M_t - M_o)}{M_o} \times 100 \tag{8}$$

where, M_t is the weight of the wet sample and M_o is the weight of the dry sample.

3 Results and Discussion

3.1 *Characterization*

3.1.1 Scanning Electron Microscope (SEM)

Figure 2 displays the SEM images used to analyze the morphology and microstructure of the ultrasound-assisted chitosan-based microspheres subjected to different power levels, this images will give an idea on how energy input alters microstructural properties. ImageJ software was also used to assess the size distribution of the chitosan microspheres.

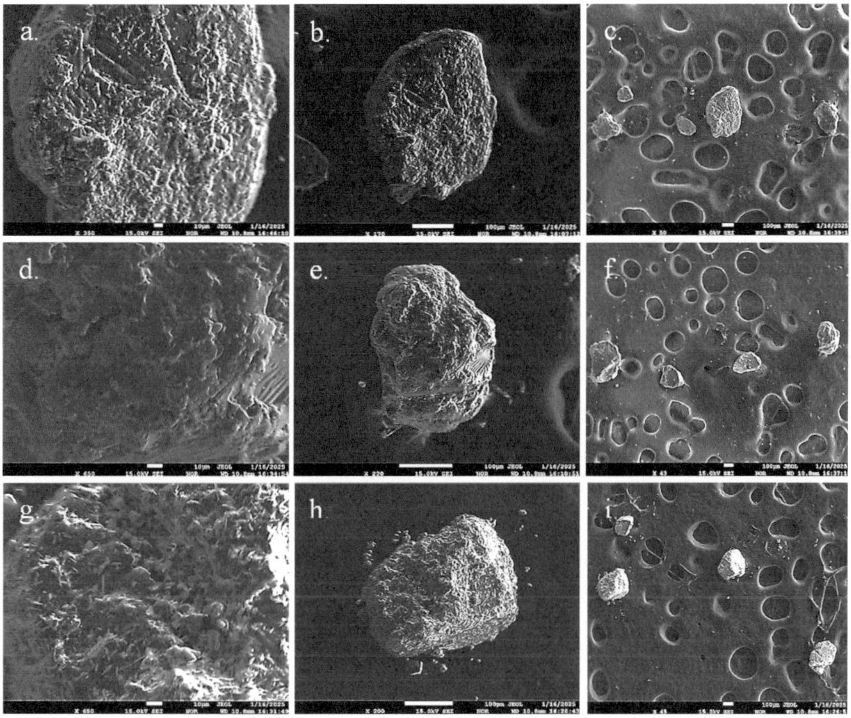

Fig. 2 SEM results of ultrasound-assisted chitosan-based microspheres at varying ultrasonic power: at 90 W (**a**–**c**), at 180 W (**d**–**f**), at 270 W (**g**–**i**)

At 90 W, the sample has a relatively smooth and compact surface with few surface irregularities with a mean particle diameter of approximately 15.8 ± 2.4 μm. At 180 W sample showed improved uniformity and more clearly defined spherical shape and moderate porosity with a reduced mean diameter of 12.6 ± 1.9 μm. Moreover, the promising results were observed in the sample subjected at 270 W where the average diameter further decreased to 10.3 ± 1.5 μm. The particles exhibited well-defined surface pores and significantly higher porosity, representing the most pronounced surface modification among all treatments. These formations of well-defined pores at higher power can be explained by the formation and collapse of chitosan bubbles during the sonication process where the cavitation improves the overall microstructural features of the microspheres, where higher ultrasound energy can facilitate intense cavitation [9].

In addition, the increase in surface porosity and reduction in particle size at higher ultrasound power are likely to enhance the surface area enhancing adsorption efficiency [10]. Furthermore, the particle sizes obtained under varying ultrasound power levels remained within the expected microscale range of 1–1000 μm [9].

3.1.2 Fourier Transform Infrared (FTIR)

FTIR spectra confirmed the structural features and crosslinking of chitosan microspheres with sodium tripolyphosphate (STP) under different ultrasound powers. Typical chitosan bands were observed at 3357 cm^{-1} (O–H and N–H stretching), 2873 cm^{-1} (C–H stretching), 1637 cm^{-1} (amide I), 1535 cm^{-1} (amide II), and 886 cm^{-1} (saccharide) presented in Table 1.

The fingerprint region between 870 and 1190 cm^{-1} confirmed the presence of the polysaccharide backbone in chitosan microspheres. Upon STP crosslinking, new absorption bands emerged at 1064 and 1027 cm^{-1}, corresponding to P–O and P–O–P vibrations. These phosphate-related peaks became progressively more intense with increasing ultrasonic power, indicating stronger electrostatic interactions between chitosan amino groups and STP phosphate groups. Such changes were absent in unmodified chitosan, highlighting the effect of both crosslinking and sonication.

Samples crosslinked under 90, 180, and 270 W ultrasonic power (Samples B–D) showed increasingly sharp and pronounced phosphate peaks. This trend reflects enhanced ionic crosslinking efficiency, as higher ultrasonic energy promotes deeper penetration of STP into the chitosan matrix. With the ultrasonication disrupts intermolecular hydrogen bonding and increases chain mobility, facilitating more uniform and efficient crosslinking [11].

The intensified P–O and P–O–P bands also suggest the formation of a denser and better-organized network within the microspheres. Such structural improvements

Table 1 FTIR spectral interpretation

Significant wavenumber (cm^{-1})	Interpretations	CS	CS90	CS180	CS270
3700–3200	–OH stretching	Broad, strong	Broad, slightly weaker	Broad, medium	Broad, medium
3000–2850	C–H stretching	Broad, medium	Medium, slightly decreased	Medium, slightly decreased	Medium, slightly decreased
1637	Amide I (C=O stretching)	Broad, medium	Medium, slightly shifted	Medium, slightly shifted	Medium, slightly shifted
1535	Amide II (N–H bending)	Broad, Medium	Slightly weak	Weak	Weak
1064	P–O asymmetric stretching	Absent	Narrow, medium	Narrow, strong	Narrow, stronger
1027	P–O–P asymmetric stretching (polyphosphate)	Absent	Weak	Narrow, medium	Sharp, strong
886	Saccharide ring/ P–O bending	Broad, medium	Slightly sharper	Narrow, medium	Narrow, medium

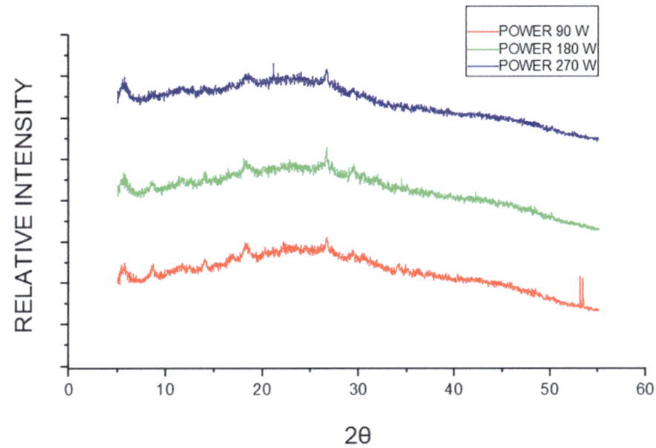

Fig. 3 XRD results of ultrasound-assisted chitosan-based microspheres

are expected to influence key properties, including microsphere stability, swelling behavior, and drug encapsulation efficiency. Overall, these findings demonstrate that ultrasonic power is a critical factor in enhancing crosslinking interactions and optimizing the functional performance of chitosan–STP microspheres.

3.1.3 X-ray Diffraction (XRD)

To ascertain the crystalline properties of three chitosan microsphere samples (1-P90, 2-P180, and 3-P270) and connect them to their possible adsorption performance, X-ray diffraction was used. Based on their intensity and sharpness, which indicate the level of crystallinity in the samples, the peaks in the diffractograms, which were captured over a 2θ range of 5–55°, were analyzed presented in Fig. 3.

XRD analysis revealed that Sample 1-P90 exhibited the most defined peaks, indicating a more ordered crystalline structure, while Sample 2-P180 showed semi-crystalline features. In contrast, Sample 3-P270 displayed broad, low-intensity peaks, confirming a largely amorphous structure. The sonication reduces chitosan crystallinity through peak softening and broadening. The trend clearly shows that crystallinity decreases as ultrasonic power increases [12]. The amorphous nature of Sample 3-P270 suggests enhanced adsorption capacity due to greater surface area and active sites. The reduced crystallinity improves dye adsorption efficiency. Overall, lower crystallinity enhances the ability of chitosan microspheres to adsorb both hydrophilic and hydrophobic contaminants, making P270 the most effective sample for adsorption applications.

3.1.4 X-ray Fluorescence (XRF)

XRF analysis (Fig. 4a–d) revealed significant elemental variations between pure 80% DD chitosan and STP-crosslinked microspheres. Major elements detected included Al, Si, P, Ca, and Fe, while others were present in negligible amounts.

Phosphorus content increased substantially with STP introduction, particularly at 180 W, confirming successful crosslinking, while calcium showed a marked increase at 270 W.

The rise in phosphorus indicating that STP enhances crosslinking and influences chitosan's swelling, zeta potential, and biocompatibility. Traces of calcium likely originated from residual minerals in chitosan, and phosphate groups may interact with Ca^{2+} to form calcium phosphate complexes. Meanwhile, Al, Si, and Fe initially present in pure chitosan diminished after ultrasonication, possibly due to removal of impurities or degradation of micropollutants. These observations suggest that ultrasonication not only promotes crosslinking but also improves material purity.

Batch adsorption experiments showed that oil removal efficiency increased with ultrasonic power, reaching 88.86% at 90 W, 94% at 180 W, and 97.14% at 270 W. Optimal adsorption occurred at 0.6 g adsorbent dosage, 0.5 g/L oil–seawater concentration, and 1 h contact time. The results indicate that 270 W provides sufficient energy for effective microsphere formation without causing structural damage, while

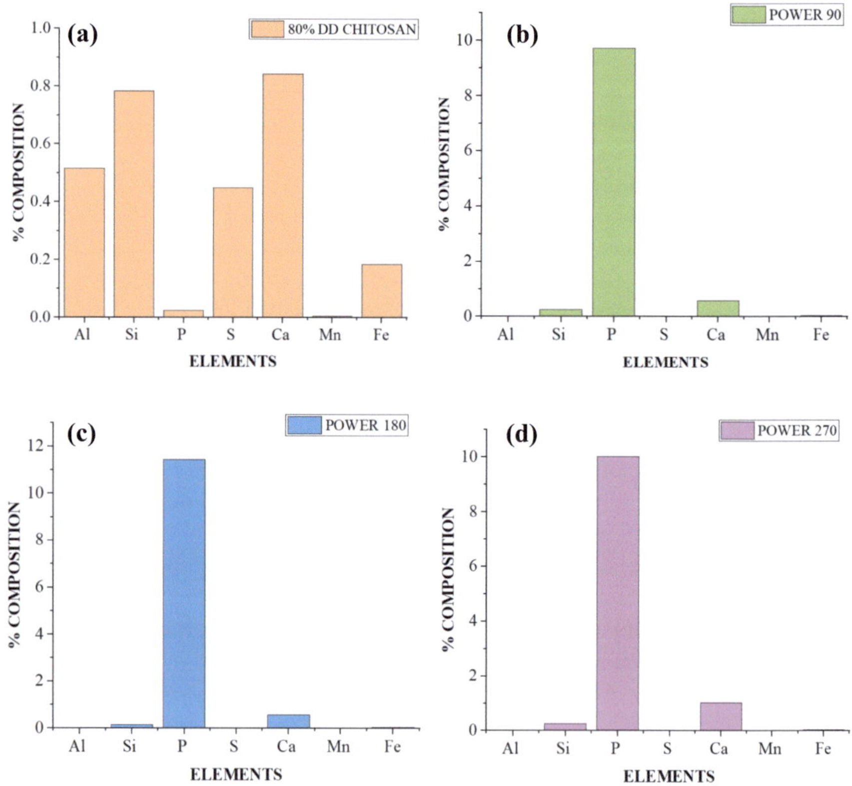

Fig. 4 XRF results of the ultrasound-assisted chitosan-based microspheres at varying power: **a** pure 80% DD chitosan, **b** power 90 W, **c** power 180 W, **d** power 270 W

excessive cavitation beyond this threshold could negatively affect chitosan structure [13].

SEM analysis supported these trends, revealing smoother, compact microspheres at 90 W, moderate porosity at 180 W, and highly porous, rough surfaces at 270 W. Increased porosity and surface roughness at higher power levels enhance dispersibility and surface area for adsorption. XRD analysis further showed a decrease in crystallinity with increasing ultrasonic power, from highly crystalline at 90 W to predominantly amorphous at 270 W. This reduction in crystallinity increases surface activity and available binding sites, supporting improved adsorption performance.

3.2 *Batch Adsorption Experiment*

3.2.1 Effect of Contact Time and Effect of Oil–Seawater Concentration

Figure 5 shows the effect of contact time on the adsorption capacity of chitosan microspheres for marine diesel oil. The results showed a distinct trend, with the amount of oil removal increasing with extended exposure until it approaches a steady state. From 0.5 to 1.5 h the oil removal efficiency increases significantly from 26.40 to 65.99%. This rapid rise during the initial stage is likely due to the accessibility of active adsorption sites available on the chitosan microspheres' surface. The rate of oil removal increases more gradually between 1.5 and 2.5 h, peaking at 83.68% during this interval. This trend suggests a reduction in the number of available active sites, consistent with the progressive filling of adsorption sites.

Adsorption experiments showed that removal efficiency stabilized between 90.33 and 90.54% after 3–4 h, indicating saturation of active sites on the chitosan microspheres. This plateau reflects equilibrium in the adsorption process, where extending

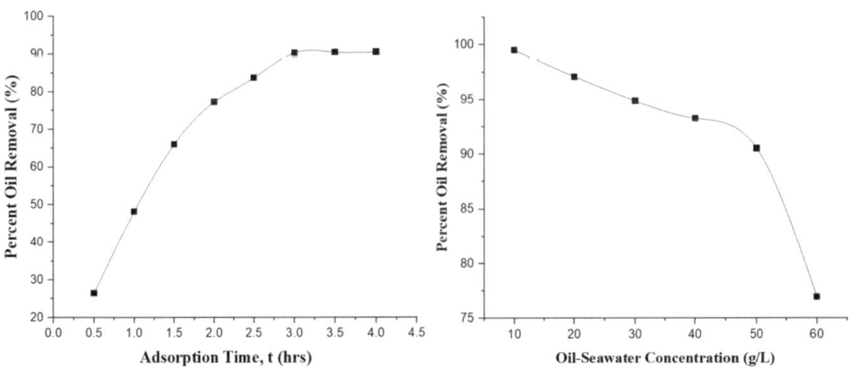

Fig. 5 Effect of adsorption time and oil–seawater concentration on the removal efficiency of ultrasound-assisted chitosan-based microspheres for marine diesel oil removal from seawater

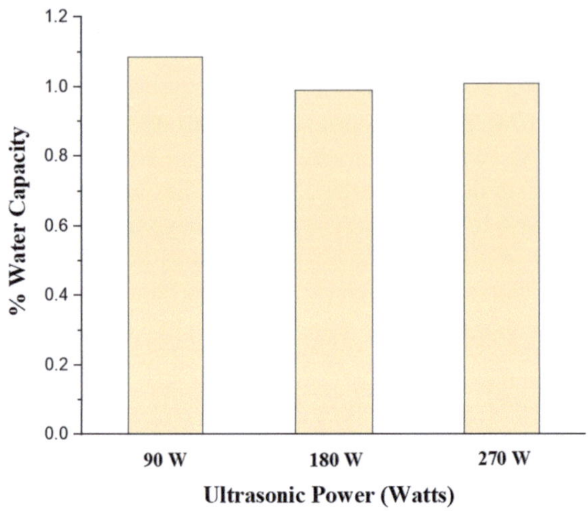

Fig. 6 Water capacity at varying ultrasonic power: 90, 180 and 270 W

contact time does not significantly improve removal efficiency. The trends that is observed in oil–saline water systems with chitosan, confirming that adsorption time is a critical factor for optimizing performance. In this study, the optimal adsorption time was approximately three hours, beyond which only marginal gains were achieved.

The effect of oil–seawater concentration on adsorption is shown in Fig. 6. High removal efficiencies were maintained at 10–50 g/L, with near-complete adsorption (~100%) at 10 g/L. Efficiency declined at concentrations above 50 g/L, reaching its lowest at 60 g/L. This decrease is attributed to active site saturation and microsphere aggregation, which limit interactions between oil droplets and the adsorbent. The insufficient active sites and particle overlap reduce adsorption efficiency at higher oil loads [14].

These results emphasize the importance of optimizing the oil-to-adsorbent ratio and contact time to maintain high removal performance in practical applications. Maintaining appropriate concentrations and exposure times ensures maximum utilization of microsphere surface area and active sites, enhancing overall adsorption efficiency.

3.2.2 Water Capacity

The results of the water sorption experiment for the ultrasound-assisted chitosan-based microspheres are summarized in Fig. 7. This was conducted to assess the hydrophilic behavior of the chitosan-based microspheres that were subjected to varying ultrasonic power levels (90, 180, 270 W) providing an insight into how ultrasonication influences the microsphere's ability to absorb water which is a factor in oil–water separation performance.

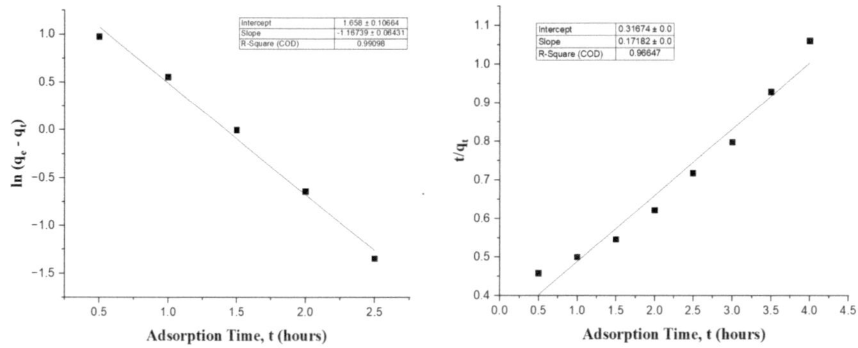

Fig. 7 **a** Pseudo-first order kinetic model and **b** pseudo-second order kinetic model plot of adsorption of marine diesel oil onto ultrasound-assisted chitosan-based microspheres

The results of the water sorption experiment for the ultrasound-assisted chitosan-based microspheres are summarized in Fig. 6. This was conducted to assess the hydrophilic behavior of the chitosan-based microspheres that were subjected to varying ultrasonic power levels (90, 180, 270 W) providing an insight into how ultrasonication influences the microsphere's ability to absorb water which is a factor in oil–water separation performance.

Among the samples, sample 1 (90 W) showed the highest average water uptake at 1.0864%. As the power increased to 180 W (sample 2), water uptake decreased significantly to 0.9899%. At 270 W (sample 3), water uptake slightly increased to 1.0098%. This trend implies that ultrasound intensity directly influences the microstructure of the chitosan. The low water uptake also indicates that there was a direct change in the hydrophilic nature of chitosan.

3.3 Modeling Adsorption Kinetics

To understand the mechanism and kinetic limitations involved in the removal of marine diesel oil from seawater using ultrasound-assisted chitosan-based microspheres, the initial parameters for the non-linear kinetic models were determined using the linearized equation forms of pseudo-first order and pseudo-second order as presented in Eqs. (2) and (3). By comparing the coefficient of determination (R^2) and kinetic parameters obtained, the most suitable kinetic model describing the adsorption process was identified.

Table 2 summarizes the kinetic parameters obtained from two independent experiments conducted at contact times ranging from 0.5 to 4 h. Adsorption capacity at equilibrium was achieved at 2.5 h where $q_{e(\text{experimental})} = 3.7726$ g/g while the $q_{e(\text{theoretical})}$ presented at the table are the predicted equilibrium adsorption capacity by fitting the data to kinetic models. Meanwhile, k_1 and k_2 are pseudo first and second order constants or the rate at which adsorption occurs.

Table 2 Pseudo-first and second order constants and coefficient of determination for the adsorption of marine diesel oil onto ultrasound-assisted chitosan-based microspheres

System	Pseudo-first order kinetic model			Pseudo-second order kinetic model		
	q_e (g/g)	k_1 (min^{-1})	R^2	qe (g/g)	k_2 (g/g min^{-1})	R^2
Oil-microsphere	5.2488	1.1674	0.991	5.8201	0.0932	0.9965

For the pseudo-first order (Fig. 7a), the linear relationship between $\ln(q_e - q_t)$ versus time is illustrated in Fig. 4.4. Based on these plots, the values ofk_1 and q_e were obtained. For the pseudo-second order, the values of q_e andk_2 were determined using the linear plot of $\frac{t}{q_t}$ versus time. The values were calculated from the respective slope and intercepts of the kinetic models [7]. As observed, the $q_{e(\text{theoretical})}$ is 5.2488 g/g. The data also presents $R^2 = 0.991$ suggesting that the adsorption behavior across various contact times can be adequately approximated by the pseudo-first order kinetic model where the system follows a physical adsorption.

The pseudo-second order (PSO) kinetic model was also applied to further evaluate the adsorption mechanism and rate-controlling steps of the system. This model assumes a chemisorption rate-limiting step. The $q_{e(\text{theoretical})} = 5.8201$ (g/g) which is higher than the experimental value obtained in the PFO model, however, the correlation of determination $R^2 = 0.9665$ is slightly lower than that of the PFO model, indicating a less precise fit for the experimental data. Furthermore, since $\frac{k_1}{k_2} \gg q_e$; $\frac{1.1674}{0.0932} \gg 5.2488$, means that complex kinetic equations are unnecessary as the system becomes mathematically equivalent to the PFO model. In addition, although the system suggests a chemisorption behavior, the better fit would be a PFO model.

3.4 *Modeling of Adsorption Isotherm*

To further understand the adsorption behavior marine diesel oil onto the chitosan microspheres, the results of the batch experiment were applied to Langmuir and Freundlich isotherm models. This was done at constant time of 3 h and varying oil–seawater concentration from 10 to 60 g/L.

Table 3 gives a comparison of isotherm constants obtained from two independent experiments as interpreted by both chosen isotherm models. q_m is the maximum adsorption capacity of the adsorbate (g/g), K_L and K_F are Langmuir and Freundlich constant (L/g) indicating the affinity of the adsorbent towards the adsorbate, and R_L is the dimensionless separation factor, and *1/n* adsorption intensity indicating how adsorption energy is distributed across the adsorbent surface [8].

In the Langmuir isotherm model, the linearized form as presented in Eq. (4), is used to plot $\frac{C_e}{q_e}$ versus C_e where C_e is the adsorbate concentration at equilibrium (g/L) and q_e is the amount of adsorbate adsorbed at equilibrium (g/g) presented in Fig. 8a. From these plots, the values of q_m and K_L were obtained using the slope and the intercept. The key characteristic of Langmuir is the R_L that indicates whether the

Table 3 Adsorption isotherm constants

System	Pseudo-first order kinetic model				Pseudo-second order kinetic model		
	Q_m (g/g)	k_L (min^{-1})	R^2	R_L	1/n	k_F (g/g)(L/g)$^{1/n}$	R^2
Oil-microsphere	4.0393	1.4946	0.9995	0.8438	0.2698	2.1417	0.9620

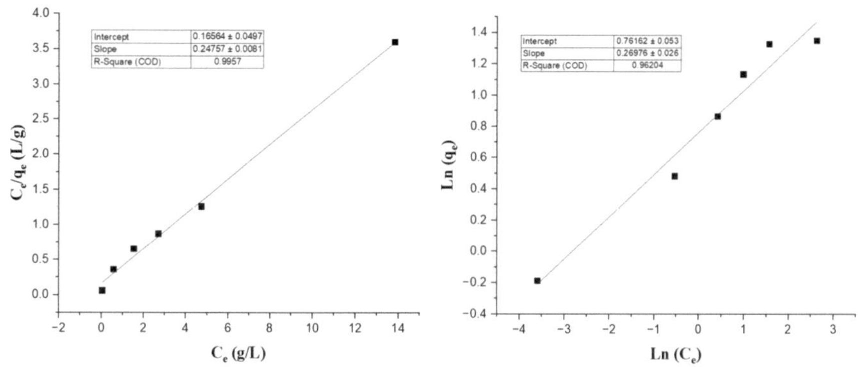

Fig. 8 **a** Langmuir adsorption isotherm and **b** Freundlich adsorption isotherm plot for the adsorption of marine diesel oil onto ultrasound-assisted chitosan-based microspheres

adsorption system is favorable or not. It can be computed using Eq. (9):

$$R_L = \frac{1}{1 + K_L C_o} \tag{9}$$

The data obtained shows a favorable adsorption with separation factor $0 < R_L = 0.8438 < 1$. On the other hand, the experimental equilibrium q_e is achieved at 3.8018 g/g and the predicted q_m of Langmuir is 4.0393 g/g. The ratio $\theta = \frac{q_e}{q_m} \approx 0.941$ indicates that there is a high affinity between the adsorbent and the contaminant with its value nearing to 1. This result alone validates the applicability of the Langmuir model, and the system follows a monolayer adsorption. This is supported by the values of $K_L = 1.4946$ L/g which is greater than 1 meaning a strong affinity is present. Furthermore, with the coefficient of determination $R^2 = 0.9957$ which implies a strong fit of the experimental data to the Langmuir model and assuming a uniform surface adsorption.

Freundlich isotherm model was also applied to further evaluate the adsorption behavior on a heterogeneous surface. This model assumes a multilayer adsorption process. The yielded adsorption intensity (*1/n*) is 0.2698 and a Freundlich constant of 2.1417. A value of $0 < \frac{1}{n} < 1$ and a K_f of greater than one means a favorable adsorption and strong affinity of the adsorbent and the adsorbate as presented in Fig. 8b. This indicates that the system likely occurs on a heterogenous surface, however, comparing the correlation of determination (R^2), the Langmuir model (R^2

= 0.9957) fits the experimental data slightly better than the Freundlich mode (R^2 = 0.9624). This suggests that while the adsorption surface may behave as multilayer adsorption, the adsorption process more closely follows a monolayer behavior.

3.5 Statistical Analysis

The ANOVA results in Table 4 confirm the significant effects of adsorbent dosage, oil–seawater concentration, and contact time on oil removal efficiency for crosslinked sodium-tripolyphosphate chitosan microspheres prepared at different ultrasonic powers (90, 180, and 270 W). The general linear ANOVA, performed on 12 replicates for each microsphere type, yielded p values of 0.000 for all factors, providing strong evidence against the null hypothesis. This indicates that all three variables significantly influence the adsorption process, with adsorbent dosage exerting the most pronounced positive effect.

Among the three samples, P-270 exhibited the best performance, achieving the highest overall removal efficiency. In contrast, oil–seawater concentration showed a consistently negative effect, confirming that higher concentrations reduce the removal percentage due to site saturation and competitive adsorption. The adjusted sum of squares of 8864.25 with negligible error variance further validates the robustness of the model, establishing a strong relationship between predictor variables and oil removal efficiency. The contour plots (Fig. 9a, c, e) illustrate the interaction between adsorbent dosage and oil–seawater concentration at a fixed contact time of 2 h. Across all samples, oil removal efficiency decreased with increasing oil–seawater concentration, while higher adsorbent dosages enhanced performance up to an optimal point.

The surface plots (Fig. 9b, d, f) reinforced this inverse correlation, highlighting the critical role of optimizing dosage relative to oil concentration. These findings confirm that P-270 microspheres offer superior adsorption characteristics, making them the most effective formulation for oil removal under the tested conditions.

Table 4 ANOVA results on the effects of adsorbent dosage, oil–seawater concentration, and contact time on marine diesel oil removal using chitosan microspheres synthesized under varying ultrasonic power

Source	F-value (90 W)	P-value (90 W)	F-value (180 W)	P-value (180 W)	F-value (270 W)	P-value (270 W)
Absorbent dosage	46,875.0	0.000	27,256.3	0.000	*	*
Oil–sea water concentration	1,976.3	0.000	4,961.3	0.000	*	*
Contact time	208.33	0.000	1,728.0	0.000	*	*

Note (*) indicates the value is not indicated in the dataset

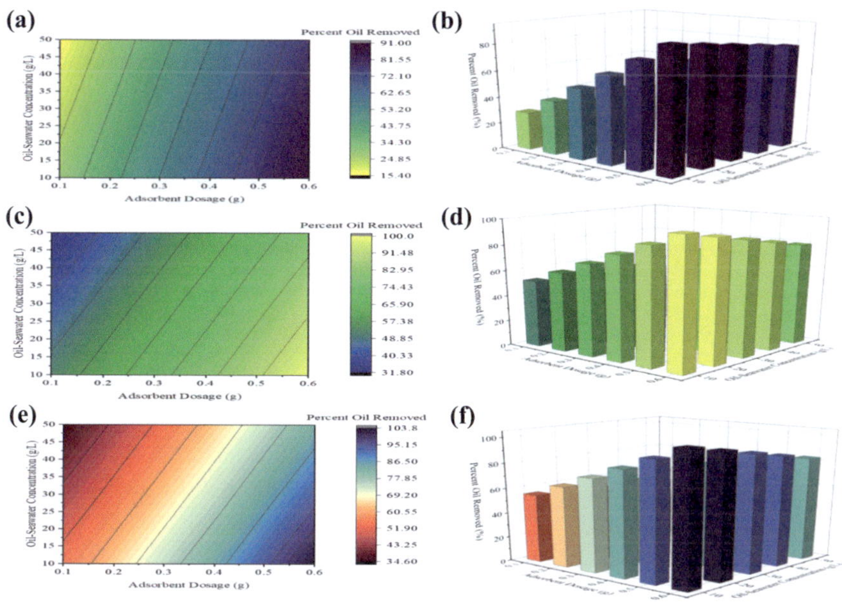

Fig. 9 Percent oil removed versus oil–seawater concentration versus adsorbent dosage: **a** contour plot 90 W; **b** surface bar plot 90 W, **c** contour plot 180 W; **d** surface bar plot 180 W, **e** contour plot 270 W; **f** surface bar plot 270 W

Oil removal efficiency increased with adsorbent dosage but declined at higher oil–seawater concentrations, consistent. The contour plots (Fig. 10a, c, e) show that, at a fixed oil–seawater concentration of 35 g/L, removal efficiency improved with both higher adsorbent dosage and longer contact time. Similarly, the surface bar plots (Fig. 10b, d, f) confirm this positive correlation, indicating that prolonged contact enhances adsorption. The increasing adsorbent dosage and contact time significantly improves oil removal efficiency [15].

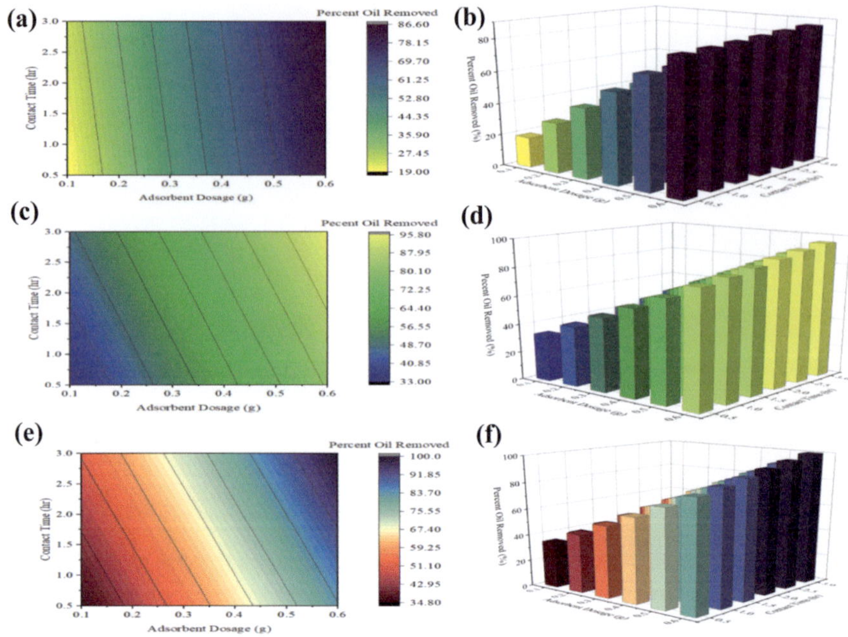

Fig. 10 Percent oil removed versus contact time versus adsorbent dosage: **a** contour plot 90 W; **b** surface bar plot 90 W, **c** contour plot 180 W; **d** surface bar plot 180 W, **e** contour plot 270 W; **f** surface bar plot 270 W

4 Conclusion

Ultrasound-assisted chitosan microspheres cross-linked with STP were successfully synthesized and showed high adsorption efficiency for marine diesel oil, with a maximum of 97.14% at 270 W, 0.6 g dosage, 10 g/L oil concentration, and 1 h contact time. SEM revealed enhanced porosity and roughness at 270 W, while XRF confirmed the elemental composition of chitosan–STP microspheres with minor impurities from processing.

Structural analysis supported these findings: XRD showed a transition from crystalline (90 W) to amorphous (270 W), indicating that higher ultrasonic power reduces crystallinity and enhances adsorption. FTIR confirmed preserved chitosan structure and stronger phosphate interactions at higher power. Overall, the 270 W sample consistently demonstrated the most favorable morphology and structural features for adsorption.

Batch adsorption tests highlighted the role of contact time and concentration. Maximum removal (90.54%) occurred at 4 h for 50 g/L oil, while the highest efficiency (99.49%) was achieved at lower oil concentration (10 g/L) with 3 h contact time. This suggests saturation of active sites beyond 3 h.

Kinetic studies followed a pseudo-first-order model ($R^2 = 0.9910$), indicating chemisorption, while isotherm analysis fit the Langmuir model ($R^2 = 0.9957$), consistent with monolayer adsorption. ANOVA confirmed that adsorbent dosage, oil concentration, and contact time significantly influenced adsorption efficiency.

Acknowledgements The proponents would like to extend their sincerest gratitude towards the Chemical Engineering Department of Adamson University and Adamson University Laboratory of Biomass, Energy and Nanotechnology (ALBEN).

References

1. Ababneh, H., Hameed, B.H.: Chitosan and chitosan composites for oil spills treatment: review of recent literature. J. Water Proc. Eng. **55**(104193), 104193 (2023)
2. Su, C., Yang, H., Zhao, H., Liu, Y., Chen, R.: Recyclable and biodegradable superhydrophobic and superoleophilic chitosan sponge for the effective removal of oily pollutants from water. Chem. Eng. J. **330**, 423–432 (2017)
3. Kadam, N.R.: Microsphere: a brief review. Asian J. Biomed. Pharm. Sci. **05**(47), 13–19 (2015)
4. Bhatt, P., Joshi, S., Urper Bayram, G.M., Khati, P., Simsek, H.: Developments and application of chitosan-based adsorbents for wastewater treatments. Environ. Res. **226**(115530), 115530 (2023)
5. Grem, I.C.d.S., Lima, B.N.B., Carneiro, W.F., Queirós, Y.G.d.C., Mansur, C.R.E.: Chitosan microspheres applied for removal of oil from produced water in the oil industry. Polímeros **23**(6), 705–711 (2013)
6. Malhas, R., Alamgir, T., Benhusain, A., Alotaibi, N., Alhaddad, N.: Oil content removal from oil water emulsion by super magnetic nanoparticles. J. Water Proc. Eng. **55**(104246), 104246 (2023)
7. William Kajjumba, G., Emik, S., Öngen, A., Kurtulus Özcan, H., Aydın, S.: Modelling of adsorption kinetic processes—errors, theory and application. In: Edebali, S. (ed.) Advanced Sorption Process Applications. IntechOpen, London, England (2019)
8. Al-Ghouti, M.A., Da'ana, D.A.: Guidelines for the use and interpretation of adsorption isotherm models: a review. J. Hazard. Mater. **393**(122383), 122383 (2020)
9. Eguico, C.S.-A. et al.: Sonophotopythochemical functionalization of graphene oxide–Al–Zn bimetal nanocomposite for corrosion inhibition. Appl. Sci. Eng. Prog. (2024)
10. Low, S.K., Tan, M.C., Chin, N.L.: Effect of ultrasound pre-treatment on adsorbent in dye adsorption compared with ultrasound simultaneous adsorption. Ultrason. Sonochem. **48**, 64–70 (2018)
11. Song, W., Yang, Z., Zhang, S., Fei, B., Zhao, R.: Properties enhancement of poly(β-hydroxybutyrate) biocomposites by incorporating surface-modified wheat straw flour: effect of pretreatment methods. Int. J. Biol. Macromol. **232**(123456), 123456 (2023)
12. Zhang, H., Yao, B., Zai, M.: Synthesis and characterization of chitosan-nanosilver microsphere by ultrasonic irradiation. Asian J. Chem. **26**(6), 1789–1791 (2014)
13. Sabnis, S.S., Raikar, R., Gogate, P.R.: Evaluation of different cavitational reactors for size reduction of DADPS. Ultrason. Sonochem. **69**(105276), 105276 (2020)
14. Olufemi, B.A., Otolorin, F.: Comparative adsorption of crude oil using mango (Mangifera indica) shell and mango shell activated carbon. Environ. Eng. Res. **22**(4), 384–392 (2017)
15. Al-Muhtaseb, R., Bhagyaraj, S., Chehimi, M.M., Krupa, I.: Adsorptive removal of emulsified crude oil from produced water using modified biowaste; its kinetics and thermodynamic study. Int. J. Environ. Sci. Technol. (Tehran) **22**(10), 9033–9048 (2025)

Parametric Analysis of Airflow and Damper Angle in a Rectangular Duct Using Solidworks CFD

Sebs Daniel D. Argarin, Basil Aloysius P. Bautista, Dexter Q. Sabusap, Allen James T. Ebay, Michael John C. Salido, and Russell Pangilinan

Abstract This study quantitatively analyzes the parametric impact of damper angles on airflow in a rectangular duct through using computational fluid dynamics (CFD) simulation. Since a volume damper is important primarily to provide proper airflow to different zones, it can be critical to maximizing performance in an HVAC system. The duct airflow is applied in practice and previous researches have explored the manual flow control system methods, but the use of simulation can assist in the setup before fabrication as well as the actual automated control in subsequent applications depending on the occupant within a zone. The main objective of this work is to investigate the impact of various damper positions on airflow velocity, pressure distribution and volume flow rate per zone inside the system. The parametric study will be performed using SolidWorks Flow Simulation. This method measures those quantities to determine the prescribed damper positions, relative to the prescribed occupancy airflow as stated in ASHRAE 62.1. Expected results indicate that the positioning of damper changes airflow behavior and causes changes in both volumetric flow rates and static pressure losses in each zone. The findings are expected to highlight the importance of adjusting damper angles for the purposes of reducing energy waste and improving ventilation. The recommendations could well be more geared toward smart damper control strategies to increase HVAC efficiency by minimizing unnecessary static pressure losses.

Keywords HVAC systems · Volume damper · Flow simulation · Airflow analysis · Damper angle · Occupancy airflow

---

S. D. D. Argarin (✉) · B. A. P. Bautista · D. Q. Sabusap · A. J. T. Ebay · M. J. C. Salido · R. Pangilinan
Adamson University, Manila, Philippines
e-mail: sebs.daniel.argarin@adamson.edu.ph

M. Awang et al. (eds.), *Proceedings of the International Conference on Mechanical, Manufacturing and Plant Engineering*, Proceedings in Technology Transfer,
https://doi.org/10.1007/978-981-95-7817-7_29

1 Introduction

1.1 Variable Air Volume (VAV) Systems

A Variable Air Volume HVAC system consists of a series of elements which work together for efficient conditioned air supply in multiple thermal zones. In Fig. 1, The Air Handling Unit (AHU) is the main core where processes such as heating, cooling, and filtering of air undergo before distributing it throughout the building. Conditioned air is delivered through a duct network, with return ducts recirculating air back to the AHU. At the zone level, VAV terminal units play a critical role by regulating the volume of conditioned air supplied to maintain desired thermal comfort and indoor air quality. Studies have identified different types of terminal units—such as throttle type, induction type, and fan-powered type—with the throttle type being the most widely used [1–3]. These units incorporate airflow rate sensors, temperature sensors, damper actuators, and sometimes valve actuators, all of which are managed by the system's control architecture. The effectiveness of these components relies on advanced control methods that continuously adjust operations to ensure efficiency and comfort [3, 4].

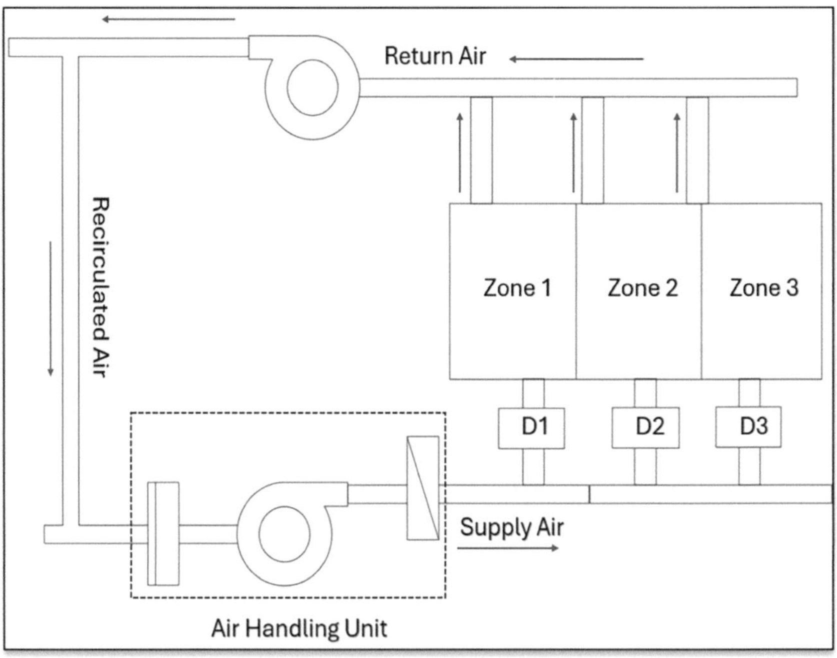

Fig. 1 Schematic diagram of a simple variable air volume (VAV) system

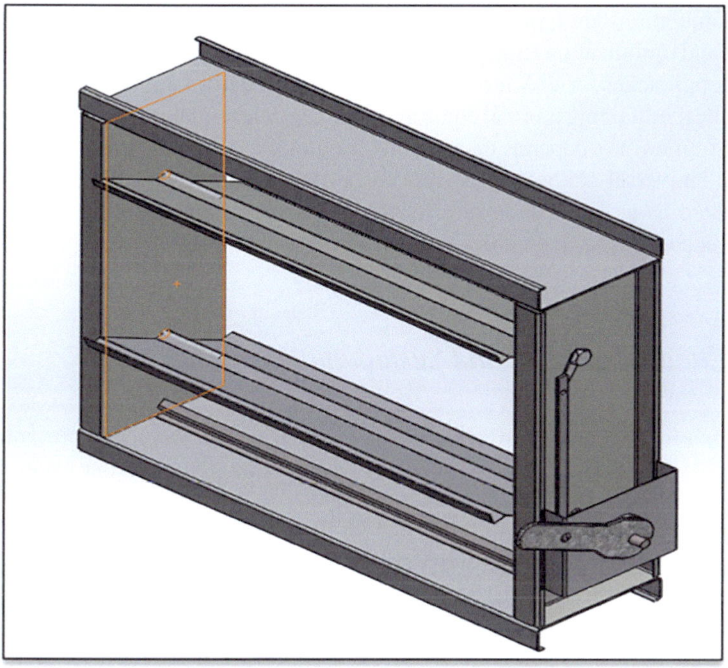

Fig. 2 3D model of volume damper

1.2 Volume and Balance Dampers

Volume dampers are central to VAV operation as they actively regulate airflow within each terminal unit. In pressure-independent configurations, damper actuators respond to control signals to adjust blade angles, thereby increasing or decreasing airflow based on zone conditions as shown in Fig. 2. Feedback from airflow and temperature sensors ensures that adjustments are continuously refined to meet set points determined by thermal load requirements [3, 5]. On the other hand, balancing dampers operate at the system level chiefly during commissioning and maintenance. They're designed to allocate airflow evenly throughout the duct system rather than in some dynamic way in those individual zones [6].

1.3 Calibration Using Fluid Simulation

While CFD simulations are rarely applied in real-time building operations, they play a valuable role in the calibration and design of HVAC control strategies. By modeling airflow patterns, pressure losses, and damper behavior, CFD provides detailed insights into system performance that inform control algorithm development.

These simulations are typically performed offline and serve as a basis for predictive models and optimization strategies that can be deployed in actual operations. Optimal control approaches for VAV terminal units are designed to minimize energy consumption while maintaining thermal comfort [4, 6]. The determination of damper positions in such frameworks depends on real-time conditions, requiring accurate sensor calibration. Universal calibration devices for airflow sensors in VAV units have been highlighted as essential to ensure reliable data, which directly influences damper control accuracy [5, 7].

1.4 Fluid Simulation and Sustainability

The integration of simulation techniques into HVAC design and operation contributes directly to sustainability goals. Research emphasizes the role of volume damper simulations in optimizing airflow distribution, thereby reducing unnecessary energy use and lowering building carbon footprints [8, 9]. These simulations also support the development of advanced control strategies that balance energy efficiency with indoor environmental quality. For instance, building energy modeling frameworks that incorporate damper simulation provide continuous opportunities for energy reduction and support decarbonization efforts across a building's lifecycle [10]. Fluid simulation is not commonly utilized for real-time application, it is a pivotal method in actual design, calibration, and optimization of HVAC systems. Computational fluid dynamics (CFD) is commonly used in the field of offline operations to model the airflow direction, pressure losses, and aerodynamic behavior of dampers. Additionally, research highlights the necessity of maintaining occupant comfort and saving energy, and demonstrates that simulation can help reach both [11]. Although full CFD simulations are still computationally expensive for real-time management, reduced-order models, and pre-computed lookup algorithms have become an option to turn accurate analyses into operational conditions [12]. These developments enhance the link between investigations of fluid dynamics and sustainable and ecologically sound building management.

2 Methodology

2.1 Design of the Test Rig

A test rig is designed and to be used for the experimental set-up to simulate a four-zone HVAC distribution system. The model consisted of a main supply duct connected to four branch ducts, each serving a separate zone. Butterfly-type volume dampers were integrated into each branch to allow modulation of airflow. The dimensions of the rectangular ducts are computed using the required volume flow rates based on the equal friction method.

In Table 1, the parameter for the prescribed airflow is selected based on the ventilation requirements for each of the four zones were established using values from ASHRAE 62.1 Table 6.1, which prescribed airflow rates are dependent either on the occupant and their activities or the floor area of the room. For this test, the typical value of 10 CFM/person is used [13]. A maximum of four (4) people per zone is used for this set-up.

Air properties such as density and viscosity were assumed under standard room conditions, while boundary conditions included inlet flow rates, outlet ambient pressure, and no-slip adiabatic wall conditions. Damper angles varied across a range between 0° for fully closed and 90° for fully opened conditions to investigate their airflow rate. In Fig. 3, the inlet and outlet boundary conditions are set which are the total supply airflow and the pressure requirement for each outlet.

Table 1 Parameters for the prescribed airflow

Occupancy type	Number of person per zone	Required airflow per zone
No occupancy	0	0 cubic feet per minute
Occupied zone	1–4	40 cubic feet per minute
Total airflow required		160 cubic feet per minute

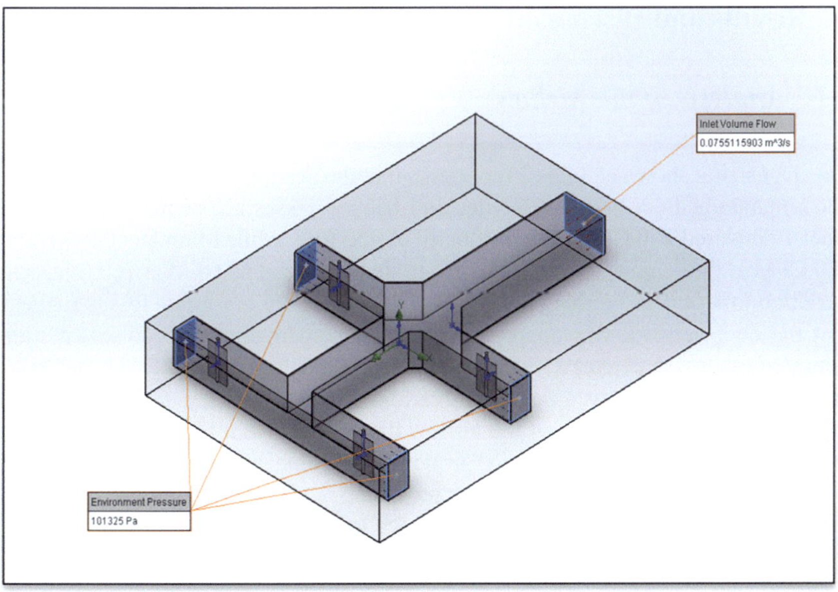

Fig. 3 3D model of the test rig with the initial conditions loaded

2.2 Parametric Study Using SolidWorks CFD

The test rig was modeled in SolidWorks, and its flow simulation module was used to conduct a what-if parametric study. For each damper angle configuration, simulations were run to predict airflow distribution, and velocity contours. The analysis considered multiple occupancy scenarios to evaluate whether the airflow delivered to each zone met the minimum ventilation requirements of ASHRAE 62.1. This approach enabled a systematic assessment of damper performance under different operating conditions.

2.3 Selection of Optimal Damper Settings

Simulation results of each iteration were compared against ASHRAE 62.1 compliance thresholds to determine the effectiveness of each configuration. The optimal damper angles were identified based on the ability to deliver the required ventilation while avoiding over-ventilation. These recommended damper settings formed the basis for an adaptive control strategy, providing guidance on how airflow can be balanced across zones.

3 Results and Discussion

3.1 Analysis of the Airflow

The duct sizing shown at Table 2 represents that the designed main ducts and branch ducts maintained consistent velocities and friction losses across the system. Main Duct 1 delivered 250 CFM with a velocity of 950 fpm, while Main Duct 2 provided 120 CFM at the same velocity. Each branch duct supplied 60 CFM at 760 fpm, with a friction loss of 0.213 mmHg per 100 ft. The uniformity in friction losses suggests that the duct network was appropriately sized to maintain balanced distribution, minimizing the risk of excessive pressure drops in localized zones. This is oversized but leaves room for overventilation or overload in the occupancy.

The visual representation of the airflow distribution from Fig. 4 under the fully opened condition validates that the system is capable of supplying the required volume flow rate across all branches, providing a reference point for evaluating the parametric study under varying occupancy loads.

Table 2 Duct size and their respective parameters

Duct	Airflow in CFM	Velocity in ft per minute	Duct Dimension (L x W)	Friction loss per 100ft (mm Hg)
Main Duct 1	250	950	8” × 6”	0.213
Main Duct 2	120	950	6” × 4”	0.213
Branch 1	60	760	4” × 4”	0.213
Branch 2	60	760	4” × 4”	0.213
Branch 3	60	760	4” × 4”	0.213

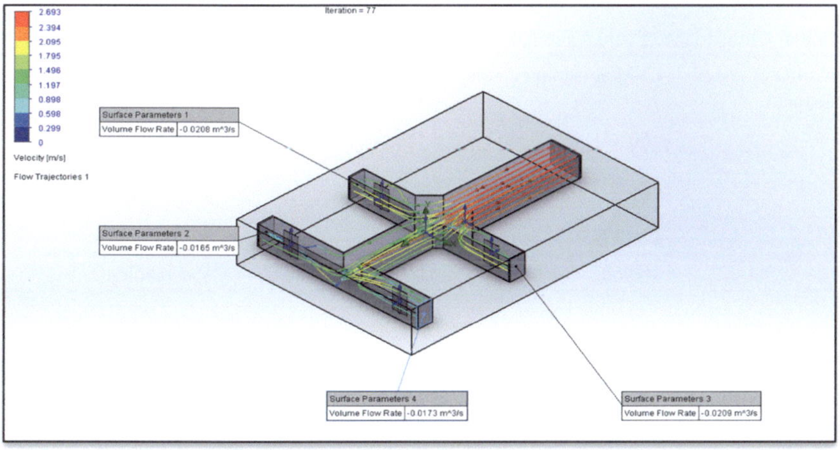

Fig. 4 Airflow at fully opened condition

3.2 Simulation Results with Occupancy

The parametric study results in Table 3 shows the simulated the performance of the system under one to four occupied zones, with damper openings adjusted accordingly leading to 24 cases to be simulated. For one occupied zone, Zone 1 received 50.165 CFM, which exceeded the ASHRAE 62.1 prescribed value of 40 CFM per zone, leading to over-ventilation by approximately 10 CFM. With two zones occupied, the system supplied a total of 81.41 CFM, with each zone receiving airflow values close to the target of 40 CFM, which is within acceptable limits. For three zones occupied, the combined supply was 119.374 CFM, closely matching the prescribed 120 CFM, indicating accurate airflow distribution. Under full occupancy of four zones, the system provided 160.607 CFM, which aligns very well with the ASHRAE requirement of 160 CFM.

The results highlighted in Table 4 shows the capability of the system to adjust damper openings to maintain compliance with ventilation requirements. The only exception occurred at single-zone occupancy, where the airflow slightly exceeded

Table 3 Result of parametric study

Number of zones occupied	Damper opening at Zone 1	CFM at Zone 1	Damper opening at Zone 2	CFM at Zone 2	Damper opening at Zone 3	CFM at Zone 3	Damper opening at Zone 4	CFM at Zone 4
1	27	50.165	0	0	0	0	0	0
2	36	40.131	36	41.279	0	0	0	0
3	72	40.999	72	39.219	72	39.156	0	0
4	90	39.219	90	41.279	90	40.122	90	39.987

Table 4 Comparison of actual simulation values versus prescribed values

Number of zones occupied	CFM at occupied zone	Prescribed CFM per zone	Remarks
1	50.165	40	Over-ventilating by 10 CFM
2	81.410	80	Acceptable
3	119.374	120	Acceptable
4	160.607	160	Acceptable

the prescribed value, leading to minor inefficiency. However, this over-ventilation remains within an acceptable tolerance range and does not significantly compromise system.

4 Conclusion

This study demonstrates the effective design and high-performance capabilities of a multi-zone ventilation system integrated with active damper control. The foundational design of the duct network proved highly successful, maintaining uniform airflow velocities and minimizing friction losses across both main and branch ducts. This leads to a consistently balanced air distribution throughout the system, effectively preventing localized excessive pressure drops and establishing a reliable operational baseline. The subsequent parametric study, which rigorously tested the system under varying occupancy loads ranging from one to four occupied zones, revealed a significant degree of adaptability and precision in airflow management, largely prescribed value for occupancy in the ASHRAE 62.1 ventilation standards.

While the system exhibited exemplary compliance with ASHRAE requirements when two, three, or four zones were occupied, a minor instance of over-ventilation was noted during single-zone occupancy. In this specific scenario, the airflow marginally exceeded the prescribed 40 CFM per zone, suggesting that a better angle is well suited. However, this deviation remained within an acceptable tolerance range, indicating that while there is room for improvement, it does not critically compromise the static pressure in a significant amount. The findings collectively validate the considerable potential of dynamic damper control strategies within multi-zone systems for achieving optimized ventilation management. Such systems can effectively balance precise air delivery, ensuring occupant comfort and air quality, with adherence to regulatory standards, thereby contributing to energy efficiency and sustainable building operations.

References

1. Motuzienė, V., Bielskus, J., Džiugaitė-Tumėnienė, R., Raudonis, V.: Occupancy-based predictive AI-driven ventilation control for energy savings in office buildings. Sustainability **17**(9), 4140 (2025)
2. Borodiņecs, A., Zemītis, J., Palcikovskis, A., Ardavs, A., Lavendelis, E.: Review of modern demand control solutions and technologies for HVAC operation. E3S Web Conf. **396**, 2020 (2023)
3. Hobson, B.W., Gunay, H.B.: Evaluating the impact sequences of operation have on the implementation of occupant-centric controls. Energy Build **266**, 112121 (2022). https://doi.org/10.1016/j.enbuild.2022.112121
4. N. Torabi, Gunay H.B., O'Brien W.: A review of common human errors in design, installation, and operation of multiple-zone VAV AHU systems. J. Phy. Conf. Ser. **2042**(1), 12130 (2021). https://doi.org/10.1088/1742-6596/2042/1/012130
5. Merabet et al. G.H.: Intelligent building control systems for thermal comfort and energy-efficiency: a systematic review of artificial intelligence-assisted techniques. Renew Sustain Energy Rev **144**. 110969 (2021). https://doi.org/10.1016/j.rser.2021.110969
6. Szekeres, S., Kostyák, A., Szodrai, F., Csáky, I.: Investigation of ventilation systems to improve air quality in the occupied zone in office buildings. Buildings **12**(4), 493 (2022). https://doi.org/10.3390/buildings12040493
7. S. Langhorst, M.M., Boughanmi, N., Schmeling, D., Wagner, C.: CFD-based comfort parameter evaluation of a flow vectoring air vent system for car cabins using a reduced order model. SAE Int. J. Passeng. Veh. Syst. **18**(2) (2025). https://doi.org/10.4271/15-18-02-0012
8. Nassif, N., Ridwana, I.: Improving building energy performance using dual VAV configuration integrated with dedicated outdoor air system. Buildings **11**(10), 466 (2021). https://doi.org/10.3390/buildings11100466
9. Kim, H., Cho, Y.-H.: Optimal control method of variable air volume terminal unit system. Energies **14**(22), 7527 (2021). https://doi.org/10.3390/en14227527.10.H
10. Zhang, H.C., Zhang, X., Cai, W., Yin, Z.: A universal calibration device for an air flow sensor of the VAV terminal unit. Sensors **22**(15), 5797 (2022). https://doi.org/10.3390/s22155797
11. Zeng, L., et al.: A variable air volume control strategy for a centralized exhaust system with multiple on-off switched terminals and flow-guide devices. J. Build. Eng. **38**, 102185 (2021). https://doi.org/10.1016/j.jobe.2021.102185
12. Nassif, N., Tahmasebi, M., Ridwana, I., Ebrahimi, P.: New optimal supply air temperature and minimum zone air flow resetting strategies for VAV systems. Buildings **12**(3), 348 (2022). https://doi.org/10.3390/buildings12030348
13. ASHRAE, ANSI/ASHRAE Standard 62.1-2022: Ventilation for Acceptable In-door Air Quality, Table 6-1, Minimum Ventilation Rates in Breathing Zone. American Society of Heating, Refrigerating and Air-Conditioning Engineers (2022)

Mechanistic Validation of a Superior Papaya Leaf Inhibitor Derived from Ultrasound-Microwave Assisted Extraction (UMAE) for High-Performance Corrosion Shielding of Mild Steel

Jozel Valenzuela, Arbee Chrystel Alera, Alexandria Bernardo, James Gabriel Galano, Rosemary Cole, Rugi Vicente Rubi, and Rich Jhon Paul Latiza

Abstract Corrosion poses significant challenges to metals used in construction and machinery. This study evaluated the Inhibitive Effect of Papaya (Carica papaya) leaf extract, obtained through Ultrasound-Microwave Assisted Extraction (UMAE), as an eco-friendly corrosion inhibitor for mild steel. Papaya leaves are rich in phytochemicals such as alkaloids, flavonoids, tannins, and saponins, which form protective layers on metal sur-faces and reduce degradation. UMAE, combining ultrasonic and microwave-assisted extraction, proved more efficient than conventional Soxhlet extraction, yielding 63.43% compared to 6.78%, while preserving bioactive compounds. Papaya leaf extract (PLE) coated mild steel strips were immersed in hydrochloric acid solutions (0.5M, 1M, 1.5M) and monitored over various exposure times. Weight loss, corrosion rates, and inhibition efficiency were measured, with the lowest weight loss of 2.1798 g under 4 coatings of PLE after 336 h and a peak efficiency of 97.63%. Characterization techniques such as SEM–EDX, FTIR, and UV–Vis confirmed the formation of corrosion-resistant barriers and identified key functional groups responsible for PLE's protective properties. These findings position PLE as a promising, sustainable solution for corrosion prevention in industrial applications.

Keywords Ultrasound-microwave assisted extraction · Corrosion · Mild steel · Weight loss · Corrosion rate · Inhibition efficiency

---

J. Valenzuela · A. C. Alera · A. Bernardo · J. G. Galano · R. Cole · R. V. Rubi · R. J. P. Latiza (✉)
Chemical Engineering Department, College of Engineering, Adamson University, Manila, Philippines
e-mail: rich.jhon.paul.latiza@adamson.edu.ph

R. V. Rubi · R. J. P. Latiza
Adamson University Laboratory of Biomass, Energy and Nanotechnology (ALBEN), Adamson University, Manila, Philippines

M. Awang et al. (eds.), *Proceedings of the International Conference on Mechanical, Manufacturing and Plant Engineering*, Proceedings in Technology Transfer,
https://doi.org/10.1007/978-981-95-7817-7_30

1 Introduction

Corrosion, a long-recognized yet persistently challenging phenomenon, refers to the electrochemical degradation of metals resulting from redox reactions with their environment. Contemporary corrosion science identifies various forms of degradation, including uniform, localized, and galvanic corrosion, which are often visually detectable [1]. More complex mechanisms such as intergranular attack, dealloying, and microbiologically influenced corrosion necessitate the use of sophisticated analytical or microscopic techniques due to their subtle manifestation and intricate pathways [2].

The National Association of Corrosion Engineers (NACE) International underscores corrosion as a significant global issue, estimating its economic impact at approximately 3–5% of a nation's gross national product (GNP) [3]. This considerable financial burden highlights the imperative for effective mitigation strategies across industrial sectors. Among these, corrosion inhibitors have emerged as a widely adopted solution, particularly those capable of interacting with chloride ions to modulate corrosion kinetics. Organic inhibitors, characterized by the presence of heteroatoms such as nitrogen, oxygen, sulfur, and phosphorus, demonstrate high efficacy due to their electron-donating capabilities. The inhibitory performance of these elements generally increases in the order $O < N < S < P$, reflecting their nucleophilicity and coordination tendencies with metal surfaces [4].

In the pursuit of sustainable and environmentally benign corrosion protection, the use of organic green corrosion inhibitors (OGCIs) sourced from plant materials has garnered significant interest [5, 6]. Papaya (Carica papaya) leaf extract (PLE) has shown notable potential as a natural corrosion inhibitor. Previous investigations have demonstrated its properties and effectiveness; however, the influence of the extraction methodology on its corrosion-inhibiting performance remains underexplored [7]. This study aims to compare conventional Soxhlet extraction with the more recent ultrasound–microwave-assisted extraction technique, focusing on their impact on the concentration of phytochemicals and bioactive compounds, and ultimately, on the efficacy of PLE as a green corrosion inhibitor.

2 Methodology

2.1 Preparation of Biomass

Papaya leaves were collected, cleaned, and prepared by removing the leafstalks and washing thoroughly to eliminate impurities. The leaves were oven-dried at 70°C for 1 h, followed by an additional drying step at 75 °C for 30 min to ensure complete moisture removal. Once dried, the leaves were finely powdered using a bullet blender, sieved through a mesh 60 to retain the undersized fraction, and stored in an airtight container for preservation [8].

2.2 Extraction of Biomass: Non-Conventional Method of Extraction (UMAE)

The ultrasound-microwave assisted extraction (UMAE) process began with ultrasound-assisted extraction (UAE), where papaya leaf powder was mixed with solvent at a 0.7689:500 ratio and sonicated in a 40 kHz, 30–40 °C ultrasonic bath for 150 min [9]. This was followed by microwave-assisted extraction (MAE) using a 700-W microwave for five minutes. The extract was then concentrated using a rotary evaporator at 59.6 °C and 60 RPM, filtered through a 0.22 μm membrane, and the yield was measured and compared to that obtained via the Soxhlet extraction method.

2.3 Screening of Phytochemical Components from Extract

Phytochemical screening of the papaya leaf extract (PLE) confirmed the presence of key bioactive compounds through qualitative tests: alkaloids were detected by a green precipitate after adding HCl and Mayer's reagent, while flavonoids produced a yellow hue when mixed with sodium hydroxide. Saponins were indicated by foam formation after shaking the extract with water, and tannins were confirmed by a green–blue or black precipitate upon adding $FeCl_3$ to a water-extract mixture.

2.4 Preparation of Mild Steel Samples

The mild steel strips were cut from an MS sheet. The sizes of the MS strips were standardized to 1 × 2 × 0.08 in. The strips obtained from the cutting procedure underwent visual inspection for possible defects. Following the inspection, the coating on the strips was removed using silicon carbide abrasive paper (P60 and P150) to expose the surface for corrosion monitoring. This was then followed by degreasing the strips with ethanol and drying them using acetone. The treated strips were subsequently stored in a moisture-free desiccator to prevent contamination before initiating corrosion studies.

2.5 Preparation and Experimental Setup of Bio-Corrosion Medium

To prepare the bio-corrosion medium, 28 airtight containers were filled with 200 mL of test solution, including two control setups Control 1 with 3.5 wt% NaCl solution and Control 2 with uninhibited solutions of 0.5 M, 1 M, and 1.5 M HCl as well as four variations of papaya leaf extract (PLE) inhibitor applications using one to four

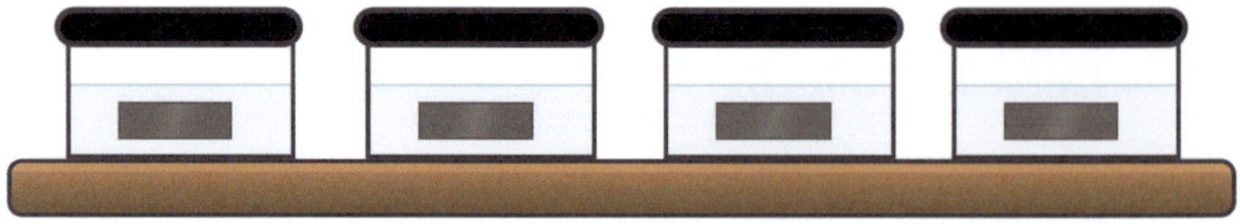

Fig. 1 Sample setup of MS strips immersed on different concentrations and number of inhibitor coatings

coatings across the same acid concentrations. This setup was duplicated using a store-bought corrosion inhibitor with matching coating levels and HCl concentrations, and all containers were kept at room temperature and observed over intervals of 24, 48, 168, 216, and 336 h, illustrated in Fig. 1.

2.6 *Weight Loss, Corrosion Rate, and Inhibition Efficiency Determination*

Weight loss assessment involved cleaning mild steel (MS) strips with 95% ethanol, air drying, etching in 5% HCl for one minute, and measuring initial weight before immersion in test solutions containing varying PLE concentrations, HCl amounts, and immersion durations, after which final weight and pH were recorded to calculate weight loss, corrosion rate (mm/yr), and inhibitor efficiency using Eqs. 1, 2, and 3.

$$\text{Weight Loss(g)} = W_f - W_i \tag{1}$$

$$\text{Corrosion Rate}\left(\frac{mm}{yr}\right) = \frac{KW}{\rho At} \tag{2}$$

$$\text{Inibition Eficiency(\%)} = \frac{CR_U - CR_I}{CR_U} \times 100\% \tag{3}$$

where, Wf and Wi are the final and initial weight of the MS strip before and after the immersion test, K, constant at 3.45 × 106, W, weight loss, ρ, density of the MS steel, A, area of specimen, t is the time duration, CR_U, corrosion rate of controlled, and CR_I corrosion rate of inhibitor.

2.7 *Characterization of Resulting Samples*

To evaluate the structural, chemical, and optical properties of the samples, Scanning Electron Microscopy (SEM) with Energy Dispersive X-ray (EDX) was used to assess surface morphology and elemental composition, while Fourier Transform Infrared

Table 1 Presence of phytochemical properties

Phytochemical properties	Results
Alkaloid	✔
Flavonoids	✔
Saponins	✔
Tannins	✔

Spectroscopy (FTIR) identified functional groups and confirmed inhibitor adsorption on the metal surface. Ultraviolet–Visible (UV–Vis) Spectroscopy analyzed the extract's optical properties and phytochemical content. These combined techniques offered comprehensive insights into the corrosion inhibition mechanism and the characteristics of the protective layers formed.

3 Results and Discussion

3.1 Screening for Phytochemical Components of the UMAE Extracted Papaya (Carica Papaya) Leaves by the Following Tests

Phytochemical testing on the PLE from UMAE was done to show that these properties make the sample a candidate to create a barrier or film layer of corrosion inhibition, presented in Table 1. The presence of phytochemicals were proven metabolites and a biomass plant, PLE, as an inhibitor attributed to the adsorption of phytochemical compounds onto the MS strips' surface, forming a protective barrier.

Alkaloids form protective layers on metal surfaces by adsorbing through their nitrogen atoms, minimizing exposure to corrosive environments like acidic solutions [10]. Flavonoids, with their polyphenolic structures, create barriers by adsorbing onto metals and forming chelate complexes, effectively blocking both anodic and cathodic corrosion processes while also leveraging antioxidant properties [11]. Saponins, due to their amphiphilic nature, develop thick, durable coatings on metal surfaces by physisorption or chemisorption, reducing the metal's contact with corrosive agents [12]. Tannins interact with metal ions to form ferric tannates or metallic tannates, acting as rust converters that stabilize and protect the metal surface [13].

3.2 Corrosion Results Analysis

The relationship between the logarithmic corrosion rate and pH under varying concentrations of PLE and HCl, as well as the controls, is illustrated in Figs. 2, 3 and 4. These graphs demonstrate how changes in pH influenced the corrosion rate.

The behavior of the curves in these figures indicates that the corrosion rate generally decreased as the pH increased, becoming more basic. However, the behavior of Control 1 is distinct, as its solution is water. In this case, iron reacts electrochemically with dissolved oxygen and water molecules, leading to corrosion.

The sample with 1.5 HCl and concentration 4 solution, as shown in Fig. 2 exhibits a prominent sigmoidal pattern, which indicates a noticeable logarithmic corrosion rate at low pH levels, signifying corrosion in acidic conditions. As the pH nears 6, a significant reduction in the corrosion rate is noted, indicating a change in the corrosion mechanism or the development of a protective effect, caused by the PLE. Following this transition, the corrosion rate stabilizes at minimal levels, signifying

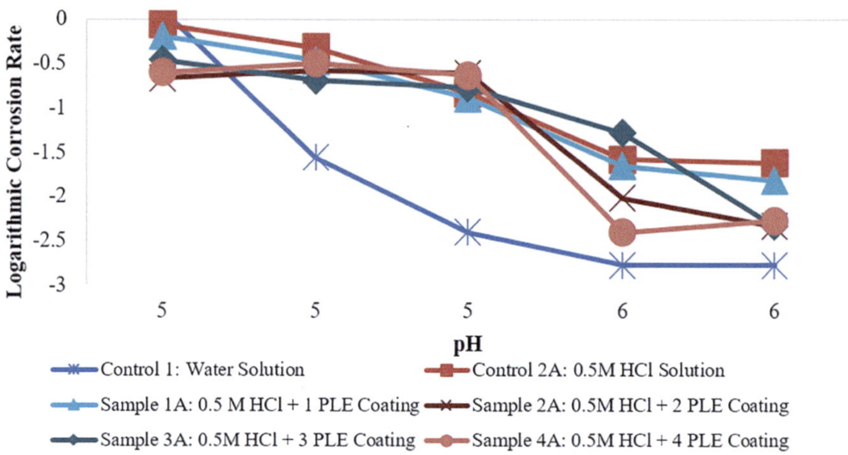

Fig. 2 Logarithm corrosion rate versus pH for samples immersed in 0.5M HCl solution

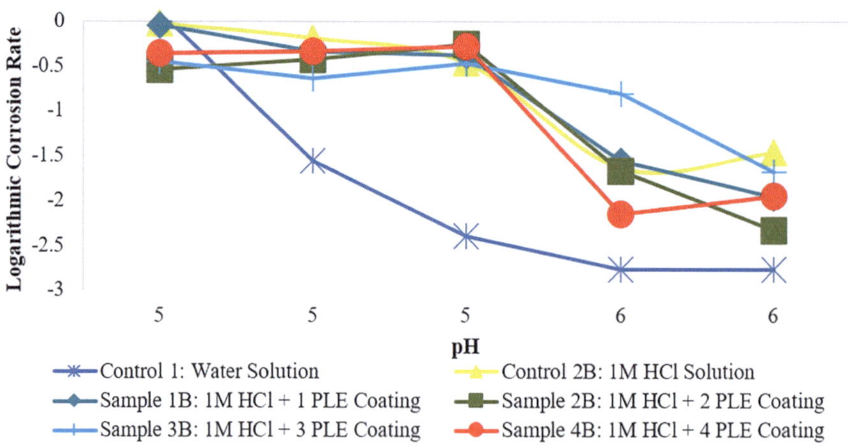

Fig. 3 Logarithm corrosion rate versus pH for samples immersed in 1M HCl solution

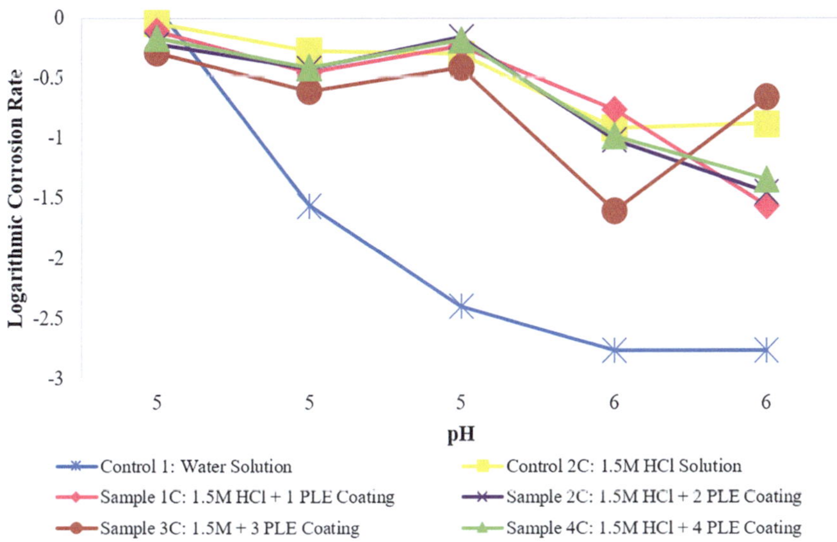

Fig. 4 Logarithm corrosion rate versus pH for samples immersed in 1.5M HCl solution

successful corrosion prevention in near neutral or less acidic environments. Such behavior highlights the pH-dependent effectiveness of the system in mitigating corrosion. Fe, iron reacts electrochemically with dissolved oxygen and water molecules, leading to corrosion. Additionally, the corrosion process of mild steel on the HCl solution involves the anodic reaction of $Fe \rightarrow Fe^{2+} + 2e^{-}$, this means that as the concentration of hydrogen ions increases, which are relevant at lower pH values, sample dissolves faster [14].

Moreover, the curves for samples coated with PLE indicate that at higher pH levels, the less aggressive acidic environment enhances the effectiveness of the extract as an inhibitor. The consistency of this behavior considers that PLE phytochemical or bioactive compounds are uniformly effective in preventing iron-oxidizing and oxygen reduction.

These bioactive compounds uniformly act by attaching onto the surface of the MS strips, obstructing the sites, and lessening the interaction of corrosive H^{+} ions in the HCl solutions [15]. This is why corrosion is moderate in a basic environment corresponding to a lower HCl concentration. This protective barrier effectively reduces the interaction between the metal and the corrosive environment, resulting in a significant decline in corrosion rates in the higher pH regions of the curves.

Corrosion rate and weight loss are directly related. As the MS strip corrodes, it loses mass due to the oxidation process. The faster the corrosion rate, the greater the weight loss over a given period, illustrations are presented in Fig. 5, 6 and 7. From Fig. 5, control 1, showed continuously higher weight loss with time, with a logarithmic weight loss at the negative side. With that, inhibited samples showed less weight loss compared to uninhibited samples. Meanwhile on Fig. 6, solutions with 1.5M such as samples 2C and 4C showed greater weight loss, with a peak log around

0.6 on 2C after 336 h, as observed by a steeper decline in the logarithm of weight loss over time, resulting to a more aggressive corrosion. This means that solutions with higher molarity of HCl demonstrated greater weight loss of the MS, as more material has been lost because of corrosive reactions for a given period and surface area, which suggests a higher corrosion rate [16, 17].

Furthermore, chloride ions $\left(Cl^{-}\right)$ are extremely aggressive and contribute significantly to the degradation of mild steel protective oxide layer. By enabling faster ion and electron $\left(H^{+}\&Cl^{-}\right)$ transport, this speeds up electrochemical reactions and

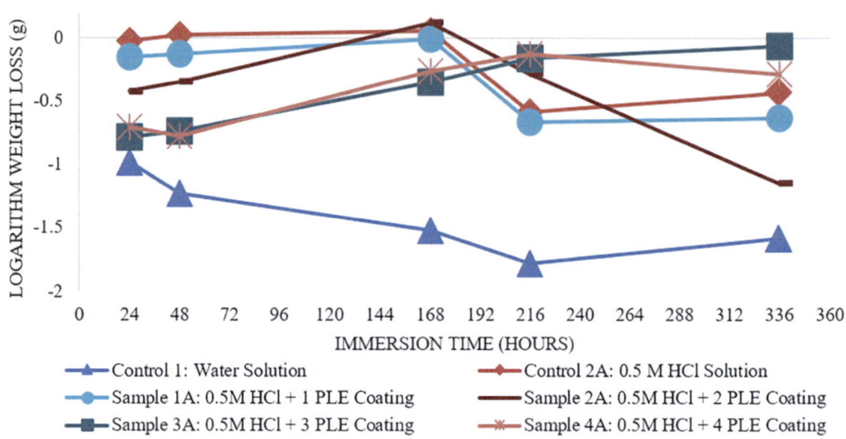

Fig. 5 Logarithm weight loss versus immersion time for samples immersed in 0.5 M HCl solution

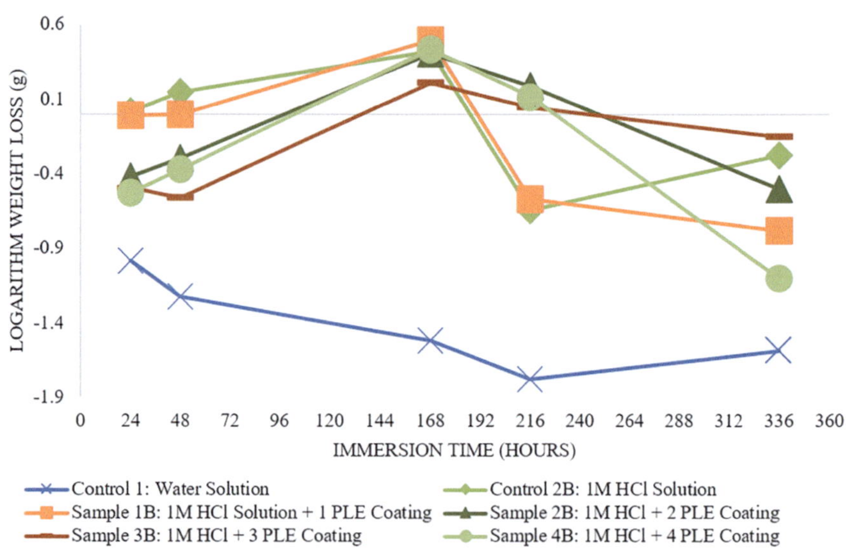

Fig. 6 Logarithm weight loss versus immersion time for samples immersed in 1 M HCl solution

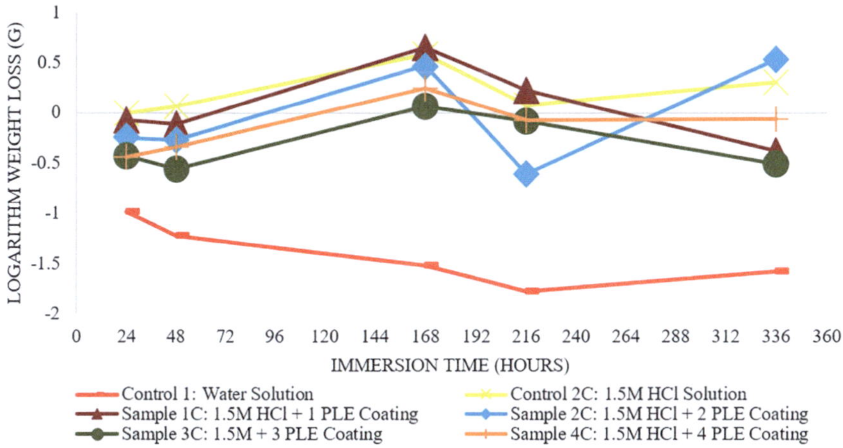

Fig. 7 Logarithm weight loss versus immersion time for samples immersed in 1.5 M HCl solution

raises the rate of corrosion. However, despite having a high molarity, control 2C in Fig. 6 have the least amount of weight loss because of the decreased oxygen availability. The overall corrosion process is slowed in highly concentrated HCl solutions because of the low solubility of oxygen caused by the high ionic strength, which prevents a cathodic reaction $2H_2O + O_2 + 4e^- \rightarrow 4OH^-$ while the anodic reaction $\left(Fe \rightarrow Fe^{2+} + 2e^-\right)$ is observed. Since corrosion involves both cathodic and anodic processes, the process will be slowed if one reaction is limited [16]. As seen in Fig. 8, the efficiency of the PLE as a corrosion inhibitor is demonstrated by the slower weight loss in inhibited samples, which correlates to a reduced corrosion rate; hence, weight loss is consistently lower, and corrosion rate decreases over time when PLE is present.

Inhibition efficiency and immersion time have a complicated relationship that changes based on several variables. The inhibition efficiency gradually increased at certain intervals (24–216 h) and then decreased after reaching the peak, indicating a dynamic interaction between the extract and the steel surface over time. Meanwhile, sample 3A in Fig. 7 obtained the peak efficiency of 97.63%. This means that the effectiveness of PLE as a corrosion inhibitor is affected by the concentration of HCl and over time, the phytochemicals in PLE create a protective barrier by adhering to the MS surface. Strong connections between the steel surface and the active functional groups, such as flavonoids that contain functional groups like hydroxyl (–OH) that enhance adsorption on the metal surface, in the phytochemicals cause interactions between the inhibitor molecules and the metal surface to intensify with immersion time, which is usually what controls this adsorption process [17].

Factors such as the desorption of inhibitor molecules, the deterioration of protective barriers, and the competition from other compounds in the environment diminished the effectiveness of the inhibitor as time progressed. This contrary evident in the behavior of sample 3B, which started at around 60% inhibition efficiency and

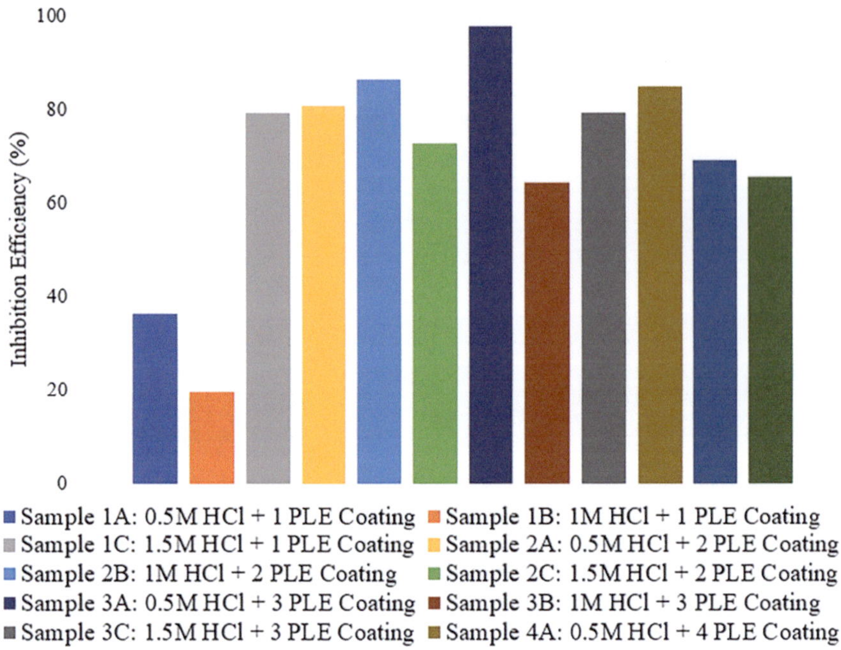

Fig. 8 Inhibition efficiency of all samples after 336 h of immersion time

gradually increased to 65% over time. The specific behavior of this relationship was influenced by factors such as the type of inhibitor, the corrosive environment, and the metal surface condition. Lastly, on Fig. 7, the efficiency after 336 h of all the samples on the same molarity of 1.5M HCl showed the highest efficiency of all the corresponding strips within the experiment time frame. This indicates that the PLE coating proved to be effective over time on reducing the corrosion rate of the samples.

3.3 Characterization of Samples

3.3.1 SEM–EDX Analysis

The SEM micrograph results (at 15 kV) presented in Fig. 9a–c compare blank mild steel (MS) strips with those coated with PLE and SB, which achieved the lowest weight loss in 0.5 M HCl with four and one coat, respectively. These micrographs reveal distinct differences between the inhibitors and the blank MS strip. Figure 9b, c show evidence of surface corrosion after immersion in an acidic environment; however, Fig. 9c (blank MS) exhibits severe corrosion, resulting in a notably rough and pitted surface. In contrast, the inhibited samples (Fig. 9a, b) display smoother surfaces, indicating a protective film formation that limits acid attack.

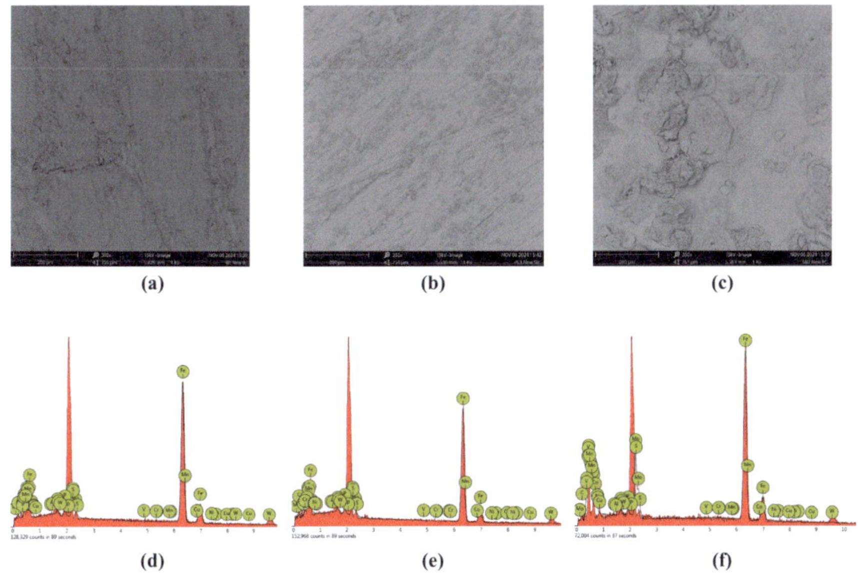

Fig. 9 SEM–EDX Results **a** SEM micrograph of Blank MS strip **b** SEM micrograph of MS strip with PLE Inhibitor **c** SEM micrograph of MS strip with SB inhibition **d** EDX micrograph of Blank MS strip **e** EDX micrograph of MS strip with PLE Inhibitor **f** EDX micrograph of MS strip with SB inhibition

The effectiveness of the organic inhibitors is supported by their ability to reduce corrosion current density and shift the corrosion potential to more positive values [18]. This behavior suggests adsorption of the inhibitor molecules on the MS surface, forming a barrier that impedes both anodic metal dissolution and cathodic hydrogen evolution. The SEM images visually confirm this protective behavior, reinforcing the electrochemical data and weight loss measurements. Furthermore, the denser and more uniform coating in the PLE-treated sample (Fig. 9a) correlates with its higher number of coatings and improved corrosion resistance, highlighting the importance of coating thickness and uniformity in inhibition performance.

Additionally, the EDX spectra results depicted in Fig. 9d–f, and summarized in Table 2, reveal the presence of oxygen on the MS strip coated with SB even after immersion. This suggests the formation of iron oxide due to oxidation reactions occurring during the corrosion process [19]. The detection of oxygen signals supports the partial breakdown of the inhibitor film, leading to localized corrosion. Moreover, a comparison of Fe content between the blank MS strip and the PLE-coated strip shows that the latter has a lower Fe signal, indicating reduced metal dissolution. This suggests that the PLE coating forms a more stable and adherent protective layer that limits corrosion progression [20, 21], further validating its superior performance in acidic environments.

The IR spectra comparison of the three samples PLE, Sample 3A, and Sample 4A is presented in Fig. 10, with the PLE spectrum referenced from Oghenerukevwe

Table 2 Weight quantification from EDX

Compound	Weight concentration percentage		
	Blank MS strip	MS strip PLE inhibitor	MS strip with SB inhibitor
Fe	76.05	65.07	67.09
O	0.42	0	1.35
Others	23.53	34.93	31.56
Total	100	100	100

[13]. The spectral profiles of Sample 3A and Sample 4A closely resemble that of PLE, indicating the presence of similar organic functional groups such as hydroxyl (–OH), carbonyl (C = O), and aromatic (C = C) structures [15]. These characteristic peaks suggest that the molecular framework of the applied inhibitors shares components with PLE, implying either direct derivation or incorporation of PLE-based compounds into Samples 3A and 4A.

The strong absorption bands observed near ~ 3400 cm^{-1} correspond to O–H stretching vibrations, typically associated with alcohols or phenolic groups. The peaks around ~ 1700 cm^{-1} indicate C = O stretching from carboxylic acids or esters, while the absorptions in the region of ~ 1600 cm^{-1} are attributed to aromatic C = C bonds, confirming the presence of conjugated systems [15]. These functional groups are known for their ability to adsorb onto metal surfaces and form protective complexes, which enhance corrosion inhibition.

The presence of these groups in all three samples supports the hypothesis that PLE is either a precursor or an integral component of the formulations used in Samples 3A and 4A. The slight variations in peak intensities or shifts in wavenumber between the samples may be due to structural modifications, additive interactions, or different processing conditions during sample preparation.

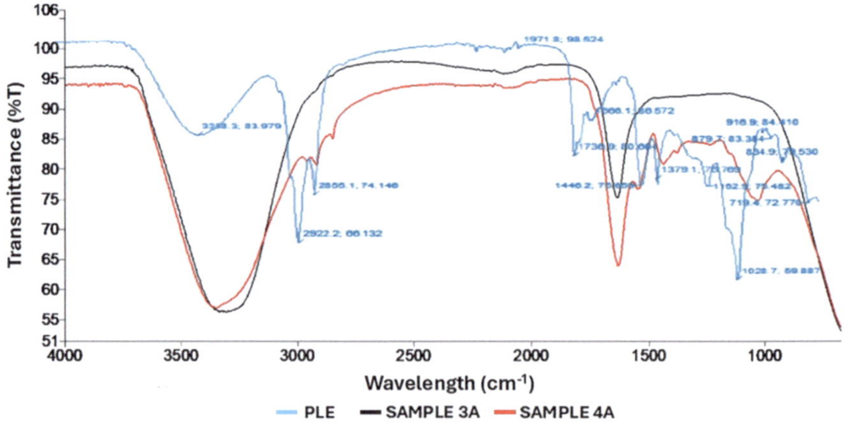

Fig. 10 FTIR results for papaya leaf extract, Sample 3A, and Sample 4A

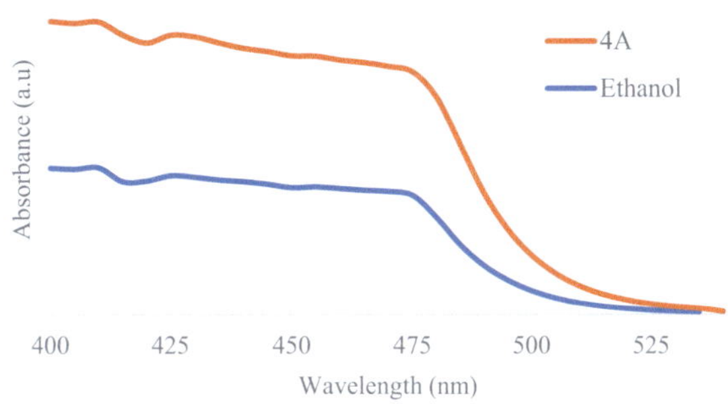

Fig. 11 Merged graph of 2 Samples (Wavelength versus Absorbance)

The chemical similarity and potential functional behavior of the inhibitors reinforce evidence the role of PLE-derived organic compounds as active corrosion-inhibiting agents through surface adsorption and complex formation with metal ions.

3.3.2 UV–Vis Analysis

The UV–Vis spectroscopy results for ethanol (blue line) and the 4A PLE-coated MS strip (green line) are shown in Fig. 11. Ethanol, serving as a control solvent, exhibited low and relatively constant absorbance across the measured wavelengths, establishing a baseline for comparison. In contrast, the 4A PLE-coated MS strip showed a significantly higher absorbance value of 3.959 at lower wavelengths around 415 nm, where ethanol displayed no observable peaks, confirming the presence of light-absorbing compounds in the coated strip. by a reduced corrosion current and a more positive corrosion potential [18].

This increased absorbance suggests a concentration of active components linked to corrosion inhibition [13]. Within the absorption band of 440–470 nm, the 4A PLE-coated MS strip displayed the highest intensity, indicating a greater concentration of active chromophores capable of absorbing UV light, which are essential for its protective properties [22–24]. These chromophores are associated with $\pi \rightarrow \pi^*$ or $n \rightarrow \pi^*$ electronic transitions, characteristics of compounds such as polyphenols, flavonoids, and other conjugated organic molecules. Such compounds, commonly found in papaya leaf extracts, are known for their antioxidant and corrosion-inhibiting properties.

4 Conclusion

This study demonstrated that Ultrasound-Microwave Assisted Extraction (UMAE) significantly outperforms Soxhlet Extraction (SE) in obtaining papaya leaf extract (PLE), yielding 63.43% compared to SE's 6.78%, due to better solvent penetration and preservation of thermosensitive compounds. Qualitative tests confirmed the presence of key bioactive compounds in PLE—alkaloids, flavonoids, saponins, and tannins—that contribute to its corrosion-inhibiting properties. PLE effectively reduced corrosion rates and weight loss in mild steel strips by slowing pH changes and forming a protective barrier, with inhibition efficiency peaking at 97.63% under optimal conditions. While increased acid concentration and excessive coatings diminished effectiveness, an optimal balance of coatings and environmental factors was identified. These findings were supported by SEM–EDX, FTIR, and UV–Vis analyses, which con-firmed the formation of protective films and the presence of functional groups responsible for corrosion inhibition, underscoring PLE's potential as a sustainable, natural corrosion inhibitor.

References

1. Ajoku, N.N.M., Fayomi, O.S.I., Olanrewaju, G., Olusanyan, D.: A mini review on the dependent factor causing corrosion crises: The pictorial perspective. IOP Conf. Ser.: Mater. Sci. Eng. **1107**(1), 012077 (2021)
2. Kadhim, A., Al-Amiery, A.A., Alazawi, R., Al-Ghezi, M.K.S., Abass, R.H.: Corrosion inhibitors: a review. Int. J. Corros. Scale Inhib. **10**(1), 54–67 (2021)
3. Abdulwahab, M., Obot, I.B., Umoren, S.A.: Natural products as corrosion inhibitors: a comprehensive review of current trends and future perspectives. J. Ind. Eng. Chem. **85**, 34–35 (2020)
4. Aribou, Z., Ouakki, M., El Hajri, F., Ech-chihbi, E., Saber, I., Benzekri, Z., Boukhris, S., Al-Sadoon, M.K., Galai, M., Cha-rafeddine, J., Touhami, M.E.: Comprehensive assessment of the corrosion inhibition properties of quinazoline derivatives on mild steel in 1.0 M HCl solution: an electrochemical, surface analysis, and computational study. Int. J. Electrochem. Sci. **19**(11), 100788 (2024)
5. Acharya, D., et al.: Alkaloid of rhynchostylis retusa as green inhibitor for mild steel corrosion in 1 M H2SO4 solution. Elec-trochem **3**(2), 211–224 (2022)
6. Latiza, R.J.P., Olay, J., Eguico, C., Yan, R.J., Rubi, R.V.: Environmental applications of carbon dots: addressing micro-plastics, air and water pollution. J. Hazard. Mater. Adv. **17**(100591), 100591 (2025)
7. Agu, P.C., Okon, K.: The inhibition of mild steel corrosion by papaya and neem extracts. Zast. Mater. **64**(3), 274–282 (2023)
8. Latiza, R.J.P., Rubi, R.V., Quitain, A.: Supercritical water gasification for hospital wastewater. J. Hazard. Mater. Adv. **18**(100651), 100651 (2025)
9. Dwivedi, M.K., Sonter, S., Mishra, S., Patel, D.K., Singh, P.K.: Antioxidant, antibacterial activity, and phytochemical characterization of Carica papaya flowers. Beni-Suef Univ. J. Basic Appl. Sci. **9**(23), 11 (2020)
10. Rubi, R.V., Olay, J., Calugay, P., Diaz, M., Dimayuga, K., Gagui, F., Tare, K.: Ultrasound-microwave assisted extraction (UMAE) of andrographolide from sinta (Andrographis paniculata) with its bioactivity assessment. J. Environ. Sci. Manage. (SE-1), 1–7 (2020)

11. de Souza Morais, W.R., da Silva, J.S., Queiroz, N.M.P., de Paiva e Silva Zanta, C.L., Ribeiro, A.S., Tonholo, J.: Green corrosion inhibitors based on plant extracts for metals and alloys in corrosive environment: a technological and scientific prospection. Appl. Sci. (Basel, Switzerland) **13**(13), 7482 (2023)
12. Ouakki, M., Galai, M., Benzekri, Z., Aribou, Z., Ech-chihbi, E., Guo, L., Dahmani, K., Nouneh, K., Briche, S., Boukhris, S., Cherkaoui, M.: A detailed investigation on the corrosion inhibition effect of by newly synthesized pyran derivative on mild steel in 1.0 M HCl: Experimental, surface morphological (SEM-EDS, DRX& AFM) and computational analysis (DFT & MD simula-tion). J. Mol. Liq. **344**(117777), 117777 (2021)
13. Al-Amiery, A.A., Isahak, W.N.R.W., Al-Azzawi, W.K.: Corrosion inhibitors: natural and synthetic organic inhibi-tors. Lubricants (Basel, Switzerland) **11**(4), 174 (2023)
14. Cullity, B.D., Stock, S.R.: Elements of X-Ray Diffraction, 3rd ed. Pearson Education Limited (2014)
15. Eguico, C.S.-A., Abanto, M.M., Cendaña, H.T., Perez Famero, D.A., Pediongco, K.B., Cruz Evangelista, A.D., Del Castillo Rubi, R.V.: Sonophotopythochemical functionalization of graphene oxide—Al–Zn bimetal nanocomposite for corrosion inhibition. Appl. Sci. Eng. Prog. **18**(2), 7613 (2024)
16. Oghenerukevwe, P.O., Ajuwa, C.I., Samuel, O.D., Adepoju, T.F.: Carica papaya leaf extract (CPLE) as green corrosion inhibitor for AISI 4140 steel in 15% hydrochloric acid medium. Case Stud. Chem. Environ. Eng. **8**(100479), 100479 (2023)
17. Salleh, S., Yusoff, A., Zakaria, S., Taib, M.A., Seman, A., Masri, M., Mohamad, M., Mamat, S., Sobri, S., Arlina, A., Teo, P.: Plant extracts as green corrosion inhibitor for ferrous metal alloys: a review. J. Clean. Prod. **304**, 127030 (2021)
18. Sithuba, T., Masia, N.D., Moema, J., Murulana, L.C., Masuku, G., Bahadur, I., Kabanda, M.M.: Corrosion inhibitory po-tential of selected flavonoid derivatives: electrochemical, molecular-Zn surface interactions and quantum chemical approaches. Results Eng. **16**(100694), 100694 (2022)
19. Miralrio, A., Espinoza Vázquez, A.: Plant extracts as green corrosion inhibitors for different metal surfaces and corrosive media: a review. Processes (Basel, Switzerland) **8**(8), 942 (2020)
20. González-Parra, J., Di Turo, F.: The use of plant extracts as sustainable corrosion inhibitors for cultural heritage alloys: a mini-review. Sustainability **16**(5), 1868 (2024)
21. Latiza, R.J.P., Rubi, R.V., Olay, J., Soriano, A.: Is the future of energy rotten? Novel perspective on tri-phase fermentation and the food waste paradox. Appl. Sci. Eng. Prog. (2025)
22. Pratiwi, R., Nandiyanto, A.: How to read and interpret UV-VIS spectrophotometric results in determining the structure of chemical compounds. Indones. J. Educ. Res. Technol. **2**(1), 1–20 (2022)
23. Latiza, R.J.P., Mustafa, A., Delos Reyes, K., Nebres, K.L., Abrogena, S.L., Rubi, R.V.: Optimization, characterization, reversible thermodynamically favored adsorption, and mechanistic insights into low-cost mesoporous Fe-doped kapok fibers for efficient caffeine removal from water. Sep. Sci. Technol., 1–16 (2025)
24. Cordero, M.T., Agustin, J.C., Alfaro, R.E., Caoile, J.R., Ercilla, A.J., Latiza, R.J.P., Evangelista, A., Raguindin, R., Rubi, R.V.: Lacustrine agroforestry waste-derived coconut shell nanobiochar impregnated with chocolate mint extract for antimicrobial textile applications. Emergent Mater. (2025)

A Comparative Analysis of Statistical Correlation and Machine Learning Feature Importance for the Building Shape Coefficient in Cooling Load Estimation from a Higher Institution in the Philippines Building

Mark John Paul T. Obsid, Basil Aloysius P. Bautista, and Dexter Lyndon Q. Sabusap

Abstract This study examines the role of the Building Shape Coefficient in cooling load estimation by comparing statistical correlation and machine-learning feature importance. A dataset of 10,000 simulated building cases was generated using HAP, EnergyPlus, and jEPlus Parametric Study, incorporating seven building and thermal parameters, including floor level, window-to-wall ratio, gross floor area, target temperature, external wall U value, window U value, and the Building Shape Coefficient. After preprocessing and outlier removal, correlation analysis and machine learning with Support Vector Machine, Random Forest, and XGBoost were applied. The statistical results show that the Building Shape Coefficient has a moderate negative correlation with cooling load, confirming that compact buildings generally require less cooling. However, features such as floor level, gross floor area, and window-to-wall ratio demonstrated stronger positive correlations and greater influence. Machine learning analysis further revealed that the Building Shape Coefficient contributed less consistently across models. Its influence was weak in Support Vector Machine, moderate in Random Forest, and strongest in XGBoost when combined with thermostat setpoints. SHAP analysis confirmed that while the Building Shape Coefficient is physically significant, its predictive role is overshadowed by other dominant variables. The findings suggest that the Building Shape Coefficient remains an important geometric indicator of cooling demand, but design strategies should prioritize floor level, gross floor area, and window-to-wall ratio as primary drivers

M. J. P. T. Obsid (✉) · B. A. P. Bautista · D. L. Q. Sabusap
Adamson University, Manila, Philippines
e-mail: mark.john.paul.obsid@adamson.edu.ph

B. A. P. Bautista
e-mail: basil.aloysius.bautista@adamson.edu.ph

D. L. Q. Sabusap
e-mail: dexter.lyndon.sabusap@adamson.edu.ph

M. Awang et al. (eds.), *Proceedings of the International Conference on Mechanical, Manufacturing and Plant Engineering*, Proceedings in Technology Transfer,
https://doi.org/10.1007/978-981-95-7817-7_31

of cooling load. The integration of statistical correlation and machine learning interpretability provides a comprehensive framework for evaluating the role of building geometry in energy performance analysis.

Keywords Building shape coefficient · Cooling load estimation · Statistical correlation · Machine learning · SHAP analysis

1 Introduction

The variation in physical features of the building, such as wall surface area and shape coefficient, has a significant influence on cooling and heating requirements across different typologies [1]. Geometric parameters such as building shape coefficients influence heat transfer rates and cooling load requirements. These suggest that the compactness of a building is a significant parameter in energy thermal design, depending on building type. Building shape has been identified as a universal strategy for reducing energy demand as a reliable metric for evaluating geometric compactness [2]. This shows the strong link between compact geometry and lower energy demand, reinforcing the importance of using shape-based measures to evaluate building performance. The role of the surface-to-volume ratio in building energy performance under future climate change scenarios, showing that more compact buildings with smaller surface areas relative to their volume reduce cooling demand in hot regions and mitigate increases in moderate climates. Building shape is a critical factor influencing energy consumption, yet its role in heating and cooling demand remains insufficiently understood, especially during the early design phase. The study emphasized that while the building shape coefficient is widely used, it has limitations in physical interpretation. Research relies heavily on simulations of a few common building forms, such as rectangular, square, L-shaped, and U-shaped layouts, with limited attention to irregular geometries or large-scale multi-parameter simulations, leaving gaps in the standardization of design tools [3]. The importance of the building shape coefficient as a practical indicator of thermal performance, while calling for more comprehensive simulation-based studies across diverse shapes and climates, for a better guide in building design. Buildings are increasingly vulnerable to the impacts of climate change, with rising greenhouse gas emissions and changing weather patterns significantly altering energy demand profiles.

A simulation-based analysis of office building prototypes revealed that medium-sized office buildings are the most vulnerable to climate change, showing an average increase of 12.5% in Energy Use Intensity by 2080, compared to 7.4% for large offices and a slight decline of 2.5% for small ones. Passive design strategies such as proper orientation, shading, natural ventilation, and high-performance envelopes, combined with efficient HVAC systems and stricter building regulations, were emphasized as critical in mitigating rising cooling demands, especially for medium-sized buildings in climate-sensitive regions [4]. Building size and form play a central role in determining resilience to future climatic conditions. While compact and well-designed building forms can significantly reduce vulnerability, studies showed that the traditional shape coefficient does not always capture real thermal performance, since solar radiation, orientation, and wall or window insulation strongly affect energy demand. A modified coefficient that integrates these factors reduced values by more than 30%, offering a more accurate prediction of actual building energy use [5, 6]. These findings suggest that simulation-based applications of the shape coefficient must incorporate thermal and climatic factors to better reflect its role in energy performance. Machine learning methods provide further insights into this relationship. A study in Dongguan, China, integrated large-scale smart meter data with 3D building footprints using XGBoost and SHAP. Results showed that the building shape coefficient was the most influential factor, with a strong positive correlation to energy use intensity. Threshold analysis revealed that when the coefficient exceeded 0.38, energy use rose sharply, underscoring the role of compactness in hot summer and warm winter climates [7]. This demonstrates the advantage of combining simulation with explainable machine learning for identifying key design thresholds. Other studies applied machine learning to predict cooling loads using architectural features. For tropical buildings, ensemble and tree-based models achieved the highest accuracy, while support vector regression performed the worst [8]. Random Forest was found most effective for predicting both energy consumption and comfort in radiant cooling systems, achieving the lowest errors [9]. In larger simulation datasets with over 9,000 building scenarios, artificial neural networks outperformed other methods with a prediction accuracy of 98%, compared to only 67% for linear regression [10]. Hybrid models have also been explored: a combination of support vector regression and group method of data handling achieved superior accuracy with correlation values above 99% [11]. While a particle swarm optimization–XGBoost model provided the best heating load predictions, identifying building height, roof, wall, and surface areas as the most influential variables [12]. Similarly, in China, a machine learning framework was developed across six building types and five climate zones. Using nine design-related features and eight different algorithms, the models achieved accuracy above 90%. Among them, gradient boosting and k-nearest neighbors were most effective, with climate zone and floor height identified as the strongest predictors of cooling load [13]. These studies provide evidence that machine learning techniques, with a focus on ensemble and hybrid models, achieve high predictive accuracy in estimating cooling loads. The findings also indicate that while the building shape coefficient is a significant parameter, its effect must be considered alongside other influential factors, including climate zone, floor level, and building envelope performance.

The influence of individual features on cooling load predictions using SHAP (SHapley Additive exPlanations) values. The study examines feature importance both locally, for a single point in time, and globally, for the entire dataset, across many various buildings with different characteristics. The results show a correlation between building characteristics and the importance of features for cooling load prediction. Therefore, building characteristics are implicitly represented in the SHAP values, enabling a purely data-driven interpretation of these characteristics. Additionally, the study demonstrates that local SHAP values provide a satisfactory description of heat flows in a room, offering data-driven insights into building heat dynamics [14]. Also, an investigation of how building characteristics influence feature importance in heating load predictions using 15,552 simulated room variants. Methods included thermal balance analysis, correlation coefficients, Random Forest, Permutation, and SHAP values. Outdoor temperature emerged as the most influential feature, and SHAP values showed the strongest correlation with thermal balance characteristics. The study demonstrates a systematic relationship between building characteristics and feature importance, highlighting that models must be interpreted in the context of specific building attributes to improve model interpretability [15]. An Explainable AI framework integrating XGBoost with SHAP and LIME to enhance the interpretability and accuracy of multi-scale household energy forecasting. SHAP values provide insights into global and local features, while LIME offers detailed explanations for individual predictions. Permutation Feature Importance identifies key determinants, and Partial Dependence Plots illustrate input–output relationships. The incorporation of Holt-Winters Exponential Smoothing captures both short- and long-term trends across various time intervals, from hourly to appliance-level consumption [16]. LightGBM and CatBoost to predict energy use intensity for U.S. apartments and single-family houses. SHAP analysis identified key drivers such as building size, heating type, climate, and age. The study provides actionable insights for energy-efficient building design [17]. A combined SHAP with data-driven model predictive control to improve the interpretability of artificial neural networks in building energy systems. Using the Hydronic Heat Pump test case, the study identified how feature characteristics affect predictions, highlighted unimportant features, and explained performance differences between models trained on base and exploring datasets. The results demonstrate that SHAP can uncover model weaknesses, guide feature selection, and support comparison across different ANN models, enhancing both transparency and control performance [18]. These studies emphasize that SHAP offers a powerful approach to interpret machine learning predictions in building performance analysis. They highlight the importance of considering individual building characteristics, such as shape, size, and envelope properties, when assessing feature influence. For studies focused on the Building Shape Coefficient (BSC) in Philippine buildings.

The predictive contribution of the Building Shape Coefficient (BSC) remains underexplored. Most studies examine BSC alongside multiple features such as floor area, floor level, envelope properties, and climate factors, making it unclear whether its effect is statistically significant but predictively overshadowed by stronger predictors. Furthermore, existing analyses are predominantly based on temperate or subtropical climates, with limited investigation in tropical contexts like the Philippines, where climate conditions may alter the influence of BSC on cooling load. Many studies also rely on simplified or prototypical building geometries and small simulation datasets, restricting the ability to generalize findings across diverse building shapes. While SHAP analyses have highlighted dominant predictors such as building features, the relative importance of BSC is often underrepresented, and its dynamic interactions or threshold effects remain insufficiently analyzed. Finally, few studies translate feature importance into actionable guidance for early stage building design, leaving the practical role of BSC in optimizing cooling loads inadequately addressed. Addressing these gaps is essential to clarify whether the BSC is statistically meaningful yet predictively overshadowed, particularly in tropical building contexts, and to support evidence-based design strategies. The primary objective of this study is to investigate the role of the Building Shape Coefficient (BSC) in cooling load estimation for Philippine buildings, using a combination of large-scale simulation data and feature importance analysis.

2 Methodology

The paper employs a quantitative research design integrating simulation-based data generation, statistical analysis, and machine learning modelling. The approach compares the statistical correlation of the Building Shape Coefficient (BSC) with cooling load against its predictive feature importance in machine learning models. The research was conducted using a Philippine Building design, which was parameterized and simulated in HAP, EnergyPlus, and jEPlus Parametric Study.

A modelling framework consisted of three key stages: (1) identifying input variables to generate a raw dataset, (2) pre-processing and correlation analysis of the dataset (3) evaluating the performance of different machine learning models. The research workflow illustrated in Fig. 1 outlines the process from dataset generation to feature importance.

A dataset with seven input variables on building and thermal features by batch simulation using Hap, EnergyPlus, and jEPlus Parametric Study. The dataset is stored in a CSV file, which was then pre-processed to remove outliers and perform correlation analysis. SVM, Random Forest, and XGBoost are introduced as algorithms to analyze and compare predictive performance. Finally, a comparative analysis was conducted to contrast the statistical correlation of Building Shape Coefficients with its machine learning feature importance.

2.1 Variable Selection for Building Cooling Load Modelling

The paper focuses on the peak cooling demand of the building. The identification of critical variables reflecting building characteristics is fundamental to building a dataset for thermal load prediction. Seven feature variables were selected in this

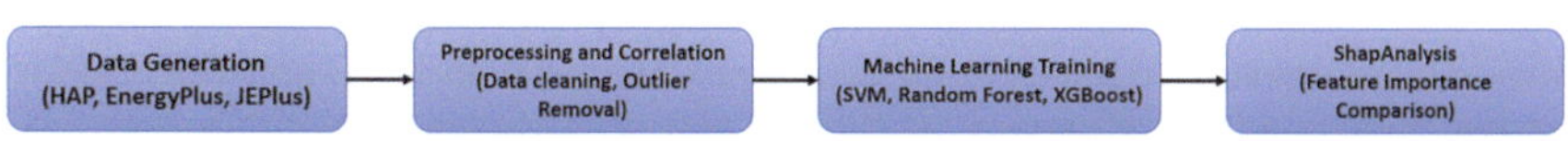

Fig. 1 Research framework showing dataset generation, preprocessing, machine learning, and SHAP analysis

Table 1 Building feature variables with corresponding value ranges

Category	Parameters	Values
Geometry	Floor level	1, 3, 4, 8, 12
	Window-to-wall ratio	0.1, 0.4, 0.6, 0.8
	Building shape coefficient	Computed (range: approx. 0.13–0.44)
	Gross floor area (m^2)	From HAP (331.60–10,484)
Envelope	Overall U-external wall (W/m·K)	0.4, 0.6, 0.8, 1.6, 2.0
	Overall U-window (W/m^2-K)	5.9, 3.1, 2.6, 2.2, 1.9
Thermostat	Target temperature	18, 19, 20, 21

study to generate the modeling dataset. These include Floor Level, Window-to-Wall Ratio (WWR), Building Shape Coefficient (BSC), Gross Floor Area (m^2), Target Temperature (°C), Overall U-Value of the External Wall (W/m^2·K), and Overall U-Value of the Window (W/m^2·K).

In this study, the independent variables comprised both controlled and derived parameters. The Building Shape Coefficient (BSC) was subsequently calculated from each model's geometric properties rather than directly assigned. This combination of controlled inputs computed parameters, and simulation-derived values ensured that the dataset reflected realistic building conditions while retaining sufficient variability for cooling load prediction. Table 1 summarizes the independent variables, indicating their classification as controlled, computed, or simulation-derived, along with the corresponding value ranges used in the modeling process.

2.2 *Parametric Model Generation and Simulation*

As shown in Fig. 2, the dataset used in this study was developed through a stepwise simulation workflow that combined manual geometric modification with automated parametric analysis. First, a base building was modeled in HAP. Once simulated, HAP automatically generated the corresponding EnergyPlus IDF file, from which the Building Shape Coefficient (BSC) was computed using the building volume and the summation of external wall and roof areas. The IDF file was then adapted for parametric analysis in jEPlus Parametric Study, where systematic variations in envelope and thermal parameters, specifically the Overall U value of the external wall, the Overall U value of the window, and the thermostat setpoint temperature.

The procedure was systematically repeated for different combinations of floor levels and window-to-wall ratio (WWR) values. Besides the building feature variables, annual weather data were used as input files in both HAP and EnergyPlus. Internal heat gain schedules, including occupancy schedules, lighting operation schedules, and equipment operation schedules, were assigned based on ASHRAE Standard 183 Peak Load Calculation. With weather data and internal heat gains

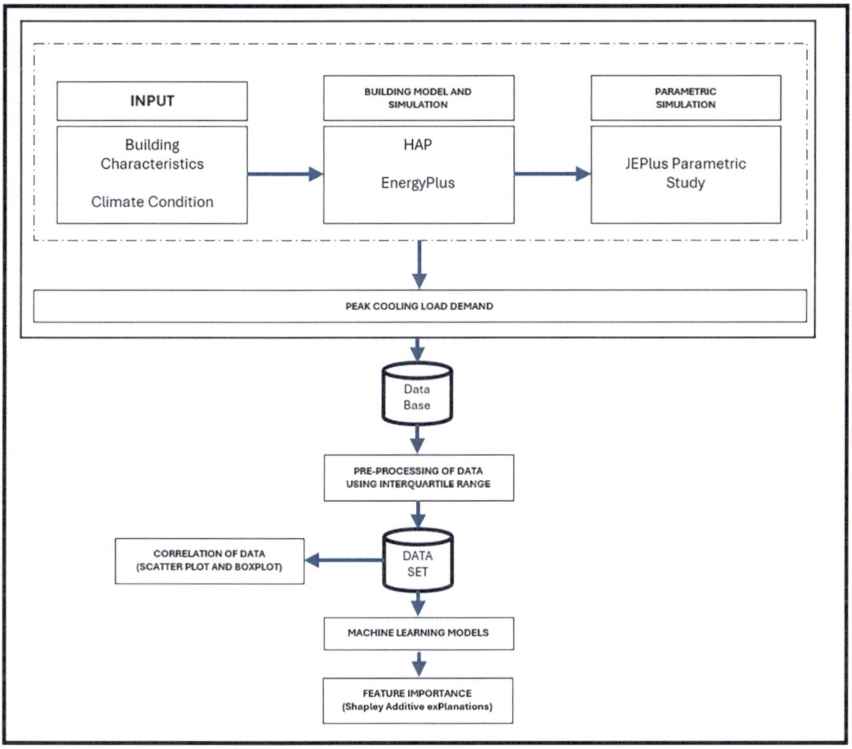

Fig. 2 Methodological framework for cooling load estimation using simulation and machine learning with SHAP-based feature importance analysis

applied, 10,000 dataset models of each building type were simulated, and their corresponding peak cooling loads were obtained. All simulation outputs were consolidated into a single CSV database, resulting in a comprehensive dataset of approximately 10,000 cases that captured both geometric and thermal variations.

According to the generation approach, each model in the dataset contains seven building feature variables as independent inputs and the corresponding cooling load as the dependent output. The simulation results were consolidated into a CSV file, which served as the primary database for preprocessing, correlation analysis, and machine learning applications. In this structure, each row represents a single building model configuration, comprising seven independent variables: floor level, window-to-wall ratio, building shape coefficient, gross floor area, target temperature, overall U-value of the external wall, and overall U-value of the window, along with the calculated cooling load. This tabular format ensured that the dataset was structured, easily accessible, and directly usable for subsequent statistical and machine learning analysis.

2.3 Data Preprocessing and Statistical Analysis

The raw dataset containing ten thousand entries was first subjected to Interquartile Range filtering to remove statistical outliers. This process reduced the dataset to 7,680 valid cases, thereby eliminating approximately 23% of the extreme values. The removal of these outliers improved the consistency of the dataset by reducing noise and bias, which strengthened its suitability for subsequent regression and machine learning analysis. Following preprocessing, exploratory statistical techniques were applied, including the Coefficient of Variation (CV) and the Coefficient of Quartile Variation (CQV), to quantify the relative variability of input variables. Finally, Pearson correlation analysis and scatter plots were used to examine the relationships between building feature variables and cooling load, providing a statistical baseline against which the results of machine learning feature importance could later be compared.

2.4 Predictive Modeling with Machine Learning

After statistical analysis, three supervised machine learning algorithms were employed to evaluate the predictive contribution of the Building Shape Coefficient (BSC) in cooling load estimation. The models used were Support Vector Machine (SVM), Random Forest (RF), and Extreme Gradient Boosting (XGBoost), which were selected for their proven ability to capture nonlinear relationships and handle relatively large datasets. Each model was trained using the seven building feature variables as independent inputs and cooling load as the dependent output.

SVM is a kernel-based method that performs regression by mapping inputs into a higher-dimensional space, where an optimal hyperplane is identified to capture nonlinear relationships. In this study, an SVM with a radial basis function (RBF) kernel was implemented, with hyperparameters. Random Forest is an ensemble method that constructs multiple decision trees and averages their predictions, thereby improving generalization and reducing overfitting. XGBoost is a gradient boosting algorithm that sequentially builds decision trees to minimize residual errors, offering strong predictive performance and computational efficiency on large datasets. All models were implemented in Python using the scikit learn and XGBoost libraries. The dataset was divided into training 80% and testing 20% subsets. These three algorithms were selected to compare statistical correlation and machine learning feature importance in assessing the role of the Building Shape Coefficient in estimating cooling loads.

Table 2 presents a summary of the model hyperparameters, and validation settings used in this study. This provides a reference for the implementation of each machine learning method. For the SVM, the penalty parameter (C), kernel coefficient (γ), and epsilon (ε) were set to capture nonlinear relationships while controlling error tolerance. The Random Forest model used 100 estimators, which provided stable

Table 2 Model hyperparameters and validation settings

Model	Key hyperparameters	Validation approach
SVM (RBF Kernel)	C = 100, $\gamma = 0.1$, $\varepsilon = 0$	1, 3, 4, 8, 12
Random forest	n_estimators = 100 random_state = 42	0.4, 0.6, 0.8, 1.6, 2.0
XGBoost	n_estimators = 100, objective = reg:squarederror, random_state = 42	18, 19, 20, 21

predictions without excessive computation. For XGBoost, the squared error objective and 100 boosting rounds were applied, allowing the model to optimize residuals effectively while avoiding overfitting. The 80–20 train-test split ensured consistent evaluation of model generalization.

2.5 *Interpretation of Machine Learning Models with SHAP Values*

The interpretability of the machine learning models, Shapley Additive exPlanations (SHAP), was applied. SHAP is based on cooperative game theory and assigns each input variable a value that reflects its contribution to the model prediction. This approach allows for a transparent examination of how the Building Shape Coefficient (BSC) and other features influence cooling load estimation. SHAP analysis provides both global explanations, which rank the overall importance of variables across the dataset, and local explanations, which quantify how individual feature values affect specific predictions. This dual capability makes SHAP particularly suitable for building energy studies, where both overall trends and case-specific influences are important for interpretation.

In this study, SHAP values were computed in Python for the trained Support Vector Machine, Random Forest, and XGBoost models. Beeswarm plots were used to present global feature importance, while dependence plots were generated to illustrate how variations in BSC interact with other features in shaping cooling load predictions. By combining statistical correlation with SHAP-based feature importance, the methodology established a comparative framework that highlights both statistical associations and predictive significance.

3 Results and Discussion

3.1 Outlier Detection and Removal Using IQR

To improve the reliability of the analysis, outlier removal was performed using the Interquartile Range (IQR) method. Outliers were identified as data points falling outside the range of the building feature variables. After applying this method, the dataset was reduced from 10,000 to 7680 entries, approximately 23% of the cases were considered outliers and removed. This preprocessing step ensured that extreme values were minimized. This produced a dataset that preserved the essential distribution of BSC relative to cooling load, thereby improving the reliability of both the statistical correlation analysis and the subsequent machine learning feature importance evaluation.

3.2 Coefficient of Variation Analysis of Building Features and Cooling Load

The Coefficient of Variation (CV) was calculated for all input and target variables to assess relative variability in the dataset. Gross Floor Area and Floor Level showed the highest variability, reflecting wide differences in building design, while Cooling Load (TR) also exhibited significant variation. The Building Shape Coefficient (CV = 0.2605) displayed relatively low variability compared to other geometric and thermal features, which may explain its reduced predictive influence in machine learning-based cooling load estimation. Target Temperature showed minimal variation, supporting its role as a stable setpoint parameter. A summary of CV values for all features is presented in Table 3. The framework for building a test model is shown in Fig. 3.

Table 3 Coefficient of variation (CV) for input and target variables, showing relative variability across building design parameters and cooling load

Coefficient of variation (CV)	Value	Interpretation
Floor level	0.73	High variability
Window-wall ratio	0.54	High variability
Building shape coefficient	0.26	Moderate variability
Gross floor area (m^2)	0.77	High variability
Target temperature (C)	0.06	Very low variability
Overall U-external wall (W/m·K)	0.57	High variability
Overall U-window ($W/m^2 \cdot K$)	0.19	Low variability
Cooling load (TR)	0.71	High variability

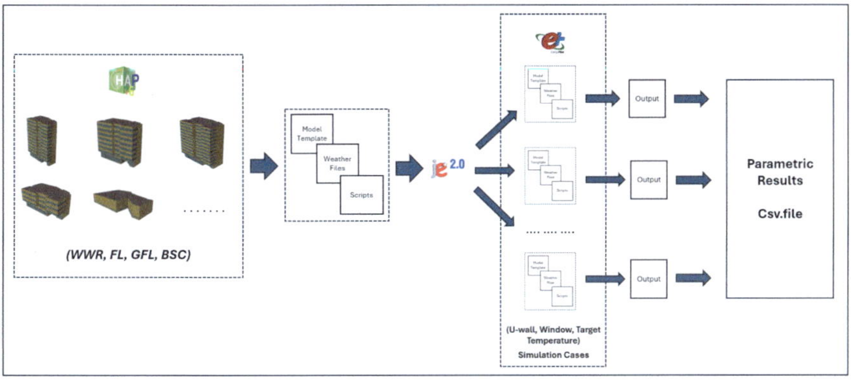

Fig. 3 Framework for building a test model

3.3 Scatter Plot and Box Plot Analysis of Input Features and Cooling Load

The scatter plot analysis shown in Fig. 4, that Floor Level, Gross Floor Area, and Window-to-Wall Ratio exhibit strong positive correlations with Cooling Load, indicating that taller buildings, larger areas, and greater glazing ratios directly increase cooling demand. In contrast, Target Temperature and the U-values of external walls and windows display weak or negligible correlations, suggesting limited influence within the simulated cases. The Building Shape Coefficient demonstrates a moderate negative correlation with Cooling Load, but its effect is smaller than that of Floor Level and Gross Floor Area, which vary more widely across the dataset. The box plot results shown in Fig. 5 reinforce these patterns by showing greater variability in cooling load for higher floors, larger buildings, and higher glazing ratios, while compact buildings with higher BSC values exhibit reduced and more stable cooling demand. This indicates that although BSC influences cooling performance, it is less dominant compared to structural parameters that capture larger variation. The Pearson correlation heatmap shown in Fig. 6 synthesizes these findings, confirming strong positive associations of Cooling Load with Floor Level ($r = 0.96$) and Gross Floor Area ($r = 0.89$), a moderate negative correlation with BSC ($r = -0.67$), and negligible effects of Target Temperature and U-values. Together, these results show that while BSC remains a meaningful geometric indicator, its predictive influence is outweighed by the stronger and more variable drivers of cooling load in Philippine building design.

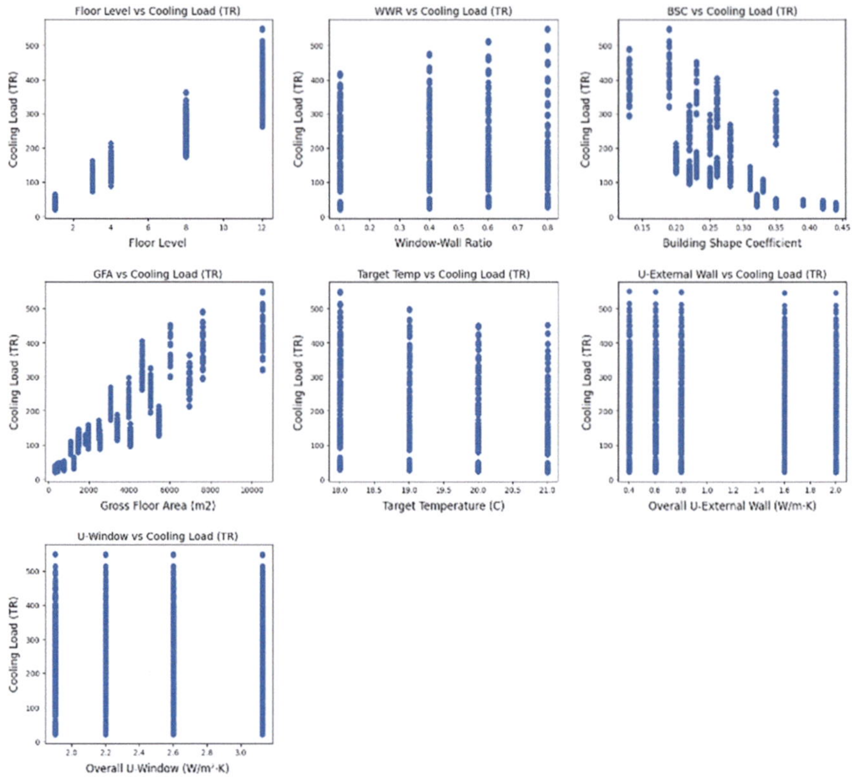

Fig. 4 Scatter plots showing the relationship between input features and cooling load (TR) for all simulated building cases

3.4 Feature Importance Analysis Using SHAP

Figure 7 presents the SHAP analysis for the three machine learning models to analyze the building features in predicting cooling load. Floor Level consistently emerges as the most dominant predictor, particularly in tree-based models, confirming that higher floors demand greater cooling capacity. Gross Floor Area shows high importance in SVM, reflecting the direct effect of size in linear models, while the Building Shape Coefficient (BSC) is especially influential in XGBoost, underscoring the impact of geometry when nonlinear interactions are considered. Target Temperature and Window-to-Wall Ratio contribute moderately, suggesting that indoor setpoints and glazing design affect cooling load but less strongly than geometry and scale. In contrast, U-values of walls and windows contribute minimally across all models, implying that within this dataset, insulation plays a smaller role than structural and geometric factors.

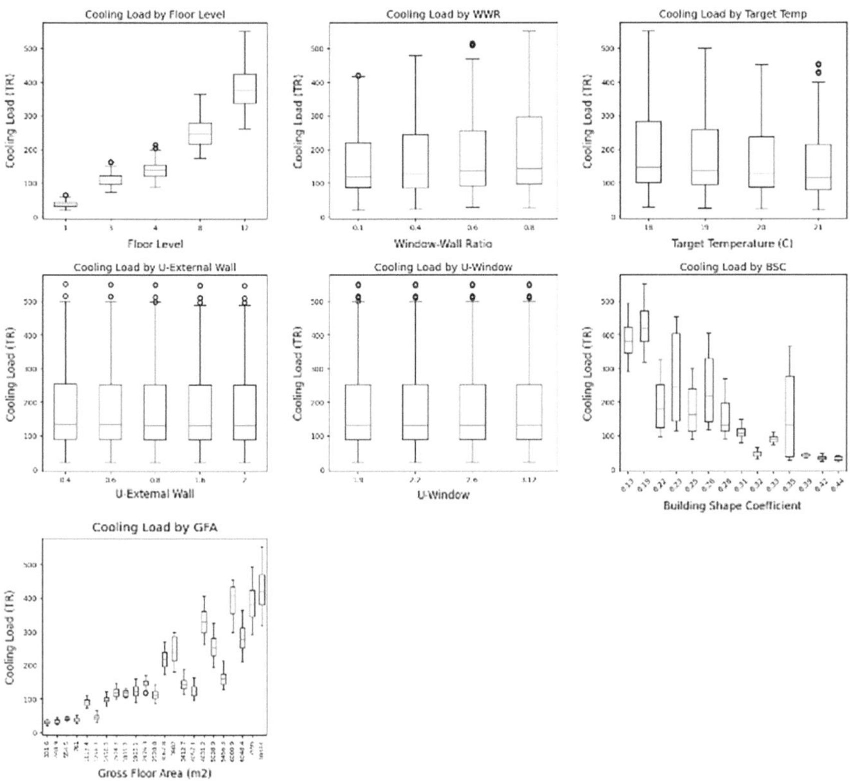

Fig. 5 Box plots showing the relationship between input features and cooling load (TR) for all simulated building cases

3.5 Dependence Plot Analysis Using SHAP

To further investigate interaction effects, SHAP dependence plots were generated for the Building Shape Coefficient (BSC) and Target Temperature (Figs. 8, 9 and 10). BSC was selected as the primary feature of interest due to its role in representing geometric compactness, while Target Temperature emerged as the most influential thermal parameter. The SVM dependence plot (Fig. 8) shows only a modest influence of BSC, with SHAP values clustering near zero and limited interaction with Target Temperature. This suggests that SVM struggles to capture nonlinear dependencies between compactness and thermostat setpoints. In contrast, Random Forest (Fig. 9) reveals a clearer pattern, where higher BSC values correspond to greater cooling load contributions. The color gradient shows that lower setpoints intensify the effect of BSC, while higher setpoints dampen it, highlighting nonlinear feature interactions. The XGBoost dependence plot (Fig. 10) captures the most complex relationships. As BSC increases, SHAP values diverge into both strong positive and negative contributions, with the interaction most pronounced at lower setpoints. This indicates that

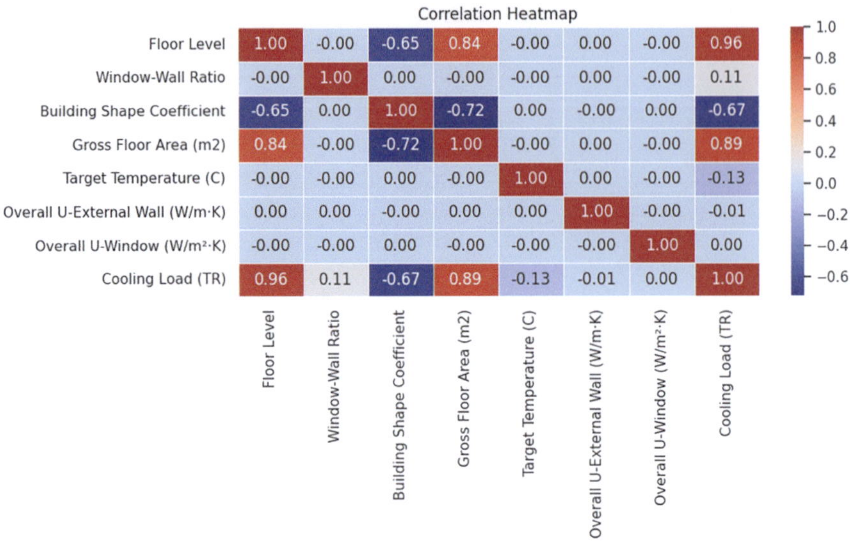

Fig. 6 Correlation heatmap of building features and cooling load (TR) using Pearson correlation coefficients

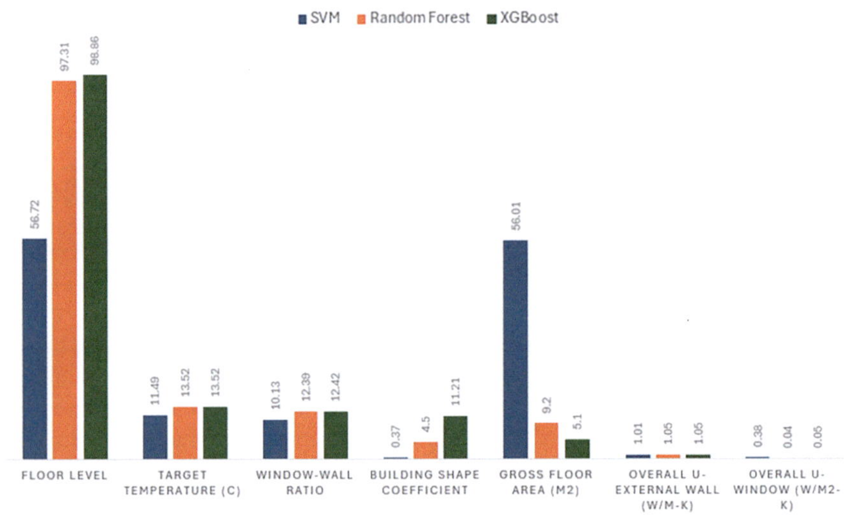

Fig. 7 Comparison of SHAP Feature Importance across models

XGBoost not only recognizes the direct impact of compactness but also captures its interaction with indoor temperature control, making it the most sensitive model to geometric–thermal interactions. The dependence plots confirm that BSC contributes meaningfully to cooling load estimation, but its influence varies by algorithm. SVM treats it as marginal, Random Forest shows moderate interaction, and XGBoost highlights its complexity and nonlinear role. These findings reinforce the conclusion that while BSC is a significant design factor, its predictive role becomes most evident when advanced ensemble models are employed.

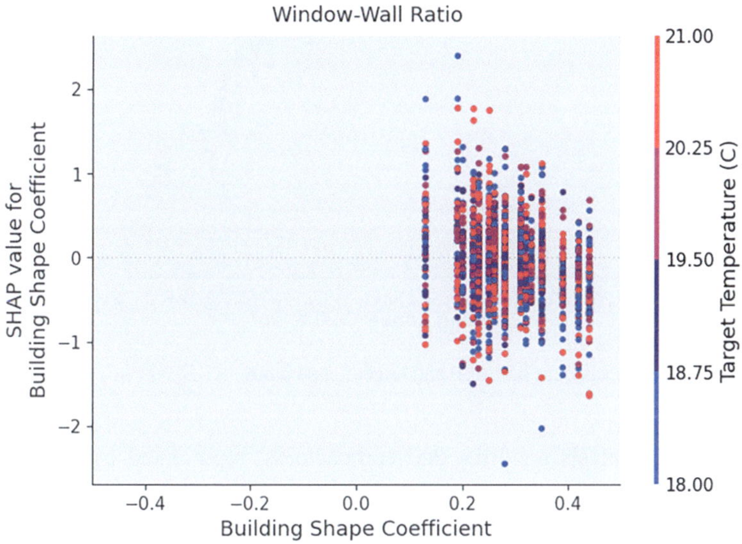

Fig. 8 Support vector machine (SVM) dependence plot BSP and target temperature

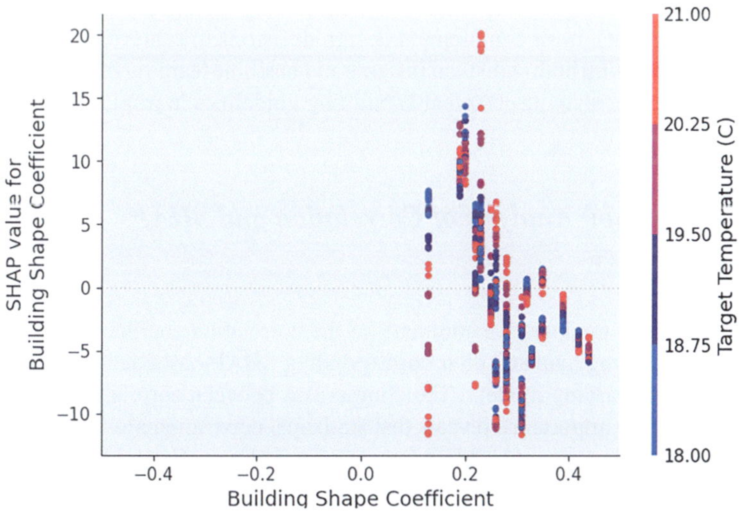

Fig. 9 Random forest dependence plot BSP and target temperature

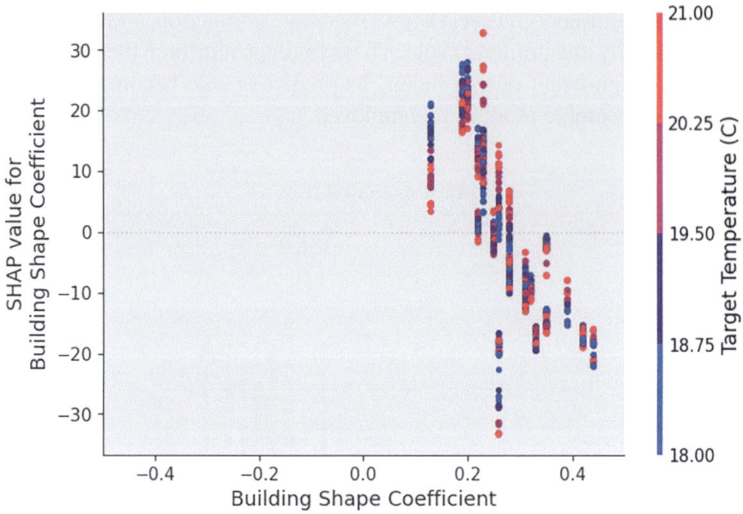

Fig. 10 XGBoost dependence plot BSP and target temperature

The SHAP analysis confirms that compactness, represented by the Building Shape Coefficient, interacts strongly with other features such as indoor setpoints, but its predictive role is most evident in models like XGBoost. For Philippine buildings, where cooling dominates cooling load design, these results imply that design strategies should prioritize floor levels, floor area, and glazing ratios while adopting more compact geometries to minimize cooling demand. Additionally, the interaction between BSC and Target Temperature highlights that operational controls, such as thermostat setpoints, must complement design decisions to achieve optimal energy efficiency. Integrating both statistical insights and machine learning findings provides a stronger basis for shaping sustainable building guidelines in tropical climates.

3.6 Comparison Analysis of Correlation and SHAP Importance

Table 4 presents a comparative summary of the correlation coefficients of building features with cooling load and their corresponding SHAP-based feature importance across machine learning models. The comparison between correlation coefficients and SHAP feature importance reveals that statistical correlation alone does not fully capture the predictive role of building features. For instance, Gross Floor Area shows a strong positive correlation with cooling load (r = 0.89), yet its SHAP importance is relatively low in Random Forest (9.2%) and XGBoost (5.1%). Conversely, Floor Level not only has a high correlation (r = 0.96) but also consistently dominates SHAP rankings across all models, exceeding 95% in Random Forest and XGBoost.

The Building Shape Coefficient exhibits a moderate negative correlation with cooling load (r = –0.67), but its SHAP importance varies widely, being negligible in SVM (0.37%) and more pronounced in XGBoost (11.21%). This highlights the distinction between correlation, a measure of linear association, and SHAP values, indicators of predictive contribution in nonlinear models.

The comparative analysis further clarifies the practical role of each feature in cooling load estimation. Floor Level and Gross Floor Area emerged as the strongest determinants, with higher floor positions and larger volumes consistently driving higher cooling demand, particularly relevant in tropical climates. The Window-to-Wall Ratio also showed moderate yet consistent influence, confirming the role of glazing design in controlling solar heat gains. The Building Shape Coefficient, while negatively correlated with cooling load, was less dominant in predictive models, suggesting that its influence is exceeded by features with greater variability, such as Floor Level and Gross Floor Area. Target Temperature demonstrated moderate importance, emphasizing the impact of setpoint control on. In contrast, the U-values of walls and windows contributed minimally, suggesting that material thermal properties were less influential in this dataset compared to geometric and design-related parameters.

From the perspective of Philippine building design, these results underscore the importance of prioritizing geometric features that strongly influence cooling demand. The dominance of Floor Level and Gross Floor Area in both correlation and SHAP-based feature importance suggests that vertical expansion and larger floor areas substantially increase cooling loads. This is highly relevant in the tropical climate of the Philippines, where cooling accounts for a major share of energy use. The consistent influence of the Window-to-Wall Ratio further emphasizes the need for careful glazing design to minimize solar heat gains while maintaining daylighting. Although the Building Shape Coefficient was less dominant in predictive models, its negative correlation with cooling demand highlights the value of compact and well-proportioned forms in reducing external surface exposure. These findings imply that Philippine building design should prioritize compact geometries, controlled glazing ratios, and optimized floor layouts alongside efficient HVAC systems and materials.

Table 4 Comparison of correlation coefficients and SHAP feature importance across models

Feature	(r)	Shap importance SVM (%)	Shap importance RF (%)	Shap importance XGBoost (%)
FL	0.96	56.72	97.31	98.66
WWR	0.11	10.13	12.38	12.42
BSP	-0.67	0.37	4.5	11.21
GFA	0.89	56.01	9.2	5.1
TT	-0.13	11.49	13.42	13.52
U-wall	-0.01	1.01	1.05	1.05
U-window	0.00	0.38	0.04	0.05

Integrating these strategies into architectural practice and policy can significantly improve the energy performance of tropical low- and mid-rise buildings.

4 Conclusion

This study conducted a comparative analysis of the Building Shape Coefficient in cooling load estimation by examining both its statistical correlation with cooling demand and its predictive feature importance in machine learning models. Using a dataset generated from parametric simulations in HAP, EnergyPlus, and jEPlus Parametric Study, seven building feature variables were analyzed with particular focus on the Building Shape Coefficient as a measure of building compactness. The results reveal that the Building Shape Coefficient shows a moderate negative correlation with cooling load, which means that compact buildings generally require less cooling. However, its predictive importance is overshadowed by more dominant variables such as floor level, gross floor area, and window-to-wall ratio. Statistical analysis confirmed the physical significance of the Building Shape Coefficient in reducing cooling demand, but its moderate variability within the dataset limited its influence in machine learning predictions.

The machine learning models provided further insights into the role of the Building Shape Coefficient. Support Vector Machine reflected minimal sensitivity to this parameter, Random Forest captured stronger nonlinear effects, and XGBoost demonstrated the most significant interaction between the Building Shape Coefficient and cooling load, particularly when combined with thermostat setpoints. SHAP analysis highlighted that although the Building Shape Coefficient consistently contributed to prediction, its relative influence varied across algorithms, which reinforces the need to evaluate building geometry together with other design parameters. The Building Shape Coefficient remains an important geometric parameter in cooling load estimation, particularly in statistical and physical terms, but its predictive role is less dominant compared to features with greater variability. The findings suggest design strategies in the Philippine context should prioritize highly influential variables such as floor level, gross floor area, and window-to-wall ratio while recognizing the complementary role of compact building geometry. This study shows that in building-specific simulations, its role is outweighed by other features, despite being physically and statistically significant. This combined approach that balances statistical correlation with machine learning interpretability provides a more comprehensive framework for understanding the contribution of the Building Shape Coefficient in building thermal demand analysis.

References

1. Yang, Y., Wang, P.: Effects of building physics form on energy consumption for buildings. J. Phys. Conf. Ser. **2186**(1), 012008 (2022)
2. Koźniewski, E.: Assessment of building compactness at initial design stage of single-family houses. Energies **18**(13), 3569 (2025)
3. Li, J., Liang, C., Zhou, W.: A review of building physical shapes on heating and cooling energy consumption. Energies **17**(22), 5766 (2024)
4. Niknia, S., Ghiai, M.: Simulation-based prediction of office buildings energy performance under RCP scenarios across all U.S. climate zones. Architecture **5**(2), 1–22 (2025)
5. Li, J., Yue, T., Zhang, Y.: Experimental and simulation study on the modified shape coefficient of building integrated thermal considerations for heating buildings in solar-enriched areas. Sol. Energy **264**, 112060 (2023)
6. Li, J., Yue, T., Zhang, Y.: Modified shape coefficient of building with thermal considerations: concept, experiment and application in solar-enriched areas. J. Asian Archit. Build. Eng. **24**(3), 1744–1761 (2025)
7. Liu, G., et al.: Assessing urban morphology effects on residential building electricity consumption via explainable machine learning: evidence from China's hot summer and warm winter zone. Energy Build. **345**, 116063 (2025)
8. Bekdaş, G., Aydın, Y., Isıkdağ, Ü., Sadeghifam, A.N., Kim, S., Geem, Z.W.: Prediction of cooling load of tropical buildings with machine learning. Sustain. **15**(11), 9061 (2023)
9. Lu, S., et al.: A comparative analysis of machine learning algorithms in predicting the performance of a combined radiant floor and fan coiling system. Buildings **14**(6), 1659 (2024)
10. Ertosun Yildiz, M., Beyhan, F.: Prediction of cooling load via machine learning on building envelope design parameters. J. Build. Eng. **100**, 111724 (2025)
11. Moradzadeh, A., Mohammadi-Ivatloo, B., Abapour, M., Anvari-Moghaddam, A., Roy, S.S.: Heating and cooling loads forecasting for residential buildings based on hybrid machine learning applications: a comprehensive review and comparative analysis. IEEE Access **10**, 2196–2215 (2022)
12. Le, L.T., Nguyen, H., Zhou, J., Dou, J., Moayedi, H.: Estimating the heating load of buildings for smart city planning using a novel artificial intelligence technique PSO-XGBoost. Appl. Sci. **9**(13), 2714 (2019)
13. Chi, B., Li, Y., Zhou, D.: A hybrid method of cooling and heating consumption prediction for six types of buildings based on machine learning. Sustain. **16**(24), 11200 (2024)
14. Neubauer, A., Brandt, S., Kriegel, M.: Explainable AI for cooling load prediction and heat flow estimation using SHAP values. In: Proceedings of BauSim 2024 10th Conference IBPSA-Germany Austria 10, pp. 152–159 (2024)
15. Neubauer, A., Brandt, S., Kriegel, M.: Relationship between feature importance and building characteristics for heating load predictions. Appl. Energy **359**, 122668 (2024)
16. Devanathan, B., Varshitha, J.J., Kumar, P.P., Lakshmanan, S.A., Prakash, K.K.: Explainable AI framework using XGBoost with SHAP and LIME for multi-scale household energy forecasting. IEEE Access **13**, 149750–149764 (2025)
17. Cui, X., Lee, M., Koo, C., Hong, T.: Energy consumption prediction and household feature analysis for different residential building types using machine learning and SHAP: toward energy-efficient buildings. Energy Build. **309**, 113997 (2024)
18. Henkel, P., Kasperski, T., Stoffel, P., Müller, D.: Interpretable data-driven model predictive control of building energy systems using SHAP. Proc. Mach. Learn. Res. **242**, 222–234 (2024)

Study of Polymer Nanocomposite Filaments for Additive Manufacturing

Chuang Lek Cham, Thai Kiat Ong, Mum Wai Yip, Joo Eng Lim, Ming Yeng Chan, Seong Chun Koay, and Huei Ruey Ong

Abstract This study investigates the development of high-performance polymer nanocomposites for additive manufacturing applications. The primary objective was to produce polymer nanocomposite filaments and evaluate their mechanical properties. Polylactic Acid (PLA) was filled with 1 wt. %, 3 wt. %, and 5 wt.% of cellulose nanocrystals (CNC) to fabricate cellulose nanocrystals reinforced polylactic acid (PLA/CNC) composite filaments. The findings reveal that incorporating 1 wt.% CNC into PLA enhanced the ultimate tensile strength and Young's modulus by approximately 5.56% and 6.68% respectively as compared to neat PLA. In contrast, composites with 5 wt.% CNC exhibited a reduction in tensile strength of 8.76% and modulus of 6.91%. However, addition of 5 wt.% CNC into PLA matrix able to increases maximum bending stress (25%) and flexural modulus (31.73%) of PLA/CNC composites. These results demonstrate the feasibility of utilizing PLA/CNC nanocomposites in additive manufacturing, offering improved mechanical performance tailored to specific applications.

Keywords Additive manufacturing · Nanocomposite · PLA · Nanocellulose

C. L. Cham · M. W. Yip · J. E. Lim
Faculty of Engineering and Technology, Tunku Abdul Rahman University of Management and Technology, Kuala Lumpur, Malaysia

T. K. Ong (✉) · M. Y. Chan
Faculty of Engineering and Technology, Centre for Advanced Materials, Tunku Abdul Rahman University of Management and Technology, Kuala Lumpur, Malaysia
e-mail: ongtk@tarc.edu.my

S. C. Koay
Department of Mechanical and Materials Engineering, Centre for Sustainable Mobility Technologies, Lee Kong Chian Faculty of Engineering and Science, Universiti Tunku Abdul Rahman, Kajang, Malaysia

H. R. Ong
Faculty of Engineering and Technology, DRB, HICOM University of Automotive Malaysia, Pekan, Pahang, Malaysia

M. Awang et al. (eds.), *Proceedings of the International Conference on Mechanical, Manufacturing and Plant Engineering*, Proceedings in Technology Transfer,
https://doi.org/10.1007/978-981-95-7817-7_32

1 Introduction

Additive manufacturing, also known as three-dimensional (3D) printing, has become a widely adopted technology in industries for producing intricate, customizable parts. Its ability to fabricate designs directly from computer-aided design (CAD) software such as SolidWorks and computer-aided three-dimensional interactive application (CATIA) offers significant advantages over conventional manufacturing, including reduced material waste, faster production times, greater flexibility, and shorter setup durations. However, a significant limitation of 3D printing is the inadequate mechanical performance of pure polymers, which restricts their use in load-bearing applications. This research aims to develop polymer composites suitable for 3D printing while exhibiting enhanced mechanical properties. By focusing on mechanical performance, this study seeks to expand the practical applications of polymer composites beyond prototyping, enabling their use in functional, load-bearing components.

2 Literature Review

2.1 Additive Manufacturing

Additive manufacturing, commonly known as 3D printing, is a layer-by-layer fabrication process that transforms 3D model data into physical objects by precisely joining materials. This technology minimizes material waste through its high-accuracy deposition method. The process begins with creating a 3D model using CAD software, which is then converted into a surface tessellation language (STL) file. The STL file is sliced into 2D layer instructions and sent to the 3D printer for fabrication [1].

According to Quek et al. (2025) [2], 3D-printed polymers have significant potential across various fields, including rapid prototyping, rapid tooling and rapid manufacturing. Many 3D-printed products are no longer limited to only prototyping; they can also use as functional components for low load-bearing applications. For example, Philips enables customers to 3D-print customized pendants and lamp covers. For high load-bearing requirements can be addressed by using polymer composites, which combine a polymer matrix with reinforcing materials to enhance strength and structural performance.

Fused filament fabrication (FFF) is among the most widely used technologies in the 3D printing industry for producing thermoplastic components. FFF systems offer a wide selection of thermoplastic materials, ranging from engineering to biodegradable plastics and general used plastics such as polylactic acid (PLA), acrylonitrile butadiene styrene (ABS), polypropylene (PP), and polycarbonate (PC). The FFF process works by extruding thermoplastic filament layer by layer onto a build platform, where each layer fuses and solidifies to form the final part. FFF is widely adopted by researchers, industry users, and hobbyists due to its ease of operation, low machine cost, and affordable material options [3].

One of the key challenges in this study is the fabrication of polymer composite filaments. Ensuring uniform dispersion of reinforcements within the polymer matrix during the extrusion process is critical. Additionally, process parameters must be optimized to eliminate voids in the composite filaments, which can compromise their structural integrity. Products fabricated using 3D printers often exhibit reduced physical properties, as the viscosity of composite materials is limited by their melting temperature. Furthermore, in FFF, the removal of structural supports used for producing complex parts presents additional challenges, requiring extra post-processing steps to achieve a finished product.

FFF offers several advantages, including low production costs, high production speeds, and the simplicity of the printer's design. FFF also enables the simultaneous deposition of diverse materials, as multiple extrusion nozzles can be loaded with different materials. This capability allows for the fabrication of parts composed of multiple materials or composites, enhancing the versatility and functionality of printed components.

2.2 Polymer Composites

Polymer composites are materials formed by combining different components, with a polymer serving as the matrix material and is strengthened by reinforcement. These composites typically exhibit superior performance and greater versatility compared to the base polymer alone. Polymer matrix composites are fabricated by incorporating particulate, fibrous, or nanoscale reinforcements into the polymer matrix to enhance its properties. Polymer nanocomposites (PNCs) are materials in which nanoparticles are dispersed within a polymer matrix. Common nanomaterials used in PNCs include carbon nanotubes, graphene, graphite, ceramics, and metal nanoparticles, which offer unique mechanical, electrical, and thermal properties. Compared to traditional fibres or powders, nanomaterials enable the development of high-performance functional composites, making PNCs a promising solution for advanced applications.

The incorporation of nanomaterials in polymer nanocomposites (PNCs) can significantly improve the mechanical properties of polymers, expand their performance capabilities and enable new applications for plastics. According to Winey et al. (2007), even small amounts (1–5%) of nanoparticles such as silicates or carbon nanotubes can enhance properties by 15–40% for load-bearing purposes [4]. Additionally, a study by Song et al. (2018) demonstrated that the addition of minimal amounts of nanofillers into a polymer matrix can notably reduce bending deflections and increase critical buckling load [5]. Shoftner et al. (2003) reported that adding 10 wt.% carbon nanofibers or multi-walled carbon nanotubes increased the tensile strength by 39% and 7.5%, respectively, compared to unfilled polymers, though these parts also became more brittle and exhibited reduced elongation [6]. In stereolithography (SLA) fabrication, Lin et al. (2015) found that graphene oxide/photopolymer composites improved both strength and ductility, with just 0.2% graphene oxide leading to a 62% increase in tensile strength and a 12.8% increase in elongation [7].

Nanocomposites are more challenging to prepare than conventional polymer composites due to the difficulty in achieving homogeneous dispersion of nanoparticles, which is critical for obtaining the desired performance in 3D printing. One of the biggest challenges is preventing nanoparticle agglomeration and ensuring good interfacial bonding between nanoparticles and the polymer matrix. According to Zheng et al. (2006), nanoparticles often require chemical surface treatment before the printing process. In their study, emulsion polymerization was used to coat nano-aluminium oxide (nano-Al_2O_3) particles with polystyrene, resulting in a denser structure and a 300% increase in tensile strength compared to untreated particles when processed with selective laser sintering (SLS) printers [8]. Another approach to improve interfacial bonding is the addition of linker molecules that cross-link with the polymer matrix on the nanoparticle surface. For instance, oxidized graphite nanoplatelets enhanced the ultimate strength and Young's modulus of SLS-printed graphite nanocomposites after nitric acid treatment. In research by Steurer et al. (2009), the optimal graphite oxide loading level in an ABS system was found to be 0.06%, offering multiple property improvements and favorable economic feasibility. At this loading level, 3D-printed nanocomposite ABS exhibited higher strain-to-failure (14% facedown, 29% upright) and toughness (20% facedown, 55% upright) [9]. Fracture strength increased by 3.5% facedown and 10% upright, while stiffness decreased by 6% facedown and 15% upright. The improvements in toughness enhance impact and fatigue resistance, as well as the damping coefficient of the printed parts [10, 11].

Biodegradable polymers are gaining popularity as a sustainable alternative to addressing the global reliance on petroleum-based products. Polylactic acid (PLA) is a promising successor due to its favorable mechanical properties and environmental sustainability [12, 13]. PLA exhibits relatively high strength, good optical transparency, and a high modulus compared to other synthetic polymers. To enhance PLA's strength for load-bearing applications in 3D printing, nanocomposites incorporating reinforcements such as cellulose nanofibers and cellulose nanocrystals are frequently utilized.

Cellulose, a naturally abundant material found in plants, marine organisms, algae, fungi, and bacteria [12], is an attractive reinforcement for polylactic acid (PLA) due to its availability and proven ability to enhance mechanical and thermal properties in 3D-printed PLA products. However, a significant challenge arises from the hydrophilic nature of cellulose and the hydrophobic nature of PLA, which can lead to cellulose aggregation and uneven dispersion within the PLA matrix [12, 13]. To address this, Iwatake et al. (2008) demonstrated that premixing cellulose fibers in an organic solvent, followed by solvent removal during kneading, achieves a uniform dispersion of cellulose nanofibers in PLA [14]. This approach resulted in a 25% increase in tensile strength and a 40% increase in Young's modulus at a fiber loading level of 10% [12, 14]. Similarly, Jonoobi et al. (2015) found that casting and melt blending using twin-screw extrusion reduces cellulose nanofiber aggregation, particularly at fiber contents of 1wt.% and 3 wt.%. At 5 wt.% CNF, the composites exhibited a 21% increase in tensile strength, highlighting the effectiveness of this method for producing cellulose-reinforced PLA composites [15].

3D printing is an ideal technique for fabricating products with polymer nanocomposites, owing to its ability to accommodate varying volume fractions of nanomaterials tailored to specific needs and applications. This flexibility in design composition enables the optimization of printed parts' properties, making it possible to achieve enhanced performance characteristics for diverse applications.

3 Materials and Methodology

3.1 Preparation of PLA/CNC Composites

Polylactic acid (PLA) (NatureWorks Ingeo™ Biopolymer 3052D) was blended with Nano Cellulose Crystal (CNC) at weight percentages of 0% (100PLA), 1% (01CNC99PLA), 3% (03CNC97PLA), and 5% (05CNC95PLA) using a Brabender plastograph® EC. The mixer temperature and rotor speed were set at 210 °C and 30 min^{-1} respectively. The compound was processed through a plastic crusher to produce smaller pieces, which were then fed into a single-screw extruder (Teach Line E20). Filaments with a diameter of 175 ± 5 mm were produced at a screw speed of 20 rpm and a roller speed of 360 rpm. The temperature settings for the extruder zones 1, 2, 3, 4, and 5 were 45 °C, 180 °C, 200 °C, 205 °C, and 197 °C, respectively.

3.2 3D Printing of PLA/CNC Composites Filament

Four rolls of PLA/CNC composite filaments with varying weight percentages were used to fabricate specimens for tensile tests and three-point bending tests. Five specimens were prepared for each test at each weight percentage. Three-point bending tests were performed using a GOTECH 2-Ton Universal Testing Machine, in accordance with ASTM D790 standard. Tensile test was carried out with a SHIMADZU AGS-X Universal Testing Machine, complying with ASTM D638 standard.

4 Results and Discussion

4.1 Tensile Properties

Refer to Table 1 and Figs. 1, 2 and 3, PLA with 1 wt. % CNC exhibited the best mechanical properties among all tested weight percentages. Specifically, PLA/CNC composite with addition of 1 wt. % of CNC (01CNC99PLA) achieved an ultimate tensile strength (UTS) of 42.91 MPa, a Young's modulus of 2672.18 MPa, and a maximum elongation of 1.83%. This phenomenon, as observed in this study, corroborates findings by Iwatake et al. (2008), who also noted improvements in mechanical properties through nanocellulose compounding [14].

The highest improvements in ultimate tensile strength and Young's modulus were observed at the 1 wt. % CNC loading, consistent with the conclusions of Winey and Vaia (2007) [4]. Similarly, Murphy et al. (2018) demonstrated that incorporating nanocellulose into PLA enhances tensile strength and modulus [12]. At a 3 wt. % CNC weight percentage, the ultimate tensile strength returned to its baseline value of approximately 40.62 MPa. A further increase to 5 wt. % CNC caused a decline in tensile strength to 37.09 MPa. Similar finding was reported by Herrera et al.'s (2015) stated that nanoparticle oversaturation compromises composite integrity [16].

Young's modulus of PLA/CNC composites with 1 wt. % of CNC (01CNC99PLA) and 3 wt. % CNC (03CNC97PLA) were recorded at 2672.18 MPa and 2642.85 MPa, respectively, both surpassing the performance of neat PLA. Additionally, incorporating CNC improved the maximum elongation across all three weight percentages. This enhancement can be attributed to the addition of CNC reinforcement improved the stiffness of the PLA nanocomposites and good dispersion of CNC within the PLA

Table 1 Tensile test results of different test samples

Sample	Young's modulus (MPa)	Ultimate tensile strength (MPa)	Max elongation (%)
100PLA	2504.87	40.65	1.78
01CNC99PLA	2672.18	42.91	1.83
03CNC97PLA	2642.85	40.62	1.80
05CNC95PLA	2331.66	37.09	1.80

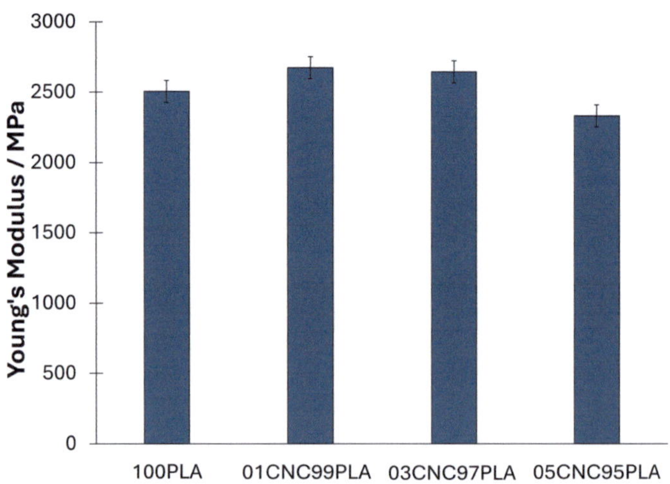

Fig. 1 Young's modulus of different test samples

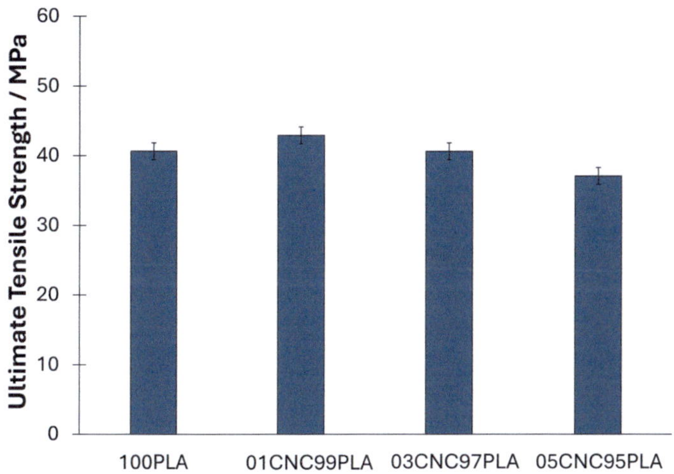

Fig. 2 Ultimate tensile strength of different test samples

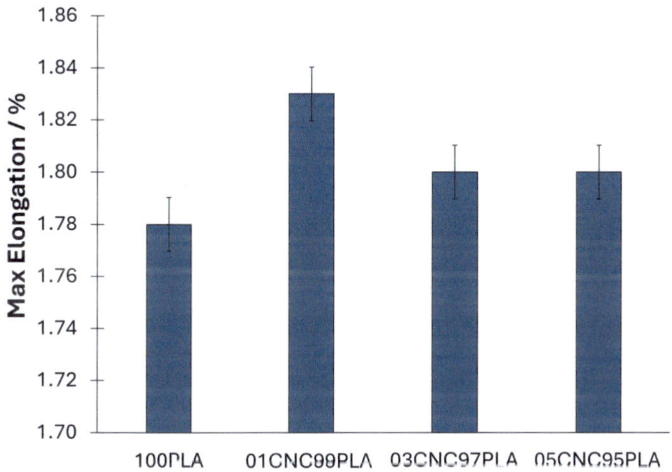

Fig. 3 Max elongation of different test samples

matrix, which enhances stress distribution during deformation and allows the material to exhibit greater ductility. These findings are consistent with the observations of Iwatake et al. (2008) [14].

4.2 Flexural Properties

The flexural test results as shown in Table 2, Figs. 4 and 5, demonstrate that incorporating nanoparticles into the PLA matrix enhances its maximum bending stress and flexural modulus. All three CNC reinforced PLA composites outperformed neat PLA. This finding aligns with the study by Song et al. (2018), which reported that small amounts of nanofillers effectively reduce bending deflections and increase critical load capacity [5]. Interestingly, while higher CNC weight percentages may not significantly improve ultimate tensile strength and Young's modulus, they are most effective in enhancing maximum bending stress and flexural modulus. CNC reinforced PLA composites with a 5 wt. % of CNC addition (05CNC95PLA) exhibited a 25% increase in maximum bending stress and a 31.7% increase in flexural modulus compared to neat PLA, with values of 55.17 MPa and 2406 MPa, respectively. The performance of 01CNC99PLA and 03CNC97PLA in maximum bending stress was also notable, measuring 53.92 MPa and 54.33 MPa, respectively, indicating that even as low as 1 wt.% CNC can significantly enhance the maximum bending stress of PLA.

These results suggest that the addition of CNC nanoparticles improves stress distribution and delays crack formation, resulting in higher maximum bending stress and

Table 2 Three-point bending test results of different test samples

Sample	Maximum bending stress (MPa)	Flexural modulus (MPa)
100PLA	44.12	1826.82
01CNC99PLA	53.92	2196.26
03CNC97PLA	54.33	2193.71
05CNC95PLA	55.17	2406.48

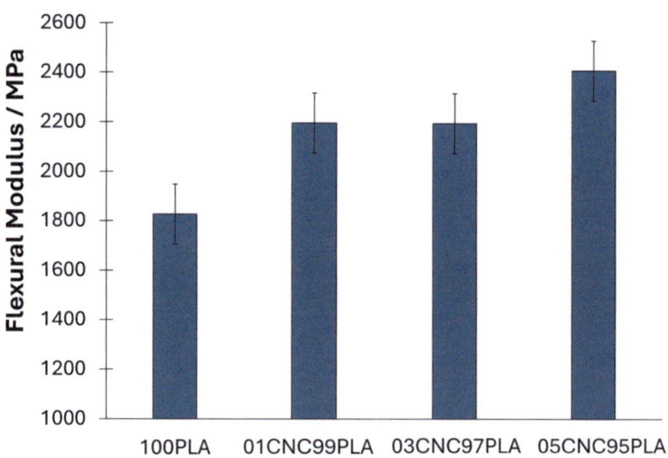

Fig. 4 Flexural modulus of different test samples

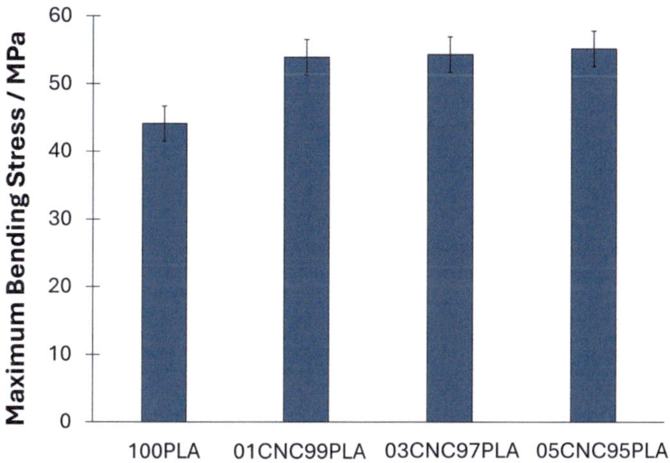

Fig. 5 Maximum bending stress of different test samples

flexural modulus. Furthermore, the trend of increasing performance with higher CNC weight percentages suggests the potential for even greater improvements at higher CNC concentrations. In contrast, the flexural modulus shows a more pronounced and consistent improvement with increasing CNC content, as the addition of rigid CNC restrict the chain mobility of PLA and improves the overall stiffness of composite materials.

5 Conclusion

In conclusion, this research represents a meaningful step toward broadening the application of additive manufacturing by addressing the challenge of creating composite filaments with enhanced mechanical properties compared to their pure counterparts. The study successfully demonstrated that addition of 1 wt.% of CNC into PLA matrix improved ultimate tensile strength by 5.56% and Young's modulus of PLA/CNC composites by 6.68%, while higher CNC weight percentages enhanced maximum bending stress and flexural modulus.

Additionally, incorporation of CNC into PLA matrix improved flexural modulus and maximum bending stress up to 25% and 31.7%, respectively. These findings confirm that the research objective to produce a 3D-printable composite filament with superior mechanical properties was achieved. This research provides valuable insights into the selection of CNC weight percentages for creating PLA/CNC composites, serving as a guideline for producing nanocomposite filaments for additive manufacturing. Moreover, it highlights the potential of recyclable materials like PLA and CNC in sustainable manufacturing. By showcasing these benefits, this study aims to inspire further exploration and adoption of environmentally friendly materials in additive manufacturing.

References

1. Tan, Y.A., Chan, M.Y., Koay, S.C., Ong, T.K.: 3D polymer composite filament development from post-consumer polypropylene and disposable chopstick fillers. J. Vinyl Add. Tech. **29**(5), 909–922 (2023)
2. Quek, C.Y., Calista, N.H.T., Chai, J.Y., Lee, Z.Y., Khaw, M.Y., Ting, C.H., Koay, S.C., Chan, M.Y.: A comparative study in fused filament fabrication using commercial and recycled high-density polyethylene. J. Phys. Sci. **36**(3), 1–14 (2025)
3. Lim, K.Y., Koay, S.C., Chan, M.Y., Ong, T.K., Ho, S.H., Ting, C.H., Pang, M.M.: Effects of corn husk fiber as filler in recycled single-use polypropylene for fused filament fabrication. J. Vinyl Add. Tech. **30**(2), 620–634 (2023)
4. Winey, K.I., Vaia, R.A.: Polymer nanocomposites. MRS Bull. **32**(4), 314–322 (2007)
5. Song, M., Yang, J., Kitipornchai, S.: Bending and buckling analyses of functionally graded polymer composite plates reinforced with graphene nanoplatelets. Compos. B **134**, 106–113 (2018)
6. Shoftner, M., Lozano, K., Rodriguez-Macias, F., Barrera, E.: Nanofiber-reinforced polymers prepared by fused deposition modeling. J. Appl. Polym. Sci. **89**(11), 3081–3090 (2003)
7. Lin, D., Jin, S., Zhang, F., Wang, C., Wang, Y., Zhou, C., Cheng, G.J.: 3D stereolithography printing of graphene oxide reinforced complex architectures. Nanotechnology **26**(43), 434003 (2015)
8. Zheng, H., Zhang, J., Lu, S., Wang, G., Xu, Z.: Effect of core–shell composite particles on the sintering behavior and properties of nano-Al_2O_3/polystyrene composite prepared by SLS. Mater. Lett. **60**(9–10), 1219–1223 (2006)
9. Steurer, P., Wissert, R., Thomann, R., Mülhaupt, R.: Functionalized graphenes and thermoplastic nanocomposites based upon expanded graphite oxide. Macromol. Rapid Commun. **30**(4–5), 316–327 (2009)
10. Yamamoto, B.E., Trimble, A.Z., Minei, B., Ghasemi-Nejhad, M.N.: Development of multifunctional nanocomposites with 3-D printing additive manufacturing and low graphene loading. J. Thermoplast. Compos. Mater. **32**(3), 383–408 (2019)
11. Bortz, D., Garcia-Heras, E., Martin-Gullon, I.: Impressive fatigue life and fracture toughness improvements in graphene oxide/epoxy composites. Macromolecules **45**(1), 238–245 (2012)
12. Murphy, C.A., Collins, M.N.: Microcrystalline cellulose reinforced polylactic acid biocomposite filaments for 3D printing. Polym. Compos. **39**(4), 1311–1320 (2018)
13. Zhu, J., Zhu, H., Njuguna, J., Abhyankar, H.: Recent development of flax fibres and their reinforced composites based on different polymeric matrices. Materials **6**(11), 5171–5198 (2013)
14. Iwatake, A., Nogi, M., Yano, H.: Cellulose nanofiber-reinforced polylactic acid. Compos. Sci. Technol. **68**(9), 2103–2106 (2008)
15. Jonoobi, M., Oladi, R., Davoudpour, Y., Oksman, K., Dufresne, A., Hamzeh, Y., Davoodi, R.: Different preparation methods and properties of nanostructured cellulose from various natural resources and residues: a review. Cellulose **22**(2), 935–969 (2015)
16. Herrera, N., Mathew, A.P., Oksman, K.: Plasticized polylactic acid/cellulose nanocomposites prepared using melt-extrusion and liquid feeding: Mechanical, thermal and optical properties. Compos. Sci. Technol. **106**, 149–155 (2015)

Author Index

M. Awang et al. (eds.), *Proceedings of the International Conference on Mechanical, Manufacturing and Plant Engineering*, Proceedings in Technology Transfer,
https://doi.org/10.1007/978-981-95-7817-7

MIX
Papier aus verantwortungsvollen Quellen
Paper from responsible sources
FSC® C105338

If you have any concerns about our products,
you can contact us on
ProductSafety@springernature.com

In case Publisher is established outside the EU,
the EU authorized representative is:
Springer Nature Customer Service Center GmbH
Europaplatz 3, 69115 Heidelberg, Germany

Printed by Libri Plureos GmbH
in Hamburg, Germany